Classical & Fluid Mechanics

Classical & Fluid Mechanics

Editor:

Olga Moreira, Ph.D.

ArclerPress

www.arclerpress.com

Classical & Fluid Mechanics

Editor: Olga Moreira, Ph.D.

708 3rd Avenue, 6th Floor
New York
NY 10017
United States of America
www.arclerpress.com

ISBN: 978-1-68094-430-3

Library of Congress Control Number: 2016951126

About the Editor

Olga Moreira, Ph.D.

Olga Moreira obtained her Ph.D. in Astrophysics from the University of Liege (Belgium) in 2010, her BSc. in Physics and Applied Mathematics from the University of Porto (Portugal). Her post-graduate travels and international collaborations with the European Space Agency (ESA) and European Southern Observatory (ESO) led to great personal and professional growth as a scientist. Currently, she is working as an independent researcher, technical writer, and editor in the fields of Mathematics, Physics, Astronomy and Astrophysics.

Contents

List of Contributors

Luis Vázquez
Departamento de Matematica Aplicada, Facultad de Inform ´ atica, Universidad Complutense de Madrid, ´ 28040 Madrid, Spain

Gülden Gün Polat
Department of Mathematics, Faculty of Science and Letters, Istanbul Technical University, Maslak, 34469 Istanbul, Turkey

Teoman Özer
Division of Mechanics, Faculty of Civil Engineering, Istanbul Technical University, Maslak, 34469 Istanbul, Turkey

J. Á. Acosta
Departamento de Ingeniería de Sistemas y Automática, Escuela Técnica Superior de Ingenieros, Camino de los Descubrimientos s/n, 41092 Sevilla, Spain

Irena Rachůnková
Department of Mathematical Analysis and Applications of Mathematics, Faculty of Science, Palacky University, t ´ r. 17. listopadu 12, 771 46 Olomouc, Czech Republic

Jan Tomeček
Department of Mathematical Analysis and Applications of Mathematics, Faculty of Science, Palacky University, t ´ r. 17. listopadu 12, 771 46 Olomouc, Czech Republic

Jakub Stryja
VSB, Technical University of Ostrava, 17. listopadu 15, 708 33 Ostrava-Poruba, Czech Republic

Martin Hermann
Fakulta¨t fu¨r Mathematik und Informatik, Friedrich-SchillerUniversita¨t Jena, 07743 Jena, Germany

Masoud Saravi
Department of Mathematics, Islamic Azad University Nour Branch, 731 Nour, Ira

Hadi Ebrahimi Khah
Fakulta¨t fu¨r Mathematik und Informatik, Friedrich-SchillerUniversita¨t Jena, 07743 Jena, Germany

Andrzej J. **Maciejewski**
Janusz Gill Institute of Astronomy, University of Zielona Góra, Licealna 9, 65-417 Zielona Góra, Poland

Maria Przybylska
Institute of Physics, University of Zielona Góra, Licealna 9, 65-417 Zielona Góra, Poland

Joseph J Smulsky
Institute of Earth's Cryosphere, Malygina Str. 86, 625000 Tyumen, Russia

Sergey V. Ershkov
Institute for Time Nature Explorations, M.V. Lomonosov's Moscow State University, Leninskie Gory 1-12, Moscow 119991, Russia

Robert D. Furber
208 43rd Street, Manhattan Beach, CA 90266-3009, USA

E. Iorfida
Surrey Space Centre, University of Surrey, Guildford GU2 7XH, UK

P. L. Palmer
Surrey Space Centre, University of Surrey, Guildford GU2 7XH, UK

M. Roberts
Department of Mathematics, University of Surrey, Guildford, UK
AIMS Tanzania, PO Box 176, Bagamoyo, Pwani, Tanzania

S. E. Abd El-Bar
Department of Applied Mathematics, Faculty of Applied Science, Taibah University, Al-Madina-Al Munawwarah, Saudi Arabia
Department of Mathematics, Faculty of Science, Tanta University, Tanta, Egypt

F. A. Abd El-Salam
Department of Mathematics, Faculty of Science, Taibah University, Al-Madina-Al Munwwarah, Saudi Arabia
Department of Astronomy, Faculty of Science, Cairo University, Cairo 12613, Egypt

José M. Ferrándiz
Department of Applied Mathematics, University of Alicante, 03080 Alicante, Spain.

Juan F. Navarr
Department of Applied Mathematics, University of Alicante, 03080 Alicante, Spain.

Alberto Escapa
Department of Applied Mathematics, University of Alicante, 03080 Alicante, Spain.

Juan Getino
Department of Applied Mathematics, Faculty of Sciences, University of Valladolid, 47011 Valladolid, Spain

F. Shojai
Department of Physics, University of Tehran, Tehran, Iran
Foundations of Physics Group, School of Physics, Institute for Research in Fundamental Sciences (IPM), Tehran

M. R. Fazel
Department of Physics, University of Tehran, Tehran, Iran

A. Stepanian
Department of Physics, University of Tehran, Tehran, Iran

M. Kohandel
Department of Sciences, Alzahra University, Tehran, Iran

M. Oliveira
Grupo de Ciências Ambientais e Recursos Naturais, Instituto Federal do Espírito Santo (IFES), Guarapari, Brazil

H. E. S. Velten
Departamento de Física, Universidade Federal do Espírito Santo (UFES), Vitória, Brazil

J. C. Fabris
Departamento de Física, Universidade Federal do Espírito Santo (UFES), Vitória, Brazil

I. G. Salako
Institut de Mathématiques et de Sciences Physiques (IMSP), Porto-Novo, Bénin

Carlos Cartes
Complex Systems Group, Universidad de los AndesChile

Orazio Descalzi
Complex Systems Group, Universidad de los AndesChile

Christopher Depcik
University of Kansas, Department of Mechanical Engineering USA

Sudarshan Loya
University of Kansas, Department of Mechanical Engineering USA

Caroline W. Kanyiri
Department of Pure and Applied Mathematics, Jomo Kenyatta University of Agriculture and Technology, Nairobi, Kenya

Mathew Kinyanjui
Department of Pure and Applied Mathematics, Jomo Kenyatta University of Agriculture and Technology, Nairobi, Kenya

Kang'ethe Giterere
Department of Pure and Applied Mathematics, Jomo Kenyatta University of Agriculture and Technology, Nairobi, Kenya

PK Kameswaran
School of Mathematics, Statistics and Computer Sciences, University of KwaZulu-Natal, Private Bag X01, Pietermaritzburg, Scottsville 3209, South Africa

M Narayana
School of Mathematics, Statistics and Computer Sciences, University of KwaZulu-Natal, Private Bag X01, Pietermaritzburg, Scottsville 3209, South Africa

P Sibanda
School of Mathematics, Statistics and Computer Sciences, University of KwaZulu-Natal, Private Bag X01, Pietermaritzburg, Scottsville 3209, South Africa

G Makanda
School of Mathematics, Statistics and Computer Sciences, University of KwaZulu-Natal, Private Bag X01, Pietermaritzburg, Scottsville 3209, South Africa

Quoc-Hung Nguyen
Mechanical Faculty, Ho Chi Minh University of Industry,Vietnam

Ngoc-Diep Nguyen
Mechanical Faculty, Ho Chi Minh University of Industry,Vietnam

Changjia Wang
School of Science, Changchun University of Science and Technology, Changchun, 130022, P.R. China

Preface

Classical mechanics (a.k.a. Newtonian Mechanics) was one of the first pillars of modern physics providing the first unification of the laws of motion and universal gravitation. Although classical mechanics has been surpassed by the relativity theory and quantum mechanics, it still provides an accurate description of macroscopic objects motion as long as their velocities are far below the speed of light.

This book is a collection of research, which the main purpose is to offer a broad glimpse of the advances of Classical & Fluid Mechanics in several areas (e.g. oscillatory motion, celestial & orbital mechanics, fluid dynamics). The first part (chapters 1 - 5) overviews modern mathematical tools and techniques used to solve and construct differential equations associated with the equation of motion and modeling of oscillatory systems (e.g. Furuta's pendulum, harmonic oscillator). The second part (chapters 6- 12) focuses on the dynamics of multi-body systems (N-body problem), planetary and orbital mechanics, including some applications to study of moons orbits, satellite motion, and Earth's rotation problem. The third includes two examples Newtonian approach to the study of stellar structure (chapters 13 - 14). The remaining chapters are dedicated to advanced topics of fluid mechanics. This includes the description of compressible and incompressible fluid dynamics (chapters: 15 - 16) as well as the modeling of Newtonian (chapters 17-18) and non-Newtonian fluids (Chapters 19-20).

Editor
Olga Moreira, Ph.D.

INTRODUCTION

Fluid mechanics is a branch of physics which concerns the study of fluids and the ways in which they interact with forces. Both liquids and gases are considered to be fluids for the purposes of this branch of science. Often, the field of fluid mechanics is divided into two more specific fields of study. These are fluid statics and fluid dynamics, which concern fluids at rest and fluids in motion, respectively. Fluid mechanics can involve highly complex mathematics, and the aid of modern computers has enhanced this science significantly.

The chronological roots of fluid mechanics go all the way back to at least the ancient Greeks. The Greek physicist and inventor Archimedes was the author of some of the first studies we know of which concern fluid statics, including the property of buoyancy. Persian philosophers in the medieval time period coupled these ancient works with their own studies of fluid dynamics that acted as an early precursor to modern fluid dynamics. Such well-known historical figures as Leonardo da Vinci and Sir Isaac Newton, as well as others, made notable contributions to our understanding of fluid mechanics.

Every type of science starts out with basic, fundamental assumptions that govern the course of their study. Fluid mechanics is typically defined as having three basic premises or assumptions at its root. The first is the conservation of mass, which means that mass can neither be spontaneously created nor destroyed, although it may change forms. The second assumption, the conservation of momentum, is somewhat similar. This law states that the total momentum in a closed system is constant, and cannot spontaneously appear or disappear.

The third basic assumption governing fluid mechanics is what is known as the continuum hypothesis. This is a way of seeing fluids that does not take into account the presence of discrete molecules. Instead, a fluid's properties are assumed to vary in a continuous way from one point to the next.

Because it ignores the actual nature of small particles of matter, the continuum hypothesis is only an approximation used as a tool in calculations. It can result in a slightly inaccurate solution, but also in solutions that are very accurate under ideal circumstances. Other, more exact methods exist, but this hypothesis is often quite useful as a preliminary assumption. Many times, it can also be assumed that a given fluid is incompressible, meaning it cannot be compressed. This is only actually true of liquids, however, and not gases.

What is Fluidics?

The application of the physical properties of liquids and gases as a fluid to perform logic operations that control other mechanical systems is called fluidics. Hydraulics and pneumatics, respectively, starting from the Industrial Revolution that began around the late 1700s, provided a foundation. Subsequent study on the dynamics of fluids — liquids in particular — developed into a theoretical model of predictive behavior. This gave engineers a framework from which to conceive switches and other logic circuits which became the forerunners of modern electronics. Although digital circuits dominate the world today, fluidic processors remain in critical use.

Fluidics is not to be confused with the compression or expansion of liquids and gases as a hydraulic or pneumatic power source. Instead, the flow of a fluid is conceived as a medium capable of changing its character, carrying this information and transmitting it to other flows. The core functioning of a fluidic device has no moving parts.

The first set of assumptions about fluid dynamics is the Newtonian physics of classical mechanics. To this is added the variables of velocity, pressure, density and temperature as functions of space and time. An additional law is especially important — the "continuum assumption," that the flow characteristics of a fluid can be described without accounting for the known fact that fluids are composed of discreet molecular particles. Both theoretical and empirical physicists continue to expand computational understanding of viscosity, turbulence and other peculiar features of a fluid in motion. Engineers have followed with increasingly sophisticated fluidic devices.

Fluidics technology did not have a full opportunity to mature. The first logic circuits, including an amplifier and a diode, were invented in the early 1960s. Concurrently, the same concepts of signal amplification and transmission were realized employing a flow of electrons, and the invention of the solid state transistor ushered in a digital revolution.

The physical flow of a fluid, of course, cannot match the speed of an electron. A fluidic signal processor typically has an operating speed of just a few kilohertz. Unlike an electron, however, the mass flow of a liquid or gas is unaffected by electromagnetic or ionic interferences. Fluidics therefore remain necessary for the control of some failure-intolerant systems, such as military avionics. Fluidics have also developed into effective processors of analog data because of the nature of fluids to flow as a wave.

One of the major challenges of fluidics is that the principles of fluid dynamics are apparently different according to scale. To be sure, climatologists have yet to fully understand how massively large bodies of water or currents of air behave. Likewise, scientists have discovered that fluids behave very differently when studied at the scale of nanotechnology. Future study and application of the latter, called nano-fluidics, pose the possibility of significantly faster and more complex circuitry, including multiple gate arrays for parallel processing.

What is Fluid Dynamics?

Fluid dynamics refers to a subcategory of the science of fluid mechanics, with the other subcategory being fluid statics. While fluid statics deals with fluids that are at rest, fluid dynamics is concerned with fluids that are in motion. Any matter in a gas or liquid state can be considered a fluid. Fluid dynamics is a discipline with many relevant applications in our modern world, most notably because it contains the study of aerodynamics, and also because it comprises part of weather prediction. A typical fluid dynamics problem may include such variables as velocity, temperature, and density.

All of the physical sciences, including fluid dynamics, are governed first and foremost by the laws of conservation. These state that the total amounts of energy, mass, and linear momentum in a closed system remain constant, and that energy and mass can neither be created nor destroyed. That they may change forms, it is true, but they cannot disappear or come from nothing. These laws constitute some of the most basic assumptions in science.

Another governing principle of fluid dynamics is the continuum assumption, also called the continuum hypothesis. While fluids are known to be composed of microscopic, discreet particles, this hypothesis states that they are continuous, and that their properties vary evenly throughout. This often serves as a helpful mathematical approximation, even though it technically ignores one of the basic characteristics of fluids.

Before the invention of powered flight and aircraft in the 20th century, the term hydrodynamics was often used interchangeably with fluid dynamics, because most of fluid mechanics was devoted to studying liquids in motion, rather than gases in motion. When traveling by aircraft became more common, the need developed for these machines to be more efficient at creating and maintaining lift, with a minimum of drag. The branch of study known as aerodynamics made leaps and bounds because of the new technology, which has also come to be applied to automobiles, to some extent, with the goal of increased fuel efficiency.

One of the more important figures in modern aerodynamics was Octave Chanute. In addition to compiling a comprehensive volume of the study of aerodynamics in the late 1800s, he personally assisted the Wright brothers in building their famous aircraft, which accomplished the first manned, powered flight in 1903. It was likely due to this help that they accomplished their goal just ahead of the next closest contender, Samuel Pierpont Langley.

What Do Fluid Engineers Do?

Fluid engineers design and maintain fluid-based mechanical systems. They rely heavily on principles of fluid mechanics, engineering, and the natural sciences as they work with mechanical, hydraulic, and pumping systems. A fluid engineer may work in the design field as an engineer or architect. He may also find employment in the drilling field, where he installs and repairs equipment used to extract water, oil, and natural gas from the earth. These professionals may also maintain and repair the equipment used in mining, construction, and similar operations.

In the design field, fluid engineers may find employment in an engineering firm, power production facility, or manufacturing plant. In these types of jobs, fluid engineers design the hydraulic systems used in major industrial equipment, buildings or power production. For example, a fluid engineering professional employed by an equipment manufacturer may be responsible for designing new equipment, or creating more effective hydraulic and fluid-based operating systems. He may design cranes, excavators and large-scale pumps, relying on his knowledge of fluid mechanics to determine how much fluid is needed and what type will produce the best performance. Fluid design engineers also develop power production systems in hydroelectric plants and similar facilities.

In the drilling arena, fluid or mud engineers set up the drilling systems used to remove oil and natural gas from the earth. Drilling fluid plays a major role in this type of operation, and is used to clear the way for the powerful drill bits as it removes cuttings and waste materials from the earth. In this setting, fluid engineers not only set up and install equipment, but also teach other members of the team how to operate and maintain these systems. They track the how fluid impacts pressure and performance deep underground, and take charge of cleaning and maintaining equipment so that it will produce maximize results.

Fluid engineers may also manage the repair and maintenance of large equipment on mining and construction sites. These professionals may be responsible for keeping washing stations or excavators operating at a gold mine, or maintaining cranes, bulldozers and other equipment used in construction and land clearing operations. In this type of position, fluid engineers perform routine maintenance like checking fluid levels and topping off the fluid as needed. They also perform major repairs and diagnostics when equipment malfunctions, which may include something as simple as replacing a damaged hose or gasket to a job as complex as rebuilding an entire hydraulic operating system.

Fluid dynamics

In physics, fluid dynamics is a sub discipline of fluid mechanics that deals with fluid flow—the science of fluids (liquids and gases) in motion. It has several sub disciplines itself, including aerodynamics (the study of air and other gases in motion) and hydrodynamics (the study of liquids in motion). Fluid dynamics has a wide range of applications, including calculating forces and moments on aircraft, determining the mass flow rate of petroleum through pipelines, predicting weather patterns, understanding nebulae in interstellar space and modelling fission weapon detonation. Some of its principles are even used in traffic engineering, where traffic is treated as a continuous fluid, and crowd dynamics.

Fluid dynamics offers a systematic structure—which underlies these practical disciplines—that embraces empirical and semi-empirical laws derived from flow measurement and used to solve practical problems. The solution to a fluid dynamics problem typically involves calculating various properties of the fluid, such as flow velocity, pressure, density, and temperature, as functions of space and time.

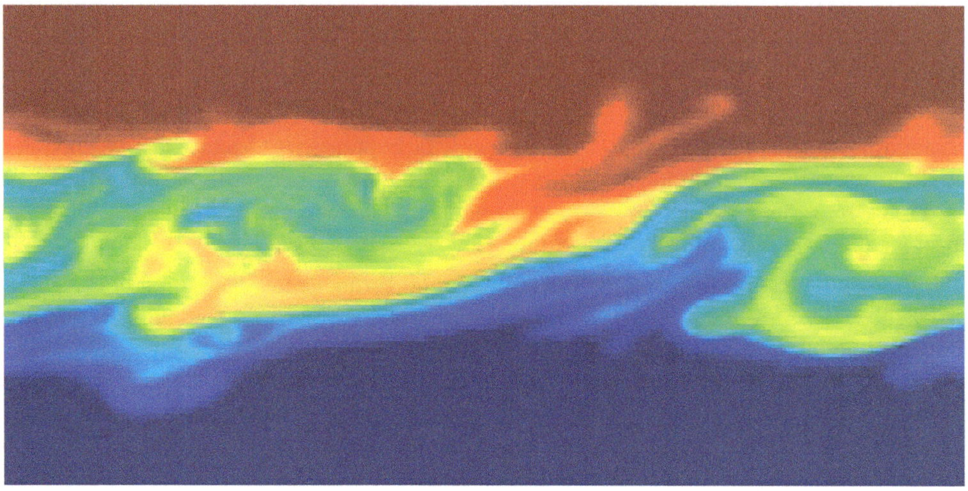

Equations of fluid dynamics

The foundational axioms of fluid dynamics are the conservation laws, specifically, conservation of mass, conservation of linear momentum (also known as Newton's Second Law of Motion), and conservation of energy (also known as First Law of Thermodynamics). These are based on classical mechanics and are modified in quantum mechanics and general relativity. They are expressed using the Reynolds Transport Theorem.

In addition to the above, fluids are assumed to obey the continuum assumption. Fluids are composed of molecules that collide with one another and solid objects. However, the continuum assumption considers fluids as continuous, rather than discrete. Consequently, properties such as density, pressure, temperature, and flow velocity are assumed well-defined at infinitesimally small points, and are assumed to vary continuously from one point to another. The fact that the fluid is made up of discrete molecules is ignored.

For fluids that are sufficiently dense to be a continuum, do not contain ionized species, and have flow velocities small in relation to the speed of light, the momentum equations for Newtonian fluids are the Navier–Stokes equations—which is a non-linear set of differential equations that describes the flow of a fluid whose stress depends linearly on flow velocity gradients and pressure. The simplified equations do not have a general closed-form solution, so they are primarily of use in Computational Fluid Dynamics. The equations can be simplified in a number of ways, all of which make them easier to solve. Some of them allow appropriate fluid dynamics problems be solved in closed form

Classical fluid mechanics

Classical fluid mechanics is a branch of continuum mechanics; that is, it proceeds on the assumption that a fluid is practically continuous and homogeneous in structure. The fundamental property which distinguishes a fluid to from other continuous media is that it cannot be in equilibrium in a state of stress such that the mutual action between two adjacent parts is oblique to the common surface. Though this property is the basis of hydrostatics and hydrodynamics, it is by itself insufficient for the description of fluid motion. In order to characterize the physical behavior of a fluid the property must be extended, given suitable analytical form, and introduced into the equations of motion of a general continuous medium, this leading ultimately to a system of differential equations which are to be satisfied by the velocity, density, pressure, etc. of an arbitrary fluid motion. In this article we shall consider these differential equations, their derivation from fundamental axioms, and the various forms which they take when more or less special assumptions concerning the fluid or the fluid motion are made. Our intent, then, is to present in a mathematically correct way, in concise form, and with more than passing attention to the foundations, the principles of classical fluid mechanics. The work includes the body of exact theoretical knowledge which accompanies the fundamental equations, and at the same time excludes relativistic and quantum effects, most of the kinetic theory, special fields such as turbulence, and all numerical or approximate work.

Vectors and tensors

The mathematical notation used in this article is that of ordinary Cartesian or Gibbsian vector analysis. This notation leads to the utmost conciseness of expression, and at the same time illuminates the physical meaning of the phenomena represented. Most of the vector operations which we use are standard, but occasionally an expression is needed which may appear unusual or ambiguous. For this reason it is convenient to define all operations in terms of vector components: then the meaning of an equation can always be made clear simply by rewriting it in component form. Another advantage accrues to this method, namely that any equation admits an immediate tensorial interpretation if so desired.

CHAPTER 1

FROM NEWTON'S EQUATION TO FRACTIONAL DIFFUSION AND WAVE EQUATIONS

Luis Vázquez

INTRODUCTION

Fractional calculus offers a very suggestive and stimulating scenario where we have the convergence of deep and fundamental mathematical questions, development of appropriate numerical algorithms, as well as the applications to modelizations in different frameworks. An illustration of the physical applications of fractional calculus is the recent following books and special issues on applications on nanotechnology and other important topics [2, 3, 4, 5, 6], among many others.

From a mathematical point of view, the modelization of the long-range dependence is associated with integrodifferential equations in a broad sense. On the other hand, in many cases such integrodifferential equations can be understood as fractional differential equations, and they can be studied in the fractional calculus framework.

Up to 1994, a very extensive review about the nonlinear nonlocal wave equations with applications in hydrodynamics, magnetohydrodynamics, and plasma can be found in the book of Naumkin and Shishmarëv [7]. The nonlocality in space is associated with the long-range interactions (many space scales) while the nonlocality in time is related to the effects with memory/delay (many time scales). The corresponding nonlocal systems are described by general integral equations. We have a large spectrum of scenarios where the integral equations occur [8, 9, 10, 11, 12, 13, 14, 15]: potential theory, signal processing, problems in geophysics, problems in electricity and magnetism, hereditary phenomena in physics and biology, problems in population growth and industrial replacement, radiation problems, optimization, and automatic control systems, communication theory, stochastic problems, and mathematical economics and financial phenomena.

Professor Magin in his recent book, *Fractional Calculus in Bioengineering* [16] established clearly the underlying mathematical context of fractional calculus.

The purpose of this book is to explore the behavior of biological systems from the perspective of fractional calculus. Fractional calculus, integration and differentiation of an arbitrary or fractional order, provides new tools that expand the descriptive power of calculus beyond the familiar integer-order concepts of rates of change and area under a curve... Fractional calculus adds new functional relationships and new functions to the familiar family of exponentials and sinusoids that arise in the realm of ordinary linear differential equations. Among such functions that play an important role, we have the Euler Gamma function, the Euler Beta function, the Mittag-Leffler functions, the Wright and Fox functions, and so forth.

From a historical point of view, the origin of fractional calculus coincides with that of differential calculus. In a letter [17] dated September 9, 1695, and addressed to G. Leibniz, M. de L'Hopital raises the question of the possible meaning of $d^n f(x)/dx^n$ if $n = 1/2$. The answer of Leibniz was:

Citation: Luis Vázquez, From Newton's Equation to Fractional Diffusion and Wave Equations, DOI: 10.1155/2011/169421.

"... $d^{1/2}x$ will be equal to $x\sqrt{dx:x}$. This is an apparent paradox from which, one day, useful consequences will be drawn". Fractional calculus arises to give a meaning to the noninteger derivatives. This concept was also analyzed by E. Euler, J. Liouville, P. S. Laplace, or Fourier. The generalization of this framework indicates that it is more appropriate to talk of *integration and differentiation of arbitrary order*. The book by Samko et al. [1] reflects the situation of fractional calculus up to 1990 with a wide bibliography and many applications. Therefore we can think in the field of Fractional Calculus as an old topic, but, from 20 years ago, the real applications of the so-called fractional models in so many branches of applied sciences and engineering have increased in a dramatic way, so we can consider the topic of fractional differential equations and their applications as a emergent field.

A nonexhaustive list of works that support the mentioned modern development of fractional calculus and its applications are in [18, 19, 20, 21, 22, 23, 24, 25]. From our point of view, the monographic work by Metzler and Klafter in 2000 [26] could be considered as the last breakpoint in developing the field of the study of fractional differential equations and their applications. Some of the more important publications on the mentioned topic during last five years, among so many references, could be the following books in [27, 28, 29, 30, 31, 32, 33, 34].

There are different definitions of fractional derivatives and, as a consequence, of fractional integrals, since the fractional derivation is the inverse operation of the fractional integration. This a generalization of what happens in ordinary calculus with the standard integer derivatives and integrals according to the fundamental theorem of calculus. The different fractional derivatives coincide in the integer case, up to a constant in some cases. We have the fractional derivatives and integrals of Riemann-Liouville, Liouville, Caputo, and so forth, and their general feature is the *nonlocal character* since they depend on the values of the function in an interval. As an illustration, we have a derivative of order β of the function $f(t)$:

$$\frac{d^\beta f(t)}{dt^\beta} = \frac{1}{\Gamma(m-\beta)}\int_0^t \frac{f^{(m)}(\tau)}{(t-\tau)^{\beta+m-1}}d\tau, \tag{1.1}$$

Where $m = 1,2,\ldots$ and $0 \leq m-1 < \beta \leq m$. This definition requires the absolute integrability of the derivative of order m.

The nonlocal character makes the fractional derivatives suitable for the modellization of systems with long-range interactions. On the other hand, the freedom in the definition of fractional derivatives allows us to incorporate different types of information. At the same time, the fractional derivatives with noninteger exponents stress which algebraic scale properties are relevant to the data analysis.

In this contribution, we present a standard approach to construct fractional differential equations from the basic equations of classical and quantum mechanics (Section 2). In Section 3, we analyze the possible bridges between fractional calculus and fractals while in Section 4, we consider the relation of fractional calculus with the modellization of relaxation processes.

FRACTIONAL DIFFERENTIAL EQUATIONS

Following the models of classical mechanics ([35]) and quantum mechanics ([36]), let us consider a possible standard path among the basic equations of physics that would allow us to interpret, in a more wide context, the level of *fractionalization* of the basic differential equations by analyzing the associated dynamics and solutions in the framework of the corresponding modellization [26, 27, 37, 38, 39].

For the sake of simplicity, let us consider systems in one space dimension. Newton›s equation for a particle of mass m moving in a one-dimensional force field F, u being the displacement, is

$$m\frac{d^2u}{dt^2} = F. \tag{2.1}$$

Let us consider now a discrete system, formed by infinite point-like masses, m , spaced a distance L , and connected by strings of the same constant k . If we name u_i the displacement of the particle i from the equilibrium, the equation of motion is given by

$$\frac{m}{L}\frac{d^2u_i}{dt^2} = kL\frac{u_{i+1}-2u_i+u_{i-1}}{L^2} \tag{2.2}$$

This equation is written in such a way as to interpret appropriately the continuous limit of the system when $L \to 0$:

$$u_i \longrightarrow u(x), \quad \frac{m}{L} \longrightarrow \rho \quad \text{(linear mass density)}$$
$$kL \longrightarrow Y \quad \text{(Young's modulus)}$$
$$\frac{u_{i+1}-2u_i+u_{i-1}}{L^2} \longrightarrow \frac{\partial^2 u}{\partial x^2}. \tag{2.3}$$

Thus, we have the equation

$$\frac{1}{c^2}\frac{\partial^2 u}{\partial t^2} - \frac{\partial^2 u}{\partial x^2} = 0, \tag{2.4}$$

where $c^2 = Y/\rho$. With this mechanical approach, we get the wave equation which appears in many contexts according to the different meanings of u and c . If we assume a damping mechanism either in the discrete system or in the continuous limit, we have the equation

$$\frac{1}{c^2}\frac{\partial^2 u}{\partial t^2} + \frac{1}{D}\frac{\partial u}{\partial t} - \frac{\partial^2 u}{\partial x^2} = 0 \tag{2.5}$$

When the damping term, $(1/D)(\partial u/\partial t)$, dominates over the inertial one, $(1/c^2)(\partial^2 u/\partial t^2)$, we have the standard diffusion equation

$$\frac{\partial u}{\partial t} - D\frac{\partial^2 u}{\partial x^2} = 0 \tag{2.6}$$

Other basic equations are obtained by considering extra linear terms. This is the case of the *Klein-Gordon equation*:

$$\frac{1}{c^2}\frac{\partial^2 u}{\partial t^2} - \frac{\partial^2 u}{\partial x^2} + \mu^2 u = 0 \tag{2.7}$$

and the *Telegraph equation*:

$$\frac{1}{c^2}\frac{\partial^2 u}{\partial t^2} + \frac{1}{D}\frac{\partial u}{\partial t} - \frac{\partial^2 u}{\partial x^2} + \mu^2 u = 0. \tag{2.8}$$

All this can be summarized in the following equation, where we also include Dirac's equation to be discussed in the following paragraphs.

Newton's equation for one particle

$\longrightarrow$ Newton's equation for one-dimensional system of particles linearly coupled

System of particles linearly coupled $\xrightarrow{\text{Continuous limit}}$ One-dimensional wave

Wave equation with damping $\longrightarrow$ Diffusion equation $\longrightarrow$ Fractional Dirac equation

Klein-Gordon equation $\longrightarrow$ Dirac equation, Telegraph equation. *(2.9)*

According to the previous statements, we have a set of very important features.

1. Fractional calculus provides a natural interpolation between different equations. For instance, we have the fractional interpolation between the wave equation (hyperbolic) and the diffusion equation (parabolic). Thus, we have the family of fractional equations

$$\frac{\partial^\beta u}{\partial t^\beta} - D\frac{\partial^2 u}{\partial x^2} = 0 \qquad (2.10)$$

When $\beta = 2$ we have the wave equation, while we have the diffusion equation for $\beta = 1$. The study of such possible interpolations has not been fully exploited yet.

2. The different equations above can be fractionalized by replacing the integer derivatives either in space or time by fractional derivatives. The range of fractional equations is very wide at the same time that the exploration of the associated dynamics and solutions is a task to be fulfilled. On the other hand, it is an open issue to analyze the relation of the different fractionalizations at the discrete and the continuous levels.

3. *The standard diffusion equation.* The diffusion equation (2.6) appears in very different physical contexts which leads to a better understanding and interpretation of the generalized diffusion equations. Such different applications allow a richer phenomenology to understand the possible meaning of the different possible fractionalizations. As an example, the equation is obtained by combining the following laws with the associated continuity equation.

4. Darcy›s law

$$\vec{q} = -K\nabla h, \qquad (2.11)$$

where q is the flux of groundwater, K is the medium property: hydraulic conductivity, and h is the potential related to the head.

5. Fourier›s law

$$\vec{Q} = -\kappa\nabla T, \qquad (2.12)$$

where Q is the flux of heat, K is the medium property: thermal conductivity and T is the potential related to the temperature.

6. Fick›s law

$$\vec{f} = -D\nabla C, \qquad (2.13)$$

where f is the flux of solute, D is the medium property: diffusion coefficient and C is the potential related to the concentration.

7. Ohm›s law

$$\vec{j} = -\sigma\nabla V, \qquad (2.14)$$

where j is the flux of charge (electrical current), σ is the medium property: electrical conductivity, and V is the potential related to the voltage.

8. *The fractional Dirac equation* (see [11, 40, 41, 42, 43]). The free Dirac equation ([36]) can be considered as the square root of the Klein-Gordon equation. This is also true for the case of $\mu = 0$ in (2.7) which corresponds to the wave equation. More precisely in this last case, we have Figure 1: where A and B are matrices satisfying the indicated algebra. The function $u(x,t)$ is scalar while $\Psi(x,t)$ is a multicomponent function with, at least, two scalar space-time components.

$$\Psi = \begin{pmatrix} \varphi(x,t) \\ \chi(x,t) \end{pmatrix} \qquad (2.15)$$

In the context of Figure 1, we can understand the following equation:

$$A\frac{\partial^{1/2}\Psi}{\partial t^{1/2}} + B\frac{\partial\Psi}{\partial x} = 0 \qquad (2.16)$$

as the square root of the standard diffusion equation (2.6). It is another way to view the possible interpolations between the hyperbolic operator of the wave equation and the parabolic one of the classical diffusion equation. Every scalar component of $\Psi(x,t)$ also satisfies the diffusion equation above. Such solutions can be interpreted as probability distributions with internal structure associated to internal degrees of freedom of the system. We could name them *diffunors* in analogy with the spinors of quantum mechanics.

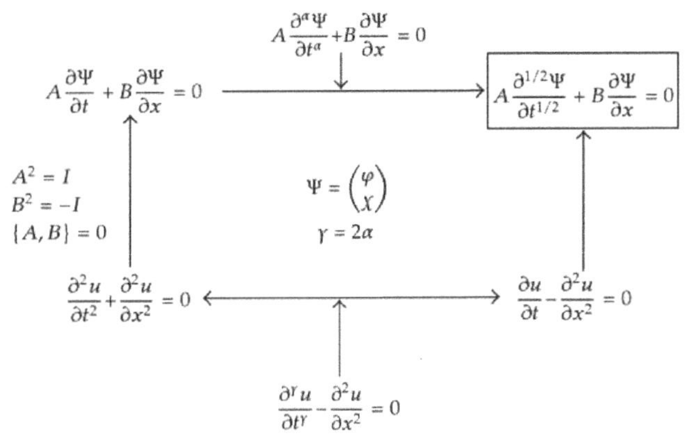

Figure 1: Fractional diffusion equation and fractional Dirac-type equation.

According to the possible representations of the Pauli algebra for A and B , we have either an uncoupled system or a coupled system of equations. An example of the first case is the representation

$$A = \begin{pmatrix} 0 & 1 \\ 1 & 0 \end{pmatrix}, \quad B = \begin{pmatrix} 0 & 1 \\ -1 & 0 \end{pmatrix} \qquad (2.17)$$

The transition between the two limit behaviours indicated in the diagram is not well understood and it deserves a deep analysis. As an illustration of the possible rich phenomenology associated to the interpolating fractional wave and Dirac equations is their behaviour with respect the discrete symmetries: time T and space inversion P (see [43]).

This analysis of the fractional wave, diffusion, and Dirac equations, fits in the general context of establish bridges between fractional calculus and the classical and quantum mechanics [44, 45]. This is also the approach of the works of Rabei and coworkers (see [46]).

FRACTIONAL CALCULUS AND FRACTALS

Fractals and fractional calculus generate parameters of intermediate order: arbitrary dimensions as well as arbitrary order of integration and differentiation [47, 48]. Let us consider the basic aspects associated with both conceptual blocks, and we will remark on possible bridges between them to be explored.

1. Fractals show self-similar structures, being the same structure at all scale levels. Such structures are introduced by using the concept of a reference structure and repeating itself over many scales, telescoping both downward and upward in scale. In general, the fractals structures are defined by an iterative process instead of an explicit mathematical formula. A simple case is the *Cantor set* which is a limiting set of points which results from discarding the middle third of each line segment in going from generation to generation, and starting from a line segment of unit length. The *fractal dimension D* is a measure of the irregularity of the fractal structure, and, in the case of the Cantor set, the fractal dimension is $D = \ln 2 / \ln 3 \approx 0.6309$.

The fractal property can be either in space or in time. In the first case we have the fractal geometric structures as the Cantor set while in the temporal context we have the structures of sequence of data generated for a system over an interval of time. A special case is the *Weierstrass function* which is an example of a function continuous everywhere but differentiable nowhere:

$$W(z) = \sum_{n=-\infty}^{\infty} \frac{(1 - e^{i\gamma^n z})e^{i\phi_n}}{\gamma^{(2-D)n}}, \tag{3.1}$$

where $\gamma > 1$, ϕ_n is an arbitrary phase and $1 < D < 2$, D being the fractal dimension of the self-similar structure defined by $W(z)$. The function can be interpreted as a superposition of harmonic terms: the first one with unit amplitude and unit frequency and the n -order term with amplitude $\gamma^{(2-D)(n-1)}$ and frequency γ^{n-1}. The lack of differentiability of $W(z)$ is related to the integer derivatives. On the other hand, it has been established that the fractional derivative (fractional integral) of the Weierstrass function is another fractal function with a greater (lesser) fractal dimension [49]. At the same time, it has been proved that the Weierstrass function is a solution of a fractional differential stochastic equation of motion. In particular, we have the following relations for the Weierstrass function.

- The fractional Riemann-Liouville integral of order β, with $0 < \beta < 1$:

$$D_{\mathrm{RL}}^{-\beta} W(z) = \frac{1}{\Gamma(\beta)} \int_{-\infty}^{z} \frac{W(u)du}{(z-u)^{1-\beta}}. \tag{3.2}$$

- The fractional Riemann-Liouville derivative of order β, with $0 < \beta < 1$:

$$D_{\mathrm{RL}}^{\beta} W(z) = \frac{1}{\Gamma(1-\beta)} \int_{-\infty}^{z} \frac{W(u)du}{(z-u)^{\beta}}. \tag{3.3}$$

- The relation between the fractal dimensions of the fractional derivative and integral of the Weierstrass function

$$\mathrm{Dim}\left[D_{\mathrm{RL}}^{-\beta} W\right] = \mathrm{Dim}[W] - \beta, \mathrm{Dim}\left[D_{\mathrm{RL}}^{\beta} W\right] = \mathrm{Dim}[W] + \beta. \tag{3.4}$$

2. One of the possible experimental contexts to apply together with the concepts of fractals and fractional calculus is related to the propagation of waves [50, 51, 52]. In the 19th Century, James Clerk Maxwell and Lord Rayleigh studied the interaction of electromagnetic waves with Euclidean regular structures (cylinders, spheres). On the other hand, there are either nonregular artificial structures or from nature that show many length scales and they are not suitable to be studied in the Euclidean context. This is the case of nonregular surfaces, disordered media, structures with specific properties of scattering, and so forth. In this context, the fundamental issue of special technological impact is *the relation between the geometrical parameters (structure descriptors) and the physical quantities that characterize electromagnetically the system*. For instance, a relevant information is the relation between the fractal dimension of the scattering surface and the scattered wave. Related to that, we have that a recent field of research has been to explore the possible links between the fractional calculus and the electromagnetic theory ([53]). The fractionalization of certain linear operators in electromagnetic theory has led to relevant solutions in radiation and scattering problems.

As a simple illustration, let us consider the propagation of a wave in fractal media. The geometrical optics cannot be applied and the eikonal equation is no longer valid since it applies for wave lengths λ □ *dimension of any change in the media*. This condition cannot be satisfied for fractal media because, in this case, there is no characteristic length of the system. We can consider the associated stationary eigenvalue problem that appears in two contexts: the wave equation in a fractal potential and the wave equation with fractal boundary conditions. In the last case, we have the general equation

$$Lu = \mu u, \tag{3.5}$$

where L is a linear differential operator on $\mathbb{R}^n$ with boundary conditions $u_0(x)$ on a nondifferentiable surface but which admits the fractional derivative D^{β} with $\beta < 1$. If we define $\Phi = D^{\beta-1}$ we have the problem $L\Phi = \mu\Phi$ with the boundary condition $\Phi_0(x)$, Φbeing differentiable. Thus, with the help of the fractional derivative, we transformed the original boundary problem into another one more suitable to be studied.

FRACTIONAL CALCULUS AND RELAXATION PROCESSES

In the study of the behavior of the materials under applied forces, we have two ideal relations between such forces and the resultant displacements in the material which are represented by linear differential equations.

1. *Linear elastic behavior* which is described by the Hooke's law (ideal solid):

$$\sigma(t) = E\varepsilon(t), \tag{4.1}$$

where

E is Young's modulus, σ = force/area is the stress and ε = elongation /original length is the strain of the corresponding mechanical model, that, in this case, is represented by a spring characterized by the constant E.

2. *Ideal viscous fluid behavior* which is described by Newton's fluid law (ideal fluid):

$$\sigma(t) = \eta \frac{d\varepsilon(t)}{dt} \tag{4.2}$$

where η is the viscosity coefficient, and, in this case, the phenomenological model is represented by a dashpot.

In general, the real materials exhibit characteristics which combine the behaviors of the ideal solids and liquids and are named *viscoelastic materials*. The basic models of such materials are based on combinations of springs and dashpots [2, 16, 54]. As an illustration, let us consider the combination of a spring and a dashpot in series. In this case, the stresses are equal for both structural parts while the strains add. This leads to the classical Maxwell model which is described by the following linear differential equation obtained from the above two equations:

$$E\frac{d\varepsilon(t)}{dt} = \frac{d\sigma(t)}{dt} + \frac{1}{\tau}\sigma(t) \tag{4.3}$$

where $\tau = \eta / E$. If we consider the Laplace transform of the above equation, we obtain the *relaxation modulus* $g(s) = \sigma(s)/\varepsilon(s) = E/((s+1)/\tau)$. Finally, the stress relaxation modulus for the Maxwell model is

$$G(t) = E\exp -\frac{t}{\tau} \tag{4.4}$$

which indicates that the Maxwell model shows an exponential stress decay, while different behaviors are observed in real materials. Fractional calculus allow us to introduce generalizations of the Maxwell model and the basic constitutive equation that leads to more general behaviors of the stress relaxation modulus.

3. The fractional constitutive equation

$$\sigma(t) = E(\tau)^{\alpha}\frac{d^{\alpha}\varepsilon(t)}{dt^{\alpha}}, \tag{4.5}$$

which is an interpolation between Hooke's law $(\alpha = 0)$ and Newton's law $(\alpha = 1)$. This fractional relation can be realized physically through hierarchical arrangements and in the limit of an infinite number of springs and dashpots [2].

4. The general fractional Maxwell equation offers a wide range of possible behaviors

$$E\tau^{\delta}\frac{d^{\delta}\varepsilon(t)}{dt^{\delta}} = \tau^{\rho}\frac{d^{\rho}\sigma(t)}{dt^{\rho}} + \sigma(t), \tag{4.6}$$

the solutions are well analyzed in [2, 16] as well as in the references therein.

5. In order to consider other generalizations and approaches for possible different applications, we can interpret the *relaxation process* either as an initial value problem or a memory problem.
   - *Initial Value Problem*. The relaxation (4.4) can be interpreted as the solution of the initial value problem

$$\tau\frac{dG(t)}{dt} = -G(t), \qquad G(0) = G_0, \tag{4.7}$$

with solution $G(t) = G_0 e^{-t/\tau}$. From (4.7), we can write as starting point for the fractional generalization:

$$G(t) - G_0 = -\tau^{-1}\int_0^t du\, G(u) = -\tau^{-1}\frac{d^{-1}G(t)}{dt^{-1}} \tag{4.8}$$

from which, we formally get the fractional generalization

$$G(t) - G_0 = -\tau^{-\alpha}\frac{d^{-\alpha}G(t)}{dt^{-\alpha}} \tag{4.9}$$

   - *Memory Problem*. Another natural generalization of (4.7) is

$$\frac{dG(t)}{dt} = \int_0^t K(t-u)G(u)du. \tag{4.10}$$

If $K(t) = K_0\delta(t)$, we recover the exponential behavior of the Maxwell model (4.4). If we have constant memory, $K = K_0$ we have the periodic behavior: $K(t) = G_0\cos\sqrt{K_0}t$. Finally, if $K(t) = K_0 t^{q-2}$ with $0 < q \le 2$ the equation (4.10) can be written as a fractional derivative and the solution can be expressed in terms of the Fox functions.

One possible application field for fractional calculus would be the modellization of the *shape memory alloys (SMA)*. These are basically materials that change shape when energy is applied or extracted from them. They exhibit hardness and elasticity properties that change radically at distinct temperatures [55]. There is a high complexity in the relation between the microscopic and macroscopic behaviors. There are no reliable constitutive models, although some have been studied by Tanaka [56], Liang and Rogers [57], and Brinson [58]. The key feature of these materials, as the name indicates, is that their dynamics is memory dependent. Thus, the possible realistic models should include the nonlocal effects in time through integrodifferential equations and, possibly, fractional differential equations.

DECLARATIONS

Acknowledgments

This paper has been partially supported by the Ministerio de Ciencia e Innovación of Spain under Grant AYA2009-14212-C05-05 and by FCT and POCI 2010 (Project MAT/58321/2004) with participation of FEDER.

REFERENCES

1. Samko SG, Kilbas AA, Marichev OI: *Fractional Integrals and Derivatives: Theory and Applications*. Gordon and Breach Science, Yverdon, Switzerland; 1993:xxxvi+976.
2. Hilfer R (Ed): *Applications of Fractional Calculus in Physics*. World Scientific, River Edge, NJ, USA; 2000:viii+463.
3. Hilfer R, Metzler R, Blumen A, Klafter J (Eds): Strange kinetics In *Chemical Physics* 2002,284(1):1-541. 10.1016/S0301-0104(02)00801-7
4. Baleanu D, Güvenç ZB, Tenreiro Machad JA: *New Trends in Nanotechnology and Fractional Calculus Applications*. Springer, New York, NY, USA; 2010:xii+531.
5. Ortigueira M, Tenreiro-Machado JA, Trujillo JJ, Vinagre B (Eds): Advances in fractional signals and

systems In *Journal of Signal Processing* 2011,91(3):349-606. 10.1016/j.sigpro.2010.08.002

6. Tenreiro-Machado JA, Luo A (Eds): Discontinuous and fractional dynamical systems In *Journal of Computational and Nonlinear Dynamics* 2008,3(2):1-125.
7. Naumkin PI, Shishmarëv IA: *Nonlinear Nonlocal Equations in the Theory of Waves, Translations of Mathematical Monographs. Volume 133.* American Mathematical Society, Providence, RI, USA; 1994:x+289.
8. Lonseth AT: Sources and applications of integral equations. *SIAM Review* 1977,19(2):241-278. 10.1137/1019039.
9. Cheney M, Borden B: *Fundamentals of Radar Imaging, CBMS-NSF Regional Conference Series in Applied Mathematics.Volume 79.* SIAM, Philadelphia, Pa, USA; 2009:xxiv+140.
10. Baillie RT, King ML (Eds): Fractional differencing and long memory processes In *Journal of Econometrics* 1996,73(1):1-324. 10.1016/0304-4076(95)01731-3
11. Vázquez L: A fruitful interplay: from nonlocality to fractional calculus. In *Nonlinear Waves: Classical and Quantum Aspects.* Edited by: Abdullaev FKh, Konotop VV. Kluwer Academic Publishers, New York, NY, USA; 2004:129-133.
12. Vázquez L, Usero D: Nonlocal equations and fractional models. *Revista de la Real Academia de Ciencias Exactas, Físicas y Naturales (España)* 2005,99(2):203-223.
13. Vázquez L, Evans WAB, Rickayzen G: Numerical investigation of a non-local sine-Gordon model. *Physics Letters A* 1994,189(6):454-459. 10.1016/0375-9601(94)91209-2
14. Alfimov GL, Usero D, Vázquez L:On complex singularities of solutions of the equation $\mathcal{H}U_x - U + U^p = 0$. *Journal of Physics A* 2000,33(38):6707-6720. 10.1088/0305-4470/33/38/305.
15. Pipiras V, Taqqu MS: Fractional calculus and its connections to fractional Brownian motion. In *Theory and Applications of Long-Range Dependence.* Edited by: Doukhan P, Oppenheim G, Taqqu MS. Birkhäuser, Boston, Mass, USA; 2003:165-201.
16. Magin RL: *Fractional Calculus and Bioengineering.* Begell House, New York, NY, USA; 2006.
17. Leibniz GW: *Oeuvres Mathematiques de Leibniz. Correpondence de leibniz avec Huygens, van Zulichem et le Marquis de L'Hospital Part 1. Volume 2.* Libr. de A. Franck, Paris, France; 1853.
18. Oldham KB, Spanier J: *The Fractional Calculus, Mathematics in Science and Engineering. Volume 11.* Academic Press, New York, NY, USA; 1974:xiii+234.
19. Miller KS, Ross B: *An Introduction to the Fractional Calculus and Fractional Differential Equations.* John Wiley & Sons, New York, NY, USA; 1993:xvi+366.
20. Oustaloup A: *La Drivation Non Entiere. Theorie, Synthese et Applications.* Hermes Science, Stanmore, UK; 1995.
21. Kiryakova V: *Generalized Fractional Calculus and Applications, Pitman Research Notes in Mathematics Series. Volume 301.* Longman Scientific & Technical, Harlow, UK; 1994:x+388.
22. Podlubny I: *Fractional Differential Equations: An Introduction to Fractional Derivatives, Fractional Differential Equations, to Methods of Their Solution, Mathematics in Science and Engineering. Volume 198.* Academic Press, San Diego, Calif, USA; 1999:xxiv+340.
23. Rubin B: *Fractional Integrals and Potentials, Pitman Monographs and Surveys in Pure and Applied Mathematics. Volume 82.* Longman, Harlow, UK; 1996:xiv+409.
24. Bonilla B, Kilbas AA, Trujillo JJ: *Cálculo Fraccionario y Ecuaciones Diferenciales Fraccionarias.* Serv. de Publicaciones de la UNED, Madrid, Spain; 2003.
25. Zaslavsky GM: Chaos, fractional kinetics, and anomalous transport. *Physics Reports* 2002,371(6):461-580. 10.1016/S0370-1573(02)00331-9
26. Metzler R, Klafter J: The random walk›s guide to anomalous diffusion: a fractional dynamics approach. *Physics Reports* 2000,339(1):77.
27. Kilbas AA, Srivastava HM, Trujillo JJ: *Theory and Applications of Fractional Differential Equations, North-Holland Mathematics Studies. Volume 204.* Elsevier Science B.V., Amsterdam, The Netherlands; 2006:xvi+523.
28. Zaslavsky GM: *Hamiltonian Chaos and Fractional Dynamics.* Oxford University Press, Oxford, UK; 2008:xiv+421.
29. Uchaikin VV: *Method of Fractional Derivatives.* Artishok-Press, Ulyanovsk, Russia; 2008.
30. Lakshmikantham V, Leela S, Vasundhara-Devi J: *Theory of Fractional Dynamic Systems.* Cambridge Scientific Publishers, Cambridge, UK; 2009.
31. Mainardi F: *Fractional Calculus and Waves in Linear Viscoelasticity: An Introduction to Mathematical Models.* Imperial College Press, London, UK; 2010:xx+347.
32. Caponetto R, Dongola G, Fortuna L, Petras I: *Fractional Order Systems: Modeling and Control Applications.* World Scientific, Singapure; 2010.
33. Tarasov VE: *Fractional Dynamics in Physics.* Higher Education Press and Springer, Berlin, Geramny; 2011.
34. Klafter J, Lim SC, Metzler R (Eds): *Fractional Dynamics: Applications of Fractional Calculus to Dynamics of Particles, Fields and Media.* World Scientific, Singapore; 2011.
35. Goldstein H: *Classical Mechanics.* Addison-Wesley Press, Cambridge, Mass, USA; 1951:xii+399.
36. Dirac PAM: *The Principles of Quantum Mechanics.* Oxford University Press, Oxford, UK; 1958.
37. Sokolov IM, Klafter J, Blumen A: Fractional kinetics. *Physics Today* 2002,55(11):48-54. 10.1063/1.1535007
38. Jiménez S, Pascual P, Aguierre C, Vázquez L: A panoramic view of some perturbed nonlinear wave equations.*International Journal of Bifurcation and Chaos* 2004,14(1):1-40. 10.1142/S0218127404009211

39. Mendes RV, Vázquez L: The dynamical nature of a backlash system with and without fluid friction. *Nonlinear Dynamics* 2007,47(4):363-366. 10.1007/s11071-006-9035-y
40. Vázquez L: Fractional diffusion equations with internal degrees of freedom. *Journal of Computational Mathematics*2003,21(4):491-494.
41. Vázquez L, Vilela Mendes R: Fractionally coupled solutions of the diffusion equation. *Applied Mathematics and Computation* 2003,141(1):125-130. 10.1016/S0096-3003(02)00326-0.
42. Kilbas AA, Pierantozzi T, Trujillo JJ, Vázquez L: On the solution of fractional evolution equations. *Journal of Physics A*2004,37(9):3271-3283. 10.1088/0305-4470/37/9/015.
43. Pierantozzi T, Vázquez L: An interpolation between the wave and diffusion equations through the fractional evolution equations Dirac like. *Journal of Mathematical Physics* 2005,46(11):113512, 12.
44. Vázquez L: Una panorámica del Cálculo Fraccionario y sus aplicaciones. *Revista de la Real Academia de Ciencias Exactas, Físicas y Naturales* 2004,98(1):17-26.
45. Turchetti G, Usero D, Vázquez L: Hamiltonian systems with fractional time derivative. *Tamsui Oxford Journal of Mathematical Sciences* 2002,18(1):31-44.
46. Rabei EM, Nawafleh KI, Hijjawi RS, Muslih SI, Baleanu D: The Hamilton formalism with fractional derivatives. *Journal of Mathematical Analysis and Applications* 2007,327(2):891-897. 10.1016/j.jmaa.2006.04.076.
47. West BJ, Bologna M, Grigolini P: *Physics of Fractal Operators*. Springer, New York, NY, USA; 2003:x+354.
48. Mandelbrot BB: *The Fractal Geometry of Nature*. W. H. Freeman, San Francisco, Calif, USA; 1982:v+460.
49. Rocco A, West BJ: Fractional calculus and the evolution of fractal phenomena. *Physica A* 1999,265(3):535-546. 10.1016/S0378-4371(98)00550-0.
50. Berry MV: Difractals. *Journal of Physics A* 1979, 12: 781-797. 10.1088/0305-4470/12/6/008.
51. Jaggard DL: On fractal electrodymanics. In *Recent Advances in Electromagnetic Theory*. Edited by: Kritikos HN, Jaggard DL. Springer, Berlin, Germany; 1990:183-224.
52. Konotop VV, Fei Z, Vázquez L: Wave interaction with a fractal layer. *Physical Review E* 1993,48(5):4044-4048. 10.1103/PhysRevE.48.4044.
53. Engheta N: Fractional paradigm in electromagnetic theory. In *Frontiers in Electromagnetism*. Edited by: Werner DH, Mittra R. IEEE Press, New York, NY, USA; 2000:523-552.
54. Vázquez L, Velasco Cebrián MP: Aplicación del Cálculo Fraccionario a los procesos de relajación. *Memorias de la Real Academia de Ciencias Exactas, Físicas y Naturales* 2009, 23: 105-115.
55. Sun QP, Hwang KC: Micromechanics modelling for the constitutive behavior of polycrystalline shape memory alloys-II. Study of the individual phenomena. *Journal of the Mechanics and Physics of Solids* 1993,41(1):19-33. 10.1016/0022-5096(93)90061-J.
56. Tanaka K: A thermo-mechanical sketch of shape memory effect: one-dimensional tensile behavior. *Res Mechanica*1986,18(3):251-263.
57. Liang C, Rogers CA: One-dimensional thermo-mechanical constitutive relations for shape memory materials.*Journal of Intelligent Material Systems and Structures* 1990,1(2):207-218. 10.1177/1045389X9000100205.
58. Brinson LC: One-dimensional constitutive behavior of shape memory alloys: thermomechanical derivation with non-constant material functions and redefined martensite internal variable. *Journal of Intelligent Material Systems and Structures* 1993,4(2):229-242. 10.1177/1045389X9300400213.

CHAPTER 2

SOLUTIONS FOR CLASSICAL MECHANICS PROBLEMS OF LIÉNARD TYPE

Gülden Gün Polat and Teoman Özer

INTRODUCTION

In classical mechanics, it is known that many important problems can be derived from Lienard equation of the form ´ $y'' + f(y)y' + g(y) = 0$. For instance, in dynamics, the Van der Pol oscillator that is a nonconservative oscillator with nonlinear damping is a physically important example of Lienard equation. Li ´ enard equation can also be considered ´ as a model for a spring-mass system where the damping force f(y) corresponds to the position (e.g., the mass might be moving through a viscous medium of varying density), and the spring constant $g(y)$ corresponds to how much the spring is stretched. It is possible to present other similar examples related to the Lienard equation. Thus, it can be said ´ that Lienard equation has an important role in mathematical ´ physics and mechanics fields.

In addition, with respect to the investigation of solution of differential equations, one of the most powerful methods for nonlinear differential equations is based on the study of Lie group of transformations. In the last century, the applications of Lie groups to the problems in mechanics, mathematics, physics, and so forth, have been carried out by many mathematicians, for example [1–12]. For the case of Lienard equation, the classical Lie point symmetries of ´ Lienard equation are investigated in detail in the study ´ [13]. However, the application of Noether theorem in the concept of theory of Lie groups to differential equations introduces that any Noether symmetry of the action of a physical system has a corresponding conservation law [1]. A conservation law means a quantity associated with a physical system that remains unchanged as the system evolves in time. In classical mechanics, the natural form of the Lagrangian is defined as the difference between the kinetic energy and potential energy of the system. An important property of the Lagrangian is that conservation laws can be easily derived from it. Furthermore, Noether theorem presents that variational symmetries are in one to one correspondence with conservation laws for the associated Euler-Lagrange equations. If the Lagrangian of a system is known, then the equations of motion of the system may be obtained by a direct substitution of the expression for the Lagrangian into the Euler-Lagrange equation. In addition, a Lagrangian of a system can be determined by making use of partial Lagrangian approach and the conservation laws can be obtained directly. The first study on this approach is carried out in [14]. In this study the Noether symmetries of Lienard equation are investigated for two different cases ´ of arbitrary functions f(y) and $g(y)$. In fact, f(y) function is considered in two different forms for corresponding to two different mechanical problems, namely, linear undamped systems and linear damped systems. For both cases, the function f(y) = k_1y+k_2 can be chosen in a linear form for constant values of k_1 and k_2. The first case for the linear undamped systems corresponds to the equation of motion of the free particle and simple harmonic oscillator, and the second case for the linear damped systems corresponds to the damped harmonic oscillator. As a result, for different choices of k_1 and k_2, the conserved forms, partial Noether symmetries can be analyzed and classified for the Lienard ´ equation.

In addition, in the literature, there is another approach called λ-symmetry, which is introduced by Muriel and Romero [15–

Citation: Gülden Gün Polat and Teoman Özer, "Solutions for Classical Mechanics Problems of Liénard Type," Advances in Mathematical Physics, vol. 2014, Article ID 107895, 16 pages, 2014. doi:10.1155/2014/107895

18]. They introduce a new prolongation formula to investigate λ-symmetries for second order differential equations and they prove that although the equation has no Lie symmetries and by using the prolongation formula, then λ-symmetry of differential equation can be obtained and so the order of equation can be reduced. For convenience they consider a specific form (linear) for λ-function such as $\lambda(x,y,y') = \lambda_1(x,y)y' + \lambda_2(x,y)$ [17]. For example, for the specific choices functions f(y) and $g(y)$, the Lienard equations correspond to the general modified ′ Emden-type equation and then its groups properties can be analyzed with linear term and constant external forcing via λ-symmetries. Moreover, it is possible to apply λ-symmetry approach to the Lienard equation by choosing other different ′ forms of f(y) function and one can examine different $g(y)$ functions corresponding to each $f(y)$ function. So we can obtain different types of Lienard equations having ′ physical meaning and highly nonlinear forms and by λ- symmetry concept new solutions, conservation laws and classification properties of the Lienard equation can be ′ investigated.

The outline of paper is as follows. In Section 2 we introduce some preliminaries about partial Noether theorem. Section 3 is devoted to determine conservation laws and partial Noether symmetries of some important problems in classical mechanics derived from Lienard equation. In ′ Section 4 we present some fundamental definitions about λ-symmetry approach and the calculation procedure of the integration factor and the conservation laws from λ- symmetries algorithmically. In Section 5 we investigate conservation laws, integration factors, and invariant solutions by using λ-symmetry concept, as an alternative approach, of some mechanics problems considered in Section 3. Some important results in the study are discussed in Section 6.

FUNDAMENTAL DEFINITIONS ABOUT NOETHER THEOREM

Let $x = (x^1,\ldots,x^n)$ be the independent variable with coordinates x^i and $y = (y^1,\ldots,y^m)$ the dependent variable with coordinates y^α. The derivatives of y with respect to x are

$$y_i^\alpha = D_i(y^\alpha), \qquad y_{ij}^\alpha = D_j D_i(y^\alpha) \tag{1}$$

where

$$D_i = \frac{\partial}{\partial x^i} + y_i^\alpha \frac{\partial}{\partial y^\alpha} + y_{ij}^\alpha \frac{\partial}{\partial y_j^\alpha}, \qquad i = 1,\ldots,n \tag{2}$$

is the operator of total differentiation. Additionally, if A is a vector space of all differential functions of all finite orders, that is called the universal space and then the operators given below can be defined in the space A.

Definition 1. For each α the Euler-Lagrange operator is defined by

$$\frac{\delta}{\delta y^\alpha} = \frac{\partial}{\partial y^\alpha} + \sum_{s\geq 1}(-1)^s D_{i_1}\cdots D_{i_s}\frac{\partial}{\partial y^\alpha_{i_1\cdots i_s}}, \qquad \alpha = 1,\ldots,m. \tag{3}$$

Definition 2. Generalized operator can be formulated as

$$X = \xi^i \frac{\partial}{\partial x^i} + \eta^\alpha \frac{\partial}{\partial y^\alpha}, \qquad \xi^i, \eta^\alpha \in \mathscr{A}, \tag{4}$$

and for convenience this operator can be written as

$$X = \xi^i \frac{\partial}{\partial x^i} + \eta^\alpha \frac{\partial}{\partial y^\alpha} + \sum_{s\geq 1}\zeta^\alpha_{i_1\cdots i_s}\frac{\partial}{\partial y^\alpha_{i_1\cdots i_s}}, \tag{5}$$

where the additional coefficients are examined by the formula

$$\zeta_i^\alpha = D_i(W^\alpha) + \xi^j y_{ij}^\alpha,$$
$$\zeta^\alpha_{i_1\cdots i_s} = D_i(W^\alpha) + \xi^j y_{ij}^\alpha, \qquad s \geq 2, \tag{6}$$

in which W^α is the Lie characteristic function defined as

$$W^\alpha = \eta^\alpha - \xi^j y_j^\alpha. \tag{7}$$

The generalized operator (5) can also be rewritten by using Lie characteristic function such as

$$X = \xi^i D_i + W^\alpha \frac{\partial}{\partial y^\alpha} + \sum_{s\geq 1} D_{i_1}\cdots D_{i_s}(W^\alpha)\frac{\partial}{\partial y^\alpha_{i_1\cdots i_s}}, \tag{8}$$

and the Noether operator associated with a generalized operator X is given by

$$N = \xi^i + W^\alpha \frac{\partial}{\partial y^\alpha} + \sum_{s\geq 1} D_{i_1}\cdots D_{i_s}(W^\alpha)\frac{\delta}{\delta y^\alpha_{ii_1\cdots i_s}},$$
$$i = 1,\ldots,n, \tag{9}$$

where the Euler-Lagrange operators with respect to derivatives of y^α are obtained from (3) by replacing y^α by the corresponding derivatives

$$\frac{\delta}{\delta y_i^\alpha} = \frac{\partial}{\partial y^\alpha} + \sum_{s\geq 1}(-1)^s D_{j_1}\cdots D_{j_s}\frac{\partial}{\partial y^\alpha_{ij_1\cdots j_s}},$$
$$i = 1,\ldots,n,\ \alpha = 1,\ldots,m. \tag{10}$$

Definition 3. Suppose that a nth-order ordinary differential equation system is given by

$$f_\alpha(x, y, y_{(1)}, y_{(2)}, \ldots, y_{(n)}) = 0, \qquad \alpha = 1,2,\ldots,m, \tag{11}$$

with maximal rank and is locally solvable.

Definition 4. $I^j = I^j(x, y, y_{(1)}, y_{(2)}, \ldots, y_{(k-1)}) \in \mathscr{A}$ which satisfies what is given below

$$D_i(I^i) = 0 \tag{12}$$

is called a conservation law of (11). Moreover, (12) exists for all solutions of (11). In addition, (11) are assumed to be as the following form:

$$f_\alpha \equiv f_\alpha^0 + f_\alpha^1 = 0, \qquad \alpha = 1,2,\ldots,m. \tag{13}$$

Let us consider differential equations of the form

$$\frac{\delta L}{\delta y^\alpha} = 0, \qquad \alpha = 1,2,\ldots,m, \tag{14}$$

where $L = (Lx,(x,y,y_{(1)},\ldots,u_{(l)})) \in \mathscr{A}, l \leq k, k$ being the order of (14), is a Lagrangian and $\delta/\delta y\alpha$ is the Euler-Lagrange operator defined by (3)

Suppose that if nonzero functions $g_\alpha^\beta \in \mathscr{A}$ are defined such that (14)

$$\frac{\delta L}{\delta y^\alpha} = g_\alpha^\beta f_\beta^1, \tag{15}$$

in which $f_\beta^1 \neq 0$ then L is called a partial Lagrangian of (13).

Definition 5. If there exists a vector $B = (B^i, \ldots, B^n)$, $B^i \in \mathcal{A}$, $B^i \neq N^i L + C^i$, C^i are constants and $W = (W^1, \ldots, W^m)$. If $W\alpha \in A$ is the characteristic of X then

$$X(L) + LD_i(\xi^i) = W^\alpha \frac{\delta L}{\delta y^\alpha} + D_i(B^i) \tag{16}$$

is called Noether-type symmetry operator corresponding to a partial Lagrangian $L \in A$. In addition, B is called gauge function. If X is a partial Noether operator corresponding to partial Lagrangian L then the gauge function B exists then the first integral is given by

$$I = \xi L + (\eta - y'\xi) L_{y'} - B. \tag{17}$$

PARTIAL NOETHER SYMMETRIES OF LIÉNARD EQUATION

Let us consider the one-dimensional nonlinear Lienard-type ´ differential equation such as

$$y'' + f(y)y' + g(y) = 0, \tag{18}$$

where $f(y)$ and $g(y)$ are smooth functions of y. In this section we analyze the partial Noether symmetries of (18).

For the Lienard equation (18), the Euler-Lagrange operator can be written as (3)

$$\frac{\delta}{\alpha y^\alpha} = \frac{\partial}{\partial y^\alpha} - D_x \frac{\partial}{\partial y_x} + D_x^2 \frac{\partial}{\partial y_{xx}}, \tag{19}$$

and the partial Lagrangian L for the Lienard equation (´ 18) is given by

$$L = \frac{1}{2} y'^2 - \int g(y)\,dy. \tag{20}$$

Therefore (18) becomes

$$\frac{\delta L}{\delta y} = -g(y) - y' = y' f(y) \tag{21}$$

The partial Noether operators corresponding to L are

$$X(L) = \xi \frac{\partial}{\partial x} + \eta \frac{\partial}{\partial y} + \eta^1 \frac{\partial}{\partial y'}, \tag{22}$$

where η^1 is equal to

$$\eta^1 = \eta_x + (\eta_y - \xi_x) y' - \xi_y y'^2. \tag{23}$$

By using the relation (16) we have

$$\begin{aligned} &\eta_x y' + (\eta_y - \xi_x) y'^2 - \xi_y y'^3 - \eta g(y) \\ &+ \frac{1}{2} y'^3 \eta_y - \xi_x \int g(y)\,dy - y'\xi_y \int g(y)\,dy \\ &= \eta f(y) y' - \xi f(y) y'^2 + B_x + y' B_y, \end{aligned} \tag{24}$$

and analyzing (24) the coefficients of derivatives y' gives the following determining equations:

$$\xi_y = 0, \tag{25}$$

$$\eta_y - \frac{1}{2}\xi_x + \xi f(y) = 0, \tag{26}$$

$$\eta_x - \eta f(y) - B_y = 0, \tag{27}$$

$$-\eta g(y) - \xi_x \int g(y)\,dy - B_x = 0. \tag{28}$$

First, (25) gives

$$\xi = a(x), \tag{29}$$

where $a(x)$ is an arbitrary function. Then substituting ξ into (26) η gives

$$\eta = \frac{1}{2} a(x)' y - a(x) \int f(y)\,dy + \beta(x) \tag{30}$$

where $\beta(x)$ is an arbitrary function. Differentiating (27) and (28), respectively, with respect to x and y we have the following equation:

$$\begin{aligned} &g'(y)\left(a'(x) y - 2a(x) \int f(y)\,dy + 2\beta(x)\right) \\ &+ g(y)\left(3a'(x) - 2a(x) f(y)\right) \\ &+ a'''(x) y - 2a''(x) \int f(y)\,dy + 2\beta''(x) \\ &- a''(x) y f(y) + 2a'(x) f(y) \int f(y)\,dy \\ &- 2\beta'(x) f(y) = 0, \end{aligned} \tag{30}$$

which is a differential equation in terms of the unknown functions $f(y)$ and $g(y)$. In the following subsections we consider some classical mechanics problems related with Lienard-type differential equation for different choices of ´ f(y) and $g(y)$ functions.

Linear Undamped Systems: f(y)=0

If we consider f(y) = 0, then (31) is equal to

$$\begin{aligned} &g'(y)\left(2\beta(x) + a'(x) y\right) + 3a'(x) g(y) \\ &+ a'''(x) y + 2\beta''(x) = 0. \end{aligned} \tag{31}$$

It is easy to see that a solution of (32) is

$$g(y) = k_1 y + k_2, \tag{32}$$

where k_1 and k_2 are arbitrary constants. We now analyze the following subcases:

(i) $k_1 = k_2 = 0$,

(ii) $k_1 = 0$ and $k_2 \neq 0$,

(iii) $k_1 \neq 0$ and $k_2 = 0$,

(iv) $k_1 \neq 0$ and $k_2 \neq 0$

Case 1 (Free particle motion ($k_1 = k_2 = 0$)). If we take $k_1 = k_2 = 0$, then (33) is equal to zero and thus the Lienard ´ equation (18) becomes

$$y'' = 0. \tag{34}$$

This form of (18) represents the equation of motion of a free particle. By using (32), (29), and (30) we obtain the partial Noether infinitesimals

$$\xi = \frac{c_1 x^2}{2} + c_2 x + c_3, \qquad \eta = \frac{(c_1 x + c_2) y}{2} + c_4 x + c_5, \tag{35}$$

where c_1, c_2, c_3, c_4, and c_5 are arbitrary constants. The corresponding partial Noether symmetry generators are

$$X_1 = \frac{x^2}{2}\frac{\partial}{\partial x} + \frac{xy}{2}\frac{\partial}{\partial y}, \qquad X_2 = x\frac{\partial}{\partial x} + \frac{y}{2}\frac{\partial}{\partial y},$$

$$X_3 = \frac{\partial}{\partial x}, \qquad X_4 = x\frac{\partial}{\partial y}, \qquad X_5 = \frac{\partial}{\partial y}. \tag{36}$$

Additionally, the gauge function is obtained from (25) and (26)

$$B(x, y) = \frac{1}{4}c_1 y^2 + c_4 y. \tag{37}$$

The corresponding conservation laws by the (17) formula are obtained in the following forms:

$$I_1 = -\frac{y^2}{4} + \frac{1}{2}xyy' - \frac{1}{4}x^2(y')^2,$$

$$I_2 = \frac{yy'}{2} - \frac{1}{2}x(y')^2, \qquad I_3 = -\frac{1}{2}x(y')^2,$$

$$I_4 = -y + xy', \qquad I_5 = y'. \tag{38}$$

It is important to mention that each conservation law satisfies the equality $D_x I = W(-y'') = 0$ with respect to the characteristic W

Case 2 (Free falling particle $(k_1 = 0, k_2 \neq 0)$). For this case of k_1 and k_2 (18) becomes

$$\xi = \frac{c_1 x^2}{2} + c_2 x + c_3,$$

$$\eta = \frac{(c_1 x + c_2)y}{2} + c_3 x + c_4 - \frac{3}{4}c_2 x^2 k_2 - \frac{1}{4}c_1 x^3 k_2,$$

$$B = \frac{1}{16}\big(16c_5 + 16c_3 y + 4c_1 y^2 - 16c_4 x k_2$$

$$- 8c_3 x^2 k_2 - 24c_2 xyk_2$$

$$-12c_1 x^2 y k_2 + 4c_2 x^3 k_2^2 + c_1 x^4 k_2^2\big). \tag{40}$$

The corresponding partial Noether symmetry generators are

$$X_1 = \frac{x^2}{2}\frac{\partial}{\partial x} + \left(\frac{xy}{2} - \frac{x^3 k_2}{4}\right)\frac{\partial}{\partial y},$$

$$X_2 = x\frac{\partial}{\partial x} + \left(\frac{y}{2} - \frac{3x^2 k_2}{4}\right)\frac{\partial}{\partial y},$$

$$X_3 = \frac{\partial}{\partial x} + y\frac{\partial}{\partial y}, \qquad X_4 = \frac{\partial}{\partial y}, \tag{41}$$

and the conservation laws corresponding to each partial Noether symmetries are

$$I_1 = \frac{1}{16}\big(- x^4 k_2^2 - 4y^2 - 4x^3 k_2 y'$$

$$-4x^2 k_2 y'^2 + 4xy\left(xk_2 + 2y'\right)\big),$$

$$I_2 = \frac{1}{4}\left(-x^3 k_2^2 - 3x^2 k_2 y' - 2xy'^2 + 2y\left(x\lambda_2 + y'\right)\right),$$

$$I_3 = \frac{1}{2}\left(x^2 k_2 - 2(1 + k_2)y + 2xy' - (y')^2\right),$$

$$I_4 = y' + xk_2.$$

Case 3 (Free linear harmonic oscillator $(k_1 \neq 0, k_2 = 0)$). If we chose k_1 as an arbitrary and $k_2 = 0$ thus the Lienard ′ equation (18) is equal to

$$y'' + k_1 y = 0, \tag{43}$$

which refers to the equation of a linear harmonic oscillator for $k1 > 0$. Then the solution of determining (25)–(28) for the f(y) = 0 and $g(y) = k_1 y$ gives

$$\xi = c_3 - \frac{c_2 \cos\left(2x\sqrt{k_1}\right)}{2\sqrt{k_1}} + \frac{c_1 \sin\left(2x\sqrt{k_1}\right)}{2\sqrt{k_1}},$$

$$\eta = c_4 \cos\left(x\sqrt{k_1}\right) + c_5 \sin\left(x\sqrt{k_1}\right)$$

$$+ \frac{1}{2}y\left(c_1 \cos\left(2x\sqrt{k_1}\right) + c_1 \cos\left(2x\sqrt{k_1}\right)\right),$$

$$B = \frac{1}{2\sqrt{k_1}}\Big(2c_6\sqrt{k_1} + 2c_5 k_1 \cos\left(x\sqrt{k_1}\right)$$

$$+ c_2\left(k + y^2 k_1\right)\cos\left(2x\sqrt{k_1}\right)$$

$$- 2c_4 y k_1 \sin\left(x\sqrt{k_1}\right)$$

$$-kc_1 \sin\left(2x\sqrt{k_1}\right) - c_1 y^2 k_1 \sin\left(2x\sqrt{k_1}\right)\Big). \tag{44}$$

The corresponding partial Noether symmetry generators are found as follows:

$$X_1 = \frac{\sin\left(2x\sqrt{k_1}\right)}{2\sqrt{k_1}}\frac{\partial}{\partial x} + \frac{\cos\left(2x\sqrt{k_1}\right)}{2}y\frac{\partial}{\partial y},$$

$$X_2 = -\frac{\cos\left(2x\sqrt{k_1}\right)}{2\sqrt{k_1}}\frac{\partial}{\partial x} + \frac{\sin\left(2x\sqrt{k_1}\right)}{2}y\frac{\partial}{\partial y},$$

$$X_3 = \frac{\partial}{\partial x}, \qquad X_4 = \cos\left(x\sqrt{k_1}\right)\frac{\partial}{\partial y},$$

$$X_5 = \sin\left(x\sqrt{k_1}\right)\frac{\partial}{\partial y},$$

and the conservation laws are

$$I_1 = \frac{1}{4}\Bigg(\sqrt{k_1}\sin\left(2x\sqrt{k_1}\right)y^2$$

$$+2yy'\cos\left(2x\sqrt{k_1}\right) - \frac{\sin\left(2x\sqrt{k_1}\right)y^2}{\sqrt{k_1}}\Bigg),$$

$$I_2 = \frac{1}{4}\Bigg(- \sqrt{k_1}\cos\left(2x\sqrt{k_1}\right)y^2$$

$$+2yy'\sin\left(2x\sqrt{k_1}\right) + \frac{\cos\left(2x\sqrt{k_1}\right)y^2}{\sqrt{k_1}}\Bigg),$$

$$I_3 = -\frac{1}{2}\left(-2k + y^2 k_1 - (y')^2\right),$$

$$I_4 = \sqrt{k_1}\sin\left(x\sqrt{k_1}\right)y + \cos\left(x\sqrt{k_1}\right)y',$$

$$I_5 = -\sqrt{k_1}\cos\left(x\sqrt{k_1}\right)y + \sin\left(x\sqrt{k_1}\right)y', \tag{46}$$

in which it is easy to see that the relation $D_x I=0$ is satisfied for each conserved form.

Case 4 (Displaced linear harmonic oscillator $(k_1 \neq k_2 \neq 0)$). If the k_1 and k_2 are assumed to be arbitrary, then we obtain the Lienard equation (18) as the following form:

$$y'' + k_1 y + k_2 = 0, \tag{47}$$

which represents a displaced simple harmonic motion. Equation (47) transforms the previous case with a change of variable such as $y \rightarrow \tilde{y} = y + (k_2/k_1)$. Therefore, this form of Lienard equation gives the similar partial Noether ′ symmetries, generators, and conservation laws to Case 3.

Linear Damped Systems f(y) = Constant = k

The other case for f(y) is to be considered as a constant. By substituting f(y) = k into the (29) and (30) then we calculate the infinitesimals

$$\xi = a(x),$$

$$\eta = \left(\frac{1}{2}a'(x) - a(x)k\right)y + \beta(x) \tag{48}$$

and rewrite (31), which includes the unknown functions $f(y)$ and $g(y)$

$$\begin{aligned} & g'(y)\left(2\beta(x) - 2ka(x)y + a'(x)y\right) \\ & \quad + g(y)\left(3a'(x) - 2ka(x)\right) \\ & \quad + a'(x)2k^2y - 2k\beta'(x) - 3kya''(x) + 2\beta''(x) \\ & \quad + ya'''(x) = 0. \end{aligned} \tag{49}$$

If we analyze the solutions of (49) then a solution of this equation for $g(y)$ can be considered as linear function of y, which is similar to the section A; that is,

$$g(y) = k_1 y + k_2, \tag{50}$$

where k_1 and k_2 are arbitrary constants. To examine all possibilities of f(y)we again consider the following subcases:

i. $k_1 = k_2 = 0$,

ii. $k_1 = 0$ and $k_2 \neq 0$,

iii. $k_1 \neq 0$ and $k_2 = 0$,

iv. $k_1 \neq 0$ and $k_2 \neq 0$.

Case 1 (Free particle in a viscous medium ($k_1 = k_2 = 0$)). For this case the Lienard equation (18) is equal to

$$y'' + ky' = 0, \tag{51}$$

which represents the equation of a free particle in a viscous medium. If we substitute $g(y) = 0$ into (49), then by using the determining (27)-(28) we find the partial Noether symmetries ξ, η and gauge function $B(x, y)$ such that

$$\begin{aligned} \xi &= \frac{e^{2kx}}{2k}c_1 + \frac{e^{2x}}{k}c_2 + c_3, \\ \eta &= -\frac{1}{2}\left(c_2e^{kx} + 23k\right)y + c_4\frac{e^{kx}}{k} + c_5, \\ B(x,y) &= -\frac{1}{2k}\left(2c_2e^{kx} + c_1e^{2kx} + 2c_5k^2y - c_3k^3y^2\right), \end{aligned} \tag{52}$$

and partial Noether symmetry generators are

$$\begin{aligned} & X_1 = \frac{e^{2kx}}{2k}\frac{\partial}{\partial x}, \qquad X_2 = \frac{e^{kx}}{k}\frac{\partial}{\partial x} - \frac{e^{kx}y}{2}\frac{\partial}{\partial y}, \\ & X_3 = -ky\frac{\partial}{\partial y}, \qquad X_4 = \frac{e^{kx}}{k}\frac{\partial}{\partial y}, \qquad X_5 = \frac{\partial}{\partial y}. \end{aligned} \tag{53}$$

By applying the definition of first integral (17) we obtain the conservation laws

$$\begin{aligned} & I_1 = -\frac{1}{4k}e^{2kx}y'^2, \qquad I_2 = -\frac{e^{kx}y'}{2k}\left(y + ky'\right), \\ & I_3 = -\frac{1}{2}\left(2 + k^2y^2 + 2kyy' - y'^2\right), \\ & I_4 = \frac{1}{k}e^{kx}y', \qquad I_5 = y + ky', \end{aligned} \tag{54}$$

which satisfy the relation $D_x I=0$.

Case 2 (Falling particle in a viscous medium $(k_2 \neq 0, k_1 = 0)$). If the functions f(y) = k and $g(y) = k_2$ are substituted into the Lienard equation (′ 18) we obtain

$$y'' + ky' + k_2 = 0, \tag{55}$$

which is the equation of a falling particle in a viscous medium. It is obvious that this form of Lienard equation ′ (55) can be transformed into previous case by using the simple transformation such as $y \rightarrow \tilde{y} = y + (k_2/k)w.$. Therefore, if this transformation can be applied for (52)–(54) then partial Noether symmetries, partial Noether generators, and conservation laws are derived, respectively, in terms of w for (55)

Case 3 (Damped linear harmonic oscillator $(k_1 \neq 0, k_2 = 0)$). Suppose that $k1$ is arbitrary and $k2$ is equal to zero and thus Lienard equation (′ 18) can be obtained as the form

$$y'' + ky' + k_1 y = 0, \tag{56}$$

which corresponds to the equation of a damped linear harmonic oscillator. By considering similar process to previous cases we get partial Noether symmetries, partial Noether generators, and gauge function

$$\begin{aligned} \xi &= c_1e^{x(k-\gamma)} + c_2e^{x(k+\gamma)} + c_3e^{kx}, \\ \eta &= -\frac{1}{2}\left(2c_4e^{1/2x(k-\gamma)} + 2c_5e^{1/2x(k+\gamma)} - e^{x(k-\gamma)}y \right. \\ & \quad \times \left(e^{x\gamma}\left(c_3k - c_2e^{x\gamma}(\gamma - k)\right)\right. \\ & \quad \left.\left. + c_1(k+\gamma)\right)\right), \\ & B(x,y) \\ & = \frac{1}{4}e^{-x\gamma}\left(-4ce^{kx}\left(c_1 + e^{x\gamma}\left(c_3 + c_2e^{x\gamma}\right)\right)\right. \\ & \quad + y\left(2c_5e^{1/2x(k+3\gamma)}\right. \\ & \quad \times \left(-k + \gamma - 2c_4e^{1/2x(k+\gamma)}\right. \\ & \quad \times (k+\gamma) + e^{kx}y \\ & \quad \times \left(c_1\left(k^2 + k\gamma - 4k_1\right) - c_2e^{2x\gamma}\right. \\ & \quad \left.\left.\left.\left.\left. \times \left(-k^2 + k\gamma + 4k_1\right)\right)\right)\right)\right)\right), \end{aligned} \tag{57}$$

where c_1, c_2, c_3, c_4, and c_5 are arbitrary constants and $\sqrt{k^2 - 4k_1; k^2 - 4k_1}$ is assumed to be positive. And the partial Noether generators are

$$X_1 = e^{x(k-\gamma)}\frac{\partial}{\partial x} - \frac{1}{2}e^{x(k-\gamma)}y(k+\gamma)\frac{\partial}{\partial y},$$

$$X_2 = e^{x(k+\gamma)}\frac{\partial}{\partial x} + \frac{1}{2}e^{x(k-\gamma)+2x\gamma}y(\gamma-k)\frac{\partial}{\partial y},$$

$$X_3 = e^{kx}\frac{\partial}{\partial x} - \frac{1}{2}e^{kx}ky\frac{\partial}{\partial y}, \qquad X_4 = e^{1/2x(k-\gamma)}\frac{\partial}{\partial y},$$

$$X_5 = e^{1/2x(k4\gamma)}\frac{\partial}{\partial y}, \tag{58}$$

and the corresponding conservation laws are

$$I_1 = -\frac{1}{4}e^{x(k-\gamma)}\left(\left(k^2+k\gamma-2k_1\right)y^2+2\left(k+\gamma yy'+2y'^2\right)\right),$$

$$I_2 = \frac{1}{4}e^{x(k+\gamma)}\left(\left(-k^2+k\gamma+2k_1\right)y^2+2\left(\gamma-k\right)yy'-2y'^2\right),$$

$$I_3 = -\frac{1}{2}e^{kx}\left(k_1y^2+kyy'+y'^2\right),$$

$$I_4 = \frac{1}{2}e^{x(k-\gamma)}\left(\left(k+\gamma\right)y+2y'\right),$$

$$I_5 = \frac{1}{2}e^{x(k+\gamma)}\left(\left(k-\gamma\right)y+2y'\right). \tag{59}$$

Case 4 (Displaced damped harmonic oscillator ($k_1 \neq k_2 \neq 0$)). For this case we get the Lienard equation (′ 18)

$$y''+ky'+k_1y+k_2=0, \tag{60}$$

which is the equation of displaced damped harmonic oscillator. By using the change of variable such as $y \rightarrow \tilde{y} = y + k_2/k_1$ this case corresponds to the previous form of Lienard ′ equation. So by applying this transformation to (57)–(59) partial Noether operators, partial Noether generators, and conservation laws can be obtained similarly

λ-SYMMETRY APPROACH FOR DIFFERENTIAL EQUATIONS

In this section we consider λ-symmetry properties of Lienard ′ equation and for this purpose we first present some fundamental definitions and theorems about λ-symmetries [15, 17]. Let us consider a nth-order ordinary differential equation

$$\tilde{\Delta}\left(x, y^{(n)}\right) = 0 \tag{61}$$

where variables (x, y) are in some open set $M \subset X \times Y \cong \mathbb{R}^2$. For $k \in \mathbb{N}$, $M^{(k)} \subset X \times Y^{(k)}$ denotes the corresponding k- jet space and the elements of (k) are denoted by $(x, {}^{(k)}) = (x, y, y', ..., y^k)$. By applying implicit function theorem to (61), as a result, this equation can be written in the explicit form

$$y^{(n)} = \phi\left(x, y^{(n-1)}\right) \tag{62}$$

The vector field

$$A = \frac{\partial}{\partial x} + y'\frac{\partial}{\partial y} + \cdots + \phi\left(x, y^{(n-1)}\right)\frac{\partial}{\partial y^{(n-1)}} \tag{63}$$

is called the vector field associated with (62). An integrating factor of (61) is assumed to be such as $(x, {}^{(k)})$ for some k and $0 \le k \le n$. If the left-hand side of (61) is multiplied by integration factor μ, total derivative of some function $\Delta(x, y^{(n-1)})$ is obtained as the form

$$\mu\left(x, y^{(k)}\right)\cdot\tilde{\Delta}\left(x, y^{(n)}\right) = D_x\left(\Delta\left(x, y^{(n-1)}\right)\right) \tag{64}$$

It is clear that λ-symmetries can be used for reduction of order of a differential equation. Namely, $(\Delta(x, y^{(n-1)})) = 0$ is an exact differential equation, which is the result of trivial reduction of order $\Delta(x, y^{(n-1)}) = C, C \in \mathbb{R}$ [18] If $\lambda \in C^{\infty}(M^{(k)}), 0 \le k \le n$, is any solution of the partial differential equation:

$$\sum_{i=0}^{n-1}(D_x+\lambda)^i(1)\frac{\partial\Delta}{\partial y^{(i)}} = 0, \tag{65}$$

when $D_x(\Delta(x, y^{(n-1)})) = 0$, then the vector field] = ∂_y is a λ-symmetry of (61).

Theorem 6 (see [17]). Assume that (61)is a nth-order ordinary differential equation that admits an integrating factor μ such that $\mu|_{\tilde{\Delta}=0} \neq 0$. If λ is any particular solution of (65), then the vector field] = ∂_y is λ-symmetry of (61).

Now let us consider the second order differential equation (62) and a vector field (63)

$$y'' = \phi\left(x, y, y'\right), \tag{66}$$

$$A = \frac{\partial}{\partial x} + y'\frac{\partial}{\partial y} + \cdots + \phi\left(x, y, y'\right)\frac{\partial}{\partial y'}; \tag{67}$$

then a conservation law (first integral) of (66) is any function such as $I(x, y, y')$ which satisfies the relation

$$A(I) = 0. \tag{68}$$

If] is assumed to be a λ-symmetry of (66) and $\omega(x, y, y')$ is a first-order invariant of][λ,(1)] and any particular solution of the equation

$$\omega_y + \lambda\left(x, y, y'\right)\omega_{y'} = 0, \tag{69}$$

then a first-order invariant reduced equation of the form $\Delta_R(x, y, y')$=0 can be obtained by using the reduction process associated with the λ-symmetry. Thus the general solution is found such as an equation of the implicit form

$$G(x, \omega) = C, \quad C \in \mathbb{R} \tag{70}$$

It is clear that $((x, \omega(x, y, y')) = 0$ is an equivalent form of (66). Consequently,

$$\mu\left(x, y, y'\right) = G_{\omega}\left(x, y, \omega\left(x, y, y'\right)\right)\cdot\omega_{y'}\left(x, y, y'\right) \tag{71}$$

is an integrating factor of (66).

Theorem 7 (see [17]). Let $y^n = (x, {}^{(n-1)})$ be a nth-order ordinary differential equation, where ϕ is an analytic function of its arguments. There exists a function $(x, {}^{(k)})$, for some $k<n$, such that the vector field v = ∂_y is a λ-symmetry of the equation

λ-SYMMETRIES, CONSERVATION LAWS, AND INTEGRATING FACTORS OF LIÉNARD EQUATION

In this section we investigate λ-symmetries of (18) for different cases of arbitrary functions of f(y) and g(y). By considering theorem (62) the vector field v = ∂_y is assumed to be as a λ-symmetry of (18). If we write explicit form of (18) corresponding to (66) we obtain

$$\phi(x, y, y') = -y' f(y) - g(y) \tag{72}$$

By applying (65) to (72) we find

$$\lambda_x + y'\lambda_y + \phi\lambda_{y'} + \lambda^2 - \phi_y - \lambda\phi_{y'} = 0, \tag{73}$$

in which λ is any particular solution of (73). For convenience a solution of λ of (73) can be assumed to be linear form such that

$$\lambda(x, y, y') = \lambda_1(x, y)\, y' + \lambda_2(x, y) \tag{74}$$

Therefore, the expansion of (73) becomes

$$\begin{aligned} &- g(y)\lambda_1 + y'^2\lambda_1^2 + f(y)\lambda_2 + 2y'\lambda_1\lambda_2 \\ &+ \lambda_2^2 + y' f'(y) + g'(y) + y'^2\lambda_{1y} \\ &+ y'\lambda_{2y} + y'\lambda_{1x} + \lambda_{2x} = 0. \end{aligned} \tag{75}$$

The usual separation of powers of derivatives of y (75) reduces to the system

$$\lambda_1^2 + \lambda_{1y} = 0, \tag{76}$$

$$2\lambda_1\lambda_2 + f'(y) + \lambda_{2y} + \lambda_{1x} = 0, \tag{77}$$

$$-g'(y) - g(y)\lambda_1 + f(y)\lambda_2 + \lambda_2^2 + \lambda_{2x} = 0. \tag{78}$$

A particular solution of (76) is found

$$\lambda_1(x, y) = \frac{1}{y} \tag{79}$$

And so (77) becomes

$$\frac{2\lambda_2}{y} + f'(y) + \lambda_{2y} = 0. \tag{80}$$

To obtain the solution of (80) different cases of f(y) should be considered. Firstly in order to compare Noether and λ- symmetry approaches we investigate the similar cases of functions f(y) and $g(y)$, which are analyzed in Section 3.

Case 1. In this case we analyze f(y) and $g(y)$ functions which correspond to the case of Section 3.1. Namely, f(y) = 0 and $g(y) = k_1 y + k_2$, which represent linear undamped systems. For this functions the determining equations (80) and (78) become

$$\frac{2\lambda_2}{y} + \lambda_{2y} = 0, \tag{81}$$

$$-k_1 - (k_1 y + k_2)\lambda_1 + \lambda_2^2 + \lambda_{2x} = 0. \tag{82}$$

The solution of (81) is given by

$$\lambda_2(x, y) = \frac{c(x)}{y} \tag{83}$$

Case 1.1 ((y) = 0 and (y) = 0 ($k_1 = k_2 = 0$)). If we substitute (83) into (82), we obtain (x) = 0 and so $\lambda_2(x, y) = 0$. Therefore, (x, y, y') is found by using (79) as follows:

$$\lambda(x, y, y') = \frac{y'}{y} \tag{84}$$

As a result, a first-order invariant (x, y, y') of $v^{[\lambda,(1)]}$ is applied to (69) then the conservation law and integration factor can be derived by using the λ-symmetry (84)

$$\omega_y + \frac{y'}{y}\omega_{y'} = 0. \tag{85}$$

A solution of (85) is

$$\omega(x, y, y') = \frac{y'}{y} \tag{86}$$

In order to express the Lienard equation (18) in terms of {x, ω, ω' }, the terms y' and y'' can be eliminated from (86) and so the reduced form becomes

$$\omega' + \omega^2 = 0, \tag{87}$$

and the general solution of (88) is

$$\omega = \frac{1}{x - G}, \quad G \in \mathbb{R}. \tag{88}$$

The integration factor corresponding to (71) can be written as

$$\mu(x, y, y') = \frac{y}{y'^2} \tag{89}$$

Here, (x, (x, y, y')) which satisfies the relation (68) is equal to conserved form

$$D_x\left(x - \frac{y}{y'}\right) = 0. \tag{90}$$

Case 1.2 (f(y) = 0 and $g(y) = k_2\ (k_1 = 0, k_2 \neq 0)$). By applying same process with the previous case, it is clear that in order to satisfy (82), $k2$ must be equal to zero. As a result, the results in this case are similar to the Case 1.1.

Case 1.3 (f(y) = 0 and $g(y) = k_1 y\ (k_1 \neq 0, k_2 = 0)$). For this form for (18), similar results can be obtained with (84)–(86). Therefore the reduced form of (18) can be written as

$$\omega' + \omega^2 + k_1 = 0. \tag{91}$$

The general solution of (91) is

$$\omega = -\sqrt{a}\tan\left(\sqrt{a}(x - G)\right), \quad G \in \mathbb{R}. \tag{92}$$

By using the relation (71) integration factor is given by

$$\mu(x, y, y') = \frac{y}{ay^2 + y'^2}. \tag{93}$$

In (92), (x, (x, y, y')) corresponds to conserved form

$$D_x\left(\frac{1}{\sqrt{a}}\left(x\sqrt{a} + \arctan\left(\frac{y'}{y\sqrt{a}}\right)\right)\right) = 0, \tag{94}$$

which is equal to original Lienard equation (18) for this case. This relation allows us to obtain new invariant solution of (18) which is

$$y(x) = c_1 \cos((K - x)\sqrt{a}) \quad (95)$$

where c_1 and K are constants.

Case 2. Now we analyze the case which is similar to the case in Section 3.2. As a reminder, in this section the Lienard equation represents linear damped systems for the choices of f(y) = k = constant and $g(y) = k_1y+k_2$. For this case of f(y) and $g(y)$ the third determining equation (78) becomes

$$k_1 - (k_1 y + k_2)\lambda_1 + k\lambda_2 + \lambda_2^2 + \lambda_{2x} = 0. \quad (96)$$

Now we evaluate different cases of k_1 and k_2.

Case 2.1 (f(y) = k and $g(y) = 0$ ($k_1 = k_2 = 0$)). By applying similar procedure and after determination of y' and y", the reduced form of Lienard equation (18) is obtained in terms of $\{x, \omega, \omega'\}$

$$\omega' + \omega^2 + k\omega = 0. \quad (97)$$

The general solution of (97) is

$$\omega = \frac{e^{kG}k}{e^{kG} - e^{kx}}, \quad G \in \mathbb{R} \quad (98)$$

According to (71) the integration factor is

$$\mu(x, y, y') = \frac{y}{kyy' + y'^2}. \quad (99)$$

The relation related to the (x, (x, y, y')) gives the corresponding conservation form

$$D_x\left(\frac{1}{k}\left(kx - \ln\left(1 + \frac{ky(x)}{y'(x)}\right)\right)\right) = 0. \quad (100)$$

Then the new invariant solution of (18) can be written as

$$y(x) = \left(1 - e^{k(C-x)}\right)c_2, \quad (101)$$

where C and c_2 are constants.

Case 2.2 (f(y) = k and $g(y)$ = $(k_1 = 0, k_2 \neq 0)$). After same manipulations it is possible to show that this case corresponds to the Case 2.1.

Case 2.3 (f(y) = k and $g(y)$ = $k_1 y\ (k_1 \neq 0, k_2 = 0)$). According to (84)–(86), the reduced form of (18) is given by

$$\omega' + \omega^2 + k\omega + k_1 = 0. \quad (102)$$

The general solution of (102) is

$$\omega = \frac{1}{2}\left(-k + \sqrt{k^2 - 4k_1}\tan\left(\frac{1}{2}\sqrt{4k_1 - k^2}(G - x)\right)\right),$$

$$G \in \mathbb{R}. \quad (103)$$

By using of the relation (71) the integration factor is obtained such that

$$\mu(x, y, y') = \frac{y}{ay^2 + y'(ky + y')} \quad (104)$$

(x, (x, y, y')) which is equal to conserved form can be examined from (102) and it can be written as

$$D_x\left(\frac{1}{4k_1 - k^2}\right.$$
$$\times\left(4k_1 x - k^2 x\right.$$
$$\left.\left. + 2\arctan\left(\frac{1}{4k_1 - k^2}\left(k + \frac{2y'}{y}\right)\right)\right)\right) = 0, \quad (105)$$

which satisfies the original Lienard equation (18) for this case. And so we can derive new invariant solution of (18) in the following form:

$$y(x) = e^{1/2(-k_1x + 2\ln(\cos(1/2)\sqrt{4k_1 - k^2}(x - C)))}c_2, \quad (106)$$

where c^2 and C are constants.

Modified Emden Equations f(y)=ky; k Is a Constant

It is clear that a particular solution of (31) for this special choice of f(y) can be given by

$$g(y) = \frac{1}{9}k^2y^3 + k_1 y + k_2.$$

In this case by the choices of the functions of f(y) and $g(y)$, Lienard equation is considered in the nonlinear form [′ 19]. We analyze four different case of k_1 and k_2. These cases are

$k_1 = k_2 = 0,$

$k_1 = 0$ and $k_2 \neq 0$

$k_1 \neq 0$ and $k_2 = 0,$

$k_1 \neq 0$ and $k_2 \neq 0$

In this study we consider the first and the third cases since one can find only lambda function for the cases in which it is assumed in the linear form.

Modified Emden Equation ($k_1=k_2=0$)

Corresponding Lienard equation for ′ $k_1 = k_2 = 0$ is

$$y'' + ky'y + \left(\frac{1}{9}\right)k^2y^3 = 0, \quad (108)$$

which is nonlinear ordinary differential equation. In order to obtain λ-symmetries the determining equations (76)–(78) should be evaluated together. So the results of these equations give λ-symmetry (79) such that

$$\lambda(x, y, y') = \frac{y'}{y} + \frac{ky}{3}, \quad (109)$$

and by substituting (109) into (69) we obtain

$$\omega_y + \left(\frac{y'}{y} + \frac{ky}{3}\right)\omega_{y'} = 0. \quad (110)$$

It is clear that a solution of this (110) is

$$\omega\left(x, y, y'\right) = \frac{ky^2 + 3y'}{3y}. \tag{111}$$

By using (111) y' and y'' can be derived in terms of w and so Lienard equation can be written as

$$\omega' + \omega^2 = 0, \tag{112}$$

which is the reduced form. The general solution of (112) is

$$\omega = \frac{1}{x - G}, \quad G \in \mathbb{R}. \tag{113}$$

Integration factor corresponding to (71) is

$$\mu\left(x, y, y'\right) = \frac{9y}{\left(ky^2 + 3y'\right)^2}. \tag{114}$$

It is clear that (x, (x, y, y')) is equivalent to conserved form

$$D_x\left(\frac{-3y + ky^2x + 3y'x}{ky^2 + 3y'}\right) = 0, \tag{115}$$

and the reduced equation of (108) can be derived as

$$\frac{-3y\,(x) + ky(x)^2x + 3y'\,(x)\,x}{ky(x)^2 + 3y(x)'} - C = 0. \tag{116}$$

The solution of (116) gives the new invariant solution of (108)

$$y\,(x) = \frac{6\,(C - x)}{2Ckx - kx^2 - 6c_1}, \tag{117}$$

where C and $c1$ are constants.

Modified Emden Equation with Linear Term ($k_1 \neq 0$ and k_2=0).

Choosing $k1$ as arbitrary $k2 = 0$ Lienard equation is ′ obtained as a nonlinear ordinary differential equation in the form

$$y'' + kyy'\frac{1}{9}k^2y^3 + k_1y \tag{118}$$

By the same way and for this case the reduced form of (18) can be obtained

$$\omega' + \omega^2 + k_1 = 0. \tag{119}$$

The general solution of (119) is

$$\omega = -\sqrt{k_1}\tan\left(\sqrt{k_1}\,(x - G)\right), \quad G \in \mathbb{R}. \tag{120}$$

According to (71) the integration factor is examined

$$\mu\left(x, y, y'\right) = \frac{9y}{\left(ky^2 + 3y'\right)^2 + 9k_1y^2}. \tag{121}$$

It is clear that (x, (x, y, y')) is equivalent to conserved form

$$D_x\left(\left(x\arctan\left(\frac{ky^2 + 3y'}{3y\sqrt{k_1}}\right)\right)\right) = 0. \tag{122}$$

The reduced equation of (118) can be derived as

$$x\arctan\left(\frac{ky^2 + 3y'}{3y\sqrt{k_1}}\right) - C = 0, \tag{123}$$

and the solution of (123) gives the new invariant solution of (108)

$$y\,(x) = \frac{3\sqrt{k_1}\cos\left((C - x)\sqrt{k_1}\right)}{3\sqrt{k_1}c_1 - k\sin\left((C - x)\sqrt{k_1}\right)}, \tag{124}$$

where C and c_1 are constants.

Alternative Approach for the Consideration of f(y) and g(y).

Alternatively, to get a classification, we first define f(y) function and then try to determine λ_1, λ_2, and $g(y)$ functions. The corresponding subcases are given below.

$f(y)=ay^n+by^{n-1}+c, (n \neq -2, n \neq -1)$

where a, b, and c are arbitrary constants. For this case of f(y) we get (80) of the form

$$\frac{2\lambda_2}{y} + \lambda_{2y} + any^{n-1} + b\,(n - 1)\,y^{n-2} = 0. \tag{125}$$

A particular solution of this equation gives the $\lambda_2(x, y)$ such that

$$\lambda_2\,(x, y) = -y^{n-2}\left(\frac{by\,(n - 1)}{n + 1} + \frac{any^2}{n + 2}\right). \tag{126}$$

Hence, λ-symmetry can be written by using (79)

$$\lambda\left(x, y, y'\right) = \frac{y'}{y} - y^{n-2}\left(\frac{by\,(n - 1)}{n + 1} + \frac{any^2}{n + 2}\right) \tag{127}$$

If we substitute λ_1 and λ_2 into (78), then the new form of differential equation is obtained in terms of unknown function f(y)

$$g'\,(y) - \frac{g\,(y)}{y} + y^{n-2}\left(\frac{by\,(n - 1)}{n + 1} + \frac{any^2}{n + 2}\right) \times\left(y^{n-2}\left(\frac{by\,(n - 1)}{n + 1} + \frac{any^2}{n + 2}\right) - ay^n - by^{n-1} - c\right) = 0, \tag{128}$$

then the general solution of (128) is given by

$$g\,(y) = yc_1 + \frac{1}{\left(n^2 + 3n + 2\right)^2} \times\left(y^{n-1}\left(b\,(n + 2) + a\,(n + 1)\,y\right) \times\left(cy\left(n^2 + 3n + 2\right) + y^n\left(b\,(n + 2) + a\,(n + 1)\,y\right)\right)\right), \tag{129}$$

where c_1 is constant. Thus the Lienard equation for these ′ special cases for f(y) and $g(y)$ (18) becomes

$$y'' + y'\left(ay^n + by^{n-1} + c\right) + yc_1 + \frac{1}{\left(n^2 + 3n + 2\right)^2}$$
$$\times \left(y^{n-1}\left(b(n+2) + a(n+1)y\right)\right.$$
$$\times \left(cy\left(n^2 + 3n + 2\right)\right.$$
$$\left.\left. + y^n\left(b(n+2) + a(n+1)y\right)\right)\right) = 0. \tag{130}$$

In order to obtain the conservation law and the integration factor by using λ-symmetry (127), a first-order invariant (x, y, y') of][λ,(1)] is obtained in the following form from the relation (69):

$$\omega_y + \left(\frac{y'}{y} - y^{n-2}\left(\frac{by(n-1)}{n+1} + \frac{any^2}{n+2}\right)\right)\omega_{y'} = 0, \tag{131}$$

which has a solution of the form

$$\omega\left(x, y, y'\right) = \frac{1}{\left(n^2 + 3n + 2\right)} \times \left(2by^n + bny^n + ay^{n+1} + any^{n+1} + 2y' + 3ny' + n^2y'\right). \tag{132}$$

So in terms of $\{x, \omega, \omega'\}$ the Lienard equation (18) can be written as

$$\omega' + \omega^2 + b\omega + c_1 = 0. \tag{133}$$

And the general solution of (133) is given by

$$\omega = \frac{1}{2}\left(-b\sqrt{4c_1 - b^2}\tan\left(\frac{1}{2}\sqrt{4c_1 - b^2}\,(G - x)\right)\right), \quad G \in \mathbb{R}. \tag{134}$$

According to (71) we obtain the integration factor as

$$\mu\left(x, y, y'\right) = \left(y^2\left(n^2 + 3n + 2\right)^2\right)$$
$$\times \left(\left(c\left(n^2 + 3n + 2\right)y\right.\right.$$
$$+ 2\left(by^n(n+2) + (n+1)\right.$$
$$\left.\left.\times\left(ay^{n+1} + (n+2)y'\right)\right)\right)^2$$
$$\left. - y^2\left(n^2 + 3n + 2\right)^2\right)^{-1}. \tag{135}$$

It is the fact that (x, (x, y, y')) is equivalent to the following conserved form

$$D_x\left(\frac{1}{b^2 - 4c_1}\right.$$
$$\times \left(b^2 - 4c_1x - 2\sqrt{4c_1 - b^2}\right.$$
$$\times \arctan\left(\frac{c}{\sqrt{4c_1 - b^2}}\right.$$
$$+ 2\left(by^n(n+2) + (n+1)\right.$$
$$\left.\times\left(ay^{n+1} + (n+2)y'\right)\right)$$
$$\times\left(\sqrt{4c_1 - b^2}\right.$$
$$\left.\left.\left.\left.\times\left(n^2 + 3n + 2\right)y\right)^{-1}\right)\right)\right) = 0 \tag{136}$$

which gives the original Lienard equation given in the form ′ (130). It is possible to say that the solution of reduced equation of (130) cannot be solved for the general n. But, if we choose specifically n=1 then the reduced form of (130) is

$$\frac{1}{b^2 - 4c_1}$$
$$\times \left(b^2 - 4c_1x - 2\sqrt{4c_1 - b^2}\right.$$
$$\left.\times \arctan\left(\frac{3by(x) + 6y(x)' + 2ay(x)^2}{3y(x)\sqrt{4c_1 - b^2}}\right)\right)$$
$$- k_1 = 0, \tag{137}$$

where k_1 is a constant. It is a fact that this type of Lienard ′ equation (130) for the case n=1 cannot be solved by known methods in the literature. But in this analysis, the solution of reduced equation (127) as a new invariant solution of (130) can be obtained as

$$y(x) = \left(6c_1\cos\left(\frac{1}{2}\sqrt{4c_1 - b^2}\,(x - k_1)\right)\right)$$
$$\times\left(6k_2e^{bx/2} - ab\cos\left(\frac{1}{2}\sqrt{4c_1 - b^2}\,(k_1 - x)\right)\right.$$
$$\left. - a\sqrt{4c_1 - b^2}\sin\left(\frac{1}{2}\sqrt{4c_1 - b^2}\,(k_1 - x)\right)\right)^{-1}, \tag{138}$$

where k_2 is a constant.

In this case, it is possible to check that two specific cases for n should be analyzed separately such that ($n = -2$, $n = -1$). We investigate these, respectively, in the following subcases.

$f(y) = (a/y^2) + (b/y^3) + c,\ n = -2.$

By using this form of f(y), (80) becomes

$$\frac{2\lambda_2}{y} + \lambda_{2y} - \frac{3b}{y^4} - \frac{2a}{y^3} = 0. \tag{139}$$

A particular solution of $\lambda_2(x, y)$ from differential equation (139) is

$$\lambda_2(x,y) = \frac{2ay\ln y - 3b}{y^3}. \tag{140}$$

Thus the λ-symmetry is obtained from the relation (79) as

$$\lambda(x,y,y') = \frac{y'}{y} + \frac{2ay\ln y - 3b}{y^3}. \tag{141}$$

The differential equation (78) can be rewritten in terms of unknown function f(y) by considering λ_1 and λ_2 in the following form:

$$g'(y) - \frac{g(y)}{y} + \frac{1}{y^4}\left(2a\ln y - \frac{3b}{y}\right)^2 + \frac{1}{y^2}\left(\left(\frac{a}{y^2} + \frac{b}{y^3} + c\right)\left(2a\ln y - \frac{3b}{y}\right)\right) = 0; \tag{142}$$

hence, the general solution of (142) is found as follows:

$$g(y) = yc_1 + \frac{1}{4y^5} \times \left((-2b + ay + 2ay\ln y) \times \left(-2b + ay + 2cy^3 + 2ay\ln y\right)\right), \tag{143}$$

where c_1 is constant. So we can rewrite the Lienard equation ′ (18) for this case of f(y) and $g(y)$

$$y'' + y'\left(\frac{a}{y^2} + \frac{b}{y^3} + c\right) + yc_1 + \frac{1}{4y^5}\left((-2b + ay + 2ay\ln y) \times \left(-2b + ay + 2cy^3 + 2ay\ln y\right)\right). \tag{144}$$

If we substitute λ-symmetry into (69), the relation

$$\omega_y + \left(\frac{y'}{y} + \frac{2ay\ln y - 3b}{y^3}\right)\omega_{y'} = 0 \tag{145}$$

is obtained. A solution of this equation is

$$\omega(x,y,y') = \frac{-2b + ay + 2y^2y' + 2ay\ln y}{2y^3}. \tag{146}$$

In order to obtain Lienard equation (18) in terms of $\{x, \omega, \omega'\}$, the equalities y' and y'' can be simplified (146). And so the reduced form is

$$\omega' + \omega^2 + b\omega + c_1 = 0, \tag{147}$$

and the general solution of (147) is given by

$$\omega = \frac{1}{2}\left(-b\sqrt{4c_1 - b^2}\tan\left(\frac{1}{2}\sqrt{4c_1 - b^2}\,(G - x)\right)\right), \quad G \in \mathbb{R}; \tag{148}$$

thus the integration factor corresponding to (71) is

$$\mu(x,y,y') = \left(y^6\right) \times \left(\left(-2b + y\left(a + cy^2 + 2yy'\right) + 2ay\ln y\right)^2 - \left(c^2 - 4c_1\right)y^6\right)^{-1}. \tag{149}$$

It is clear that $(x, (x, y, y'))$ is equal to the conserved form

$$D_x\left(\frac{1}{b^2 - 4c_1} \times \left(c^2 - 4c_1x - 2\sqrt{4c_1 - b^2} \times \arctan\left(\frac{cy^3 - 2b + ay + 2y^2y' + 2ay\ln y}{y^3\sqrt{4c_1 - b^2}}\right)\right)\right) = 0, \tag{150}$$

which gives the original Lienard equation such as the form ′ (143).

$f(y)=(a/y)+(b/y^2)+c,\ n = -1$

Equation (80) is found in the following form:

$$\frac{2\lambda_2}{y} + \lambda_{2y} - \frac{2b}{y^3} - \frac{a}{y^2} = 0. \tag{151}$$

A particular solution of (151) is equal to

$$\lambda_2(x,y) = \frac{ay + 2b\ln y}{y^2}. \tag{152}$$

By substituting (79) and (152) into (74) we have

$$\lambda(x,y,y') = \frac{y'}{y} + \frac{ay + 2b\ln y}{y^2}. \tag{153}$$

If λ_1 and λ_2 are inserted to (78) the differential equation is obtained in terms of f(y)

$$g'(y) - \frac{g(y)}{y} + \frac{(ay + 2b\ln y)^2}{y^4} + \frac{1}{y^2}\left(\frac{a}{y} + \frac{b}{y^2} + c\right)(ay + 2b\ln y) = 0; \tag{154}$$

the general solution of (154) is

$$g(y) = yc_1 + \frac{1}{4y^3} \times \left((b + 2ay + 2b\ln y)\left(2b\ln y + b + 2cy^2 + 2ay\right)\right), \tag{155}$$

where c_1 is constant. For this case of f(y) and $g(y)$ Lienard ′ equation (18) corresponds to

$$y'' + y'\left(\frac{a}{y} + \frac{b}{y^2} + c\right) + yc_1 + \frac{1}{4y^3}$$
$$\times \left(\left(b + 2ay + 2b\ln y\right)\left(2b\ln y + b + 2cy^2 + 2ay\right)\right)$$
(156)

According to (69) we obtain

$$\omega_y + \left(\frac{y'}{y} + \frac{ay + 2b\ln y}{y^2}\right)\omega_{y'} = 0. \quad (157)$$

A solution of (157) is

$$\omega\left(x, y, y'\right) = \frac{b + 2ay + 2yy' + 2b\ln y}{2y^2}. \quad (158)$$

By using (158) the reduced form of Lienard equation (18) in terms of $\{x, \omega, \omega'\}$ can be written in the form

$$\omega' + \omega^2 + c\omega + c_1 = 0, \quad (159)$$

and the general solution of (159) is

$$\omega = \frac{1}{2}\left(-b\sqrt{4c_1 - c^2}\tan\left(\frac{1}{2}\sqrt{4c_1 - c^2}\,(G - x)\right)\right), \quad G \in \mathbb{R}. \quad (160)$$

According to (71) the integration factor is equal to

$$\mu\left(x, y, y'\right) = \left(4y^3\right)$$
$$\times\left(4y^4c_1 + \left(b + 2y\left(a + y'\right) + 2b\ln y\right)\right.$$
$$\left.\times\left(b + 2y\left(a + cy + y'\right) + 2b\ln y\right)\right)^{-1}. \quad (161)$$

$(x, (x, y, y'))$ that is reduced to (160) corresponds to conserved form of (18)

$$D_x\left(\frac{1}{c^2 - 4c_1}\right.$$
$$\times\left(c^2 - 4c_1x - 2\sqrt{4c_1 - c^2}\right.$$
$$\left.\left.\times\arctan\left(\frac{c + 2b + 4ay + 4yy' + 4b\ln y}{\sqrt{4c_1 - c^2}}\right)\right)\right) = 0, \quad (162)$$

which is equal to the original Lienard equation of the form ′ (156).

$f(y) = a\cosh y + b.$

According to (80) the differential equation in terms of λ_2 is given by

$$\frac{2\lambda_2}{y} + \lambda_{2y} + a\sinh y = 0. \quad (163)$$

The solution of above equation is found as follows:

$$\lambda_2\left(x, y\right) = \frac{a\left(2y\sinh y - \left(2 + y^2\right)\cosh y\right)}{y^2}. \quad (164)$$

The λ-symmetry considered as in the linear form of functions of λ_1 and λ_2 is equal to

$$\lambda\left(x, y, y'\right) = \frac{y'}{y} + \frac{a\left(2y\sinh y - \left(2 + y^2\right)\cosh y\right)}{y^2}. \quad (165)$$

Thus, (78) that includes the (y) function is obtained as the form

$$g'\left(y\right) - \frac{g\left(y\right)}{y}$$
$$- \frac{a\left(a\cosh y + b\right)\left(\left(2 + y^2\right)\cosh y - 2y\sinh y\right)}{y^2}$$
$$+ \frac{\left(\left(2 + y^2\right)\cosh y - 2y\sinh y\right)^2 a^2}{y^4} = 0. \quad (166)$$

The general solution of (166) is

$$g\left(y\right) = yc_1 + \frac{1}{y^3}$$
$$\times\left(a\left(y\sinh y - \cosh y\right)\right.$$
$$\left.\times\left(y\left(a\sinh y + by\right) - a\cosh y\right)\right), \quad (167)$$

where c_1 is a constant. Consequently, Lienard equation (18) for this case of f(y) and $g(y)$ is found as below

$$y'' + y'\left(a\cosh y + b\right) + yc_1$$
$$+ \frac{1}{y^3}\left(a\left(y\sinh y - \cosh y\right)\right.$$
$$\left.\times\left(y\left(a\sinh y + by\right) - a\cosh y\right)\right). \quad (168)$$

By consideration of the relation (69) the equation

$$\omega_y + \left(\frac{y'}{y} + \frac{a\left(2y\sinh y - \left(2 + y^2\right)\cosh y\right)}{y^2}\right)\omega_{y'} = 0 \quad (169)$$

is derived. A solution of (169) is equal to

$$\omega\left(x, y, y'\right) = \frac{y'y - a\cosh y + ay\sinh y}{y^2}. \quad (170)$$

In order to find the reduced form of Lienard equation (18), y’ and y” can be examined (170). Therefore we can write the following equation:

$$\omega' + \omega^2 + c\omega + c_1 = 0, \quad (171)$$

and the general solution of (171) is equal to

$$\omega = \frac{1}{2}\left(-c\sqrt{4c_1 - c^2}\tan\left(\frac{1}{2}\sqrt{4c_1 - c^2}\,(G - x)\right)\right), \quad G \in \mathbb{R}. \quad (172)$$

The integration factor corresponding to (71) is

$$\mu(x,y,y') = (y^3) \times \left(c_1 y^4 + \left(-a\cosh y + y\left(y' + a\sinh y\right)\right) \times \left(-a\cosh y + y\left(by + y' + -a\cosh y\right)\right)\right)^{-1} \quad (173)$$

The conserved form (x, (x, y, y')) can be derived from (172)

$$D_x\left(\frac{1}{b^2-4c_1} \times \left(b^2 - 4c_1 x - 2\sqrt{4c_1 - b^2} \times \arctan\left(\frac{1}{\sqrt{4c_1-b^2}} \times \left(b+2 \times \left(\frac{y'y - a\cosh y + ay\sinh y}{y^2}\right)\right)\right)\right)\right) = 0, \quad (174)$$

which gives the original Lienard equation of the form (′ 168).

$f(y)=ae^{by}$

For this case of f(y) the differential equation in terms of λ_2 is written as follows:

$$\frac{2\lambda_2}{y} + \lambda_{2y} + abe^{by} = 0. \quad (175)$$

A particular solution of (175) is

$$\lambda_2(x,y) = -\frac{ae^{by}\left(2 - 2by + b^2y^2\right)}{b^2y^2}. \quad (176)$$

By substituting (79) and (176) into (74) we obtain

$$\lambda(x,y,y') = \frac{y'}{y} - \frac{ae^{by}\left(2 - 2by + b^2y^2\right)}{b^2y^2}. \quad (177)$$

By applying λ_1 and λ_2 to (78) the differential equation which depends on unknown function f(y) is derived

$$g'(y) - \frac{g(y)}{y} - \frac{a^2e^{2by}\left(2 - 2by + b^2y^2\right)}{b^2y^2} + \frac{a^2e^{2by}\left(2 - 2by + b^2y^2\right)}{b^4y^4} = 0, \quad (178)$$

and the general solution of (178) is equal to

$$g(y) = yc_1 + \frac{a^2e^{2by}(by-1)^2}{b^4y^3}, \quad (179)$$

where c_1 is constant. The Lienard equation (18) for this case of f(y) and g(y) corresponds to the following form:

$$y'' + y'ae^{by} + yc_1 + \frac{a^2e^{2by}(by-1)^2}{b^4y^3} = 0. \quad (180)$$

Equation (69) can be written as

$$\omega_y + \left(\frac{y'}{y} - \frac{ae^{by}\left(2 - 2by + b^2y^2\right)}{b^2y^2}\right)\omega_{y'} = 0. \quad (181)$$

A solution of this equation is found as below

$$\omega(x,y,y') = \frac{-ae^{by} + abe^{by}y + b^2yy'}{b^2y^2}. \quad (182)$$

The reduced form of Lienard equation (18) in terms of variables {x, ω, ω' } becomes

$$\omega' + \omega^2 + c_1 = 0; \quad (183)$$

the general solution of (183) is given by

$$\omega = -\sqrt{c_1}\tan\left(\sqrt{c_1}(x - G)\right), \quad G \in \mathbb{R}. \quad (184)$$

As a result, the integration factor is obtained in the form by using (71)

$$\mu(x,y,y') = \frac{b^4y^3}{c_1b^4y^4 + \left(ae^{by}(by-1) + b^2yy'\right)^2}. \quad (185)$$

The conserved form associated with (x, (x, y, y')) is

$$D_x\left(\frac{1}{\sqrt{c_1}}\left(x\sqrt{c_1}\arctan\left(\frac{-ae^{by} + abe^{by}y + b^2yy'}{b^2\sqrt{c_1}y^2}\right)\right)\right) = 0, \quad (186)$$

which gives the original Lienard equation such as the form ′ (180).

CONCLUSION

In this study we apply partial Noether symmetry and λ-symmetry approaches to Lienard equation, which has an ′ important role in terms of physics and mechanics. Based on the studies of partial Noether operators, we derive partial Lagrangian of Lienard equation, which leads to obtain partial ′ Noether symmetries. Since the Lienard equation includes ′ two arbitrary functions such as f(y) and g(y), we evaluate different cases of these functions for the classification. Partial Noether symmetries are investigated for two different problems related to the linear damped and undamped systems. And we derive new first integrals (conserved forms) corresponding to each partial Noether symmetry.

Alternatively, in the literature the λ-symmetry concept is used to study integration factors, first integrals, and invariant solutions of ordinary differential equations. Hence, the other part of this study is based on the relation between integrating factors and λ-symmetries for Lienard equation related to ′ some mechanics problems. In fact, in this approach the determining equation for the differential equation includes not only infinitesimal functions but also λ function, which are unknown functions but it is not possible to get a solution of determining equations easily. In order to overcome this difficulty for determining equations, the λ function is assumed to be in a linear form. In this way the system of determining equations which include partial differential equations can be solved easily.

In the first part of the study we consider same functions f(y) and $g(y)$ considered in the case of partial Noether symmetries. Based on these functions we study λ- symmetries and first integrals. Then we conclude that it is possible to obtain new first integral by using λ-symmetry approach, which is different from partial Noether symmetry approach. Furthermore, we analyze only some specific cases of arbitrary function f(y) to determine λ-symmetries. As a result of determining equations we find different f(y) functions corresponding to each of $f(y)$ functions and obtain new λ-symmetries. By using λ-symmetries we obtain new reduced forms, integration factors, and conserved forms of highly nonlinear Lienard type equation. We present that the ′ conserved forms lead to new invariant solutions of corresponding classical mechanics problems, which are considered by different approaches in the literature.

REFERENCES

1. E. Noether, "Invariante Variationsprobleme," Nachrichten von der Gesellschaft der Wissenschaften zu Göttingen, Mathematisch-Physikalische Klasse, vol. 2, pp. 235–257, 1918, English Translation in Transport Theory and Statistical Physics, vol. 1, no. 3, pp. 186–207, 1971.
2. G. W. Bluman and S. Kumei, Symmetries and Differential Equations, vol. 81 of Applied Mathematical Sciences, Springer, New York, NY, USA, 1989.
3. L. V. Ovsjannikov, Group Analysis of Differential Equations, Nauka, Moscow, Russia, 1978.
4. N. H. Ibragimov, Ed., CRC Handbook of Lie Group Analysis of Differential Equations, vol. 1, CRC Press, New York, NY, USA, 1993.
5. P. J. Olver, Applications of Lie Groups to Differential Equations, Springer, New York, NY, USA, 1986.
6. H. Stephani, Differential Equations: Their Solutions Using Symmetries, Cambridge University Press, Cambridge, UK, 1989.
7. E. Yaşar and T. Özer, "On symmetries, conservation laws and invariant solutions of the foam-drainage equation," International Journal of Non-Linear Mechanics, vol. 46, no. 2, pp. 357–362, 2011.
8. T. Özer, N. Antar, and D. Sahin, "Lie group analysis of gravity currents," Nonlinear Analysis: Real World Applications, vol. 11, no. 2, pp. 978–994, 2010.
9. T. Özer and E. Yaşar, "Conservation laws for one-layer shallow water wave systems," Nonlinear Analysis: Real World Applications, vol. 11, no. 2, pp. 838–848, 2010.
10. T. Özer, "Symmetry group classification for one-dimensional elastodynamics problems in nonlocal elasticity," Mechanics Research Communications, vol. 30, no. 6, pp. 539–546, 2003. · · ·
11. T. Ozer, "On symmetry group properties and general similarity forms of the Benney equations in the Lagrangian variables," Journal of Computational and Applied Mathematics, vol. 169, no. 2, pp. 297–313, 2004.
12. T. Özer, "Symmetry group classification for two-dimensional elastodynamics problems in nonlocal elasticity," International Journal of Engineering Science, vol. 41, no. 18, pp. 2193–2211, 2003.
13. S. N. Pandey, P. S. Bindu, M. Senthilvelan, and M. Lakshmanan, "A group theoretical identification of integrable cases of the Liénard-type equation x..+f(x) x.+g(x)=0. I. Equations having nonmaximal number of Lie point symmetries," Journal of Mathematical Physics, vol. 50, no. 8, Article ID 082702, 19 pages, 2009.
14. A. H. Kara, F. M. Mahomed, I. Naeem, and C. Wafo Soh, "Partial Noether operators and first integrals via partial Lagrangians," Mathematical Methods in the Applied Sciences, vol. 30, no. 16, pp. 2079–2089, 2007.
15. C. Muriel and J. L. Romero, "First integrals, integrating factors and λ-symmetries of second-order differential equations," Journal of Physics A: Mathematical and Theoretical, vol. 42, no. 36, Article ID 365207, 17 pages, 2009.
16. C. Muriel and J. L. Romero, "New methods of reduction for ordinary differential equations," IMA Journal of Applied Mathematics, vol. 66, no. 2, pp. 111–125, 2001.
17. C. Muriel and J. L. Romero, "Integrating factors and λ-symmetries," Journal of Nonlinear Mathematical Physics, vol. 15, no. 3, pp. 300–309, 2008.
18. C. Muriel and J. L. Romero, "C∞-symmetries and reduction of equations without lie point symmetries,"Journal of Nonlinear Mathematical Physics, vol. 13, no. 1, pp. 167–188, 2003.
19. E. Yaşar, "Integrating factors and first integrals for Liénard type and frequency-damped oscillators,"Mathematical Problems in Engineering, vol. 2011, Article ID 916437, 10 pages, 2011.

CHAPTER 3

FURUTA'S PENDULUM: A CONSERVATIVE NONLINEAR MODEL FOR THEORY AND PRACTISE

J. Á. Acosta

INTRODUCTION

The control of underactuated mechanical systems has attracted the interest of the control engineering community for many years. Underactuated mechanical systems are control systems with fewer control inputs than degrees of freedom. This lack of inputs is, precisely, the main difficulty to control them. Moreover, the control of underactuated mechanical systems is currently an active field of research due to their broad applications in Robotics, Aerospace, and Marine Vehicles. Some examples of underactuated mechanical systems include flexible-link robots, mobile robots, walking robots, robots on mobile platforms, cars, locomotive systems, snake-type and swimming robots, acrobatic robots, aircraft, spacecraft, helicopters, satellites, surface vessels, and underwater vehicles.

One of the most well-known underactuated mechanical systems is the pendulum which is an excellent benchmark for many control problems, attracting the attention of many researchers during years. The available version of the pendulum is the rotary pendulum shown in Figure 1, and this survey is focused on that laboratory pendulum where many control strategies have been designed and successfully tested. The rotary pendulum is one of the most complex two-degrees-of-freedom pendulum since its configuration provides centrifugal torques, which are quadratic in velocities, giving rise to many interesting and difficult control problems.

Figure 1: Available Furuta's pendulum.

The rotary pendulum is also known as Furuta's pendulum. Furuta was the inventor, and has been studied by many authors such as Furuta himself [1] (see also [2–4] among others). It is a non-linear underactuated mechanical system that is unstable at the desired upright position. In fact, it shows, mainly, two different and very interesting control problems. The first one is to swing the pendulum up from the rest (down) to the upright (up) position, which is commonly solved with energy control strategies. Once the pendulum is close to the desired upright position, at low speed, a stabilization or balancing strategy "catches" it there. Other control problems are also quite interesting as the stabilisation of autonomous oscillations and control through bifurcation analysis. All of them are briefly commented in the survey focusing on all the designed and experimentally tested in the available laboratory pendulum.

Citation: J. Á. Acosta, "Furuta's Pendulum: A Conservative Nonlinear Model for Theory and Practise," Mathematical Problems in Engineering, vol. 2010, Article ID 742894, 29 pages, 2010, doi:10.1155/2010/742894

The paper is organized as follows. In Section 2 the non-linear model for the available pendulum is derived in detail, including a dynamical-friction model. In Section 3, a survey of the non-linear strategies designed and tested on the available pendulum is given. A literature review showing the relations with other approaches are also briefly commented. Finally, a conclusion section and two appendices, one describing the whole experimental control system and another with the MATLAB codes used for the non-trivial identification of parameters of the dynamical-friction model.

NOTATION AND DEFINITIONS

Throughout the paper it is assumed that all functions and vector functions are sufficiently smooth. Moreover, concerning functions, vector functions, and matrices which are functions of some variables, we will write explicitly this dependence the first time they are defined. The norm of a vector $x \in \mathbb{R}^n$ will be denoted by $\|x\|$ and the square of the Euclidean norm by $\|x\|_2^2 = x^\top x$. Acronyms used: w.r.t. means "with respect to"; i.e. means "that is."

Consider a dynamical system affine in control as $\dot{x}(t) = f(x(t)) + g(x(t))u$, where $x \in \mathbb{R}^n$ is the state space vector, $u \in \mathbb{R}^m$ is the control input vector and where we assume that the origin is at $x = 0$ with $f(0) = 0$ and $g(0) \neq 0$. Thus,

(i) the open-loop system means the unforced system, that is, with u=0;

(ii) a (smooth) static state-feedback means a control input of the form $u = \phi(x)$, also called controller, control law and stabiliser when a stabilisation problem is considered;

(iii) the closed-loop system means the system with a specified static state-feedback, i.e. $\dot{x} = f(x) + g(x)\phi(x)$;

(iv) linearisation means a non-linear static state-feedback that linearises fully or partially the dynamics;

(v) the Jacobian linearisation means the first approximation of the system dynamics through the Jacobian at the origin.

The schematic of Furuta's pendulum is described in Figure 2, where the two degrees of freedom are shown: the angular positions of the arm (ϕ) and the pendulum (θ). The arm shaft (corresponding to the angle ϕ referred to the x-axis) is subject to a torque, while no torque is applied directly to the pendulum shaft (angle θ of pendulum referred to the z-axis). Thus, it is an underactuated mechanical system because it has only one control input and two degrees of freedom. Rigorously, the configuration space is a 2-torus, that is $(\theta, \varphi) \in \mathcal{S}^1 \times \mathcal{S}^1$, and the corresponding tangent space is $(\theta, \varphi, \dot{\theta}, \dot{\varphi}) \in \mathcal{S}^1 \times \mathcal{S}^1 \times \mathbb{R}^2$.

We define the physical parameters of the system as

m: mass of the pendulum,

2l: total length of the pendulum,

r: length of the arm (radius),

g: acceleration of gravity,

$\mathcal{J}$: moment of inertia of the pendulum w.r.t. the pivot,

$\mathcal{J}_a$: moment of inertia of the arm and the motor,

K: constant torque,

u: control input,

τ: control torque (τ := Ku)

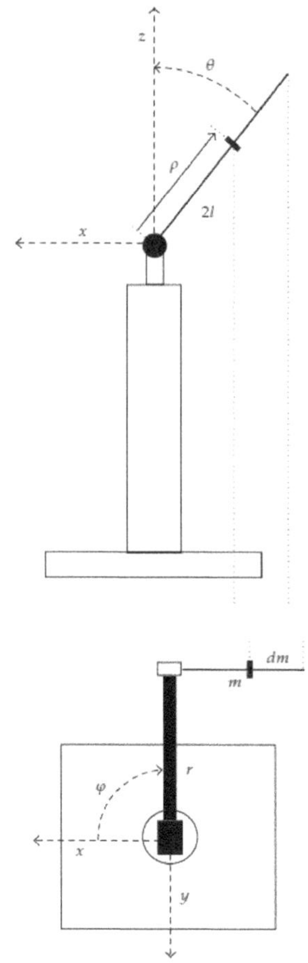

Figure 2: Reference frame of coordinates.

In what follows, energy methods of Classical Mechanics are used to derive the dynamical equations of motion. So, defining the reference frame of coordinates as in Figure 2 and assuming a linear distribution of the mass of the pendulum then, the coordinates of a differential mass element, namely ρ-mass element (see Figure 2), of the pendulum referred to that reference frame read

$$\begin{aligned} x_\rho &= r\cos\varphi - \rho\sin\theta\sin\varphi, \\ y_\rho &= r\sin\varphi + \rho\sin\theta\cos\varphi, \\ z_\rho &= \rho\cos\theta. \end{aligned} \tag{2.1}$$

Since it is a simple mechanical system the total energy is composed by the sum of the kinetic and the potential energies. Thus, on the one hand the instantaneous kinetic energy of the system is composed by the kinetic energies of the pendulum and the arm (including the motor) and becomes

$$\mathcal{T}(\theta, \dot{\theta}, \dot{\varphi}) = \frac{1}{2}\int_0^{2l} v_\rho^2(\rho)dm + \frac{1}{2}\mathcal{J}_a\dot{\varphi}^2, \tag{2.2}$$

where v_ρ is the module of the velocity of the ρ-mass element of the pendulum, which is obtained differentiating (2.1) w.r.t. the time and adding the squares. Thus, it becomes

$$v_\rho^2(\rho) = r^2\dot{\varphi}^2 + \rho^2\dot{\varphi}^2\sin^2\theta + 2r\rho\dot{\theta}\dot{\varphi}\cos\theta + \rho^2\dot{\theta}^2. \tag{2.3}$$

Since by assumption $dm = m/2l\, d\rho$ then, the kinetic energy from (2.2) and (2.3) becomes

$$\mathcal{T}(\theta, \dot{\theta}, \dot{\varphi}) = \frac{1}{2}\begin{bmatrix}\dot{\theta}\\ \dot{\varphi}\end{bmatrix}^\top \begin{bmatrix}\mathcal{J} & mrl\cos\theta \\ mrl\cos\theta & \mathcal{J}_a + mr^2 + \mathcal{J}\sin^2\theta\end{bmatrix}\begin{bmatrix}\dot{\theta}\\ \dot{\varphi}\end{bmatrix}. \tag{2.4}$$

On the other hand, the potential energy of the system w.r.t. the given reference frame is only the one of the pendulum and, integrating for the ρ-mass element, becomes

$$\mathcal{V}(\theta) = mgl \ (\cos\theta - 1), \tag{2.5}$$

which is zero at the upright position of the pendulum. Now, with the energy of the whole system completely defined we use both standard and well-known energetic approaches, Euler-Lagrange and Hamilton, to derive the dynamical equations.

Euler-Lagrange Equations

Let define the vectors $q := (\theta, \varphi)^\top \in \mathcal{S}^1 \times \mathcal{S}^1$ and $\dot{q} := (\dot{\theta}, \dot{\varphi})^\top \in \mathbb{R}^2$ and, just for compactness, another different set of physical parameters as

$$\alpha := \frac{mrl}{\mathcal{J}}, \qquad \beta := \frac{\mathcal{J}_a + mr^2}{\mathcal{J}}, \qquad \gamma := \frac{K}{mgl}, \qquad \omega_0^2 := \frac{mgl}{\mathcal{J}}. \tag{2.6}$$

Thus, to derive the Euler-Lagrange equations we calculate the Lagrangian of the system as the difference between the kinetic and potential energies of the whole system from equations 2.4 and 2.5, respectively. Then, the Lagrangian can be written as

$$\mathcal{L}(q,\dot{q}) = \frac{1}{2}\dot{q}^\top M(q)\dot{q} - \mathcal{V}(q), \tag{2.7}$$

where we have introduced the inertia matrix M(q) that from (2.4) and with the new set of parameters (2.6) becomes

$$M(q) = \mathcal{J}\begin{bmatrix} 1 & \alpha\cos q_1 \\ \alpha\cos q_1 & \beta + \sin^2 q_1 \end{bmatrix} \tag{2.8}$$

The well-known Euler-Lagrange equations [5] for this system take the form

$$\frac{d}{dt}\left(\frac{\partial\mathcal{L}}{\partial\dot{q}_1}\right) - \frac{\partial\mathcal{L}}{\partial q_1} = 0,$$
$$\frac{d}{dt}\left(\frac{\partial\mathcal{L}}{\partial\dot{q}_2}\right) - \frac{\partial\mathcal{L}}{\partial q_2} = \tau, \tag{2.9}$$

which can be rewritten as

$$M(q)\ddot{q} + C(q,\dot{q})\dot{q} + \frac{\partial\mathcal{V}}{\partial q}(q) = \begin{bmatrix} 0 \\ \tau \end{bmatrix} \tag{2.10}$$

where C(q, q˙) is the matrix of Coriolis and gyroscopic torques. Thus, the matrices defined in 2.10are given by

$$C(q,\dot{q}) = \mathcal{J}\begin{bmatrix} 0 & -\dot{q}_2\sin q_1\cos q_1 \\ -\alpha\dot{q}_1\sin q_1 + \dot{q}_2\sin q_1\cos q_1 & \dot{q}_1\sin q_1\cos q_1 \end{bmatrix}$$
$$\frac{\partial\mathcal{V}}{\partial q}(q) = \mathcal{J}\begin{bmatrix} -\omega_0^2\sin q_1 \\ 0 \end{bmatrix}, \tag{2.11}$$

and so (2.10) reads

$$\ddot{q}_1 + \alpha\cos q_1\ddot{q}_2 - \frac{1}{2}\sin 2q_1\dot{q}_2^2 - \omega_0^2\sin q_1 = 0, \tag{2.12}$$

$$\alpha\cos\theta\ddot{q}_1 + \left(\beta + \sin^2 q_1\right)\ddot{q}_2 - \alpha\sin q_1\dot{q}_1^2 + \sin 2q_1\dot{q}_1\dot{q}_2 = \gamma\omega_0^2 u. \tag{2.13}$$

Notice that, if the matrix C(q, q) is derived through Christoffel symbols [5, 6] (which is the case), then the matrix M˙ (q) – 2C(q, q) is skew-symmetric.

Since, the inertia matrix M(q) is positive definite and so invertible, then these equations can be rewritten as an affine-in-control system, just isolating the accelerations from (2.10) as

$$\ddot{q} = -M(q)^{-1}\left(C(q,\dot{q})\dot{q} + \frac{\partial\mathcal{V}}{\partial q}(q)\right) + M(q)^{-1}\begin{bmatrix} 0 \\ \tau \end{bmatrix} \tag{2.14}$$

Denoting the determinant of the inertia matrix as Δ(q), then (2.14) becomes

$$\Delta\ddot{q}_1 = \left(-\alpha^2\dot{q}_1^2 + \left(\beta + \sin^2 q_1\right)\dot{q}_2^2 + 2\alpha\cos q_1\dot{q}_1\dot{q}_2\right)\sin q_1\cos q_1$$
$$+ \beta\omega_0^2\sin q_1 + \omega_0^2\sin^3 q_1 - \alpha\gamma\omega_0^2\cos q_1 u, \tag{2.15}$$

$$\Delta\ddot{q}_2 = \alpha\left(\dot{q}_1^2 - \cos^2 q_1\dot{q}_2^2\right)\sin q_1 - \frac{1}{2}\left(2\dot{q}_1\dot{q}_2 + \alpha\omega_0^2\right)\sin 2q_1 + \omega_0^2\gamma u, \tag{2.16}$$

with $\Delta(q_1) = \mathcal{J}(\beta + \sin^2 q_1 - \alpha^2\cos^2 q_1)$. Notice that (2.15)-(2.16) is a fourth-order non-linear system strongly coupled and the coordinate q2 is cyclic. Remind that, a coordinate is said to be cyclic when it does not appear in the Lagrangian [7], and therefore the corresponding momentum is a constant of motion for the unforced system $(u = 0)$. The latter means that the system is completely integrable. Roughly speaking, a completely integrable system is a system with the same number of integrals of motion and degrees of freedom.

Partial Linearisation

A common practise in control design, in order to reduce the complexity of the equations of motion, is to introduce a (non-linear) change in the control input to linearise the actuated equation (2.13), or (2.16), in the acceleration of the q_2-coordinate. Thus, from (2.16) and defining

$$\omega_0^2\gamma u := \Delta\overline{u} - \alpha\left(\dot{q}_1^2 - \cos^2 q_1\dot{q}_2^2\right)\sin q_1 + \frac{1}{2}\left(2\dot{q}_1\dot{q}_2 + \alpha\omega_0^2\right)\sin 2q_1, \tag{2.17}$$

system (2.12)-(2.13) becomes

$$\ddot{q}_1 - \frac{1}{2}\sin 2q_1\dot{q}_2^2 - a\sin q_1 = -b\cos q_1\overline{u},$$
$$\ddot{q}_2 = \overline{u}, \tag{2.18}$$

where $a := \omega_0^2, b := \alpha$, and $\overline{u}$ is the new control input. Notice that $(q,\dot{q},u) \rightsquigarrow (q,\dot{q},\overline{u})$ is a globally defined change of coordinates and control. The system in $(q,\dot{q},\overline{u})$-coordinates can be seen as an Euler-Lagrange system where, through the Routh transformation [7], the energy is an integral of motion of the unforced system $(\overline{u} = 0)$ and reads

$$\mathcal{E}(q,\dot{q}) = \frac{1}{2}\|\dot{q}\|_2^2 + \frac{1}{4}\cos 2q_1\dot{q}_2^2 + a(\cos q_1 - 1) \tag{2.19}$$

Notice that the effective potential has been added because ˙q2 is the another integral of motion. Therefore 2.18 it is a new conservative system, completely integrable [5].

Hamilton Equations

Formally the Hamilton equations can be derived through the Legendre transformation [5] defining the conjugate momenta as $p := \partial\mathcal{L}/\partial\dot{q}$˙. Thus, the Hamilton function becomes

$$\mathcal{H}(q,p) := \dot{q}^\top p - \mathcal{L}(q,p) \tag{2.20}$$

Notice that for a Lagrangian given by (2.7) then $p = M(q)\dot{q}$ with $\mathcal{L}(q,p)$ being the Lagrange function in the Hamiltonian set of coordinates $(q,p) \in \mathcal{S}^1 \times \mathcal{S}^1 \times \mathbb{R}^2$. . Nevertheless, since the system is autonomous then the Hamiltonian function coincides with the internal energy of the system, and so it can be easily computed as

$$\mathcal{H}(q,p) = \frac{1}{2}p^{\top}M(q)^{-1}p + \mathcal{V}(q). \tag{2.21}$$

The Hamilton equations [5] are

$$\dot{q} = \frac{\partial \mathcal{H}}{\partial p}, \quad \dot{p} = -\frac{\partial \mathcal{H}}{\partial q} + \begin{bmatrix} 0 \\ \tau \end{bmatrix}, \tag{2.22}$$

and for Furuta's pendulum become

$$\dot{q}_1 = \frac{1}{\Delta}\left[\left(\beta + \sin^2 q_1\right)p_1 - \alpha p_2 \cos q_1\right], \tag{2.23}$$

$$\dot{q}_2 = \frac{1}{\Delta}\left[p_2 - \alpha p_1 \cos q_1\right], \tag{2.24}$$

$$\begin{aligned}\dot{p}_1 = {} & \frac{1}{2\Delta^2}\mathcal{J}\left[\left(\beta + \sin^2 q_1\right)p_1^2 + p_2^2 - 2\alpha p_1 p_2 \cos q_1\right] \cdot \left[\left(1 + \alpha^2\right)\sin 2q_1\right] \\ & + \mathcal{J}\omega_0^2 \sin q_1 - \frac{1}{2\Delta}\left[p_1^2 \sin 2q_1 + 2\alpha p_1 p_2 \sin q_1\right],\end{aligned} \tag{2.25}$$

$$\dot{p}_2 = \mathcal{J}\gamma\omega_0^2 u, \tag{2.26}$$

With $\Delta(q_1) = \mathcal{J}(\beta + \sin^2 q_1 - \alpha^2\cos^2 q_1)$

Table 1: Physical parameter values of the laboratory pendulum

Parameter	Value	Unit
Mass of the pendulum	0.0679	Kg
Length of the pendulum	0.14	m
Length of the arm	0.235	m
Mass of the arm	0.2869	Kg
Inertia of the motor	0.0012	Kg m^2
Constant torque	7.4	N m/v

Non-Conservative Friction Torques

In the previous sections a conservative mechanical model was derived using Classical Mechanics. In this section, we add the non-conservative torques due to friction in the real system. To this end, we focus the analysis in our available laboratory Furuta's pendulum, which was shown in Figure 1. The physical values of the parameters of this pendulum are shown in Table 1 (see 8) for details. The friction phenomena in this pendulum come up in the joints. The pendulum joint is composed by a small ball bearing in which friction can be well modelled by a damping constant. Otherwise, in the arm joint a DC-motor and a slip ring are directly coupled, and therefore the friction is not a damping-like torque, moreover it is strongly non-linear. In what follows of this section we derive the friction models designed in both joints.

Pendulum Joint

The non-conservative torques due to a natural damping is introduced through Rayleigh's dissipation function, namely D(q). So, calling cp to the damping constant then, the non-conservative torque, namely τd, becomes

$$\tau_d = \frac{d\mathfrak{D}(\dot{q}_1)}{d\dot{q}_1} = \frac{d}{d\dot{q}_1}\left(\frac{1}{2}\bar{c}_p\dot{q}_1^2\right) = \bar{c}_p\dot{q}_1. \tag{2.27}$$

This constant is easily estimated measuring the damping of a free-motion experiment yielding $\bar{c}_p = 0.0000226$. The small value is due to the small ball bearing and so it could be neglected.

Arm Joint

A carefully experimental study reveals that the friction torque in the motor shaft is not only a natural damping [8]. In fact, static and dynamic friction torques have to be taken into account in order to have a good estimation of the friction torque in this joint. To this end, a dynamic friction model, called LuGre from Lund Institute and Grenoble Laboratory(model 9–11), is used. The dynamical equations of this model are

$$\frac{dz}{dt} = v - \frac{\sigma_0}{g(v)}z|v|, \tag{2.28}$$

$$g(v) = \tau_c + (\tau_s - \tau_c)e^{-(v/v_s)^2}, \tag{2.29}$$

$$\tau_r = \sigma_0 z + \sigma_1\frac{dz}{dt} + \sigma_2 v, \tag{2.30}$$

where the parameters are defined as

τ_r : estimated friction torque,

τ_c: Coulomb friction torque,

τ_s: static friction torque,

$\sigma_{0,1}$: internal model parameters,

σ_2: dynamic friction constant,

v_s: Stribeck's velocity,

z: internal state of the model,

v: velocity ($q_2 = \dot{\phi}$).

Next, we give all the necessary steps to identify the parameters of the model$^{(2.28)-(2.30)}$. The physical meaning of these parameters is thoroughly described in [9]. Other kind of dynamical models for friction compensation can be also considered as in [12, 13]. Nevertheless, as we will show further, this model gave us a precise estimation. This good estimation allows to compensate the friction in this joint and, therefore, it gives us the chance of designing control strategies as if it was a conservative system. To this end, in some identification methods are suggested but, with the base given in [9] and our experience we have designed our own identification methodology. Thus, the identification of these parameters can be made in four steps corresponding to four different experiments, respectively. Obviously, to identify all the parameters the pendulum is decoupled from the system, i.e. it only remains the arm plus motor. We briefly comment the four experiments designed. St

Step 1 (identification of τ_c and σ_2). We use here a standard least-square algorithm to identify τ_c and σ_2. We consider the first-order system given by

$$\mathcal{J}_a \frac{dv}{dt} = \tau - \tau_c \operatorname{sgn}(v) - \sigma_2 v. \tag{2..31}$$

A side-effect parameter $\mathcal{J}_a$ can also be estimated with this experiment. Notice that, $^{(2.31)}$ becomes linear in the parameters. Thus, we introduce in the input τ a square signal (openloop) and then we measure the angular position of the arm, as shown in Figure 3 (top-left). The results after applying a least-square algorithm to (2.31) are shown in Table 2. Notice, from Table 2, the asymmetry which means that the identification is different for positive and negatives velocities.

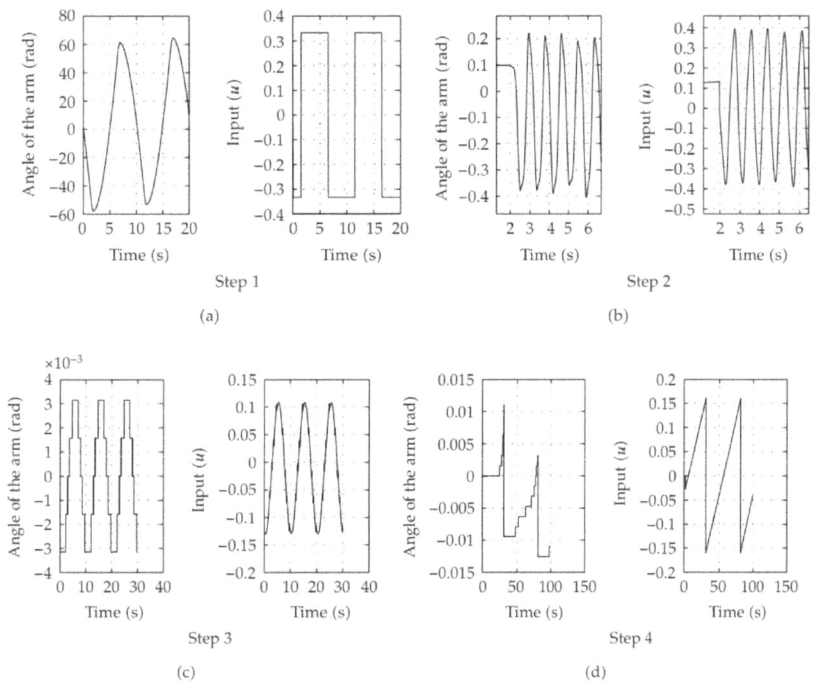

Figure 3: Experiments for Steps 1–4.

Step 2 (identification of τ_s). The static friction is identified forcing a limit cycle in the system. To this end, we feedback the angle of the motor with a slight proportional integral (PI) action, of the form $u = k_1 e + k_2 \int_0^t e(\tau)d\tau$, with $k_{1,2}$ gains and where e is the error between the measured state and the desired reference. The values are directly identified from Figure 3 (top-right) and the results are in Table 2. Notice again the asymmetric behaviour of the system.

$$\frac{dz}{dt} \approx 0 \Longrightarrow z \approx \frac{g(v)}{\sigma_0} \frac{v}{|v|}, \tag{3.32}$$

Table 2: Parameters of the LuGre model identified

Experiment	Parameter	(+)	(−)
	$\mathcal{J}_a$	0.0003	0.0004
Step 1	τ_c/K	0.2238	0.2308
	σ_2	0.0022	0.0020
Step 2	τ_s/K	0.3960	0.3789
Step 3	v_s	0.9297	0.9725
Step 4	σ_0	30	
	σ_1	0.2389	

and then, from (2.30) the torque can be approximated by

$$\tau_r \approx \left(\tau_c + (\tau_s - \tau_c)e^{-(v/v_s)^2}\right)\operatorname{sgn}(v), \tag{2.33}$$

which additionally allows to linearise the equation

$$J_a \frac{dv}{dt} = \tau - \tau_r, \tag{2.34}$$

as

$$v^2(t) = v_s^2 \ln \gamma(t), \tag{2.35}$$

where γ is given by

$$\gamma = \frac{\tau_s - \tau_c}{(\tau - J_a(dv/dt))\operatorname{sgn}(v) - \tau_c}. \tag{2.36}$$

The approximation is only valid if $\tau_s > \tau_c$ and γ_1. A sinusoidal input torque (openloop) allows us to estimate the value of γ and so vs. The experiment is shown on Figure 3 (bottom-left) and the results are shown in Table 2 yielding, again, an asymmetric behaviour. The MATLAB code for positive values of $v = q_2$.

Step 4 (identification of σ_0 and σ_1). The parameter σ_0 is estimated forcing a ramp in the input (open-loop) such that the angle cross by zero. The maximal and minimal values of the ramp should be around $\pm 0.8\,\tau_s$ and the slope small enough. The value of $\sigma 0$ can be estimated with $\sigma_0 \approx \Delta\tau/\Delta\varphi.$. The experiment is shown in Figure 3 (bottom-right).

To estimate σ_1 we assume small displacements and $|z| \ll g(v)$. Thus, $\dot z = v = \dot q_2$, and then we use a linear second-order model given by

$$J_a \ddot q_2 + (\sigma_1 + \sigma_2)\dot q_2 + \sigma_0(q_2 + q_2(0)) = \tau. \tag{2.37}$$

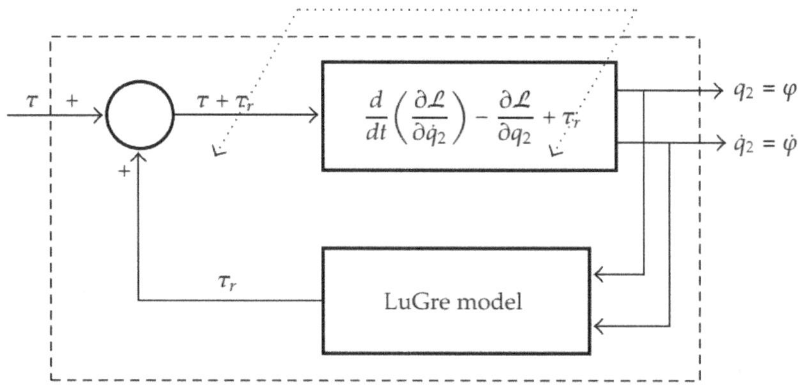

Figure 4: Friction compensation loop.

Therefore, since it is a second-order linear system the value of σ_1 is easily estimated imposing

$$\sigma_1 = 2\zeta\sqrt{\sigma_0 J_a} - \sigma_2, \tag{2.38}$$

where ζ is the damping constant of the linear model and σ_2 and J_a are known from Step 1.

The Complete Model

Once the friction torques have been estimated we add them to the model in the Lagrangian framework, being dual for the Hamiltonian case. Thus, the Euler-Lagrange equations with the non-conservative torques included become

$$\frac{d}{dt}\left(\frac{\partial \mathcal{L}}{\partial \dot q_1}\right) - \frac{\partial \mathcal{L}}{\partial q_1} + \tau_d = 0, \tag{2.39}$$

$$\frac{d}{dt}\left(\frac{\partial \mathcal{L}}{\partial \dot q_2}\right) - \frac{\partial \mathcal{L}}{\partial q_2} + \tau_r = \tau, \tag{2.40}$$

where again we normalise dividing by J with $c_p := \bar c_p / J$. Then, the differences from (2.10) are, on one hand, the matrix $C(q,\dot q)$ which turns into

$$C(q,\dot q) = J\begin{bmatrix} c_p & -\dot q_2 \sin q_1 \cos q_1 \\ -\alpha \dot q_1 \sin q_1 + \dot q_2 \sin q_1 \cos q_1 & \dot q_1 \sin q_1 \cos q_1 \end{bmatrix} \tag{2.41}$$

and on the other hand the torque is reduced in $\tau_r = K u_r$, where τ_r is estimated by equations (2.28)–(2.30) with $v := \dot q_2$.

After adding the non-conservative friction torques the model is a better approximation of reality. Nevertheless, to easy the computations of the non-linear control laws, we use the friction models to compensate the friction instead of modelling it. In fact, this is the success of our model in practise. However, remind that the system has two degrees of freedoms and just one actuation torque, which means that we can compensate the friction torques only in the actuated coordinate q2. This can be easily seen in (2.39)-(2.40). Thus, in (2.40) the friction torque τ_r can be compensated adding it to the input torque (positive feedback). Otherwise, in the unactuated equation (2.39) we cannot compensate anything at all, since we do not have an independent input. Recall that our objective was to approximate in practise as well as possible to the conservative model given by the equations (2.9), or equivalently (2.22). In Figure 4 we show how the added torque compensates the friction. In the pendulum joint, as commented above, we cannot compensate τ_d in (2.39). Nevertheless, by construction the friction torque can be neglected, i.e. $\tau_d \approx 0$, because we use a small enough ball bearing for the joint (recall that $\bar c_p \approx 2.26 \times 10^{-5}$).

Finally, after compensating the non-conservative torque then, the completed model is a good approximation to the conservative model given by (2.9). The reader will realise this good approximation in the next section, where most of all the non-linear controllers were designed with (2.9), or equivalently

(2.22). Moreover, in most of the controllers conservative quantities were used for the design stage.

A SURVEY OF NON-LINEAR CONTROL STRATEGIES

In this section we survey all of non-linear strategies designed and experimentally tested in the available laboratory pendulum shown in Figure 1. We recall that the paper is focused on the design and experimental results on our available Furuta's pendulum. Thus, other results will be referenced and briefly discussed because they were not experimentally tested on our pendulum.

Swing-Up Strategies

As commented in the introduction by swing-up we mean the non-linear control problem to swing the pendulum up from the rest to the upright position, the so-called inverted position. The control strategy proposed in [14–16] was based on the Speed-Gradient method, in fact just the finite form of the Speed-Pseudogradient Algorithm proposed in [17, 18] was used. Briefly, we comment the basic idea. Consider the time-invariant, affine-in-control system $\dot{x} = f(x) + g(x)u$ (in our system either (2.15)-(2.16) or (2.23)-(2.26)) with output $y = h(x)$ where $x(t) \in \mathbb{R}^n$, $y(t) \in \mathbb{R}^l$ and the input $u(t) \in \mathbb{R}^m$ Consider also the control objective $y(t) \to 0$ when $t \to \infty$ this control objective can be written as an objective function $Q(x) = (1/2)\|h(x)\|^2$ For this objective function the Speed-Pseudogradient Algorithm in finite form reads

$$u \propto -\left(\frac{\partial h(x)}{\partial x} g(x)\right)^{\top} h(x), \tag{3.1}$$

which can be derived calculating Q˙ (x) along trajectories of the system and noting that h (x) are independent invariants for the unforced system (u = 0). Stability properties of algorithm (3.1) are described in [17]. In essence, the objective Q(x) plays the role of a Lyapunov function with a closed trajectory as a limit set and then, roughly speaking, if $\dot{Q}(x) \leqslant 0$ the goal will be achieved. Some examples of the application can be seen in [17, 19, 20].

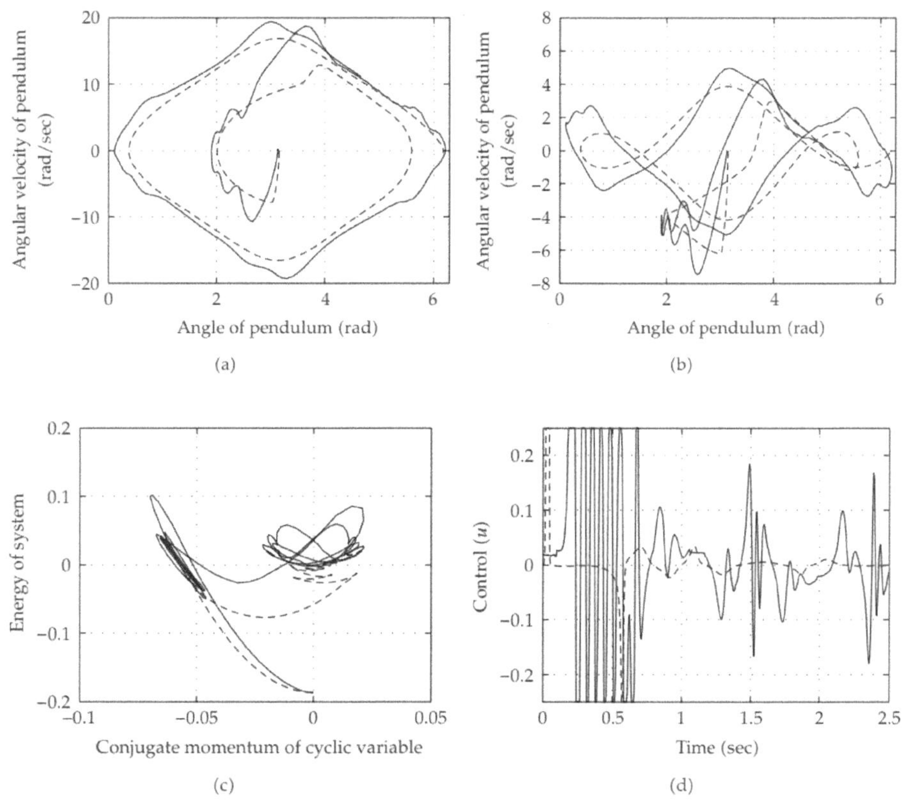

Figure 5: "Solid" experiment. "Dashed" simulation.

In the case of study, selecting the objective function as the square of the total system energy $Q := Q_1 = 1/2(\gamma_1 \mathcal{H})^2$ then, the trajectories of the mechanical system will tend towards the surface (limit set) $\{(q_1, p_1, p_2) \in \mathcal{S}^1 \times \mathbb{R}^2 \mid \mathcal{H} = 0\}$ as in [18]. This aim does not guarantee that the pendulum will pass through the origin. The objective function was modified adding a new objective function Q_2 so that the trajectories tend towards the closed trajectory (limit set) $\{Q_1 = 0\} \cap \{Q_2 = 0\}$. The Hamiltonian structure of the system helped us to find the Q_2 function needed. In fact, since the system is symmetric w.r.t. the angle q2, the conjugate momentum p_2 is a constant of motion for the unforced system, which is straightforward from (2.26). Recall that q_2 is a cyclic coordinate, which is closely related with Noether's theorem [5]. Thus, it is reasonable to choose the objective function as $Q := 1/2(\gamma_1 \mathcal{H})^2 + 1/2(\gamma_2 p_2)^2$, where $\gamma_{1,2}$ are positive control gains and p_2 is the conjugate momentum of the coordinate q_2. The output vector reads $y = h(x) = (\gamma_1 \mathcal{H}, \gamma_2 p_2)^{\top}$ The objective closed trajectory $\{Q_1 = 0\} \cap \{Q_2 = 0\}$ is the homoclinic orbit for Furuta's pendulum [21]. Thus, the objective function for Fradkov's Speed-Gradient method includes not only the energy but also the arm momentum, which are both conserved quantities for the unforced system. To show the fidelity of the model proposed we show Figure 5 from [15, 16], where we compare the experiment (solid line) with the simulation (dashed line) with the controller proposed. Another proof of the fidelity of the conservative model was shown in [14] where we developed the strategy through Routh's function [7], which embedded both conservative quantities given before. A video of the experiment can be displayed from http://www.esi2.us.es/~jaar/investiga e.htm. Remark 3.1. We underscore that the success of the proposed strategy relies on the fact that the experimental model is quasi-conservative. In fact, the control strategy is based on both conservative quantities, energy and momentum. Without that conservative property the control strategy would fail.

Literature Review

The control problem of swinging the pendulum up was firstly solved by Åström and Furuta with a reduced system model [1, 4]. The control strategy was based on neglecting the reaction torques from the pendulum to the arm. This allows greatly simplify to a second-order model in (q_1,q_2)-coordinates, and with the help of Fradkov's Speed-Gradient method the desired energy injection was easily computed [22]. However, we remind that it was based on a simplifying assumption. Moreover, the arm speed must be low for the success of any stabiliser at the switching time, but notice that this speed is not considered in the dimension 2 control law, i.e. designed with the second-order model. Fortunately, for standard pendula with the usual initial conditions (the lower position with no velocity) the dimension 2 control law behaves well. Nevertheless, if the reaction torques from the pendulum to the arm cannot be neglected, for example, either due to the mass of the pendulum or to the low friction on the arm, Åström and Furuta's law fails. Some conditions of this failed attempt case were shown in [23], where conditions to induce the approximation to fail, were tested. Moreover, our proposed control law keeps the nice properties of Åström-Furuta's one when compared with other swing-up control laws based on a fourth-order model, i.e. our swing-up control law is able to accomplish the goal with arbitrary small control signal magnitude (unlike [24, 25]) and, the control signal converges to zero as the homoclinic orbit is reached (unlike [26]).

Non-Linear Stabilisers

In the previous section we show the non-linear strategies to swing the pendulum up to the upright position. Those strategies stabilise the homoclinic orbit, which passes through the origin of the state space. Unfortunately, the origin is a saddle fixed point in closed loop and then, to solve the whole problem, that is, swing-up and stabilisation, a balancing strategy (stabiliser) has to be designed to "catch" the pendulum at the upright position. Fortunately, the swing-up strategy was designed such that the pendulum passes through the origin with $(q_1(t), \dot{q}_1(t), \dot{q}_2(t)) \rightarrow 0$, and $u(t) \rightarrow 0$. The latter allows to design the balancing stabiliser separately and then, to propose a switching strategy between both, i.e. the swing-up controller and the stabiliser. Again, theconservative model of the pendulum allowed to design non-linear stabilisers using invariants and/or conserved quantities. In this section we describe all the non-linear stabilisers designed using the conservative model and, moreover, successfully tested through experiments.

Energy Shaping

A conservative system allows, in a natural fashion, to be controlled by energy-shaping methods. In fact, a non-linear stabiliser was designed through the energy-shaping methodology called Interconnection and Damping Passivity-Based Control. This methodology was introduced in [27] to regulate the position of underactuated mechanical systems of the form (2.22) with total energy given by (2.21). The main idea is to force the closed-loop structure of the system to a desired one. This procedure is commonly called "matching." In fact, for mechanical systems the proposed closed-loop structure takes the Port-Controlled Hamiltonian form [6] given by

$$\begin{bmatrix} \dot{q} \\ \dot{p} \end{bmatrix} = \begin{bmatrix} 0 & M^{-1}M_d \\ -M_d M^{-1} & J_2 \end{bmatrix} \begin{bmatrix} \frac{\partial \mathscr{H}_d}{\partial q} \\ \frac{\partial \mathscr{H}_d}{\partial p} \end{bmatrix} \tag{3.2}$$

where in our case of study $(q,p) \in \mathcal{S}^1 \times \mathcal{S}^1 \times \mathbb{R}^2$, $M_d(q)$ is the desired inertia matrix, $\mathcal{V}_d(q)$ is the desired potential function and so, the total energy function of the closed loop reads $\mathscr{H}_d = (1/2)p^{\top} M_d^{-1}(q)p + \mathcal{V}_d(q)$. Subindex "d" denotes "desired" which means in closedloop. Finally $J_2(q,p) \in \mathbb{R}^{2\times 2}$ is a skew-symmetric matrix free for the designer. The design concludes after a damping injection stage. Stability properties require a positive definite Md in a neighborhood of the desired equilibrium $q^\star$, with $q^\star = \arg\min \mathcal{V}_d(q)$, and then $(q^\star, 0)$ is a stable equilibrium point of the closed-loop system (3.2), with Lyapunov function $\mathscr{H}_d$.

Since the system is underactuated the "matching" procedure provides on one hand, a set of algebraic constraints (in the actuated part) and, on the other hand, a set of Partial Differential Equations (PDEs) on M_d and V_d (in the unactuated part). The PDEs arise for M_d are called kinetic energy PDEs and the ones for V_d are called potential energy PDEs. Clearly, the success of this methodology relies on the possibility of solving those PDEs, which is usually a daunting task. In fact, for the case of the Furuta's pendulum a new mathematical machinery was needed to solve them. First, the reparametrization of the free skew-symmetric matrix J2 given in [28] allows to transform the kinetic energy PDE in a set of first-order ordinary differential equations depending only on the q1-coordinate. Second, the change of coordinates proposed in [29, 30] allows to simplify, drastically, the non-homogeneous term of the kinetic energy PDEs. The complete solution was reported in [31] and the experimental test will be commented further in comparison with others strategies.

Remark 3.2. Again the success of the proposed strategy relies on the fact that the model for the design is conservative. The extension of this methodology considering natural damping was given in [32, 33]. Unfortunately, in Furuta's pendulum case and considering natural damping torques no solutions were found so far.

Literature Review

Other different theoretical solutions have been reported using energy-shaping approaches. In [2] the Controlled Lagrangians methodology was used for stabilisation, through just a kinetic energy shaping (see also 34]). In [35] a way to transform the PDEs was proposed providing another solution for Furuta's pendulum. Closely related it is the non-linear controller based on a Lyapunov control approach given in [36]. All of these different solutions have not been experimentally tested, so far.

Forwarding

Cascade non-linear control techniques have attracted the interest of many researchers (see [37–39] and references therein), mainly because they provide great simplifications and/or reductions in the non-linear control design stage. Forwarding is a passivity approach applied to cascade non-linear systems with feedforward paths. In [38] the system is treated as a feedforward connection of two cascade subsystems through a non-linear vector function. Assuming that each subsystem has an associated Lyapunov function, separately, the main difficulty is the need to construct an additive cross term to get a Lyapunov function for the whole cascade. The cross term is constructed through the solution of a PDE. The solution of the PDE provides an invariant manifold decomposition. The manifold allows together with a change of coordinates to construct a compound scaled Lyapunov function solving the stabilisation problem in closed form [37]. So, to solve another PDE is needed. Finally, noting that the system is in Jurdjević-Quinn form [40], then a damping stabiliser solves the whole control problem. We apply this technique to the conservative model (2.18) and the design and experiments were reported in [41, 42]. It is worth noting that the main result of [41] was not the non-linear stabiliser designed through Forwarding. The main result was that the Lyapunov function associated to theForwarding design was used together with the one of the swing-up controller, to design a Lyapunov-like switching strategy to solve the whole problem, i.e. the almost global stabilisation of the inverted position, both in theory and practise. The usual switching strategy to solve this global control problem is, to compound the swing-up controller with a Linear Quadratic Regulator (LQR) to stabilise the origin [8, 43], i.e. a state-feedback $u := Kx$ with state $x \in \mathbb{R}^n$ and a constant gain matrix . In our available pendulum we realised that the Lyapunov function associated to the LQR cannot be used to design the switching strategy because the estimate of the Domain of Attraction (DoA) is too small. By too small we means that the level of the noise from the measurements is higher than the size of the largest admissible level set of the estimate of the DoA, and then the switching controller fails to commute.

Remark 3.3. The Forwarding needs the solution of a PDE. The solution of the PDE is an invariant which provides and invariant manifold decomposition for the closed-loop system. Thus, that conservative property was crucial again.

Literature Review

In [3] another solution was given using the non-linear cascade technique called Backstepping [44]. The approximated solution given relies on a nontrivial change of coordinates and it has not been experimentally tested so far.

Input-Output Feedback Linearisation

This technique was a step ahead after the development of the well-known State Feedback Linearisation, which linearises the state of the system from the control input through a non-trivial change of coordinates and control. This change of coordinates comes up from the solution of a set of PDEs. The extension to this approach was made through the Input-Output Feedback Linearisation which linearises the input-output response of the system [45, 46]. In the cases in which it is not possible to linearise the whole system by a static state-feedback, a partial feedback linearisation can still be possible. This partial feedback linearisation transforms the system dynamics into two parts: the external dynamics, which become linear; the internal non-linear dynamics (locally characterised by the zero dynamics [45]. As proposed in 45) a possibility to control non-linear systems is to find an output, called "fictitious" such that the system linearised through that output is minimum phase [45]. Thus, in [47, 48] we proposed a nontrivial and constructive "fictitious" output through which it is possible to achieve the asymptotic stability with (locally) exponential convergence of the origin, for a class of underactuated mechanical systems with underactuation degree one. The generalisation to underactuation degree larger than one is given in [49]. The approach was based on classical Input-Output Feedback Linearisation and Lyapunov redesign 50], yielding excellent results in theory and experiments. In particular, for Furuta's pendulum the proposed output, namely y, reads y :

$$y := \dot{q}_2 + k_1 \cos q_1 \dot{q}_1 + k_2 \sin q_1$$

where $k_{1,2}$ are positive control gains. The model used for the design was (2.18). In fact, the proposed output makes the system to be exponential minimum-phase [46], i.e. the equilibrium of the zero dynamics is hyperbolic. As it is well-known this property provides the system with robustness (to unmodelled dynamics, disturbances) [50]. The experiments in the available Furuta's pendulum equipment were reported in [51].

Remark 3.4. We would like to underscore that the exponential minimum phase property is not achieved by most of non-linear control design techniques. For example in energy-shaping and Forwarding techniques the closed-loop system is weakly minimum phase w.r.t. its output, inherent fact of those methodologies by their passivity properties (see [52] for further details).

Remark 3.5. As in the previous cases, a crucial conservative quantity allows to design the stabiliser. In this case the energy of the zero dynamics was the conserved quantity.

Literature Review

Close to Input-Output Feedback Linearisation is the Flatness approach. In the case of a system with a single control input then, it is said to be differentially flat if and only if it is fully linearisable via static state-feedback (see e.g. [53, 54]). A Flatness controller for a linearised model of Furuta's pendulum was given in [55] but it was based on a linear model, through the Jacobian, since the non-linear model is not differentially flat w.r.t. any output.

Singular Perturbations

A deep and fair review of the use of this technique to control systems with different time scales is given in [50] and references therein. A singularly perturbed system is a system in which its own structure makes different coordinates evolve in different time scales. In the beginning, it was a technique to analyse the stability of equilibria, because the time-scale separation is a property of the own structure of the system. Nevertheless, if the system has not this time-scale separation itself, we can be able to propose a stabiliser to force that separation-like property and then, to analyse the stability of the equilibrium. This is the case proposed in [56,57] for Furuta's pendulum. In fact, we even make relaxations to the Input-Output Feedback Linearisationcontrol law described above, and then we use the Singular Perturbation theory to analyse the stability. The theoretical and experimental result was reported in [56, 57], where a fair comparison with another techniques was also reported. Moreover, we designed an experiment to measure the largest attraction basin. In fact, to the best of our knowledge, this is the largest region of attraction experimental tested so far, to stabilise this kind of pendulum. We reproduce, and comment briefly, here Figure 6 from [57], where the results of the experiments were collected.

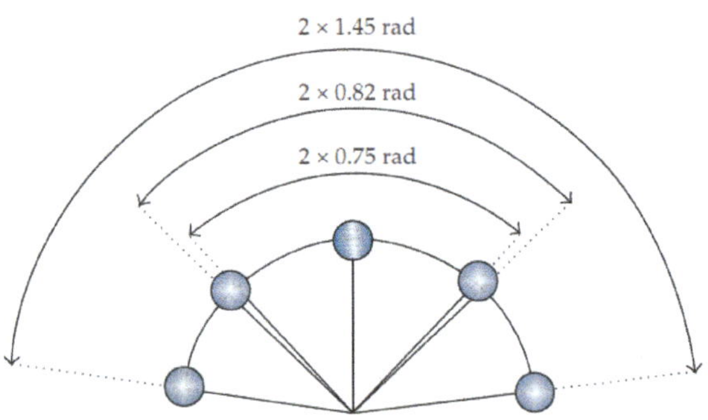

Figure 6: Experimental region of attraction.

In the experiment we forced to the system to start from an initial condition as $(\theta(0),\dot{\theta}(0),\varphi(0),\dot{\varphi}(0)) = (\theta^*,0,0,0)$. The idea was to find out the largest θ^* for which the origin (the upright position of the pendulum) was stabilised with a given controller. The controller design shows that the maximum theoretical value of θ for the controller based on singular perturbations is defined through a condition in the controller gains, and therefore can be tuned (see [57]). In 2, 31) the maximum theoretical θ depends on physical parameters of the own pendulum, and therefore the formulas are not tunable, giving rise in our pendulum to a maximum approximately the half of the value of the one given by the singularly perturbed approach. The best experimental results for each controller are collected in Figure 6 and give rise to $\theta^* = 1.45$ rad with the singular-perturbed approach; $\theta^* = 0.82$ rad with the energy-shaping approach; $\theta^* = 0.75$ rad with a Linear Quadratic Regulator. Unfortunately, the system saturates and it was not possible to enlarge even more this practical region of attraction (see [57]). Some videos of the experiments can be displayed from http://www.esi2.us.es/~jaar/investiga e.htm.

Global Stabiliser: A Challenge

In the previous sections we have shown the non-linear strategies to swing-up and stabilisation of the pendulum at the upright position. To solve the whole problem, i.e. swing-up and stabilization, a switching strategy was designed to "catch" the pendulum at the upright position properly. As commented before, fortunately, the swing-up strategy was designed such that the pendulum passes through the origin of the state space. This is the practical engineering solution to the whole problem. But one might wonder, if it would be possible to solve the whole problem with only one control strategy. We cannot close this section without describing this mathematical and still open problem that has attracted the attention of many researchers. Thus, suppose that we are interested in solving the whole problem, designing only one smooth state-feedback controller. By smooth we means only with one control strategy without any switch between strategies. Of course, from a practical engineering point of view the problem is actually solved as it has been shown in the previous sections, but from a mathematical point of view it is a challenge problem. On one hand there is an obstacle inherent to the own configuration space of the system, remind that $q \in S^1 \times S^1$, which means that we cannot globally stabilise an isolated equilibrium in the whole state space. All we can do is to stabilise the upper equilibrium from almost everywhere in the state space, except possibly a set of zero Lebesgue measure. On the other hand, since the upper equilibrium of the unforced system is unstable then we have to design a static statefeedback that makes that equilibrium asymptotically stable and all the remaining equilibria unstable.

The latter idea, for an underactuated mechanical system, was proposed in [58] in a pendulum-on-a-cart system where almost global stability of the equilibrium was proved. For Furuta's pendulum case, we have found a promising controller that makes the work but it was not possible to prove the almost global stability of the equilibrium, so far. As it was reported in [8] the static state-feedback reads $u := k_1 \sin(2q_1) + k_2 \cos(q_1)\dot{q}_1 + k_3\dot{q}_2$, with k_i, $i = 1,\dots,3$, control gains, and where we underscore the unsigned definite damping-like term. To prove almost global stability of the equilibrium, if so, is still a mathematical open problem. However we tested the proposed controller through experiments. In Figure 7 we show a successful experiment, where in the top figure we show the time histories of the position and the velocity of the pendulum and, in the bottom figure the time histories of the control input. Notice that there is no any switch. A video of the experiment can be displayed from http://www.esi2.us.es/~jaar/investiga e.htm.

Autonomous Oscillations

Another field of interest for the control community is the generation of "artificial" and autonomous oscillations, i.e. by means of an appropriate control law. To accomplish this in [59] we proposed a methodology based on matching the open-loop system with a generalised hamiltonian system, which is able to exhibit robust oscillations. Thus, the closedloop system displays robust oscillations associated with a limit cycle. The methodology mimics the energy-shaping approach but here, the objective is to reach a limit cycle (invariant limit set), which provides a stable oscillating behaviour, instead of a fixedpoint attractor. The desired limit cycle is born through a supercritical Hopf bifurcation on an approximated second-order dynamical model and, therefore, for certain values of a control parameter the closed-loop system has a fixed-point attractor, changing to a limit cycle through other values of that control parameter. Successful experiments were reported in [59] and we reproduce in Figure 8 a sequence of snapshots of the experiment for a period of the oscillations of 0.5 second. A video of the experiment can be displayed from http://www.esi2.us.es/~jaar/investiga e.htm.

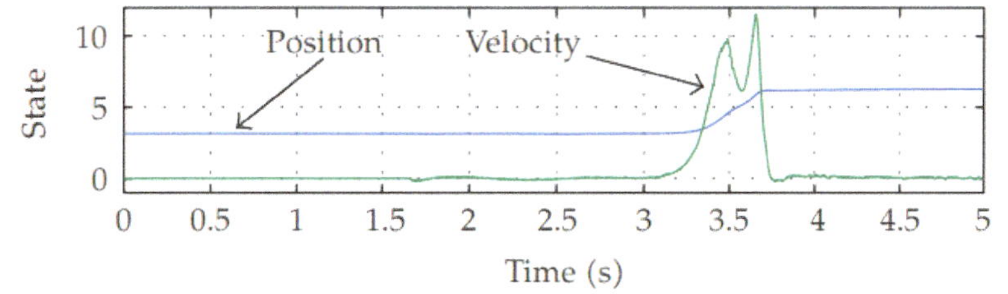

(a)

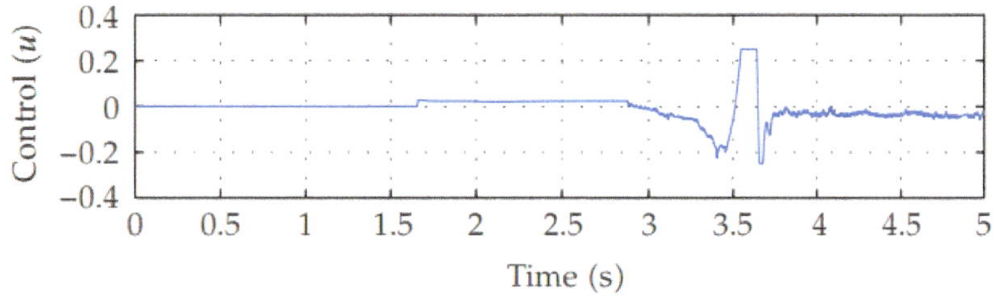

(b)

Figure 7: Experimental result with the challenging non-linear stabiliser.

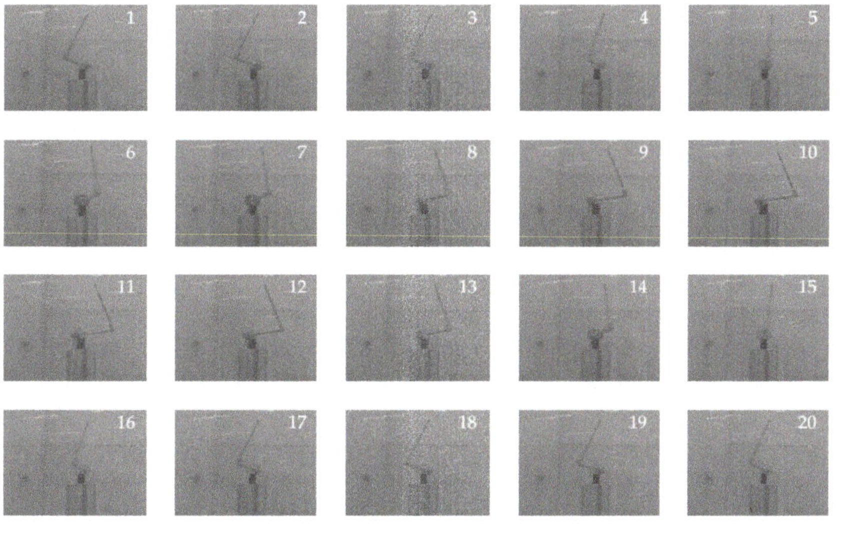

Figure 8: Sequence of snapshots showing one period of the oscillations.

Literature Review

Another approach to generate autonomous oscillations on Furuta's pendulum has been reported in [13, 60] based on virtual holonomic constraints. In our approach we design a closed-loop structure which associated Lyapunov function has a closed trajectory as its limit set (the limit cycle). In [13, 60] a closed trajectory is generated in closed loop and then only local exponential convergence can be guaranteed.

Bifurcations

This section is devoted to comment briefly some works regarding with bifurcations which have been also analysed and experimentally tested in our laboratory pendulum. The works commented here are mainly focused in the analysis of some non-linear behaviour of the controlled pendulum, rather than in control design. In fact, the controller used to analyse the behaviour is the linear one, i.e. a linear state-feedback stabiliser designed through the Jacobian. In [61, 62] a bifurcation map was constructed for Furuta's pendulum with a saturated linear controller, where the bifurcation parameters were the controller gains. Detailed analysis, simulations, and experiments in our available pendulum were reported in [63]. Finally, in [64] global bifurcation analysis was made for Furuta's pendulum. Nevertheless, we underscore that the dynamic friction model presented here was not used. Only a static model to compensate Coulomb's friction was used instead.

Literature Review

In [21] a complete bifurcation map for the Furuta's pendulum is given. Nevertheless, the bifurcation map is for the unforced pendulum and only through simulations with continuation methods.

CONCLUSION

A detailed dynamical model of the available laboratory Furuta's pendulum is provided. The completed model is based on Classical Mechanics augmented with a dynamical model that allows to compensate all those non-conservatives torques, arising to a quasi-conservative model in practise. A survey of all the non-linear controllers designed with that quasi-conservative model and successfully tested through experiments is also reported.

APPENDICES

A. Experimental Control System

The schematic representation of the full control system is depicted in Figure 9. The laboratory electromechanical system is composed by three subsystems: measurement, power, and control supervision. A brief description of them is given in this appendix.

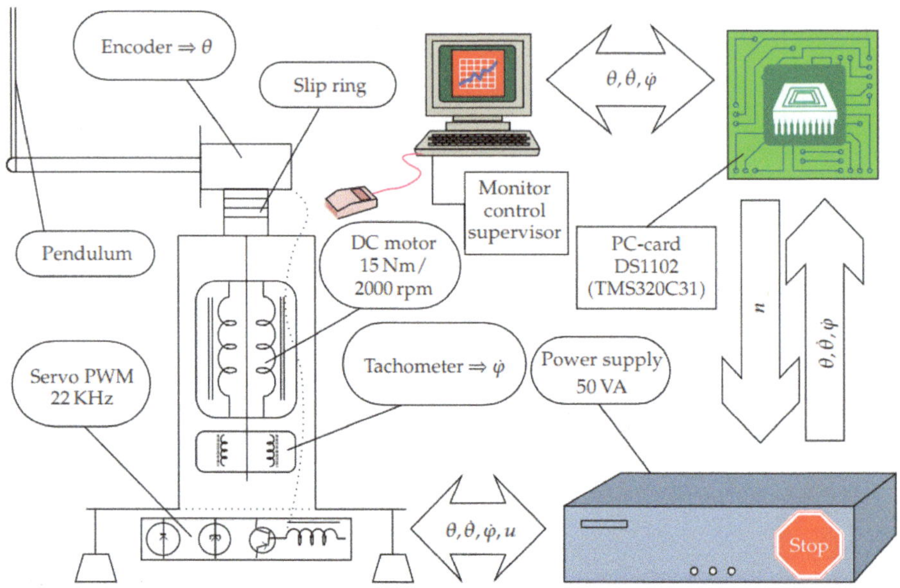

Figure 9: Schematic of the experimental control system.

To avoid problems of aliasing, mainly caused by noise, an analog filter (antialiasing) is used. The cutoff frequency is at 200 Hz, to filter the signal for the speed sensor (variable $\dot{\phi}$) before being sampled.

Power Subsystem

This subsystem provided the necessary power to drive the motor. The DC-motor provides a maximum torque of 15 N m at 2000 r.p.m., with a regulated source of 50 VA. The motor is controlled by current through a PWM servo-amplifier at 22 KHz (see Figure 9).

Control and Supervision Subsystem This subsystem closed the loop. It is composed by a PC and a data card Dspace DS1102, based on a Digital Signal Processors (DSP) TMS320C31. Among others, We have interfaz AD/DA, digital inputs/ouputs, counters, incremental encoders interface, digital filters, and communication interface. The real-time software is provided by Dspace running on Matlab/Simulink. The minimal sample time allowed is 1 msec.

MATLAB

For the sake of completeness, in this appendix we collect the detailed MATLAB code for identification of the dynamic friction model (LuGre).

B1:Code for Step 1

```
function theta = step1p(x, u, sample)
% 1st experiment of LuGre model
% theta = step1p(position, control, sample)
% Out: theta = [J sigma2 Fc+]
v_min=1; n=length(sample); k=0;          % Initialization
t = sample';                             % Time
u = u'; x = x';                          % Control, Position
T = mean(diff(t));                       % Sample time
v = diff(x)/T; v(n) = v(n-1);            % Velocity
v_dot = diff(v)/T; v_dot(n)=v_dot(n-1); % Acceleration
% Filter
tau = 0.01;                              % Time constant
[num_d, den_d] = c2dm(1, [tau, 1], T, 'tustin');
u_f = filter(num_d, den_d, u);
v_f=filter(num_d, den_d, v);
v_dot_f = filter(num_d,den_d, v_dot);
% Data selection
for i=1:n;
   if v_f(i) > v_min
      k=k+1;
      v(k) = v_f(i);
      v_dot(k) = v_dot_f(i);
      u(k) = u_f(i);
   end;
end; clear u_f v_f v_dot_f
% Least square (Moore-Penrose)
phi_p = [v_dot(1:k),v(1:k),sign(v(1:k))];
theta_p = pinv(phi_p)*u(1:k);
theta = theta_p';                        % Out
```

B.2. Code for Step 2

```
function Fs = step2(u)
% 2nd experiment of LuGre model
% Fs = step2(control)
% Out: Fs = [Fs+,Fs-]
Fs = [max(u) min(u)];                    % Out
```

B.3. Code for Step 3

```
function vs_p = step3p(Fc_p, Fs_p, J, x, u, sample)
% 3rd experiment of LuGre model
% vs+ = step3p(Fc+, Fs+, position, control, sample)
% Out: vs_p
n = length(sample); k=0; q=0;            % Initialization
u=u'; x=x';                              % Control, Position
t = sample';                             % Time
T = mean(diff(t));                       % Sample time
v=diff(x)/T; v(n)=v(n-1);                % Velocity
v_dot=diff(v)/T; v_dot(n)=v_dot(n-1);    % Acceleration
% Data selection
for i=1:n;
   if u(i)>0
      k=k+1;
      v(k)=v(i);
      v_dot(k)=v_dot(i);
      u(k)=u(i);
   end;
end;
% gamma>=1
for i=1:k
   gamma(i)=((Fs_p-Fc_p)/(u(i)-J*v_dot(i))*sign(v(i))-Fc_p);
   if gamma(i)>1
      q=q+1;
      gamma(q)=gamma(i);
        v(q)=v(i)^2;
   end;
end;
% Least square (Moore-Penrose)
phi_p = [log(gamma(1:q))'];
theta_p = pinv(phi_p)*v(1:q);
vs_p = sqrt(theta_p);                    % Out
```

B.4. Code for Step 4

```
function sigma = step4(x, u)
% 4th experiment of LuGre model
% sigma = step4(position, control)
% Out: sigma
global Inertia sigma2
delta = input('Microscopic damping:');  % Initialization
sigma0 = (u(2)-u(1))/(x(2)-x(1));
sigma1 = 2*delta*sqrt(sigma0*Inertia)-sigma2;
sigma = [sigma0 sigma1];               % Out
```

ACKNOWLEDGMENT

This work was supported by the Consejería de Innovación, Ciencia y Empresa of the Junta de Andalucía under IAC programme.

REFERENCES

1. K. J. Åström and K. Furuta, "Swinging up a pendulum by energy control," Automatica, vol. 36, no. 2, pp. 287–295, 2000.
2. A. M. Bloch, N. E. Leonard, and J. E. Marsden, "Stabilization of the pendulum on a rotor arm by the method of controlled lagrangians," in Proceedings of the 39th IEEE Conference on Robotics and Automation, vol. 1, pp. 500–505, Detroit, Mich, USA, May 1999.
3. R. Olfati-Saber, "Fixed point controllers and stabilization of the cart-pole system and the rotating pendulum," in Proceedings of the 38th IEEE Conference on Decision and Control, pp. 1174–1181, Phoenix, Ariz, USA, December 1999.
4. M. Wiklund, A. Kristenson, and K. J. Åström, "A new strategy for swinging up an inverted pendulum," inProceedings of the 12th IFAC World Congress, vol. 9, pp. 151–154, Sydney, Australia, 1993.
5. H. Goldstein, Classical Mechanics, Addison-Wesley, New York, NY, USA, 2nd edition, 1980.
6. A. J. Van der Schaft, L_2-gain and Passivity Techniques in Nonlinear Control, Springer, New York, NY, USA, 1989.
7. D. R. Merkin, Introduction to the Theory of Stability, vol. 24 of Texts in Applied Mathematics, Springer, New York, NY, USA, 1996.
8. J. Á. Acosta, Underactuated nonlinear control systems, Ph.D. thesis, Department of Automatic Control and Systems Engineering, University of Seville, Seville, Spain, 2004.
9. F. Altpeter, Friction modeling, identification and compensation, Ph.D. thesis, École Polythechnique Fédérale de Lausanne, 1999.
10. C. Canudas de Wit, "Control of system with dynamic friction," Tech. Rep. INPG-ENSIEG-CNRS, Laboratory of Automatic Control of Grenoble, Grenoble, France, 1999.
11. C. Canudas de Wit, H. Olsson, K. J. Åström, and P. Lischinsky, "A new model for control of systems with friction," IEEE Transactions on Automatic Control, vol. 40, no. 3, pp. 419–425, 1995.
12. M. Gafvert, "Dynamic model based friction compensation on the Furuta pendulum," in Proceedings of the IEEE Conference on Control Applications, vol. 2, pp. 1260–1265, Kohala Coast, Hawaii, USA, 1999.
13. A. S. Shiriaev and A. Sörensen, "Friction compensation in the Furuta pendulum for stabilizing rotational modes," in Proceedings of the IEEE Conference on Decision and Control, vol. 4, pp. 3772–3777, Orlando, FLa, USA, February 2001.
14. J. Á. Acosta, J. Aracil, and F. Gordillo, "On swinging the Furuta pendulum up based on Routh's reduction and SG method," in Proceedings of the 9th Mediterranean Conference on Control and Automation, vol. 1, Dubrovnik, Croatia, June 2001.
15. J. Á. Acosta, F. Gordillo, and J. Aracil, "A new SG law for swinging the Furuta pendulum up," inProceedings of the 5th IFAC Symposium on Nonlinear Control Systems, pp. 818–823, Saint-Petersbourg, Russia, July 2001.
16. F. Gordillo, J. Á. Acosta, and J. Aracil, "A new swing-up law for the Furuta pendulum," International Journal of Control, vol. 76, no. 8, pp. 836–844, 2003.
17. A. L. Fradkov and A. Yu. Pogromsky, Introduction to Control of Oscillations and Chaos, vol. 35, World Scientific, River Edge, NJ, USA, 1998.
18. A. L. Fradkov, O. P. Tomchina, and O. L. Nagibina, "Swinging control of rotating pendulum," inProceedings of the 3rd IEEE Mediterranean Symposium on Control and Automation, pp. 347–351, Limassol, Cyprus, July 1995.
19. A. L. Fradkov, "Swinging control of nonlinear oscillations," International Journal of Control, vol. 64, no. 6, pp. 1189–1202, 1996.
20. A. P. Konjukhov, O. L. Nagibina, and O. P. Tomchina, "Energy based double pendulum control in periodic and chaotic mode," in Proceedings of the 3rd International Conference on Motion and Vibration Control, pp. 99–104, Chiba, Japan, 1996.
21. F. J. Muñoz-Almaraz, E. Freire, and J. Galán-Vioque, "Bifurcation behavior of the Furuta pendulum,"International Journal of Bifurcation and Chaos, vol. 17, no. 8, pp. 2571–2578, 2000. ·
22. P. Andrievskii, P. Guzenko, and A. L. Fradkov, "Control of nonlinear vibrations of mechanical systems via the method of velocity gradient," Automation and Remote Control, vol. 57, pp. 456–467, 1996.
23. J. Á. Acosta, F. Gordillo, and J. Aracil, "Swinging up the Furuta pendulum by speed gradient method," inProceedings of the European Control Conference, pp. 469–474, Porto, Portugal, September 2001.

24. R. Olfati-Saber, "Cascade normal forms for underactuated mechanical systems," in Proceedings of the 39th IEEE Conference on Decision and Control, Sydney, Australia, December 2000.

25. A. Shiriaev, A. Pogromsky, H. Ludvigsen, and O. Egeland, "On global properties of passivity-based control of an inverted pendulum," International Journal of Robust and Nonlinear Control, vol. 10, no. 4, pp. 283–300, 2000.

26. I. Fantoni and R. Lozano, "Stabilization of the Furuta pendulum around its homoclinic orbit,"International Journal of Control, vol. 75, no. 6, pp. 390–398, 2002.

27. R. Ortega, M. W. Spong, F. Gómez-Estern, and G. Blankenstein, "Stabilization of underactuated mechanical systems via Interconnection and damping assignment passivity based control," IEEE Transactions on Automatic Control, vol. 47, no. 8, pp. 1218–1233, 2002.

28. J. Á. Acosta, R. Ortega, A. Astolfi, and A. M. Mahindrakar, "Interconnection and damping assignment passivity—based control of mechanical systems with underactuation degree one," Transactions on Automatic Control, vol. 50, no. 12, pp. 1936–1955, 2005.

29. G. Viola, R. Ortega, R. N. Banavar, J. Á. Acosta, and A. Astolfi, "Total energy shaping control of mechanical systems: simplifying the matching equations via coordinate changes," in Proceedings of the 3rd IFAC Workshop on Lagrangian and Hamiltonian Methods for Nonlinear Control, vol. 3, pp. 87–92, Nagoya, Japan, July 2006.

30. G. Viola, R. N. Banavar, J. Á. Acosta, and A. Astolfi, "Some remarks on interconnection and damping assignment passivity-based control of mechanical systems," in Taming Heterogeneity and Complexity of Embedded Control, pp. 721–735, ISTE, 2007.

31. G. Viola, R. Ortega, R. Banavar, J. Á. Acosta, and A. Astolfi, "Total energy shaping control of mechanical systems: simplifying the matching equations via coordinate changes," IEEE Transactions on Automatic Control, vol. 52, no. 6, pp. 1093–1099, 2007.

32. F. Gómez-Estern and A. J. Van der Schaft, "Physical damping in IDA-PBC controlled underactuated mechanical systems," European Journal of Control, vol. 10, no. 5, pp. 451–468, 2004.

33. F. Gómez-Estern, A. J. van der Schaft, and J. Á. Acosta, "Passivation of underactuated systems with physical damping," in Proceedings of the 6th IFAC Symposium on Nonlinear Control Systems, pp. 1–3, Stuttgart, Germany, September 2004.

34. A. M. Bloch, N. E. Leonard, and J. E. Marsden, "Controlled lagrangians and the stabilization of mechanical systems I: the first matching theorem," IEEE Transactions on Automatic Control, vol. 45, no. 12, pp. 2253–2270, 2000.

35. D. Auckly, L. Kapitanski, and W. White, "Control of nonlinear underactuated systems," Communications on Pure and Applied Mathematics, vol. 3, pp. 354–369, 2000.

36. C. Aguilar-Ibáñez, O. F. Gutierrez, and H. Sossa-Azuela, "Control of the Furuta pendulum by using a Lyapunov function," in Proceedings of the 45th IEEE Conference on Decision and Control, pp. 6128–6132, San Diego, Calif, USA, December 2006.

37. F. Mazenc and L. Praly, "Adding integrations, saturated controls, and stabilization for feedforward systems," IEEE Transactions on Automatic Control, vol. 41, no. 11, pp. 1559–1578, 1996.

38. R. Sepulchre, M. Janković, and P. Kokotović, Constructive Nonlinear Control, Communications and Control Engineering Series, Springer, London, UK, 1997.

39. A. R. Teel, "A nonlinear small gain theorem for the analysis of control systems with saturation," IEEE Transactions on Automatic Control, vol. 41, no. 9, pp. 1256–1270, 1996.

40. V. Jurdjevic and J. P. Quinn, "Controllability and stability," Journal of Differential Equations, vol. 28, no. 3, pp. 381–389, 1978.

41. J. Aracil, J. Á. Acosta, and F. Gordillo, "A controller for swinging-up and stabilizing the inverted pendulum," in Proceedings of the 17th IFAC World Congress, Seoul, Korea, July 2008.

42. J. Aracil, J. Á. Acosta, and F. Gordillo, "A nonlinear hybrid controller for swinging-up and stabilizing the Furuta pendulum," submitted.

43. J. Á. Acosta, J. Aracil, and F. Gordillo, "Nonlinear control strategies for the Furuta pendulum," Journal in Control and Intelligent Systems, vol. 29, pp. 101–107, 2001.

44. M. Krstić, I. Kanellakopoulos, and P. Kokotović, Nonlinear Adaptative Control Design, John Wiley & Sons, New York, NY, USA, 1995.

45. A. Isidori, Nonlinear Control Systems, Communications and Control Engineering Series, Springer, Berlin, Germany, 3rd edition, 1995.

46. S. Sastry, Nonlinear Systems: Analysis, Stability and Control, vol. 10 of Interdisciplinary Applied Mathematics, Springer, New York, NY, USA, 1999.

47. J. Á. Acosta and M. López-Martínez, "Constructive feedback linearization of mechanical systems with friction and underactuation degree one," in Proceedings of the 45th IEEE European Control Conference, vol. 1, pp. 4255–4260, Kos, Greece, July 2007.

48. M. López-Martínez and J. Á. Acosta, "Linealización por realimentación constructiva de sistemas mecánicos con grado de subactuación 1 inestables con fricción," Revista Iberoamericana de Automática e Informática Industrial, vol. 4, no. 4, pp. 70–79, 2007.

49. J. Á. Acosta and M. López-Martínez, "Input—output linearization on underactuated mechanical systems," submitted.

50. H. K. Khalil, Nonlinear Systems, Prentice-Hall, Upper Saddle River, NJ, USA, 2002.

51. J. Á. Acosta and M. López-Martínez, "Constructive feedback linearization of underactuated mechanical systems with 2-DOF," in Proceedings of the IEEE Conference on Decision and Control and European

Control Conference, pp. 4909–4914, Seville, Spain, December 2005.

52. C. I. Byrnes, A. Isidori, and J. C. Willems, "Passivity, feedback equivalence, and the global stabilization of minimum phase nonlinear systems," IEEE Transactions on Automatic Control, vol. 36, no. 11, pp. 1228–1240, 1991.

53. M. Fliess, J. Lévine, P. Martin, and P. Rouchon, "A lie-backlund approach to equivalence and flatness of nonlinear systems," IEEE Transactions on Automatic Control, vol. 44, no. 5, pp. 922–937, 1999. · ·

54. M. V. Nieuwstadt, M. Rathinam, and R. M. Murray, "Differential flatness and absolute equivalence," inProceedings of the 33rd Conference on Decision and Control, Lake Buena Vista, Fla, USA, 1995.

55. C. Aguilar-Ibáñez and H. Sira-Ramírez, "A linear differential flatness approach to controlling the Furuta pendulum," IMA Journal of Mathematical Control and Information, vol. 24, no. 1, pp. 31–45, 2007.

56. J. Á. Acosta and M. López-Martínez, "On singular perturbations of unstable underactuated mechanical systems with underactuation degree ≥1," in Proceedings of the 17th IFAC World Congress, Seoul, Korea, July 2008.

57. J. Á. Acosta and M. López-Martínez, "A nonlinear strategy to control unstable underactuated mechanical systems with underactuation ≥1. Applications to control augmentations," Automation & Control Systems Journal, vol. 2, pp. 13–20, 2009.

58. D. Angeli, "Almost global stabilization of the inverted pendulum via continuous state feedback,"Automatica, vol. 37, pp. 1103–1108, 2001.

59. J. Aracil, F. Gordillo, and J. Á. Acosta, "Stabilization of oscillation in the inverted pendulum," inProceedings of the 15th IFAC World Congress, Barcelona, Spain, July 2002.

60. L. B. Freidovich, A. Robertsson, R. Johansson, A. Sandberg, and A. S. Shiriaev, "Virtual-holonomic-constraints-based design of stable oscillations of Furuta pendulum: theory and experiments," IEEE Transactions on Robotics, vol. 23, no. 4, pp. 827–832, 2007. ·

61. D. Pagano, L. Pizarro, and J. Aracil, "Modern applied mathematics techniques in circuits, systems and control," in Bifurcation Analysis in a Rotational Inverted Pendulum, pp. 144—149, World Scientific, Singapore, 1999.

62. D. Pagano, L. Pizarro, and J. Aracil, "Local bifurcation analysis in the Furuta pendulum via normal forms," International Journal of Bifurcation and Chaos, vol. 10, no. 5, pp. 981–995, 2000.

63. D. Pagano, Bifurcaciones en los sistemas de control, Ph.D. thesis, Escuela Superior de Ingenieros. Universidad de Sevilla, Sevilla, Spain, 1999.

64. J. Aracil, K. J. Åström, and D. J. Pagano, "Global bifurcations in the Furuta pendulum," in Proceedings of the Nonlinear Control Systems Design Symposium, vol. 1, pp. 35–40, Enschede, The Netherlands, July 1998.

CHAPTER 4 OSCILLATORY SOLUTIONS OF SINGULAR EQUATIONS HYDRODYNAMICS

Irena Rachůnková, Jan Tomeček, Jakub Stryja

INTRODUCTION

We study the equation

$$(p(t)u'(t))' = p(t)f(u(t)) \tag{1.1}$$

on the half-line $[0,\infty)$, where

$$f \in \mathrm{Lip}_{\mathrm{loc}}(\mathbb{R}), \quad p \in C^1(0,\infty) \cap C[0,\infty), \tag{1.2}$$

$$p(0) = 0, \quad p'(t) > 0, \quad t > 0, \quad \lim_{t\to\infty} \frac{p'(t)}{p(t)} = 0. \tag{1.3}$$

Equation (1.1) is singular at $t = 0$ because $p(0) = 0$. If f in (1.1) fulfils moreover assumptions

$$\text{there exists } L > 0 \text{ such that } f(x) = 0 \quad \text{for } x \geq L, \tag{1.4}$$

$$xf(x) < 0 \quad \text{for } x \in (-\infty,0) \cup (0,L), \tag{1.5}$$

$$\text{there exists } \overline{B} < 0 \text{ such that } \int_{\overline{B}}^{L} f(z)\mathrm{d}z = 0, \tag{1.6}$$

then (1.1) generalizes equations which appear in hydrodynamics or in the nonlinear field theory [1–5].

Definition 1.1.

A function $u \in C^1[0,\infty)$ which has continuous second derivative on $(0,\infty)$ and satisfies (1.1) for all $t \in (0,\infty)$ is called a *solution* of (1.1).

Consider $B < 0$ and the initial conditions

$$u(0) = B, \quad u'(0) = 0. \tag{1.7}$$

The initial value problem (1.1), (1.7) has been investigated, for example, in [6–12]. In particular in [10] it was proved that for each negative B there exists a unique solution of problem (1.1), (1.7) under the assumptions (1.2)–(1.6). Consider such solution u and denote

$$u_{\sup} = \sup\{u(t) : t \in [0,\infty)\}. \tag{1.8}$$

Definition 1.2.

If $u_{\sup} < L$ ($u_{\sup} = L$ or $u_{\sup} > L$), then u is called a *damped* solution (a *homoclinic* solution or an*escape* solution) of problem (1.1), (1.7).

In [10, 12] these three types of solutions of problem (1.1), (1.7) have been studied, and the existence of each type has been proved for sublinear or linear asymptotic behaviour of f near $-\infty$. In [11], f has been supposed to have a zero $L_0 < 0$. Here we generalize and extend the results of [10–12] concerning damped solutions. We prove their existence under weaker assumptions than in the above papers. Moreover, we bring conditions under which each damped solution is oscillatory; that is, it has an unbounded set of isolated zeros.

We replace assumptions (1.4)–(1.6) by the following ones.

There exist $L_0 < 0, L > 0, C_L > 0$ such that

$$xf(x) < 0 \quad \text{for } x \in (L_0,0) \cup (0,L), \tag{1.9}$$

$$0 \leq f(x) \leq C_L \quad \text{for } x \geq L \tag{1.10}$$

($L_0 = -\infty$ is possible).

Citation: Irena Rachůnková, Jan Tomeček, Jakub Stryja, Oscillatory Solutions of Singular Equations Arising in Hydrodynamics, doi:10.1155/2010/872160

DAMPED SOLUTIONS

Theorem 2.1 (Existence and uniqueness).

Assume that (1.2), (1.3), (1.9), and (1.10) hold and let $B \in (L_0, 0)$. Then problem (1.1), (1.7) has a unique solution u, and moreover the solution u satisfies

$$u(t) \geq B \quad for\ t \in [0,\infty). \tag{2.1}$$

Proof.

Step 1.

Put

$$f_B(x) = \begin{cases} f(x) & \text{for } x \geq B, \\ f(B) & \text{for } x < B. \end{cases} \tag{2.2}$$

We will study the auxiliary differential equation:

$$(p(t)u'(t))' = p(t)f_B(u(t)) \tag{2.3}$$

By virtue of (1.2) we find the Lipschitz constant $K > 0$ for f on $[B-1, |B|+1]$ and due to (1.2), (1.10), and (2.2), we find $M_B > 0$ such that

$$|f_B(x)| \leq M_B \quad \text{for } x \in \mathbb{R}. \tag{2.4}$$

Put $\varphi(t) = \int_0^t p(s)ds/p(t)$ for $t > 0$. Having in mind (1.3), we see that p is increasing and so

$$0 < \varphi(t) \leq t \quad \text{for } t > 0, \ \lim_{t \to 0+} \varphi(t) = 0. \tag{2.5}$$

Consequently we can choose $\eta > 0$ such that

$$\int_0^\eta \varphi(t)dt \leq \min\left\{\frac{1}{2K}, \frac{1}{M_B}\right\} \tag{2.6}$$

Consider the Banach space $C[0,\eta]$ (with the maximum norm) and define an operator F: $C[0,\eta] \to C[0,\eta]$ by

$$(\mathcal{F}u)(t) = B + \int_0^t \frac{1}{p(s)} \int_0^s p(\tau)f_B(u(\tau))d\tau\, ds. \tag{2.7}$$

Using (2.4) and (2.6), we have

$$\|\mathcal{F}u\|_{C[0,\eta]} \leq |B| + M_B \int_0^\eta \varphi(s)ds \leq |B| + 1; \tag{2.8}$$

that is F maps the ball

$\mathcal{B}(0, |B|+1) = \{u \in C[0,\eta] : \|u\|_{C[0,\eta]} \leq |B|+1\}$ to itself. Due to (2.2) and the choice of K, we have for $u_1, u_2 \in \mathcal{B}(0, |B|+1)$

$$\|\mathcal{F}u_1 - \mathcal{F}u_2\|_{C[0,\eta]} \leq \int_0^\eta \frac{1}{p(s)} \int_0^s p(\tau)\left|f_B(u_1(\tau)) - f_B(u_2(\tau))\right|d\tau\, ds$$

$$\leq K\|u_1 - u_2\|_{C[0,\eta]} \int_0^\eta \varphi(s)ds \leq \frac{1}{2}\|u_1 - u_2\|_{C[0,\eta]}. \tag{2.9}$$

Hence F is a contraction on $\mathcal{B}(0, |B|+1)$, and the Banach fixed point theorem yields a unique fixed point $u \in \mathcal{B}(0, |B|+1)$ of $\mathcal{F}$.

Step 2.

The fixed point u of Step 1 fulfils

$$u(0) = B, \qquad u'(t) = \frac{1}{p(t)} \int_0^t p(s)f_B(u(s))ds, \quad t \in (0,\eta]. \tag{2.10}$$

Hence u satisfies (2.3) on $(0,\eta]$. Finally, (2.4) and (2.5) yield

$$\lim_{t \to 0+} |u'(t)| \leq M_B \lim_{t \to 0+} \varphi(t) = 0. \tag{2.11}$$

Consequently u fulfils (1.7). Choose an arbitrary $b > \eta$. Then, by (2.5) and (2.10),

$$|u'(t)| \leq M_B b, \quad |u(t)| \leq |B| + M_B b^2, \quad t \in [0,b]. \tag{2.12}$$

Having in mind that $f_B \in \mathrm{Lip}_{\mathrm{loc}}(\mathbb{R}), u$, can be (uniquely) extended as a function satisfying (2.3) onto $[0,b]$. Since b is arbitrary, u can be extended onto $[0,\infty)$ as a solution of (2.3). We have proved that problem (2.3), (1.7) has a unique solution.

Step 3.

According to Step 2 we have

$$u''(t) + \frac{p'(t)}{p(t)}u'(t) = f_B(u(t)) \quad \text{for } t \in (0,\infty) \tag{2.13}$$

Multiplying (2.13) by u' and integrating between 0 and t, we get

$$\frac{u'^2(t)}{2} + \int_0^t \frac{p'(s)}{p(s)}u'^2(s)ds = \int_0^t f_B(u(s))u'(s)ds, \quad t \in (0,\infty). \tag{2.14}$$

Put

$$F_B(x) = -\int_0^x f_B(z)dz, \quad x \in \mathbb{R}. \tag{2.15}$$

So,(2.14) has the form

$$\frac{u'^2(t)}{2} + \int_0^t \frac{p'(s)}{p(s)}u'^2(s)ds + F_B(u(t)) = F_B(B), \quad t \in (0,\infty). \tag{2.16}$$

Let

$u(t_1) \in (L_0, B)$ for some $t_1 > 0$. Then (2.16) yields $F_B(u(t_1)) \leq F_B(B)$ which is not possible because F_B is decreasing on $(L_0, 0)$ by (1.9) and (2.2). Therefore $u(t) \geq B$ for $t \in [0,\infty)$. Consequently, due to (2.2), u is a solution of (1.1).

Step 4.

Assume that there exists another solution $\tilde{u}$ of problem (1.1), (1.7). Then we can prove similarly as in Step 3 that $\tilde{u}(t) \geq B$ for $t \in [0,\infty)$. This implies that $\tilde{u}$ is also a solution of problem (2.3), (1.7) and by Step 2, $\tilde{u} \equiv u$. We have proved that problem (1.1), (1.7) has a unique solution.

Lemma 2.2.

Let $C \in \{0, L\}$ and let u be a solution of (1.1). Assume that there exists a>0 such that

$$u(a) = C, \qquad u'(a) = 0. \tag{2.17}$$

Then $u(t) = C$ *for all* $t \in [0,\infty)$.

Proof.

We see that the constant function $\tilde{u} \equiv C$ is a solution of (1.1). Let u be a solution of (1.1) satisfying (2.17) and let $u(t) \neq \tilde{u}(t)$ for some $t \in [0,\infty)$. Then the regular initial problem (1.1), (2.17) has two different solutions u and $\tilde{u}$, which contradicts (1.2).

Remark 2.3.

Let us put

$$F(x) = -\int_0^x f(z)dz \quad \text{for } x \in \mathbb{R}. \tag{2.18}$$

Due to (1.2) and (1.9) we see that F is continuous on $\mathbb{R}$, decreasing and positive on $(L_0, 0)$, increasing and positive on $(0, L)$. Therefore we can define $\overline{B} < 0$ by

$$\overline{B} = \inf\{B_0 \in (L_0, 0) : F(B) < F(L)\ \forall B \in (B_0, 0)\} \tag{2.19}$$

($\overline{B} = -\infty$ is possible.)

Theorem 2.4 (Existence of damped solutions).

Assume that (1.2), (1.3), (1.9), and (1.10) hold. Let $\overline{B}$ be given by (2.19), and assume that u is a solution of problem (1.1), (1.7) with $B \in (\overline{B}, 0)$. Then u is a damped solution.

Proof.

Since $B \in (\overline{B}, 0)$, we can find $\epsilon > 0$ such that

$$F(B) \le F(L - \epsilon). \tag{2.20}$$

Assume on the contrary that u is not damped, that is,

$$\sup\{u(t) : t \in [0, \infty)\} \ge L. \tag{2.21}$$

Then, according to Lemma 2.2, there exists $\theta \in (0, \infty)$ such that

$$u(\theta) = 0, \quad u'(\theta) > 0, \quad u(t) \in [B, 0) \quad \text{for } t \in [0, \theta). \tag{2.22}$$

By (1.1), (1.3), and (1.9) we have $(pu')' > 0$ on $(0, \theta]$. So, pu' is increasing and positive on $(0, \theta]$ and hence $u' > 0$ on $(0, \theta]$. Assumption (2.21) implies that there exists $b \in (\theta, \infty)$ such that

$$u(b) = L - \epsilon, \quad u(t) \in [B, L - \epsilon) \quad \text{for } t \in [0, b). \tag{2.23}$$

Since u fulfils (1.1), we have

$$u''(t) + \frac{p'(t)}{p(t)} u'(t) = f(u(t)) \quad \text{for } t \in (0, \infty). \tag{2.24}$$

Multiplying (2.24) by u' and integrating between 0 and b we get

$$0 < \frac{u'^2(b)}{2} + \int_0^b \frac{p'(s)}{p(s)} u'^2(s)ds = F(B) - F(L - \epsilon). \tag{2.25}$$

This contradicts (2.20).

OSCILLATORY SOLUTIONS

In this section we assume that, in addition to our basic assumptions (1.2), (1.3), (1.9), and (1.10), the following conditions are fulfilled:

$$\lim_{x \to 0-} \frac{f(x)}{x} < 0, \qquad \lim_{x \to 0+} \frac{f(x)}{x} < 0, \tag{3.1}$$

$$p \in C^2(0, \infty), \quad \limsup_{t \to \infty} \left| \frac{p''(t)}{p'(t)} \right| < \infty \tag{3.2}$$

Then the next lemmas can be proved.

Lemma 3.1.

Let u be a solution of problem (1.1), (1.7) with $B \in (L_0, 0)$. Then there exists $\theta > 0$ such that

$$u(\theta) = 0, \quad u'(t) > 0 \quad \text{for } t \in (0, \theta]. \tag{3.3}$$

Proof.

Step 1.

Assume that such θ does not exist. Then

$$u(t) < 0 \quad \text{for } t \in [0, \infty). \tag{3.4}$$

Hence (1.1), (1.7), and (1.9) yield $(pu')' > 0$ and $u' > 0$ on $(0, \infty)$. Therefore u is increasing on $(0, \infty)$ and

$$\lim_{t \to \infty} u(t) = \ell \in (B, 0] \tag{3.5}$$

Multiplying (2.24) by u' and integrating between 0 and t, we get due to (2.18)

$$\frac{u'^2(t)}{2} + \int_0^t \frac{p'(s)}{p(s)} u'^2(s)ds = F(B) - F(u(t)), \quad t \in (0, \infty). \tag{3.6}$$

Letting $t \to \infty$, we get

$$\lim_{t \to \infty} \frac{u'^2(t)}{2} = -\lim_{t \to \infty} \int_0^t \frac{p'(s)}{p(s)} u'^2(s)ds + F(B) - F(\ell). \tag{3.7}$$

Since the function $\int_0^t p'(s)/p(s) u'^2(s)ds$ is positive and increasing, it follows that there exists $\lim_{t\to\infty} u'(t) \ge 0$. If $\lim_{t\to\infty} u'(t) > 0$, then $\lim_{t\to\infty} u(t) = \infty$ contrary to (3.5). Consequently,

$$\lim_{t \to \infty} u'(t) = 0. \tag{3.8}$$

Letting $t \to \infty$ in (2.24), we get by (1.3), (1.9), and (3.5)

$$\lim_{t \to \infty} u''(t) = f(\ell) \ge 0. \tag{3.9}$$

Due to (3.8), we conclude that $f(\ell) = 0$ and hence $\ell = 0$. We have proved that if $\theta > 0$ fulfilling (3.3) does not exist, then

$$\lim_{t \to \infty} u(t) = 0, \qquad \lim_{t \to \infty} u'(t) = 0. \tag{3.10}$$

Step 2.

We define a function

$$v(t) = \sqrt{p(t)}u(t), \quad t \in [0, \infty). \tag{3.11}$$

By (1.3) and (3.2), we have $v \in C^2(0, \infty)$,

$$v'(t) = \frac{p'(t)u(t)}{2\sqrt{p(t)}} + \sqrt{p(t)}u'(t), \tag{3.12}$$

$$v''(t) = v(t)\left[\frac{1}{2}\frac{p''(t)}{p(t)} - \frac{1}{4}\left(\frac{p'(t)}{p(t)}\right)^2 + \frac{f(u(t))}{u(t)}\right], \quad t \in (0, \infty), \tag{3.13}$$

$$\lim_{t \to \infty} \frac{p''(t)}{p(t)} = \lim_{t \to \infty} \frac{p''(t)}{p'(t)} \frac{p'(t)}{p(t)} = 0. \tag{3.14}$$

Due to (1.3), (3.1), (3.10) and (3.14) there exist $\omega > 0$ and $R > 0$ such that

$$\frac{1}{2}\frac{p''(t)}{p(t)} - \frac{1}{4}\left(\frac{p'(t)}{p(t)}\right)^2 + \frac{f(u(t))}{u(t)} < -\omega \quad \text{for } t \ge R. \tag{3.15}$$

Due to (3.4), (3.11), (3.13), and (3.15), we get

$$v''(t) > -wv(t) > 0 \quad \text{for } t \ge R. \tag{3.16}$$

Thus, v' is increasing on $[R,\infty)$ and has the limit

$$\lim_{t\to\infty} v'(t) = V. \tag{3.17}$$

If $V > 0$, then $\lim_{t\to\infty} v(t) = \infty$, which contradicts (3.4) and (3.11). If $V \le 0$, then $v' < 0$ on $[R,\infty)$ and

$$v(t) \le v(R) < 0 \quad \text{for } t \in [R,\infty). \tag{3.18}$$

In view of (3.16) we can see that

$$0 < -\omega v(R) \le -\omega v(t) < v''(t) \quad \text{for } t \in [R,\infty). \tag{3.19}$$

We get $\lim_{t\to\infty} v'(t) = \infty$ which contradicts $V \le 0.$. The obtained contradictions imply that (3.4) cannot occur and hence $\theta > 0$ satisfying (3.3) must exist.

Corollary 3.2.

Let u be a solution of problem (1.1), (1.7) with $B \in (L_0,0).$. Further assume that there exist $b_1 > 0$ *and* $B_1 \in (B,0)$ such that

$$u(b_1) = B_1, \qquad u'(b_1) = 0. \tag{3.20}$$

Then there exists $\theta_1 > b_1$ such that

$$u(\theta_1) = 0, \quad u'(t) > 0 \quad \text{for } t \in (b_1,\theta_1]. \tag{3.21}$$

Proof.

We can argue as in the proof of Lemma 3.1 working with b_1 and B_1 instead of 0 and B.

Lemma 3.3.

Let u be a solution of problem (1.1), (1.7) with $B \in (L_0,0).$. Further assume that there exist $a_1 > 0$ *and* $A_1 \in (0,L)$ such that

$$u(a_1) = A_1, \qquad u'(a_1) = 0. \tag{3.22}$$

Then there exists $\delta_1 > a_1$ such that

$$u(\delta_1) = 0, \quad u'(t) < 0 \quad \text{for } t \in (a_1,\delta_1]. \tag{3.23}$$

Proof.

We argue similarly as in the proof of Lemma 3.1.

Step 1.

Assume that such $\delta_1 > a_1$ does not exist. Then

$$u(t) > 0 \quad \text{for } t \in [a_1,\infty) \tag{3.24}$$

By (1.1), (1.7), and (1.9) we deduce $u' < 0$ on (a_1,∞) and

$$\lim_{t\to\infty} u(t) = \ell_1 \in [0,A_1). \tag{3.25}$$

Multiplying (2.24) by u', integrating between a_1 and t, and using (2.18), we obtain

$$\frac{u'^2(t)}{2} + \int_{a_1}^{t} \frac{p'(s)}{p(s)} u'^2(s)\,ds = F(A_1) - F(u(t)), \quad t \in (a_1,\infty), \tag{3.26}$$

and we derive as in the proof of Lemma 3.1 that (3.10) holds.

Step 2.

We define v by (3.11) and get (3.13) for $t \in (a_1,\infty)$. As in the proof of Lemma 3.1 we find $\omega > 0$ and $R > 0$ satisfying (3.15). Due to (3.24), (3.11), (3.13), and (3.15) we get

$$v''(t) < -\omega v(t) < 0 \quad \text{for } t \ge R. \tag{3.27}$$

So, v' is decreasing on $[R,\infty)$ and $\lim_{t\to\infty} v'(t) = V$. If $V < 0$, then $\lim_{t\to\infty} v(t) = -\infty$ which contradicts (3.24) and (3.11). If $V \ge 0$, then $v' > 0$ on $[R,\infty)$ and

$$v(t) \ge v(R) > 0 \quad \text{for } t \in [R,\infty). \tag{3.28}$$

In view of (3.27) we can see that

$$v''(t) < -\omega v(t) \le -\omega v(R) < 0 \quad \text{for } t \in [R,\infty). \tag{3.29}$$

We get $\lim_{t\to\infty} v'(t) = -\infty$ contrary to $V \ge 0.$. The obtained contradictions imply that (3.24) cannot occur and that $\delta_1 > a_1$ satisfying (3.23) must exist.

Theorem 3.4.

Assume that (1.2), (1.3), (1.9), (1.10), (3.1), and (3.2) hold. Let u be a solution of problem (1.1), (1.7) with $B \in (L_0,0)$. If u is a damped solution, then u is oscillatory and its amplitudes are decreasing.

Proof.

Let u be a damped solution. By (2.1) and Definition 1.2, we can find $L_1 \in (0,L)$ such that

$$B \le u(t) \le L_1 \quad \text{for } t \in [0,\infty). \tag{3.30}$$

Step 1.

Lemma 3.1 yields $\theta > 0$ satisfying (3.3). Hence there exists a maximal interval (θ,a_1) such that $u' > 0$ on (θ,a_1). Let $a_1 = \infty$. Then, by (3.30), we get $u \in (0,L)$, $u' > 0$ on (θ,∞) and

$$\lim_{t\to\infty} u(t) = \ell_0 \in (0,L). \tag{3.31}$$

By (1.1), (1.3), and (1.9), we have $(pu')' < 0$ on (θ,∞). So pu' and u' are decreasing on (θ,∞) and, due to (3.31),

$$\lim_{t\to\infty} u'(t) = 0. \tag{3.32}$$

Letting $t \to \infty$ in (2.24) and using (1.3), (1.9), and (3.31), we get

$$\lim_{t\to\infty} u''(t) = f(\ell_0) < 0, \tag{3.33}$$

which contradicts (3.32). Therefore $a_1 < \infty$ and there exists $A_1 \in (0,L)$ such that (3.22) holds. Lemma 3.3 yields $\delta_1 > a_1$ satisfying (3.23). Therefore u has just one positive local maximum $A_1 = u(a_1)$ between its first zero θ and second zero δ_1.

Step 2.

By (3.23) there exists a maximal interval (δ_1, b_1), where $u' < 0$. Let $b_1 = \infty$. Then, by (3.30), we have $u \in [B, 0)$, $u' < 0$ on (δ_1, ∞), and

$$\lim_{t\to\infty} u(t) = \ell_1 \in [B, 0). \tag{3.34}$$

By (1.1), (1.3), and (1.9), we get $(pu')' > 0$ on (δ_1, ∞) and so pu' is increasing on (δ_1, ∞). Since $u' < 0$, we deduce that u' is increasing on (δ_1, ∞) and, by (3.34), we get (3.32). Letting $t \to \infty$ in (1.1) and using (1.3), (1.9), and (3.34), we get

$$\lim_{t\to\infty} u''(t) = f(\ell_1) > 0, \tag{3.35}$$

which contradicts (3.32). Therefore $b_1 < \infty$ and there exists $B_1 \in [B, 0)$ such that (3.20) holds. Corollary 3.2 yields $\theta_1 > b_1$ satisfying (3.21). Therefore u has just one negative minimum $B_1 \in [B, 0)$ between its second zero a_1 and third zero θ_1.

Step 3.

We can continue as in Step 1 and Step 2 and get the sequences $\{A_n\}_{n=1}^{\infty} \subset (0, L)$ and $\{B_n\}_{n=1}^{\infty} \subset [B, 0)$ of local maxima and local minima of u attained at a_n and b_n, respectively. Now, put $x_1(t) = u(t)$, $x_2(t) = u'(t)$ and write (1.1) as a system

$$x_1'(t) = x_2(t), \qquad x_2'(t) = -\frac{p'(t)}{p(t)} x_2(t) + f(x_1(t)). \tag{3.36}$$

Consider F of (2.18) and define a Lyapunov function V by

$$V(x_1, x_2) = F(x_1) + \frac{x_2^2}{2} \quad \text{for } (x_1, x_2) \in D, \tag{3.37}$$

where $D = (L_0, L) \times \mathbb{R}$. By Remark 2.3, we see that $V(0,0) = 0$ and $V(x_1, x_2) > 0$ on $D \setminus \{(0,0)\}$. By (3.6) and (3.37), we have

$$V(u(t), u'(t)) = \frac{u'^2(t)}{2} + F(u(t)) = F(B) - \int_0^t \frac{p'(t)}{p(t)} u'^2(s)\mathrm{d}s,$$

$$\dot V(t) = \frac{\mathrm{d}V(u(t), u'(t))}{\mathrm{d}t} = -\frac{p'(t)}{p(t)} u'^2(t) \le 0 \quad \text{for } t \in (0, \infty). \tag{3.38}$$

Therefore

$$\dot V(t) < 0 \quad \text{for } t \in (0, \infty),\ t \ne a_n, b_n,\ n \in \mathbb{N}. \tag{3.39}$$

By (3.30), $(u(t), u'(t)) \in D$ for $t \in [0, \infty)$. We see that $V(u(t), u'(t))$ is positive and decreasing (for the damped solution u) and hence

$$\lim_{t\to\infty} V(u(t), u'(t)) = c_B \ge 0. \tag{3.40}$$

So, sequences $\{F(A_n)\}_{n=1}^{\infty}$ and $\{F(B_n)\}_{n=1}^{\infty}$ are decreasing:

$$F(A_n) = V(u(a_n), u'(a_n)), \qquad F(B_n) = V(u(b_n), u'(b_n)) \tag{3.41}$$

for $n \in \mathbb{N}$ and

$$\lim_{n\to\infty} F(A_n) = \lim_{n\to\infty} F(B_n) = c_B. \tag{3.42}$$

Further, due to Remark 2.3, the sequence $\{A_n\}_{n=1}^{\infty}$ is decreasing and the sequence $\{B_n\}_{n=1}^{\infty}$ is increasing. Consequently,

$$\lim_{n\to\infty} A_n \in [0, L), \qquad \lim_{n\to\infty} B_n \in (B, 0]. \tag{3.43}$$

Remark 3.5.

There are two cases for the number C_B from the proof of Theorem 3.4: $c_B = 0$ and $c_B > 0$. Denote

$$\lim_{n\to\infty} A_n = A_\infty, \qquad \lim_{n\to\infty} B_n = B_\infty. \tag{3.44}$$

If

$c_B = 0$, then $F(A_\infty) = F(B_\infty) = 0$ and hence $A_\infty = B_\infty = 0$, that is, $\lim_{t\to\infty} u(t) = 0$.

Let $c_B > 0$. Consider an arbitrary sequence $\{t_n\}_{n=1}^{\infty}$ such that $\lim_{n\to\infty} t_n = \infty$. By (3.40) we have $\lim_{n\to\infty} V(u(t_n), u'(t_n)) = c_B$. By (3.30) and (3.6), the sequence $\{(u(t_n), u'(t_n))\}_{n=1}^{\infty}$ is bounded and so there exists a subsequence

$$\{(u(t_{m_n}), u'(t_{m_n}))\}_{n=1}^{\infty} \tag{3.45}$$

such that

$\lim_{n\to\infty} (u(t_{m_n}), u'(t_{m_n})) = (x_1^B, x_2^B)$, where (x_1^B, x_2^B) is a point of the level curve:

$$F(x_1) + \frac{x_2^2}{2} = c_B \tag{3.46}$$

Note that

$$c_B = 0 \quad \text{iff} \quad \int_0^\infty \frac{p'(s)}{p(s)} u'^2(s)\mathrm{d}s = F(B),$$

$$c_B > 0 \quad \text{iff} \quad \int_0^\infty \frac{p'(s)}{p(s)} u'^2(s)\mathrm{d}s < F(B). \tag{3.47}$$

Theorem 3.6 (Existence of oscillatory solutions).

Assume that (1.2), (1.3), (1.9), (1.10), (3.1), and (3.2) hold. Let $\bar B$ be given by (2.19) and let u be a solution of problem (1.1), (1.7) with $B \in (\bar B, 0)$. Then u is an oscillatory solution with decreasing amplitudes.

Proof.

The assertion follows from Theorems 2.4 and 3.4.

Remark 3.7.

The assumption (1.10) in Theorem 3.6 can be omitted, because it has no influence on the existence of oscillatory solutions. It follows from the fact that (1.10) imposes conditions on the function values of the function f for arguments greater than L however, the function values of oscillatory solutions are lower than this constant L. This condition (used only in Theorem 2.1) guaranteed the existence of solution of each problem (1.1), (1.7) for each $B < 0$ on the whole half-line, which simplified the investigation of the problem.

ACKNOWLEDGMENT

This work was supported by the Council of Czech Government MSM 6198959214.

REFERENCES

1. Dell'Isola F, Gouin H, Rotoli G: Nucleation of spherical shell-like interfaces by second gradient theory: numerical simulations. *European Journal of Mechanics* 1996,15(4):545-568.
2. Derrick GH: Comments on nonlinear wave equations as models for elementary particles. *Journal of Mathematical Physics* 1964, 5: 1252-1254. 10.1063/1.1704233.
3. Gouin H, Rotoli G: An analytical approximation of density profile and surface tension of microscopic bubbles for Van Der Waals fluids. *Mechanics Research Communications* 1997,24(3):255-260. 10.1016/S0093-6413(97)00022-0
4. Kitzhofer G, Koch O, Lima P, Weinmüller E: Efficient numerical solution of the density profile equation in hydrodynamics. *Journal of Scientific Computing* 2007,32(3):411-424. 10.1007/s10915-007-9141-05.
5. Lima PM, Konyukhova NB, Sukov AI, Chemetov NV: Analytical-numerical investigation of bubble-type solutions of nonlinear singular problems. *Journal of Computational and Applied Mathematics* 2006,189(1-2):260-273. 10.1016/j.cam.2005.05.0046.
6. Berestycki H, Lions P-L, Peletier LA:An ODE approach to the existence of positive solutions for semilinear problems in $\mathbb{R}^N$. *Indiana University Mathematics Journal* 1981,30(1):141-157. 10.1512/iumj.1981.30.300127.
7. Bonheure D, Gomes JM, Sanchez L: Positive solutions of a second-order singular ordinary differential equation. *Nonlinear Analysis: Theory, Methods & Applications* 2005,61(8):1383-1399. 10.1016/j.na.2005.02.0298.
8. Conti M, Merizzi L, Terracini S:Radial solutions of superlinear equations on $\mathbb{R}^N$. I. A global variational approach. *Archive for Rational Mechanics and Analysis* 2000,153(4):291-316. 10.1007/s0020500500159.
9. Koch O, Kofler P, Weinmüller EB: Initial value problems for systems of ordinary first and second order differential equations with a singularity of the first kind. *Analysis* 2001,21(4):373-389.10.
10. Rachůnková I, Tomeček J: Bubble-type solutions of nonlinear singular problems. *Mathematical and Computer Modelling* 2010,51(5-6):658-669. 10.1016/j.mcm.2009.10.04211.
11. Rachůnková I, Tomeček J: Strictly increasing solutions of a nonlinear singular differential equation arising in hydrodynamics. *Nonlinear Analysis: Theory, Methods & Applications* 2010,72(3-4):2114-2118. 10.1016/j.na.2009.10.011
12. Rachůnková I, Tomeček J: Homoclinic solutions of singular nonautonomous second-order differential equations. *Boundary Value Problems* 2009, 2009:-21.

CHAPTER 5

ANALYTICAL STUDY OF NONLINEAR OSCILLATORY SYSTEMS USING THE HAMILTONIAN APPROACH TECHNIQUE

Martin Hermann, Masoud Saravi, Hadi Ebrahimi Khah

Keywords: Dynamical systems, Strongly nonlinear differential equations, Analytical approximate techniques, Hamiltonian approach method, Numerical method, Runge–Kutta method

INTRODUCTION

Most of the dynamical systems facing engineers, physicists and applied mathematicians today exhibit certain essential features which preclude exact analytical solutions. Some of these features are nonlinearities, variable coefficients, complex boundary shapes, and nonlinear boundary conditions at known or, in some cases, unknown boundaries. Even if the exact solution of a problem can be found explicitly, it may be useless for mathematical and physical interpretation or numerical evaluation. Examples of such problems are Bessel functions of large argument and large-order and doubly periodic functions. Therefore, to obtain information about solutions of equations, we are forced to resort to approximate and numerical solutions, or combinations of both.

Recently, considerable attention has been directed toward the analytical approximate solutions for the strongly nonlinear differential equations of dynamical systems. The traditional perturbation methods have many shortcomings, and they are not valid for strongly nonlinear dynamical systems.

To overcome the shortcomings, many techniques have appeared in open literature, for example, parameter-expanding method [1–3], variational iteration method [4–7], energy balance method [8–14], variational approach method [15–19], amplitude–frequency formulation [20–22], homotopy perturbation method [23–29], and the other analytical approximate solutions [30].

Citation: Martin Hermann, Masoud Saravi, Hadi Ebrahimi Khah Analytical study of nonlinear oscillatory systems using the Hamiltonian approach technique, doi:10.1007/s40094-014-0133-9.

The solution of differential equations in physics and engineering, especially some oscillation equations are nonlinear, and in most cases it is difficult to solve such equations, especially analytically. Previously, He had introduced the energy balance method based on collocation and the Hamiltonian. This method can be seen as a Ritz method and leads to a very rapid convergence of the solution, and can be easily extended to other nonlinear oscillations.

This approach is very simple but strongly depends upon the chosen location point. Recently, He [31] has proposed the Hamiltonian approach to overcome the shortcomings of the energy balance method. This approach is a kind of energy method with a vast application in conservative oscillatory systems. Application of this method can be found in many literatures [31–35].

The Hamiltonian approach method is the subject of this article, as one of the analytical approximate techniques. In this article, we present two examples to illustrate the applicability, accuracy and effectiveness of the Hamiltonian approach method as one of the analytical approximate techniques.

As the first example in this paper, we investigate the nonlinear differential equation of the motion of a rigid rod rocking back on the circular surface without slipping [36]. In the second example, we investigate the nonlinear differential equation of the free vibrations of an autonomous conservative oscillator with inertia and static-type fifth-order nonlinearities [37].

To illustrate the accuracy of the Hamiltonian approach method, in first example, we compare the approximate solution result with exact solution. In second example, because there is no exact solution, we compare the approximate solution result with Runge–Kutta method as one of the known numerical methods.

The description of Hamiltonian approach method

To descript the He's Hamiltonian approach method, we consider the following general oscillator [31]:

$$\frac{d^2}{dt^2}u(t)+f(u(t))=0 \tag{1}$$

with initial conditions:

$$u(0)=A,\ \frac{d}{dt}u(0)=0. \tag{2}$$

It is easy to establish a variational principle for Eq. (1), which reads [31]:

$$J(u(t))=\int_0^{T/4}\left\{-\frac{1}{2}\left(\frac{d}{dt}u(t)\right)^2+F(u(t))\right\}dt \tag{3}$$

where T is the period of the oscillator $\partial F/\partial u = f(u)$.

In the functional (3), $-\frac{1}{2}\left(\frac{d}{dt}u(t)\right)^2$ is kinetic energy, and F(u(t)) is potential energy, so the functional (3) is the least Lagrangian action, from which we can immediately obtain its Hamiltonian, which reads:

$$H=\frac{1}{2}\left(\frac{d}{dt}u(t)\right)^2+F(u(t))=\text{constant}=H_0 \tag{4}$$

or:

$$R(t)=\frac{1}{2}\left(\frac{d}{dt}u(t)\right)^2+F(u(t))-F(A) \tag{5}$$

Equation (4) implies that the total energy keeps unchanged during the oscillation.

We use the following trial function to determine the angular frequency ω.

$$u(t)=A\cos\omega t \tag{6}$$

where ω is the frequency. Submitting Eq. (6) into Eq. (5) results in a residual:

$$R(t)=\frac{1}{2}A^2\omega^2\sin^2\omega t+F(A\cos\omega t)-F(A) \tag{7}$$

If, by chance, the exact solution had been chosen as the trial function, then it would be possible to make R zero for all values of t by appropriate choice of ω. Since u(t) = A cos ωt is an approximation to the exact solution, R cannot be made zero everywhere. According to the energy balance method [2], locating at some a special point, that is, ωt = π/4 and settingR(t = π/4ω) = 0, we can obtain an approximate frequency–amplitude relationship of the studied nonlinear oscillator. Such treatment is much simple and has been widely used by engineers [3–7]. The accuracy of such location method, however, strongly depends upon the chosen location point. To overcome the shortcoming of the energy balance method, in this paper, we apply a new approach based on Hamiltonian [8, 9].

From Eq. (7), we have:

$$\frac{\partial H}{\partial A}=0. \tag{8}$$

Introducing a new function, H¯(u(t)), defined as:

$$\bar{H}(u(t))=\int_0^{T/4}\left\{\frac{1}{2}\left(\frac{d}{dt}u(t)\right)^2+F(u(t))\right\}dt=\frac{1}{4}TH \tag{9}$$

It is obvious that:

$$\frac{\partial\bar{H}}{\partial T}=\frac{1}{4}H \tag{10}$$

Equation (10) is, then, equivalent to the following one:

$$\frac{\partial}{\partial A}\left(\frac{\partial\bar{H}}{\partial T}\right)=0 \tag{11}$$

or

$$\frac{\partial}{\partial A}\left(\frac{\partial\bar{H}}{\partial(1/\omega)}\right)=0. \tag{12}$$

From Eq. (12), we can obtain approximate frequency–amplitude relationship of a nonlinear oscillators.

The application of Hamiltonian approach method for nonlinear dynamical systems

The motion of a rigid rod rocking back on the circular surface without slipping

In this section, we consider the motion of a rigid rod rocking back on the circular surface without slipping. The governing equation of this motion is in the following form [36]:

$$\left(\frac{d^2}{dt^2}u(t)\right)+\frac{3}{4}u(t)^2\left(\frac{d^2}{dt^2}u(t)\right)+\frac{3}{4}u(t)\left(\frac{d}{dt}u(t)\right)^2+3\frac{g}{l}u(t)\cos u(t)=0 \tag{13}$$

with the following initial conditions [36]:

$$u(0)=A,\ \frac{d}{dt}u(0)=0 \tag{14}$$

where g and l are the positive constants. For this problem, we have:

$$f(u(t))=\frac{3}{4}u(t)^2\left(\frac{d^2}{dt^2}u(t)\right)+\frac{3}{4}u(t)\left(\frac{d}{dt}u(t)\right)^2+3\frac{g}{l}u(t)\cos u(t) \tag{15}$$

and:

$$F(u(t))=\frac{3}{8}u(t)^2\left(\frac{d}{dt}u(t)\right)^2+\frac{3g(\cos u(t)+u(t)\sin u(t))}{l}. \tag{16}$$

Its Hamiltonian can be easily obtained, which reads:

$$H=\frac{1}{2}\left(\frac{d}{dt}u(t)\right)^2+\frac{3}{8}u(t)^2\left(\frac{d}{dt}u(t)\right)^2+\frac{3g(\cos u(t)+u(t)\sin u(t))}{l}. \tag{17}$$

Integrating Eq. (17) with respect to t from 0 to T/4, we have:

$$\bar{H}(u(t))=\int_0^{T/4}\left\{\frac{1}{2}\left(\frac{d}{dt}u(t)\right)^2+\frac{3}{8}u(t)^2\left(\frac{d}{dt}u(t)\right)^2+\frac{3g(\cos u(t)+u(t)\sin u(t))}{l}\right\}dt. \tag{18}$$

Assuming that the solution can be expressed as u(t) = A cos ωt and substituting it to Eq. (18), we obtain:

$$\bar{H}(u(t)) = \int_0^{T/4} \left\{ \begin{array}{l} \frac{1}{2}A^2\omega^2\sin^2\omega t + \frac{3}{8}A^4\omega^2\cos^2\omega t\sin^2\omega t \\ + \frac{3g(\cos(A\cos\omega t) + A\cos\omega t\cdot\sin(A\cos\omega t))}{l} \end{array} \right\} dt$$

$$= \int_0^{\pi/2} \left\{ \frac{1}{2}A^2\omega\sin^2 t + \frac{1}{\omega}\left[\begin{array}{l} \frac{3}{8}A^4\omega^2\cos^2 t\sin^2 t \\ + \frac{3g(\cos(A\cos t) + A\cos t\cdot\sin(A\cos t))}{l} \end{array} \right] \right\}$$

$$= \frac{1}{2}A^2\omega\cdot\frac{\pi}{4} + \frac{1}{\omega}\left[\frac{3}{8}A^4\omega^2\cdot\frac{\pi}{16} + \frac{\pi}{2}\cdot\frac{3g(\mathrm{Bessel\,J}(0,A) + A\cdot\mathrm{Bessel\,J}(1,A))}{l}\right]. \quad (19)$$

In Eq. (19), we have two Bessel functions of the first kind. The Bessel functions are one of the special functions in mathematics. We can expand these Bessel functions of the first kind in the following form:

$$\text{Bessel J}(0, A) = 1 - \frac{1}{4}A^2 + \frac{1}{64}A^4 \quad (20)$$

and:

$$\text{Bessel J}(1, A) = \frac{1}{2}A - \frac{1}{16}A^3 + \frac{1}{384}A^5 \quad (21)$$

By substituting Eqs. (20) and (21) into Eq. (19), we obtain:

$$\bar{H}(u(t)) = \frac{1}{2}A^2\omega\cdot\frac{\pi}{4} + \frac{1}{\omega} \times \left\{ \frac{3}{8}A^4\omega^2\cdot\frac{\pi}{16} + \frac{\pi}{2}\cdot\frac{3g}{l}\left[\begin{array}{l} \left(1 - \frac{1}{4}A^2 + \frac{1}{64}A^4\right) \\ + A\cdot\left(\frac{1}{2}A - \frac{1}{16}A^3 + \frac{1}{384}A^5\right) \end{array} \right] \right\} \quad (22)$$

Setting:

$$\frac{\partial}{\partial A}\left(\frac{\partial\bar{H}}{\partial(1/\omega)}\right) = -A\omega^2\cdot\frac{\pi}{4} - \frac{3}{8}A^3\omega^2\cdot\frac{\pi}{4} + \frac{3g}{l}\cdot\frac{\pi}{2}\left(-\frac{1}{2}A + \frac{1}{16}A^3 + A - \frac{1}{4}A^3 + \frac{1}{64}A^5\right) = 0 \quad (23)$$

We obtain the following frequency–amplitude relationship:

$$\omega_{\mathrm{HAM}} = \frac{\sqrt{(3g/l)(32 + 12A^2)(32 - 12A^2 + A^4)}}{32 + 12A^2}. \quad (24)$$

Its period can be written in the following form:

$$T_{\mathrm{HAM}} = \frac{2\pi}{\frac{\sqrt{(3g/l)(32+12A^2)(32-12A^2+A^4)}}{32+12A^2}}. \quad (25)$$

To illustrate the accuracy of the Hamiltonian approach method, we compare the approximate solution results with exact solution. For this dynamical system, the exact period is in the following form [36]:

$$T_{\mathrm{Exact}} = 4\left(\frac{l}{3g}\right)^{1/2}\int_0^{\pi/2} \times \left(\frac{(4 + 3A^2\sin^2\varphi)A^2\cos^2\varphi}{8[A\sin A + \cos A - A\sin\varphi\sin(A\sin\varphi) - \cos(A\sin\varphi)]}\right)^{1/2} d\varphi. \quad (26)$$

A comparison of obtained results from the approximate period and the exact one is tabulated in Table 1 for g = 1.00, l = 1.00 and different values of A. From Table 1, the maximum relative error of the approximate periods is 1.9344 % for g = 1.00, l = 1.00 and A = 0.45π.

Table 1: Comparison of approximate periods with exact periods for g = 1.00, l = 1.00 and different values of A

A	**Hamiltonian approach method**	**Exact solution**	**Error percentage**
0.05π	3.66126	3.66109	0.0054
0.10π	3.76398	3.76397	0.0008
0.15π	3.94065	3.94086	0.0056
0.20π	4.20182	4.20292	0.0264
0.25π	4.56433	4.56948	0.1129
0.30π	5.05832	5.07728	0. 3734
0.35π	5.76741	5.79770	0.5224
0.40π	6.73586	6.81564	1.1705
0.45π	8.67226	8.84333	1.9344

Also, we present the comparison results of analytical approximate solution of u(t) based on twith exact solution with g = 1.00, l = 1.00 and different values of A in Figs. 1 and 2.

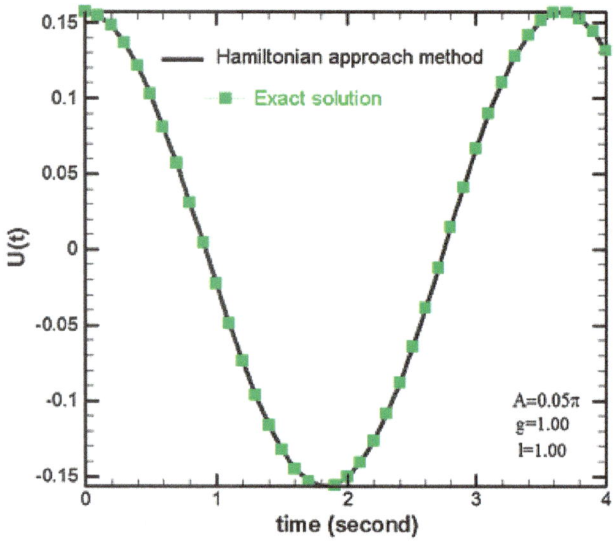

Figure 1: Comparison of Hamiltonian approach method solution of u(t) based on t with exact solution for g = 1.00, l = 1.00, A = 0.05π

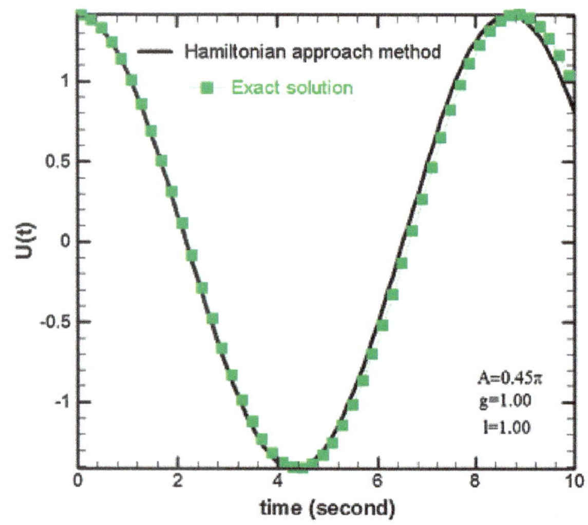

Figure 2: Comparison of Hamiltonian approach method solution of u(t) based on t with exact solution for g = 1.00, l = 1.00, A = 0.45π.

Also, to investigate on the behavior of this dynamical system, the effect of parameters g and l on the frequency corresponding to different values of amplitude (A) has been studied in Figs. 3 and4.

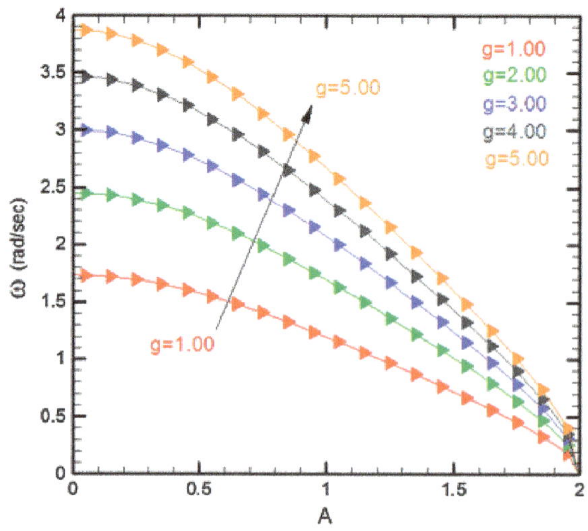

Figure 3: Comparison of frequency corresponding to different values of amplitude (A) and l = 1.00.

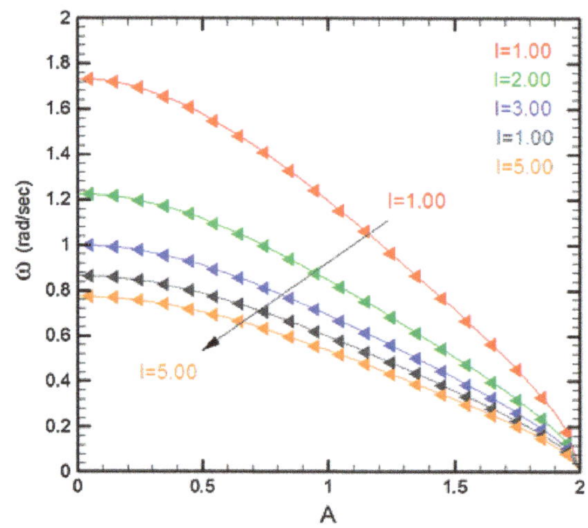

Figure 4: Comparison of frequency corresponding to different values of amplitude (A) and g = 1.00

The free vibrations of an autonomous conservative oscillator with inertia and static-type fifth-order nonlinearities

In this section, we consider the free vibrations of an autonomous conservative oscillator with inertia and static-type fifth-order nonlinearities. The differential equation of this dynamical system is the following form [37]:

$$\left(\frac{d^2}{dt^2}u(t)\right)+\varepsilon_1 u(t)^2\left(\frac{d^2}{dt^2}u(t)\right)+\varepsilon_2 u(t)^4\left(\frac{d^2}{dt^2}u(t)\right)+\varepsilon_1 u(t)\left(\frac{d}{dt}u(t)\right)^2 + 2\varepsilon_2 u(t)^3\left(\frac{d}{dt}u(t)\right)^2+\lambda u(t)+\varepsilon_3 u(t)^3+\varepsilon_4 u(t)^5=0,\quad -1\le\lambda\le 1 \tag{27}$$

with the following initial conditions [37]:

$$u(0)=A,\ \frac{d}{dt}u(0)=0 \tag{28}$$

where ε_1, ε_2, ε_3 and ε_4 are positive parameters and λ is an integer which may take values of −1, 0 and 1 [37]. For this problem, we have:

$$f(u(t)) = \varepsilon_1 u(t)^2\left(\frac{d^2}{dt^2}u(t)\right) + \varepsilon_2 u(t)^4\left(\frac{d^2}{dt^2}u(t)\right) + \varepsilon_1 u(t)\left(\frac{d}{dt}u(t)\right)^2 + 2\varepsilon_2 u(t)^3\left(\frac{d}{dt}u(t)\right)^2 + \lambda u(t) + \varepsilon_3 u(t)^3 + \varepsilon_4 u(t)^5 \tag{29}$$

and:

$$F(u(t)) = \frac{1}{2}\left(\frac{d}{dt}u(t)\right)^2 \varepsilon_1 u(t)^2 + \frac{1}{2}\left(\frac{d}{dt}u(t)\right)^2 \varepsilon_2 u(t)^4 + \frac{1}{2}\lambda u(t)^2 + \frac{1}{4}\varepsilon_3 u(t)^4 + \frac{1}{6}\varepsilon_4 u(t)^6. \tag{30}$$

Its Hamiltonian can be easily obtained, which reads:

$$H = \frac{1}{2}\left(\frac{d}{dt}u(t)\right)^2 + \frac{1}{2}\left(\frac{d}{dt}u(t)\right)^2 \varepsilon_1 u(t)^2 + \frac{1}{2}\left(\frac{d}{dt}u(t)\right)^2 \varepsilon_2 u(t)^4 + \frac{1}{2}\lambda u(t)^2 + \frac{1}{4}\varepsilon_3 u(t)^4 + \frac{1}{6}\varepsilon_4 u(t)^6. \tag{31}$$

Integrating Eq. (31) with respect to t from 0 to T/4, we have:

$$\bar{H}(u(t)) = \int_0^{T/4} \left\{ \begin{array}{l} \frac{1}{2}\left(\frac{d}{dt}u(t)\right)^2 + \frac{1}{2}\left(\frac{d}{dt}u(t)\right)^2 \varepsilon_1 u(t)^2 + \frac{1}{2}\left(\frac{d}{dt}u(t)\right)^2 \varepsilon_2 u(t)^4 \\ + \frac{1}{2}\lambda u(t)^2 + \frac{1}{4}\varepsilon_3 u(t)^4 + \frac{1}{6}\varepsilon_4 u(t)^6 \end{array} \right\} dt \tag{32}$$

Assuming that the solution can be expressed as u(t) = A cos ωt and substituting it to Eq. (32), we obtain:

$$\begin{aligned} \bar{H}(u(t)) &= \int_0^{T/4} \left\{ \begin{array}{l} \frac{1}{2}A^2\omega^2 \sin^2 \omega t + \frac{1}{2}\varepsilon_1 A^4\omega^2 \sin^2 \omega t \cos^2 \omega t + \frac{1}{2}\varepsilon_2 A^6\omega^2 \sin^2 \omega t \cos^4 \omega t \\ + \frac{1}{2}\lambda A^2 \cos^2 \omega t + \frac{1}{4}\varepsilon_3 A^4 \cos^4 \omega t + \frac{1}{6}\varepsilon_4 A^6 \cos^6 \omega t \end{array} \right\} dt \\ &= \int_0^{\pi/2} \left\{ \frac{1}{2}A^2\omega \sin^2 t + \frac{1}{\omega} \left[\begin{array}{l} \frac{1}{2}\varepsilon_1 A^4\omega^2 \sin^2 t \cos^2 t + \frac{1}{2}\varepsilon_2 A^6\omega^2 \sin^2 t \cos^4 t \\ + \frac{1}{2}\lambda A^2 \cos^2 t + \frac{1}{4}\varepsilon_3 A^4 \cos^4 t + \frac{1}{6}\varepsilon_4 A^6 \cos^6 t \end{array} \right] \right\} dt \\ &= \frac{1}{2}A^2\omega \cdot \frac{\pi}{4} + \frac{1}{\omega} \left[\begin{array}{l} \frac{1}{2}\varepsilon_1 A^4\omega^2 \cdot \frac{\pi}{16} + \frac{1}{2}\varepsilon_2 A^6\omega^2 \cdot \frac{\pi}{32} + \frac{1}{2}\lambda A^2 \cdot \frac{\pi}{4} \\ + \frac{1}{4}\varepsilon_3 A^4 \cdot \frac{3\pi}{16} + \frac{1}{6}\varepsilon_4 A^6 \cdot \frac{5\pi}{32} \end{array} \right]. \end{aligned} \tag{33}$$

Setting:

$$\frac{\partial}{\partial A}\left(\frac{\partial \bar{H}}{\partial(1/\omega)}\right) = -A\omega^2 \cdot \frac{\pi}{4} - 2\varepsilon_1 A^3\omega^2 \cdot \frac{\pi}{16} - 3\varepsilon_2 A^5\omega^2 \cdot \frac{\pi}{32} + \lambda A \cdot \frac{\pi}{4} + \varepsilon_3 A^3 \cdot \frac{3\pi}{16} + \varepsilon_4 A^5 \cdot \frac{5\pi}{32} = 0. \tag{34}$$

We obtain the following frequency–amplitude relationship:

$$\omega_{\text{HAM}} = \sqrt{\frac{6\varepsilon_3 A^2 + 5\varepsilon_4 A^4 + 8\lambda}{8 + 4\varepsilon_1 A^2 + 3\varepsilon_2 A^4}} \tag{35}$$

Its period can be written in the following form:

$$T_{\text{HAM}} = \frac{2\pi}{\sqrt{\frac{6\varepsilon_3 A^2 + 5\varepsilon_4 A^4 + 8\lambda}{8 + 4\varepsilon_1 A^2 + 3\varepsilon_2 A^4}}} \tag{36}$$

For this dynamical system, because there is no exact solution, we compare the approximate solution results with Runge–Kutta method, as one of the known numerical methods. The numerical solution with Runge–Kutta method for this nonlinear differential equation is:

$$\frac{d}{dt}u_1(t) = u_2(t), \ u_1(0) = A \tag{37}$$

and

$$\frac{d}{dt}u_2(t) = -\frac{1}{1+\varepsilon_1 u_1(t)^2+\varepsilon_2 u_1(t)^4}\begin{pmatrix}\lambda u_1(t)+\varepsilon_1 u_1(t)u_2(t)^2\\ +2\varepsilon_2 u_1(t)^3 u_2(t)^2\\ +\varepsilon_3 u_1(t)^3+\varepsilon_4 u_1(t)^5\end{pmatrix}. \tag{38}$$

Motion is assumed to start from the position of maximum displacement with zero initial velocity. λ is an integer which may take values of −1, 0 and 1, and ε_1, ε_2, ε_3 and ε_4 are positive parameters. The values of parameters ε_1, ε_2, ε_3 and ε_4 associated for a mode are shown in Table 2.

Table 2: Values of dimensionless parameters ε_i for a mode

Mode	ε_1	ε_2	ε_3	ε_4
Mode-1	0.326845	0.232598	0.129579	0.087584
Mode-2	1.642033	0.913055	0.313561	0.204297
Mode-3	4.051486	1.665232	0.281418	0.149677

To illustrate the accuracy of the Hamiltonian approach method solution, we present the comparison results of analytical approximate solution of u(t) based on t with the numerical solution which solved by Runge–Kutta method as one of the known numerical methods in Figs. 5 and 6 for λ = 1.00, A = 1.00 and various modes (mode-1 and mode-3).

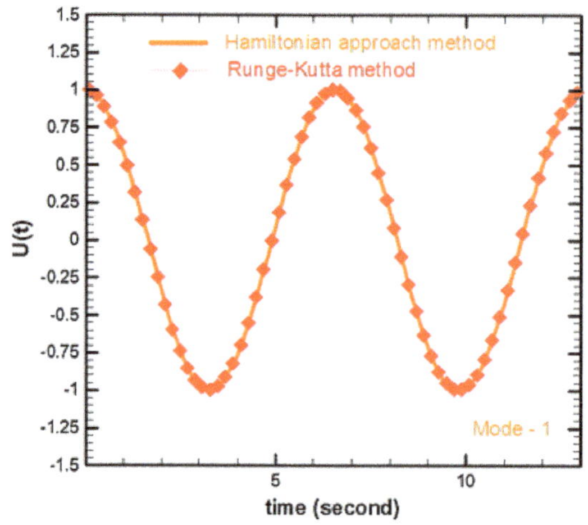

Figure 5: Comparison of Hamiltonian approach method solution of u(t) based on t with numerical solution (Runge–Kutta method) for λ = 1.00, A = 1.00in mode-1.

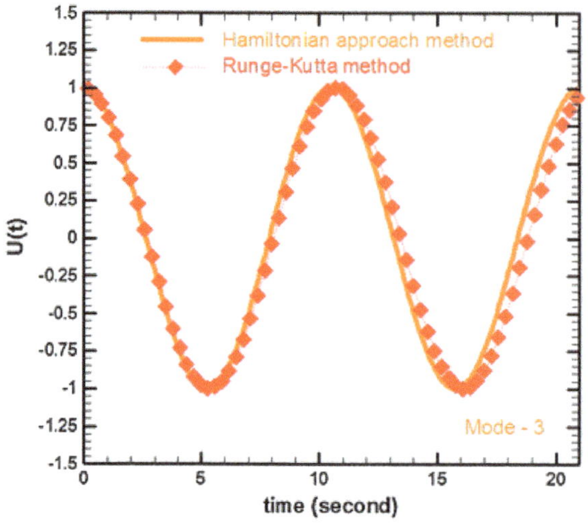

Figure 6: Comparison of Hamiltonian approach method solution of u(t) based on t with numerical solution (Runge–Kutta method) for λ = 1.00, A = 1.00 in mode-3.

Figures 7 and 8 present the high accuracy of Hamiltonian approach method solution in comparison with numerical solution for different values of ε_1, ε_2, ε_3 and ε_4, and show the phase–space curves (du/dt versus u(t) curve) for amplitude A = 1.00 and λ = 1.00.

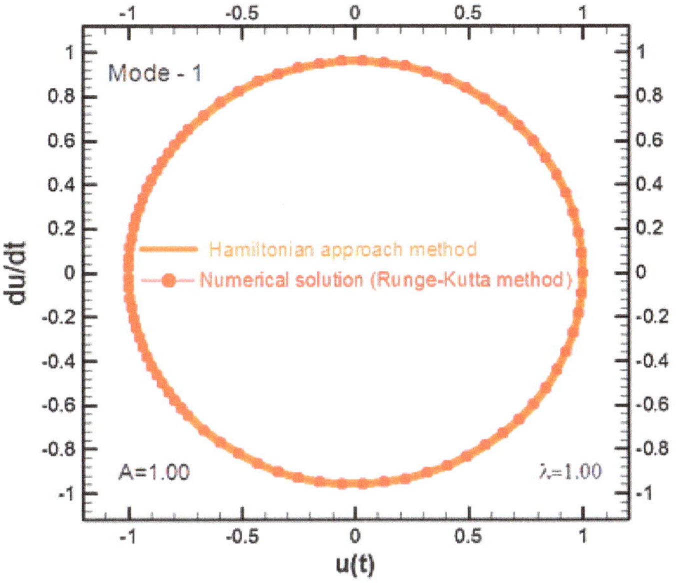

Figure 7: Comparison of analytical approximate solution of du/dt based on u(t) with the with numerical solution (Runge–Kutta method) for λ = 1.00, A = 1.00 in mode-1

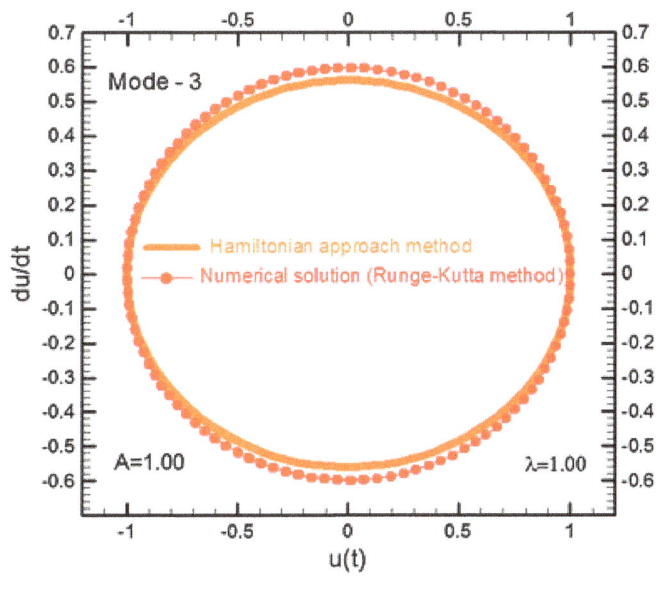

Figure 8: Comparison of analytical approximate solution of du/dt based on u(t) with the with numerical solution (Runge–Kutta method) for λ = 1.00, A = 1.00 in mode-3.

It can be observed that the phase–space curves generated from approximate solution are close to that of the numerical curves. The phase plots show the behavior of the dynamical system in modes 1 and 3. It is periodic with a center at (0, 0).

Also, to investigate on the behavior of this dynamical system, the effect of parameters ε_1, ε_2, $\varepsilon_3$and ε_4 on the frequency corresponding to different values of amplitude (A) has been studied in Fig. 9.

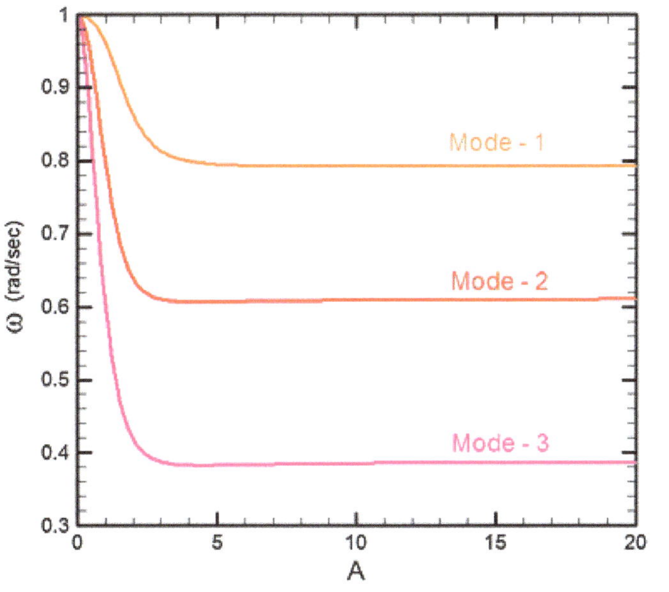

Figure 9: Comparison of frequency corresponding to different values of amplitude (A) and various modes, mode-1, mode-2 and mode-3.

It is evident that Hamiltonian approach method shows excellent agreement with the exact and numerical solutions and is quickly convergent and valid for a wide range of vibration amplitudes and initial conditions. The accuracy of the results shows that the Hamiltonian approach method can be potentiality used for the analysis of strongly nonlinear oscillation problems accurately.

CONCLUSION

In this article, we applied Hamiltonian approach method as one of the analytical approximate techniques, for studying the strongly nonlinear dynamical systems such as the motion of a rigid rod rocking back on the circular surface without slipping and the free vibrations of an autonomous conservative oscillator with inertia and static-type fifth-order nonlinearities.

Comparison of the results which are obtained by this method with the obtained result by the exact and numerical solutions reveal that the Hamiltonian approach method is very effective and convenient and does not require linearization or small perturbation and can be easily extended to other nonlinear dynamical systems and can therefore be found widely applicable in engineering and other sciences.

AUTHORS CONTRIBUTIONS

MH conceived the study and participated in its design and coordination. MS carried out the numerical solution of equations and participated in drafting the manuscript. HEK carried out the software work and the solution of equations using analytical approximate methods. All authors read and approved the final manuscript.

ACKNOWLEDGMENTS

This work was financially supported by the research deputy of Fakultät für Mathematik und Informatik, Friedrich-Schiller-Universität Jena, Germany. Authors would like to acknowledge them because of their support.

REFERENCES

1. Shou, D.H., He, J.H.: Application of parameter-expanding method to strongly nonlinear oscillators. Int. J. Nonlinear Sci. Numer. Simul. 8(1), 121–124 (2007)
2. Xu, L.: Determination of limit cycle by He's parameter-expanding method for strongly nonlinear oscillators. J. Sound Vib. 302(1), 178–184 (2007)
3. Xu, L.: Application of He's parameter-expanding method to an oscillation of a mass attached to a stretched elastic wire. Phys. Lett. A 368(3), 259–262 (2007)
4. He, J.H., XH, W.: Variational iteration method: new development and applications. Comput. Math. Appl. 54(7), 881–894 (2007)
5. Shou, D.H., He, J.H.: Beyond Adomian method: the variational iteration method for solving heat-like and wave-like equations with variable coefficients. Phys. Lett. A 372(3), 233–237 (2008)
6. Shou, D.H., He, J.H.: Variational iteration method for two-strand yarn spinning. Nonlinear Anal. Theory, Methods Appl. 71(12), 830–833 (2009)
7. He, J.H.: A short remark on fractional variational iteration method. Phys. Lett. A 375(38), 3362–3364 (2011)
8. He, J.H.: Preliminary report on the energy balance for nonlinear oscillations. Mech. Res. Commun. 29(2), 107–111 (2002)
9. Ganji, D.D., Afrouzi, G.A., Talarposhti, R.A.: He's energy balance method for nonlinear oscillators with discontinuities. Int. J. Nonlinear Sci. Numer. Simul. 10(3), 301–304 (2009)
10. Zhang, H.L., Xu, Y.G., Chang, J.R.: Application of He's energy balance method to a nonlinear oscillator with discontinuity. Int. J. Nonlinear Sci. Numer. Simul. 10(2), 207–214 (2009)
11. Ganji, D.D., Ganji, S.S., Karimpour, S.: He's energy balance and He's variational methods for nonlinear oscillations in engineering. Int. J. Mod. Phys. B 23(3), 461–471 (2009)
12. Ebrahimi Khah, H., Ganji, D.D.: A study on the motion of a rigid rod rocking back and cubic-quintic duffing oscillators by using He's energy balance method. Int. J. Nonlinear Sci. 10(4), 447–451 (2010)
13. Ebrahimi Khah, H., Ayazi, A., Lale Arefi, S.H.: Approximate analytical solution to study of nonlinear oscillator with discontinuity using energy balance method. In: Proceedings of 42th Annual Iranian Mathematics, Conference, 5–8 Sept 2011, Vali-e-Asr University of Rafsanjan, Iran
14. Ebrahimi Khah, H., Ayazi, A., Ganji, D.D.: A study on the forced vibration of strongly nonlinear oscillator using He's energy balance method. J. Adv. Res. Mech. Eng. 2(1), 27–32 (2011)
15. He, J.H.: Variational approach for nonlinear oscillators. Chaos Solitons Fractals 34(5), 1430–1439 (2007)
16. Shou, D.H.: Variational approach to the nonlinear oscillator of a mass attached to a stretched wire. Phys. Scr. 77(4), 045006 (2008)
17. Shou, D.H.: Variational approach for nonlinear oscillators with discontinuities. Comput. Math. Appl. 58(11), 2416–2419 (2009)
18. Ebrahimi Khah, H., Ganji, D.D.: Application of He's variational approach method for strongly nonlinear oscillators. J. Adv. Res. Mech. Eng. 1(1), 30–34 (2010)
19. Ebrahimi Khah, H., Ayazi, A., Daie, M.: A study on the free oscillation of pendulum using variational approach method and comparison with exact solution. In: Proceedings of 41th Iranian International Conference on Mathematics, 12–15 Sept 2010, University of Urmia, Urmia, Iran
20. He, J.H.: An improved amplitude-frequency formulation for nonlinear oscillators. Int. J. Nonlinear Sci. Numer. Simul. 9(2), 211–212 (2008)
21. He, J.H.: Asymptotic methods for solitary solutions and compactions. In: Proceedings of Abstract and Applied Analysis. Hindawi Publishing Corporation, Hindawi (2012). doi:10.1155/2012/916793
22. Ebrahimi Khah, H., Ayazi, A., Ganji, D.D.: Determination of the amplitude frequency for strongly nonlinear oscillator with using two approximate analytical techniques. J Theor. Appl. Phys. (2013). doi:10.1186/2251-7235-7-38
23. He, J.H.: Homotopy perturbation technique. Comput. Methods Appl. Mech. Eng. 178, 257–262 (1999)
24. He, J.H.: Recent development of the homotopy perturbation method. Topol. Methods Nonlinear Anal. 31(2), 205–209 (2008)
25. He, J.H.: An elementary introduction to the homotopy perturbation method. Comput. Math. Appl. 57(3), 410–412 (2009)
26. Shou, D.H.: The homotopy perturbation method for nonlinear oscillators. Comput. Math. Appl. 58(11), 2456–2459 (2009)
27. Ebrahimi Khah, H., Ayazi, A., Ganji, D.D.: The investigation and application of two approximate analytical methods for the solution of nonlinear differential equation of beam elastic deformation. J. Theor. Appl. Phys. 5(2), 53–58 (2011)
28. Ebrahimi Khah, H., Hermann, M., Saravi, M.: The comparison of homotopy perturbation method with finite difference method for determination of maximum beam deflection. J. Theor. Appl. Phys. (2013). doi:10.1186/2251-7235-7-8
29. Ebrahimi Khah, H., Ayazi, A., Lale Arefi, S.H.: The solution of beam deformation equation using homotopy perturbation method.In: Proceedings of 42th Annual Iranian Mathematics Conference, 5–8 Sept 2011, Vali-e-Asr University of Rafsanjan, Iran
30. He, J.H.: Some asymptotic methods for strongly nonlinear equations. Int. J. Mod. Phys. B 20(10), 1141–1199 (2006)
31. He, J.H.: Hamiltonian approach to nonlinear oscillators. Phys. Lett. A 374(23), 2312–2314 (2010)ADS
32. He, J.H., Zhong, T., Tang, L.: Hamiltonian approach to duffing-harmonic equation. Int. J. Nonlinear Sci. Numer. Simul. 11(1), 43–46 (2010)

33. Xu, L., He, J.H.: Determination of limit cycle by Hamiltonian approach for strongly nonlinear oscillators. Int. J. Nonlinear Sci. Numer. Simul. 11(12), 1097–1101 (2010)
34. He, J.H., Kong, H.Y., Yang, R.R., et al.: Review on fiber morphology obtained by bubble electro spinning and blown bubble spinning. Thermal Sci. 16(5), 1263–1279 (2012)
35. He, J.H.: Asymptotic methods for solitary solutions and compactions. In: Proceedings of Abstract and Applied Analysis. Hindawi Publishing Corporation, Hindawi (2012). doi:10.1155/2012/916793
36. Wu, B.S., Lim, C.W., He, J.H.: A new method for approximate analytical solutions to nonlinear oscillations of non-natural systems. Nonlinear Dyn. 32(1), 1–13 (2003)
37. Hamden, M.N., Shabaneh, N.H.: Study on the large amplitude free vibrations of a restrained uniform beam carrying an intermediate lumped mass. J. Sound Vib. 199(6), 711–736 (1997)

CHAPTER 6

DYNAMICS OF MULTIBODY CHAINS IN CIRCULAR ORBIT: NON-INTEGRABILITY OF EQUATIONS OF MOTION

Andrzej J. Maciejewski, Maria Przybylska

Keywords: Multibody systems, Integrability, Morales-Ramis theory, Planar 2-chain problem

INTRODUCTION

Investigations of multibody systems dynamics are relevant for space missions such as space robots, tethered systems, and space stations. In the simplest models, systems of points connected with joints having specific properties are considered. For example, in the perfect dumb-bell model, two point masses are joined by a massless rigid rod. An example of investigations of motion of a dumb-bell in an orbit can be found in Celletti and Sidorenko (2008). This model is in fact a simplification of a system describing the motion of a symmetric rigid body in an orbit. One can also consider a spring dumb-bell in an orbit; see, e.g., papers (Burov and Kosenko 2013; Burov 2011; Sidorenko and Celletti 2010) and references therein. These kinds of systems can serve as models with a deformable body.

Other models of deformable extended bodies consist of point masses joined by massless rigid rods with spherical hinges. The simplest of them are just open chains. A system of (n+1) points can be joined one after another to form an open chain with n links. It is called n-chain. The most general description of this type of systems in an orbital environment was done by Guerman (2003, (2006). Among other things, for such systems certain families of equilibria were found.

A peculiar dynamical property of linear chains is that if n>2, then equations of their free motion are not integrable; see, e.g., (Szumiński 2014). Moreover, an n-chain with one fixed end is just a multiple pendulum. Amazingly, even in the absence of the gravity field, the system is not integrable for n>2; see Salnikov (2013). Free systems of coupled planar rigid bodies were investigated also by Sreenath et al. (1988) and by Oh et al. (1989).

In this paper we derive equations of motion for an open chain in an arbitrary potential field. Next, we obtain equations of motion for a chain whose centre of mass moves in a circular orbit. We show that equations for 2-chains are not integrable. To obtain this result we investigate variational equations along two families of particular solutions. Thanks to this, we were able to prove the non-integrability for all allowable values of the system parameters.

EQUATIONS OF MOTION AND MAIN RESULT

We consider a system of (n+1) points P_0, ..., P_n. Point P_i is connected with point P_{i+1} by a massless perfectly rigid rod and the junction is a spherical hinge. Thus, the points form an open chain with n links. Example of three points with two links forming 2-chain is given in Fig. 1.

Let m_i and q_i denote the mass and the radius vector of point P, respectively. The rigid rods put n holonomic constraints

$$\|\boldsymbol{q}_i - \boldsymbol{q}_{i-1}\| = l_i > 0, \quad \text{for} \quad i = 1, \ldots, n, \tag{2.1}$$

on the system. We can set

$$\boldsymbol{q}_i = \boldsymbol{q}_{i-1} + l_i \boldsymbol{e}_i = \boldsymbol{q}_0 + \sum_{k=1}^{i} l_k \boldsymbol{e}_k = \boldsymbol{q}_0 + \sum_{k=1}^{n} \theta_{i,k} l_k \boldsymbol{e}_k, \tag{2.2}$$

where

$$\boldsymbol{e}_i = \frac{1}{l_i}\left(\boldsymbol{q}_i - \boldsymbol{q}_{i-1}\right), \quad \text{for} \quad i = 1, \ldots, n. \tag{2.3}$$

Citation: Andrzej J. Maciejewski, Maria Przybylska, Dynamics of multibody chains in circular orbit: non-integrability of equations of motion, DOI: 10.1007/s10569-016-9696-x.

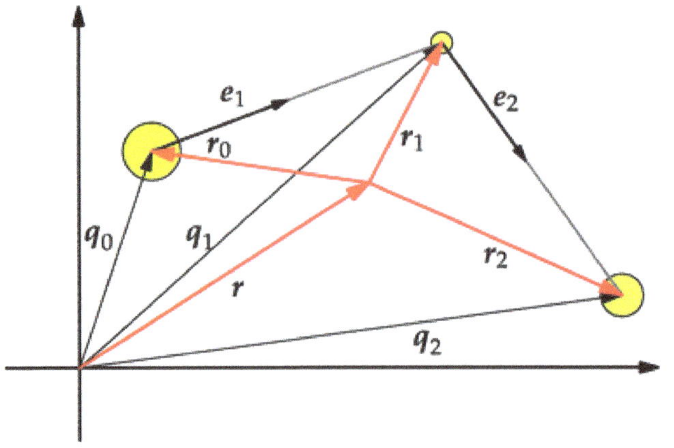

Figure 1: Geometry of the linear 2-chain.

and

$$\theta_{i,j} := \begin{cases} 1 & \text{for } j \le i, \\ 0 & \text{otherwise.} \end{cases} \tag{2.4}$$

Clearly, $(q_0, e_1, \ldots, e_n)$ specify the configuration of the system. Hence, its configuration space is $M^{2n+3} = \mathbb{R}^3 \times (\mathbb{S}^2)^n$, or, in the planar case $M^{2+n} = \mathbb{R}^2 \times (\mathbb{S}^1)^n = \mathbb{R}^2 \times \mathbb{T}^n$. Notice that instead of q_0 we can take an arbitrary point q_i with $0 \le i \le n$. However, it is natural to specify a configuration by $(r, e_1, \ldots, e_n)$, where r is the radius vector of the centre of the mass

$$\boldsymbol{r} = \frac{1}{m}\sum_{i=0}^{n} m_i \boldsymbol{q}_i, \quad \text{where} \quad m = \sum_{i=0}^{n} m_i. \tag{2.5}$$

Then, from (2.2) we obtain

$$\boldsymbol{r} = \boldsymbol{q}_0 + \frac{1}{m}\sum_{j=1}^{n} \sigma_j l_j \boldsymbol{e}_j, \quad \text{with} \quad \sigma_j = \sum_{k=j}^{n} m_k. \tag{2.6}$$

Hence, we can write

$$\boldsymbol{q}_i = \boldsymbol{r} + \boldsymbol{r}_i, \quad \text{where} \quad \boldsymbol{r}_i = \sum_{j=1}^{n} \left(\theta_{i,j} - \frac{\sigma_j}{m}\right) l_j \boldsymbol{e}_j, \tag{2.7}$$

for i=0,…,n. Obviously, we have

$$\sum_{i=0}^{n} m_i \boldsymbol{r}_i = \boldsymbol{0}. \tag{2.8}$$

Application of this equality gives

$$\sum_{i=0}^{n} m_i \boldsymbol{q}_i \cdot \boldsymbol{q}_i = \sum_{i=0}^{n} m_i (\boldsymbol{r} + \boldsymbol{r}_i) \cdot (\boldsymbol{r} + \boldsymbol{r}_i) = m \boldsymbol{r} \cdot \boldsymbol{r} + \sum_i \tag{2.9}$$

and similarly, we have

$$\sum_{i=0}^{n} m_i \|\dot{\boldsymbol{q}}_i\|^2 = m\|\dot{\boldsymbol{r}}\|^2 + \sum_{i=0}^{n} m_i \dot{\boldsymbol{r}}_i \cdot \dot{\boldsymbol{r}}_i. \tag{2.10}$$

Next, we introduce a symmetric n×n matrix $I=[I_{\alpha,\beta}]$ defined by

$$\sum_{i=0}^{n} m_i \boldsymbol{r}_i \cdot \boldsymbol{r}_i = \sum_{\alpha,\beta=1}^{n} I_{\alpha,\beta} \boldsymbol{e}_\alpha \cdot \boldsymbol{e}_\beta. \tag{2.11}$$

Using the definition of rri, see (2.7), we find that

$$I_{\alpha,\beta} = \sum_{i=0}^{n} m_i \left(\theta_{i,\alpha} - \frac{\sigma_\alpha}{m}\right) l_\alpha \left(\theta_{i,\beta} - \frac{\sigma_\beta}{m}\right) l_\beta. \tag{2.12}$$

Here, we notice that

$$\sum_{i=0}^{n} m_i \theta_{i,\alpha} = \sigma_\alpha, \quad \theta_{i,\alpha}\theta_{i,\beta} = \theta_{i,\gamma}, \quad \text{with} \quad \gamma = \max(\alpha, \beta). \tag{2.13}$$

Using the above relations, one can find

$$I_{\alpha,\beta} = l_\alpha l_\beta \left(\sigma_\gamma - \frac{\sigma_\alpha \sigma_\beta}{m}\right) \quad \text{with} \quad \gamma = \max(\alpha, \beta). \tag{2.14}$$

The kinetic energy has the form

$$T = \frac{1}{2} m \|\dot{\boldsymbol{r}}\|^2 + \frac{1}{2} \sum_{\alpha,\beta=1}^{n} I_{\alpha,\beta} \dot{\boldsymbol{e}}_\alpha \cdot \dot{\boldsymbol{e}}_\beta. \tag{2.15}$$

Let $V(\boldsymbol{r}, \boldsymbol{e}_1, \ldots, \boldsymbol{e}_n)$ be the potential of an external forces. Then the Lagrange function is

L=T−V, (2.16)

and the equations of motion have the form

$$m\ddot{\boldsymbol{r}} + \frac{\partial V}{\partial \boldsymbol{r}} = \boldsymbol{0},$$
$$\sum_{\beta=1}^{n} I_{\alpha,\beta} \ddot{\boldsymbol{e}}_\beta + \frac{\partial V}{\partial \boldsymbol{e}_\alpha} = \lambda_\alpha \boldsymbol{e}_\alpha, \quad \alpha = 1, \ldots, n, \tag{2.17}$$

where λ_α are the Lagrange multipliers. Their dependence on dynamical variables r, e_1, ..., een and velocities ˙r, ˙e_1, ..., ˙e_n can be deduced from constraints

$$\boldsymbol{e}_\alpha \cdot \boldsymbol{e}_\alpha = 1, \quad \boldsymbol{e}_\alpha \cdot \dot{\boldsymbol{e}}_\alpha = 0, \quad \alpha = 1, \ldots, n. \tag{2.18}$$

Example 2.1

Let us assume that the chain moves in a constant gravity field. The potential energy of a point mass m is given by $V = -m\,\boldsymbol{n} \cdot \boldsymbol{q}$, where $\boldsymbol{n}$ is a constant vector of a gravity field intensity. Thus, the potential energy of the chain is

$$V = -\sum_{i=0}^{n} m_i \boldsymbol{n} \cdot \boldsymbol{q}_i = -\sum_{i=0}^{n} m_i \boldsymbol{n} \cdot (\boldsymbol{r} + \boldsymbol{r}_i) = -m\,\boldsymbol{n} \cdot \boldsymbol{r}. \tag{2.19}$$

It implies that the motion of the centre of the mass of the chain moves like a point in a constant gravity field. Nevertheless, the motion of the chain is highly complicated. If n>2, the system is not integrable.

Remark 2.1

If in the above example we fix one end of the chain, then we obtain just a multiple pendulum. This modification changes the dynamics of the system. In fact, even for n=2 the system is not integrable. This problem was investigated by many authors; see, e.g., (Burov and Nechaev 2002; Moauro and Negrini 1998; Paul and Richter 1994).

Now, let us assume that the chain moves around a central body with mass M along an orbit. The potential energy of the chain is

$$V = -\sum_{i=0}^{n} \frac{GMm_i}{\|\boldsymbol{r} + \boldsymbol{r}_i\|}. \tag{2.20}$$

Let us assume that lengths of the chain links are small in comparison to the dimension of the orbit of its mass centre. It means that the following quantity

$$\varepsilon := \max_{0 \le i \le n} \frac{\|\boldsymbol{r}_i\|}{\|\boldsymbol{r}\|}, \tag{2.21}$$

is small. It is customary to approximate the potential by its truncated form. We take expansion

$$\frac{1}{\|\boldsymbol{r} + \boldsymbol{r}_i\|} = \frac{1}{\|\boldsymbol{r}\|} - \frac{\boldsymbol{r} \cdot \boldsymbol{r}_i}{\|\boldsymbol{r}\|^3} + \frac{3\,(\boldsymbol{r} \cdot \boldsymbol{r}_i)^2 - \|\boldsymbol{r}\|^2 \|\boldsymbol{r}_i\|^2}{2\|\boldsymbol{r}\|^5} + \cdots, \tag{2.22}$$

where dots denote terms of order higher than ε2, which we neglect. Then using formula (2.13) we obtain

$$V = -\frac{GMm}{\|\boldsymbol{r}\|} - \frac{GMm}{2\|\boldsymbol{r}\|^3} \sum_{\alpha,\beta=1}^{n} I_{\alpha,\beta} \left[3(\boldsymbol{e}_r \cdot \boldsymbol{e}_\alpha)(\boldsymbol{e}_r \cdot \boldsymbol{e}_\beta) - \boldsymbol{e}_\alpha \cdot \boldsymbol{e}_\beta\right], \tag{2.23}$$

where $e_r = r/\|r\|$.

Now we fix the orbit. More precisely, we assume that the centre of mass of the chain moves in a circular Keplerian orbit of radius a. The motion of the chain is conveniently investigated in its orbital frame. The first axis of this frame is directed along the radius vector, and the third axis is normal to the plane of the orbit and has the direction of the orbital angular momentum. The orbital frame rotates around the origin with constant angular velocity ω defined by

$$\omega := \sqrt{\frac{GMm}{a^3}}. \tag{2.24}$$

In our notation, $e_i = (x_i, y_i, z_i)^T$ denotes the coordinates of unit vector

$$\bar{e}_i := \frac{\overline{P_{i-1}P_i}}{|P_{i-1}P_i|} \tag{2.25}$$

in the chosen inertial frame. Coordinates of this vector in the orbital frame are denoted by $s_i = (X_i, Y_i, Z_i)^T$. The relation between these two sets of coordinates is given by $e_i = \boldsymbol{A}(t)\boldsymbol{s}_i$, where $\boldsymbol{A}(t)$ is the rotation matrix. We can assume with no loss of generality that the orbit plane coincides with the (x, y)-plane of the inertial frame and then

$$\boldsymbol{A}(t) = \begin{bmatrix} \cos\omega t & -\sin\omega t & 0 \\ \sin\omega t & \cos\omega t & 0 \\ 0 & 0 & 1 \end{bmatrix} \tag{2.26}$$

The time derivative of e_i is given by

$$\dot{\boldsymbol{e}}_i = \boldsymbol{A}(t)\left(\dot{\boldsymbol{s}}_i + \boldsymbol{\omega} \times \boldsymbol{s}_i\right), \tag{2.27}$$

where

$$\boldsymbol{\omega} = \omega\,(0, 0, 1)^T \tag{2.28}$$

is the vector of the orbital angular velocity. Using the above identities, we can rewrite the Lagrange function in the form

$$L = \frac{1}{2} \sum_{\alpha,\beta=1}^{n} I_{\alpha,\beta} \left(\dot{\boldsymbol{s}}_\alpha + \boldsymbol{\omega} \times \boldsymbol{s}_\alpha\right) \cdot \left(\dot{\boldsymbol{s}}_\beta + \boldsymbol{\omega} \times \boldsymbol{s}_\beta\right) - V, \tag{2.29}$$

where

$$V = -\frac{1}{2}\omega^2 \sum_{\alpha,\beta=1}^{n} I_{\alpha,\beta} \left[3(\boldsymbol{s} \cdot \boldsymbol{s}_\alpha)(\boldsymbol{s} \cdot \boldsymbol{s}_\beta) - \boldsymbol{s}_\alpha \cdot \boldsymbol{s}_\beta\right] \tag{2.30}$$

and $s=(1,0,0)^T$. Let us notice that we neglected terms describing the motion of the mass centre. By a proper choice of the unit of time, we can achieve that ω=1.

Planar 2-chain problem

In the rest of this paper, we investigate the motion of a 2-chain, which moves in the plane of the orbit, see in Fig. 1. As coordinates we take two angles (φ_1, φ_2) such that

$$s_1 = (\cos\varphi_1, \sin\varphi_1, 0)^T, \quad s_2 = (\cos\varphi_2, \sin\varphi_2, 0)^T \tag{2.31}$$

In these coordinates, the Lagrange function reads

$$\begin{aligned} L = &\frac{1}{2}\left[I_{1,1}(\dot\varphi_1 + 1)^2 + I_{2,2}(\dot\varphi_2 + 1)^2\right] \\ &+ I_{1,2}\cos(\varphi_1 - \varphi_2)(\dot\varphi_1 + 1)(\dot\varphi_2 + 1) - V(\varphi_1, \varphi_2), \end{aligned} \tag{2.32}$$

where

$$V(\varphi_1, \varphi_2) = -\frac{1}{2}\sum_{\alpha,\beta=1}^{2} I_{\alpha,\beta}\left[3\cos\varphi_\alpha\cos\varphi_\beta - \cos(\varphi_\alpha - \varphi_\beta)\right] \tag{2.33}$$

Explicit forms of entries of the tensor of inertia for a 2-chain are as follows

$$I_{11} = \frac{m_0(m_1+m_2)l_1^2}{m_0+m_1+m_2}, \quad I_{22} = \frac{m_2(m_0+m_1)l_2^2}{m_0+m_1+m_2}, \quad I_{12} = I_{21} = \frac{m_0 m_2 l_1 l_2}{m_0+m_1+m_2},$$

where l_1 and l_2 are lengths of arms between masses m_0 and m_1, and m_1 and m_2, respectively.

For further considerations, it is convenient to introduce relative coordinates $x_1=\varphi_1$ and $x_2=\varphi_2-\varphi_1$. Using these coordinates, we rewrite Lagrange equations in a form of the following system of first-order differential equations

$$\begin{aligned} \dot x_1 &= y_1, \\ \dot y_1 &= \frac{1}{D}\{\lambda_2 \sin x_2 [3 + 2(y_1 + y_2)(y_1 + y_2 + 2) + 3\cos(2(x_1 + x_2))] \\ &\quad + y_1(2 + y_1)\sin 2x_2 + 3\sin(x_1 + 2x_2)\cos x_1 + \frac{3}{2}[1 - 2\lambda_1\lambda_2]\sin 2x_1\bigg\}, \\ \dot x_2 &= y_2, \\ \dot y_2 &= -\frac{\sin x_2}{D}\{\lambda[3 + 2(y_1 + y_2)(y_1 + y_2 + 2)] - 2\lambda_1 y_2(2y_1 + y_2 + 2) \\ &\quad + 2[3 + 2y_1(y_1 + 2) + y_2(2y_1 + y_2 + 2)]\cos x_2 \\ &\quad + 3\lambda_1\cos 2x_1 + 3\lambda_2[\cos(2(x_1 + x_2)) + 2\lambda_1\cos(2x_1 + x_2)]\}, \end{aligned} \tag{2.34}$$

where

$$\lambda_1 = \frac{I_{1,1}}{I_{1,2}}, \quad \lambda_2 = \frac{I_{2,2}}{I_{1,2}}, \quad \lambda = \lambda_1 + \lambda_2, \tag{2.35}$$

$$D = 2\left(\lambda_1\lambda_2 - \cos^2 x_2\right) \tag{2.36}$$

The above system is a Hamiltonian; however, coordinates (x_1, y_1, x_2, y_2) are not canonical. The Hamiltonian function in these variables reads

$$\begin{aligned} H = &\frac{1}{2}\left[(\lambda + 2\cos x_2)y_1^2 + 2y_1y_2(\lambda_2 + \cos x_2) + \lambda_2 y_2^2\right] \\ &- \frac{3}{4}[\lambda + \lambda_1\cos(2x_1) + \lambda_2\cos(2(x_1 + x_2)) + 2\cos x_2 + 2\cos(2x_1 + x_2)]. \end{aligned} \tag{2.37}$$

Parameters λ_1 and λ_2 are positive, and moreover,

$$\lambda_1\lambda_2 - 1 = \frac{m_1}{m_0 m_2}(m_0 + m_1 + m_2) > 0, \tag{2.38}$$

so the set of physically admissible values of (λ_1,λ_2) is given by

$$\Lambda := \{(\lambda_1, \lambda_2) \in \mathbb{R}^2 \mid \lambda_1 > 0, \lambda_2 > 0, \lambda_1\lambda_2 - 1 > 0\}.$$

Main Theorem

Numerical tests (see Fig. 2) show that, in general, the system (2.34) is not integrable. However, such tests do not exclude the possibility that the system is integrable for certain values of $(\lambda_1,\lambda_2)\in\Lambda$. Our aim is to prove that no such values exist.

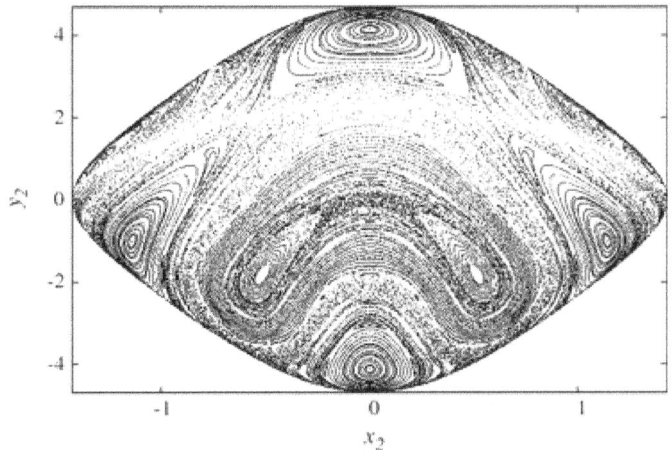

Figure 2: Poincaré cross section for system (2.34) with parameters $(\lambda_1,\lambda_2)=(2,2)$ in the level H=−3.5 of the first integral (2.37). The surface of the cross section is $x_1=0$ and the local coordinates on it are (x_2,y_2)

For the considered system, the configuration space is a two-dimensional torus T, with coordinates (x_1,x_2)\mathchoicemod2π. The phase space is $T^{\star}\mathbb{T}^2 \simeq \mathbb{T}^2 \times \mathbb{R}^2$, and (x_1, x_2, y_1, y_2) are coordinates on it.

Functions $f(x_1,x_2,y_1,y_2)$ defined on the phase space are periodic with respect to the first two arguments. We extend our system to the complex phase space, so considered functions are complex functions of complex variables $(x_1,x_2,y_1,y_2)\in C_4$.

The main result of this paper is the following theorem.

Theorem 2.1

The complex system (2.34) does not have a first integral which is meromorphic and functionally independent of H.

PROOF

In this section, we prove our main theorem by analysing properties of the differential Galois group of variational equations for certain particular solutions of the system. Our considerations are based on the general theorem of Morales Ruiz (1999), Morales-Ruiz and Ramis (2001) that formulates necessary integrability conditions using properties of the differential Galois group of variational equations.

Theorem 3.1

(Morales–Ramis). Assume that a Hamiltonian system is meromorphically integrable in the Liouville sense in a neighbourhood of a phase curve Γ. Then the identity component of the differential Galois group of the variational equations along Γ is Abelian.

The main steps of the proof are as follows. At first we find two families of particular solutions of our system. They describe oscillations (and rotations) of the chain. These solutions describe the motion of the chain when its two parts are either parallel or anti-parallel. Then we derive the variational equations and transform their normal subsystems into the form of second-order equations with rational coefficients. These equations are Fuchsian, and their differential Galois groups are subgroups of SL(2,C).

Local analysis of variational equations shows that solutions of normal variational equations have logarithmic terms. Hence, it is impossible that all their solutions are algebraic. This fact simplifies further consideration. To complete the proof of non-integrability, it is enough to show that the equations do not have exponential solutions. At this point, we apply a part of the Kovacic algorithm.

Particular solutions and variational equations

System (2.34) has two invariant manifolds

$$\mathcal{N}_k = \left\{(x_1, x_2, y_1, y_2) \in \mathbb{C}^4 \mid x_2 = (k-1)\pi, \quad y_2 = 0\right\}, \quad \text{where } k = 1, 2,$$

and its restriction to N_k reads

$$\dot{x} = y, \quad \dot{y} = -\frac{3}{2}\sin(2x), \quad (x, y) = (x_1, y_1). \tag{3.1}$$

These equations have the energy integral

$$h = \frac{y^2}{2} - \frac{3}{4}\cos(2x). \tag{3.2}$$

Let $\Gamma(\eta) \in \mathbb{C}^2$ denote a phase curve of system (3.1) lying on the level $h(x, y) = \eta$, and $\Gamma_k(\eta) \subset \mathcal{N}_k$ with $k = 1, 2$ are the respective phase curves of system (2.34).

If we denote by $[X_1, Y_1, X_2, Y_2]^T$ variations of phase variables $[x_1, y_1, x_2, y_2]^T$, then the variational equations along $\Gamma_k(\eta)$ have the form

$$\begin{bmatrix}\dot{X}_1\\ \dot{Y}_1\\ \dot{X}_2\\ \dot{Y}_2\end{bmatrix} = \begin{bmatrix}0 & 1 & 0 & 0\\ -3\cos(2x) & 0 & a_k(x, y) & 0\\ 0 & 0 & 0 & 1\\ 0 & 0 & b_k(x, y) & 0\end{bmatrix}\begin{bmatrix}X_1\\ Y_1\\ X_2\\ Y_2\end{bmatrix} \tag{3.3}$$

where $(x,y)\in\Gamma(\eta)$ and

$$a_k(x, y) = \frac{1-\varepsilon_k\lambda_2}{2(\lambda_1\lambda_2-1)}\left[3 + 2y(2+y) + 3\cos(2x)\right],$$

$$b_k(x, y) = c_k\left(3 + 2y(2+y)\right) + 3(c_k - 1)\cos(2x),$$

and

$$c_k = \frac{\varepsilon_k\lambda - 2}{2(\lambda_1\lambda_2-1)}, \quad \varepsilon_k = (-1)^k$$

for k=1,2.

The last two equations in (3.3) form a closed subsystem, which is called the normal variational equations. The equivalent second-order differential equation reads

$$\ddot{X} - b_k(x, y)X = 0, \quad (x, y) \in \Gamma_k(\eta), \quad k = 1, 2. \tag{3.4}$$

For both phase curves $\Gamma_1(\eta)$ and $\Gamma_2(\eta)$, we fix the same level of integral h, so

$$\eta = \frac{2}{3}y^2 - \cos(2x) \quad \text{for} \quad (x, y) \in \Gamma(\eta). \tag{3.5}$$

For further considerations, it is convenient to make the following change of independent variable

$$t \longmapsto z = \frac{1}{\sqrt{3}} y(t). \tag{3.6}$$

Then from (3.5), we obtain

$$\cos(2x) = 2z^2 - \eta,$$

and

$$(\dot{z})^2 = -\frac{3}{4}\left[\left(\eta - 2z^2\right)^2 - 1\right], \quad \ddot{z} = 3z\left(\eta - 2z^2\right)$$

Using the above formulae, we transform equations (3.4) to the following form

$$X'' + P(z)X' + Q_k(z)X = 0, \tag{3.7}$$

where prime denotes the derivative with respect to z, and rational coefficients of these equations are as follows

$$P(z) = \frac{\ddot{z}}{(\dot{z})^2} = \frac{2z}{2z^2 - \eta + 1} + \frac{2z}{2z^2 - \eta - 1},$$

$$Q_k(z) = -\frac{b_k}{(\dot{z})^2} = \frac{4\left[-6z^2 + c_k(3 + 4z(\sqrt{3} + 3z) - 3\eta) + 3\eta\right]}{3\left[\left(\eta - 2z^2\right)^2 - 1\right]}. \tag{3.8}$$

Next, we make the following change of dependent variable

$$X = w \exp\left[-\frac{1}{2}\int^{z} P(v)dv\right] \tag{3.9}$$

which convert equations (3.7) into its reduced form

$$w'' = r_k(z)w, \quad r_k(z) = \frac{1}{2}P'(z) + \frac{1}{4}P(z)^2 - Q_k(z) \tag{3.10}$$

Moreover, we fix the energy η=1 for both particular solutions. It is clear that for this value of η two singular points of equation (3.10) collapse into one at z=0. Then, the coefficient rk is given by

$$r_k(z) = \frac{3/4}{z^2} - \frac{3/16}{(z-1)^2} - \frac{3/16}{(z+1)^2} + \frac{4c_k}{\sqrt{3}z} + \frac{39 - 32c_k(3 + \sqrt{3}z)}{24(z^2 - 1)}. \tag{3.11}$$

It is worth to notice here that rk is a symmetric function of λ1 and λ2. In fact, we have

$$c_k = \frac{\varepsilon_k \lambda - 2}{2\sigma}, \quad \text{where} \quad \sigma := \lambda_1\lambda_2 - 1, \quad \lambda := \lambda_1 + \lambda_2. \tag{3.12}$$

The admissible values of parameters (λ,σ) are distinguished by the following inequalities

$$\lambda > 0, \quad \sigma > 0 \quad \text{and} \quad \lambda^2 - 4(\sigma + 1) \geq 0 \tag{3.13}$$

see Fig. 3.

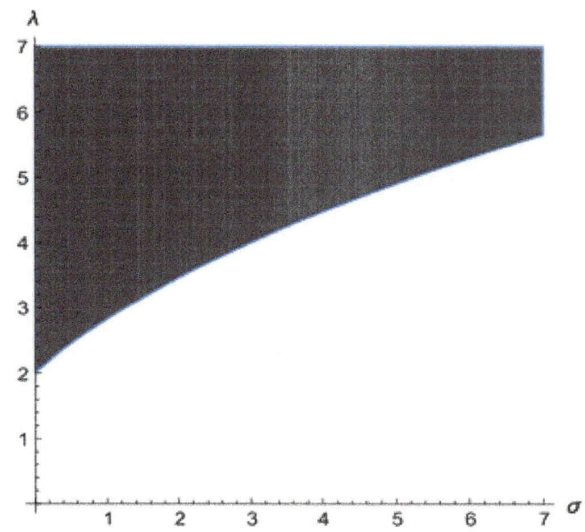

Figure 3: Range of parameters satisfying physical restrictions.

Equation (3.10) with the above coefficient r_k is a Fuchsian equation with four regular singular points $z_0 = 0$, $z_{\pm 1} = \pm 1$ and $z_\infty = \infty$. The respective differences of exponents at these points are as follows

$$\Delta_0 = 2, \quad \Delta_{\pm 1} = \frac{1}{2}, \quad \Delta_\infty = \delta_k, \tag{3.14}$$

where

$$\delta_k = \sqrt{1 + 8\frac{(\lambda_1 - \varepsilon_k)(\lambda_2 - \varepsilon_k)}{(\lambda_1\lambda_2 - 1)}}, \quad \text{for} \quad k = 1, 2, \tag{3.15}$$

respectively. In terms of σ and λ, these quantities read

$$\delta_1 = \sqrt{9 + 8\frac{2+\lambda}{\sigma}}, \quad \delta_2 = \sqrt{9 + 8\frac{2-\lambda}{\sigma}}. \tag{3.16}$$

Moreover, rational functions rk(z), see (3.11), can be rewritten in the following form

$$r_k(z) = \frac{9 - 42z^2 + 24z^4 + (\delta_k^2 - 9)z(\sqrt{3} + 3z)(z^2 - 1)}{12z^2(z^2 - 1)^2}. \tag{3.17}$$

Logarithmic Terms

Since the difference of exponents Δ0=2 is an integer, local solutions near z=0 can have logarithmic terms; see, e.g., (Whittaker and Watson 1935, Chap. 10) and (Maciejewski et al.2013, App. B). More precisely, two linearly independent local solutions w1 and w2 of (3.10) in a neighbourhood of z=0 have the following form

$$w_1(z) = z^{3/2} f(z), \quad w_2(z) = g w_1(z) \ln z + z^{-1/2} h(z), \tag{3.18}$$

where f(z) and h(z) are holomorphic at z=0 and f(0)≠0. Coefficient g multiplying the logarithmic term is

$$g = \frac{1}{6}\left(8c_k^2 - 6c_k + 3\right) \tag{3.19}$$

Since g>0 for arbitrary $c_k \in \mathbb{R}$, the logarithmic term appears in local solutions of the variational equation for all values of the parameters.

A continuation of the matrix of fundamental solutions along a small loop γ encircling the origin z=0 counterclockwise gives rise to a triangular monodromy matrix

$$\begin{bmatrix} w_1(z) & w_2(z) \\ w_1'(z) & w_2'(z) \end{bmatrix} \underset{\gamma}{\longrightarrow} \begin{bmatrix} w_1(z) & w_2(z) \\ w_1'(z) & w_2'(z) \end{bmatrix} \begin{bmatrix} -1 & -2\pi i \\ 0 & -1 \end{bmatrix} \tag{3.20}$$

Hence, the monodromy group of equation (3.10) contains matrix

$$M_0 := \begin{bmatrix} -1 & -2\pi i \\ 0 & -1 \end{bmatrix} \tag{3.21}$$

which is not diagonalisable.

Differential Galois group of the variational equation

The variational equation (3.10) is a parameterised family of second-order differential equations of the following form

$$y'' = r(z)y, \quad r(z) \in \mathbb{C}(z) \tag{3.22}$$

For such equations, their differential Galois group G is an algebraic subgroup of SL(2,C). The following lemma describes all possible types of G and relates these types to forms of solution of (3.22); see (Kovacic 1986; Morales Ruiz 1999).

Lemma 3.1

Let G be the differential Galois group of equation (3.22). Then one of four cases can occur.

G is reducible (it is conjugated to a subgroup of a triangular group); in this case equation (3.22) has an exponential solution of the form $y = \exp\int \omega$, where $\omega \in \mathbb{C}(z)$,

G is conjugated with a subgroup of

$$D^\dagger = \left\{\begin{bmatrix} c & 0 \\ 0 & c^{-1} \end{bmatrix} \middle| c \in \mathbb{C}^*\right\} \cup \left\{\begin{bmatrix} 0 & c \\ c^{-1} & 0 \end{bmatrix} \middle| c \in \mathbb{C}^*\right\}$$

in this case equation (3.22) has a solution of the form $y = \exp\int \omega$, where ω is algebraic over C(z) of degree 2,

G is primitive and finite; in this case all solutions of equation (3.22) are algebraic,

G=SL(2,C) and equation (3.22) has no Liouvillian solution.

For the notion of Liouvillian solutions, see, e.g., Kovacic (1986) and references therein. Now, let us return to our variational equation (3.10). Let M_k and G_k denote its monodromy and differential Galois group, respectively. It is known that $M_k \subset G_k$. We showed that its monodromy group contains non-diagonalisable matrix M_0. Thus, G_k is not a finite subgroup of SL(2,C). In this way, we exclude the third case of Lemma 3.1.

By the same reason, G_k cannot be also a subgroup of the dihedral group D† because it cannot contain a non-diagonalisable triangular matrix. Thus, the differential Galois group G_k of the variational equation is either a triangular subgroup of SL(2,C) or SL(2,C).

Elimination of the first case of Lemma 3.1

Let us assume that for our variational equation (3.10) the first case of Lemma 3.1 occurs. Then this equation has a nonzero solution of the form

$$w = P(z)\exp\left[\int^z \omega(s)\mathrm{d}s\right], \quad P \in \mathbb{C}[z], \quad \omega(z) \in \mathbb{C}(z) \tag{3.23}$$

see Kovacic (1986). If such a solution exists, then polynomial P(z) of degree d and rational function ω(z) can be found by application of Case 1 of the Kovacic algorithm, see Kovacic (1986). In order to construct this solution, first we calculate for each singularity auxiliary sets

$$\mathcal{E}_i = \left\{\frac{1}{2}(1 \pm \Delta_i)\right\}, \quad \text{for} \quad i \in \{-1, 0, 1, \infty\}$$

Next, we select from the Cartesian product

$$\mathcal{E} = \mathcal{E}_{-1} \times \mathcal{E}_0 \times \mathcal{E}_1 \times \mathcal{E}_\infty,$$

these elements $\rho = (\rho_{-1}, \rho_0, \rho_1, \rho_\infty) \in \mathcal{E}$ for which

$$d(\rho) := \rho_\infty - \sum_{i=-1}^{1} \rho_i \tag{3.24}$$

is a non-negative integer. The integer d(ρ) is the degree of polynomial P(z) entering into solution (3.23). For each selected element ρ∈E, we define the corresponding rational function

$$\omega = \sum_{i=-1}^{1} \frac{\rho_i}{z - z_i}. \tag{3.25}$$

Moreover, polynomial P must satisfy equation

$$P'' + 2\omega P' + (\omega' + \omega^2 - r_k)P = 0. \tag{3.26}$$

For our normal variational equations, the auxiliary sets are the following

$$\mathcal{E}_0 = \left\{-\frac{1}{2}, \frac{3}{2}\right\}, \quad \mathcal{E}_1 = \mathcal{E}_{-1} = \left\{\frac{1}{4}, \frac{3}{4}\right\}, \quad \mathcal{E}_\infty = \left\{\frac{1}{2}(1-\delta_k), \frac{1}{2}(1+\delta_k)\right\}$$

For possible choices of ρ, we have that

$$d(\rho) = \frac{1}{2}(1-\delta_k), \quad \text{or} \quad d(\rho) = \frac{1}{2}(\delta_k - m),$$

with m∈{−1,0,1,3,4,5}. We require that d(ρ) is a non-negative integer for k=1 and k=2. This implies that δ_1 and δ_2 are integers, which we denote by $n_k=\delta_k$. Using equation (3.16), one can express parameters (λ_1,λ_2) in terms of (n_1,n_2) and symmetric functions λ and σ defined in Eq. (3.12) take the forms

$$\lambda = \frac{2(n_1^2 - n_2^2)}{n_1^2 + n_2^2 - 18}, \quad \sigma = \frac{32}{n_1^2 + n_2^2 - 18}. \tag{3.27}$$

Now the last inequality in (3.13) expressed in terms of (n1,n2) reads

$$16\frac{63 + n_1^2 + n_2^2 - n_1^2 n_2^2}{(n_1^2 + n_2^2 - 18)^2} \geq 0. \tag{3.28}$$

Hence, n_1 and n_2 are non-negative integers, which satisfy three inequalities

$$63 + n_1^2 + n_2^2 - n_1^2 n_2^2 \geq 0, \quad n_1^2 > n_2^2, \quad n_1^2 + n_2^2 > 18. \tag{3.29}$$

From the last two conditions, it follows that $n_1^2 > 9$. Putting $n_1^2 = 9+a$, with a certain a>0, from the first condition (3.29) we obtain

$$9 > \frac{72 + a}{8 + a} > n_2^2 \quad \text{for} \quad a \in (0, \infty). \tag{3.30}$$

Thus, we conclude that there is only a finite number of admissible values of n_2, namely n_2∈{0,1,2}. We investigate these three cases using the variational equation corresponding to the second particular solution, i.e., we fix k=2.

If $n_2 = \delta_2 = 0$, then $\mathcal{E}_\infty = \{1/2\}$. There are two possible choices of ρ∈E for which d(ρ) is a non-negative integer, namely

$$\text{for} \quad \rho_1 = \left(\frac{1}{4}, -\frac{1}{2}, \frac{3}{4}, \frac{1}{2}\right), \quad d(\rho_1) = 0,$$

$$\text{for} \quad \rho_2 = \left(\frac{3}{4}, -\frac{1}{2}, \frac{1}{4}, \frac{1}{2}\right), \quad d(\rho_2) = 0.$$

So, the polynomial P(z) is a nonzero constant polynomial, and we can assume that P(z)=1. Then equation (3.26) reduces to the following one

$$\omega' + \omega^2 - r_2(z) = 0, \tag{3.31}$$

with $\omega=\omega(z)$ given by (3.25). In the considered case, this function reads

$$\omega = \frac{z^2 \pm z + 1}{2z(z^2-1)}, \tag{3.32}$$

with an upper sign for ρ_1. Moreover, as $\delta_2=0$, coefficient r_2 in (3.17) simplifies to

$$r_2 = \frac{3 - 14z^2 + 8z^4 - 3z(\sqrt{3}+3z)(z^2-1)}{4z^2(z^2-1)^2}$$

Hence,

$$\omega' + \omega^2 - r_2(z) = \frac{3\sqrt{3} \mp 2}{4z(z^2-1)} \neq 0. \tag{3.33}$$

In other words, this case is excluded.

If $\delta_2=1$, then $E_4=\{1,0\}$, and

$$r_2(z) = \frac{9 - 42z^2 + 24z^4 - 8z(\sqrt{3}+3z)(z^2-1)}{12z^2(z^2-1)^2}.$$

One can find that there are three admissible vectors $\rho \in E$, namely

$$\text{for} \quad \rho_1 = \left(\frac{3}{4}, -\frac{1}{2}, \frac{3}{4}, 1\right), \quad d(\rho_1) = 0,$$

$$\text{for} \quad \rho_2 = \left(\frac{1}{4}, -\frac{1}{2}, \frac{1}{4}, 0\right), \quad d(\rho_2) = 0,$$

$$\text{for} \quad \rho_3 = \left(\frac{1}{4}, -\frac{1}{2}, \frac{1}{4}, 1\right), \quad d(\rho_3) = 1.$$

For these vectors, the respective rational functions ω are as follows

$$\text{for} \quad \rho_1, \quad \omega = \frac{1+2z}{2z(z^2-1)},$$

For ρ_1 and ρ_2, we can set $P(z) = 1$, and we have to check if equation (3.31) is satisfied. But for both vectors, we obtain

$$\omega' + \omega^2 - r_2(z) = \frac{2}{\sqrt{3}z(z^2-1)} \neq 0.$$

For ρ_3, we can set P(z)=z+a with a certain $a \in C$. This polynomial must fulfill equation (3.26), which reduces to

$$\frac{3 + 2\sqrt{3}(z+a)}{3z(z^2-1)} \neq 0,$$

for an arbitrary $a \in C$. Thus, this case is also excluded.

If $\delta_2=2$, then $E_4=\{-1/2,3/2\}$, and

$$r_2(z) = \frac{9 - 42z^2 + 24z^4 - 5z(\sqrt{3}+3z)(z^2-1)}{12z^2(z^2-1)^2}.$$

Now, we have two possible choices for ρ. Namely,

$$\text{for} \quad \rho_1 = \left(\frac{1}{4}, -\frac{1}{2}, \frac{3}{4}, \frac{3}{2}\right), \quad d(\rho_1) = 1,$$

$$\text{for} \quad \rho_2 = \left(\frac{3}{4}, -\frac{1}{2}, \frac{1}{4}, \frac{3}{2}\right), \quad d(\rho_2) = 1.$$

The respective functions ω are given by (3.32), and we can put P=z+a. Then equation (3.26) reduces to equality

$$\frac{12 + a(\mp 6 + 5\sqrt{3}) + (\pm 6 + 5\sqrt{3} - 12a)z}{12z(z^2 - 1)} = 0,$$

which cannot be fulfilled for any a∈C. Thus, this case is also excluded.

In summary, it is impossible that for both variational equations their differential Galois groups are triangular subgroups of SL(2,C). At the same time, this shows also that at least for one particular solution the identity component of the differential Galois group of the respective normal variational equation is not Abelian. Thus, by Theorem 3.1, the system is non-integrable.

ACKNOWLEDGMENTS

This work was partially supported by the National Science Centre of Poland under grants DEC-2012/05/B/ST1/02165 and DEC-2013/09/B/ST1/04130 and by the EC Marie Curie Network for Initial Training Astronet-II, Grant Agreement No. 289240.

REFERENCES

1. Burov, A.A.: Oscillations of a vibrating dumbbell in an elliptic orbit. Dokl. Akad. Nauk. 437(2), 186–189 (2011)
2. Burov, A.A., Kosenko, I.I.: On the existence and stability of "orbitally uniform" rotations of a vibrating dumbbell in an elliptical orbit. Dokl. Akad. Nauk. 451(2), 164–167 (2013)
3. Burov, A.A., Nechaev, A.N.: On nonintegrability in the problem of the motion of a heavy double pendulum. In: Problems in the investigation of the stability and stabilization of motion, pp. 128–135, Ross. Akad. Nauk, Vychisl. Tsentr im. A. A. Dorodnitsyna, Moscow (in Russian) (2002). http://www.ccas.ru/depart/mechanics/TUMUS/z_SBORNIKI/Issues.html
4. Celletti, A., Sidorenko, V.: Some properties of the dumbbell satellite attitude dynamics. Celest. Mech. Dyn. Astron., 101(1–2), 105–126 (2008)
5. Guerman, A.D.: Equilibria of multibody chain in orbit plane. J. Guid. Control Dyn. 26(6), 942–948 (2003)
6. Guerman, A.D.: Spatial equilibria of multibody chain in a circular orbit. Acta Astronautica58(1), 1–14 (2006). doi:10.1016/j.actaastro.2005.05.002ADSCrossRef
7. Kovacic, J.J.: An algorithm for solving second order linear homogeneous differential equations. J. Symbol. Comput. 2(1), 3–43 (1986)
1. Maciejewski, A.J., Przybylska, M., Simpson, L., Szumiński, W.: Non-integrability of the dumbbell and point mass problem. Celest. Mech. Dyn. Astron. 117(3), 315–330 (2013)
8. Moauro, V., Negrini, P.: Chaotic trajectories of a double mathematical pendulum. Prikl. Mat. Mekh. 62(5), 892–895 (1998)
9. Morales Ruiz, J.J.: Differential Galois theory and non-integrability of Hamiltonian systems, volume 179 of Progress in Mathematics. Birkhäuser Verlag, Basel (1999)
10. Morales-Ruiz, J.J., Ramis, J.P.: Galoisian obstructions to integrability of Hamiltonian systems. I. Methods Appl. Anal. 8(1), 33–95 (2001)
11. Oh, Y.-G., Sreenath, N., Krishnaprasad, P.S., Marsden, J.E.: The dynamics of coupled planar rigid bodies. II. Bifurcations, periodic solutions, and chaos. J. Dyn. Differ. Equ. 1(3), 269–298 (1989)
12. Paul, A., Richter, P.H.: Application of Greene's method and the MacKay residue criterion to the double pendulum. Z. Phys. B 93(4), 515–520 (1994)
13. Salnikov, V.: On numerical approaches to the analysis of topology of the phase space for dynamical integrability. Chaos Solitons Fractals 57, 155–161 (2013)
14. Sidorenko, V.V., Celletti, A.: A "spring-mass" model of tethered satellite systems: properties of planar periodic motions. Celest. Mech. Dyn. Astron. 107(1–2), 209–231 (2010)
15. Sreenath, N., Oh, Y.-G., Krishnaprasad, P.S., Marsden, J.E.: The dynamics of coupled planar rigid bodies. I. Reduction, equilibria and stability. Dyn. Stab. Syst. 3(1–2), 25–49 (1988)
16. Szumiński, W.: Dynamics of multiple pendula without gravity. Chaotic Model. Simul. (CMSIM)1, 57–67 (2014)
17. Whittaker, E.T., Watson, G.N.: A Course of Modern Analysis. Cambridge University Press, London (1935)

CHAPTER 7

EXACT SOLUTION TO THE PROBLEM OF N BODIES FORMING A MULTI-LAYER ROTATING STRUCTURE

Joseph J Smulsky

Keywords: Problem of N bodies, Multi-layer rotating structure, Exact solution, Homographic and planar central configurations

BACKGROUND

For some configurations of N material points, there exist exact solutions to the problem of the Newtonian gravitational interaction of the material points. For instance, for N bodies axisymmetrically arranged along a circumference, a complete solution to the problem was previously obtained in (Smulsky 1999, 2004; Smul'skii 2003). In this problem, a central body can be located at the center of the circumference. Depending on their initial velocities, the peripheral bodies can move along an ellipse, a parabola, or a hyperbola. Apart from such single-layer configurations, in a number of studies the matter of exact solutions for configurations involving several layers was addressed (Silushik 2003; Grebenikov et al. 2006; Diarova et al.2006; Gutsu et al. 2007; Grebenikov 2010). Traditionally, the configurations of interest were treated as systems formed by mutually embedded polygons rotating, as an entity, at angular velocity ω. At polygon vertices, material points interacting among themselves by the Newton law of gravitation are placed. Within the context of the problem under consideration, example solutions for embedded triangles, rhombs, squares, pentagons, and hexagons were reported in a generalizing study by Grebenikov (Grebenikov 2010). The vertices in neighbor polygons can lie either in one radius or in radii passing through the middle points of the sides of neighbor polygons.

The above problem was solved for several layers of such polygons: for triangles, up to four layers; for squares, up to three layers, and for pentagons and hexagons, up to two layers (Grebenikov 2010). The largest number of the interacting bodies located at polygon vertices amounted to 12 bodies. Within the Hamiltonian dynamics, the problem can be reduced to a system of algebraic equations that can be solved by computer-algebra methods (Grebenikov2010; Grebenikov et al. 2002). Each particular problem requires special consideration, and much effort has to be spent on its solution. Earlier, such problems for interacting material points were treated as homographic-dynamics problems (Grebenikov 2010) or problems of planar central configurations (Saari 1980; Perko and Walter 1985; Diacu 1990; Xia 1991; Albouy 1996; Bang and Elmabsout 2001; Moeckel 1990; Hampton and Moeckel 2006; Albouy et al. 2008; Shi and Xie 2010). In turn, homographic dynamics itself has emerged as a new field in space dynamics (Grebenikov 2010).

In the present publication, we consider a somewhat different approach to solving the problems described above. This approach presents further development of the method that was previously used in tackling the axisymmetric problem (Smulsky 1999, 2004; Smul'skii 2003) and in treating multi-layered ring structures (Smul'skii 2011). The motion of the bodies is investigated considering the forces acting between the bodies. Instead of polygons, whose imaginary sides connect the bodies, in the present study we deal with the circumferences along which the bodies are located. The authors of all the above-mentioned publications mainly studied theoretical aspects of the Hamiltonian dynamics of the systems of interest such as, for instance, the solution existence problem for a specific form of central configurations. Finding stable central configurations

Citation: Joseph J Smulsky, Exact solution to the problem of N bodies forming a multi-layer rotating structure, DOI: 10.1186/s40064-015-1141-1.

is very difficult a problem (Yu and Zhang 2015). The present work is devoted to obtaining all exact solutions, calculation of structures, and possible use of results obtained.

Statement of the problem

Consider a multi-layer axisymmetric structure comprising several material points whose interaction is governed by the Newton law of gravitation (see Figure 1). The structure involvesN_2 circumferences with N_3 bodies located in each circumference. We will call the system of bodies whose centers are arranged along a circumference a body ring, or a body layer. The rings are enumerated with numbers j = 1, 2 … N_2, and the bodies in each ring, with numbers l = 1, 2 …N_3. In the plane x_0y_0, in which the bodies are disposed, for a body of mass $m_{j,l}$ we introduce a polar radius $r_{j,l}$ and a polar angle $\varphi_{j,l}$. For simplicity, in what follows the symbol $m_{j,l}$ will also be used to denote the body itself. All bodies in a ring have identical radii $r_{j,l} = r_j$, where r_j is the radius of the ring, and their masses are also identical, i.e. $m_{j,l} = m_j$. The polar angle that defines the angular position of the first body in each ring, $\varphi_{j,1}$, specifies the particular form of the structure. In what follows, this angle will be assumed a specified parameter. The polar angles of all other peripheral bodies are given by the formula

$$\varphi_{j,l} = \varphi_{j,1} + (l-1)\cdot\Delta\varphi_0, \tag{1}$$

where $\Delta\varphi_0 = 2\pi/N_3$ is the angular separation between the bodies in each ring.

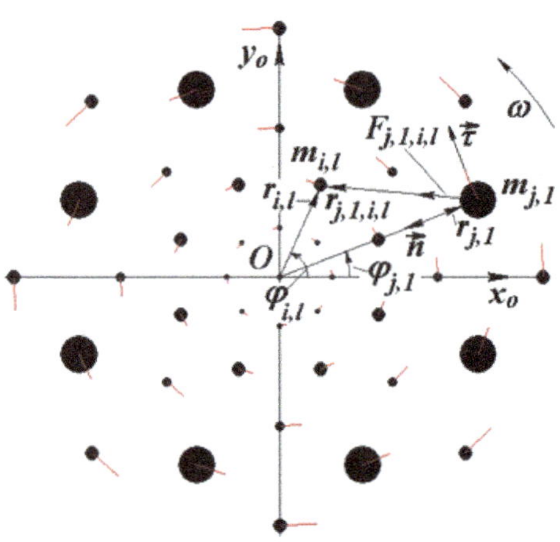

Figure 1: Geometric characteristics of the multi-layer axisymmetric structure with $N_2 = 5$ and$N_3 = 8$ in which the angles of the first body in neighbor rings $\varphi_{j,1}$ alternate one another. For clarity, the bodies are shown as full circles whose radii vary in proportion to the bodies' masses, and the line segments at the circles show the velocity vectors of the bodies.

To summarize, the geometry of a multi-layered axisymmetric structure is defined by the total number of the involved rings N_2, by the number of the bodies in each ring N_3, by the ring radiir_j, and by the angles defining the angular position of the first body in each ring $\varphi_{j,1}$. The mass of each body in the j-th ring is m_j and, in the presence of a central body of mass m_0, the mass of the whole system is

$$m_{SS} = m_0 + N_3\cdot\sum_{j=1}^{N_2} m_j \tag{2}$$

The whole system revolves at angular velocity ω. A multi-layer structure is specified by the set quantities N_2, N_3, $\varphi_{j,1}$, m_0 and ω, while the ring radii r_j and the body masses m_j present unknown quantities.

The forces acting between the bodies

Consider the forces that act on the first body in the j-th ring, whose mass is $m_{j,1}$, from the side of all other bodies (see Figure 1). To the body $m_{j,1}$, we attach a natural coordinate system (n,τ) in which n and τ are the normal and tangent lines to the trajectory. The gravity force $F_{j,1,i,l}$ due to the body $m_{i,l}$ in the i-th ring that acts on the body $m_{j,1}$ is $F_{j,1,i,l} = G\cdot m_{j,1}\cdot m_{i,l}/r_{j,1,i,l}^2$, where G is the gravitation constant and $r_{j,1,i,l}$ is the distance from body $m_{j,1}$ to body $m_{i,l}$. Then, the projections of the force $F_{j,1,i,l}$ onto the axes $\vec{n}$ n→ and $\vec{\tau}$ τ→ are

$$F_{n,j,1,i,l} = \frac{G\cdot m_{j,1}\cdot m_{i,l}\cdot n_{j,1,i,l}}{r_{j,1,i,l}^3}; \tag{3}$$

$$F_{\tau,j,1,i,l} = \frac{G\cdot m_{j,1}\cdot m_{i,l}\cdot \tau_{j,1,i,l}}{r_{j,1,i,l}^3}, \tag{4}$$

with $n_{j,1,i,l}$ and $\tau_{j,1,i,l}$ being the projections of the distance $r_{j,1,i,l}$, respectively onto the n- and τ-axis.

In the triangle $Om_{i,l}m_{j,1}$ (see Figure 1), the angle between the body radii r_i and r_j is

$$\Delta\varphi_{j,1,i,l} = \varphi_{i,l} - \varphi_{j,1}. \tag{5}$$

By the cosine theorem, the distance between the bodies is

$$r_{j,1,i,l}^2 = r_j^2 + r_i^2 - 2r_ir_j\cdot\cos\Delta\varphi_{j,1,i,l}. \tag{6}$$

Then, the projections of the latter distance onto the n- and τ-axis are respectively

$$F_{n,j,1,0} = \frac{G\cdot m_{j,1}\cdot m_0\cdot n_j}{r_j^3}. \tag{7}$$

Apart from the peripheral bodies, the body m_j is additionally acted upon by the central body of mass m_0 located at the point O (see Figure 1). The projection of the force due to the latter body onto the τ-axis is zero, and the projection of that force onto the n–axis can be expressed by analogy with formula (3):

$$F_{n,j,1,0} = \frac{G\cdot m_{j,1}\cdot m_0\cdot n_j}{r_j^3}.$$

In view of formula (7), in which we assume $r_i = 0$, for the central body the latter formula yields$n_j = r_j$.

After substitution of (6) and (7) into expressions (3) and (4), and after summation of the forces over all bodies in the system, for the projections of the forces acting on the body $m_{j,1}$ from the side of the rest bodies we obtain:

$$F_{n,j,1} = G \cdot m_j \left[\frac{m_0}{r_j^2} + \sum_{i \neq j}^{N_2} \left[m_i \cdot \sum_{l=1}^{N_3} \frac{r_j - r_i \cdot \cos \Delta\varphi_{j,1,i,l}}{\left(r_j^2 + r_i^2 - 2r_j \cdot r_i \cdot \cos \Delta\varphi_{j,1,i,l}\right)^{3/2}} \right] + \frac{m_j}{r_j^2} \sum_{l=2}^{N_3} \frac{0.5}{\left(2 - 2 \cdot \cos \Delta\varphi_{j,1,j,l}\right)^{1/2}} \right]; \tag{8}$$

$$F_{\tau,j,1} = G \cdot m_j \left[\sum_{i \neq j}^{N_2} \left[m_i \cdot \sum_{l=1}^{N_3} \frac{r_i \cdot \sin \Delta\varphi_{j,1,i,l}}{\left(r_j^2 + r_i^2 - 2r_j \cdot r_i \cdot \cos \Delta\varphi_{j,1,i,l}\right)^{3/2}} \right] + \frac{m_j}{r_j^2} \sum_{l=2}^{N_3} \frac{\sin \Delta\varphi_{j,1,i,l}}{\left(2 - 2 \cdot \cos \Delta\varphi_{j,1,j,l}\right)^{3/2}} \right] \tag{9}$$

In (8) and (9), m_j is mass of each of the bodies $m_{j,1}$, and m_i is the mass of each of the bodies $m_{i,l}$.

For excluding the self-action force of the body $m_{j,1}$, in formulas (8) and (9) the action due to all other bodies in the j-th ring is separated out from the total sum, and that action is then written as a last term. This action readily results if we replace the subscript i in the previous term with the subscript j. The exclusion of the j-th ring from the expression under the summation sign $\sum\sum$is denoted as $i \neq j$.

We consider such configurations of rotating structures for which expressions (8) and (9) yield one and the same force value for each body in the j-th ring. The latter is only possible if, as then-axis passes through any body in the j-th ring, the geometric positions of all other bodies acting on this body remain unchanged. The latter condition is satisfied providing that the initial angle of ring bodies assumes either a value $\varphi_{j,1} = 0$ or a value $\varphi_{j,1} = 0.5 \cdot \Delta\varphi_0$. The arrangement of a structure in which the initial angles $\varphi_{j,1}$ assume sequentially alternating values in neighbor rings is shown in Figure 1. Structures with arbitrary alternation pattern of initial angles $\varphi_{j,1}$ also satisfy the above condition.

Note that it is the above conditions that define the term «axisymmetric» as used in the present article. Some structure presents an axisymmetric structure if its geometric and dynamic characteristics remain unchanged as the structure revolves through the angle $\Delta\varphi_0$.

For the configurations considered above, the normal n presents a symmetry axis (see Figure 1). That is why, in view of (5), the angles of deflection of the interacting bodies from the n-axis,$\Delta\varphi_{j,1,i,l}$, have values pairwise equal in magnitude and opposite in sign. Hence, the sines in the nominators of formula (9) are also pairwise equal in magnitude and opposite in sign. The cosines of those angles in the denominators being identical, the tangential forces vanish. With the total number of the bodies N_3 being an even number, another body can be found in the n-axis, this body being located symmetrically about the center O. Since the angle $\Delta\varphi_{j,1,i,l}$ of the latter body is π, then the interaction force due to this body in (9) is also zero. Thus, the projections of all the forces onto the tangent axis are zero, $F\tau,j,1=0F\tau,j,1=0$. That is why the force acting on each body in the ring with number j from the side of all other bodies in the multi-layer axisymmetric structure is directed along the normal n to the trajectory, i.e. towards the centerO, and this force is defined by expression (8).

According to (1), the difference between the angles (5) in the j-th ring is

$$\Delta\varphi_{j,1,j,l} = \varphi_{j,l} - \varphi_{j,1} = 2\pi(l-1)/N_3. \tag{10}$$

Then, the expression in the denominator of the last term in formula (8) can be written as

$$2 \cdot [1 - \cos(2\pi(l-1)/N_3)] = 4 \cdot \sin^2(\pi(l-1)/N_3). \tag{11}$$

After substitution of (11) into (8), for the force acting on arbitrary body in the j-th ring from the side of all other bodies we obtain the expression

$$F_{n,j} = \frac{G \cdot m_j}{r_j^2} \left[m_0 + \sum_{i \neq j}^{N_2} \left[m_i \cdot \sum_{l=1}^{N_3} \frac{1 - r_{i,j} \cdot \cos \Delta\phi_{j,1,i,l}}{\left(1 + r_{i,j}^2 - 2r_{i,j} \cdot \cos \Delta\phi_{j,1,i,l}\right)^{3/2}} \right] + m_j \cdot f_{n3} \right] \tag{12}$$

where

$$f_{n3} = 0.25 \sum_{l=2}^{N_3} \frac{1}{\sin\left[\pi(l-1)/N_3\right]}; \tag{13}$$

$r_{i,j} = r_i/r_j$ is the ratio between the radii of the rings with numbers i and j; and the angle difference $\Delta\varphi_{j,1,j,l}$ is given by formula (10). Force (12) is directed towards the center O.

Motion equations of the rotating structure

By the force (12), the body $m_{j,1}$, whose mass is m_j, executes an accelerated motion (see Figure 1). In the natural coordinate system (n, τ), the force (12) acts along the n-axis; along the same axis, the normal acceleration $w_n = v^2/\rho$ is directed (here, v is the tangential velocity of the body $m_{j,1}$, and ρ is the radius of curvature of the body's trajectory. Hence, the differential equation of motion for the body can be written as

$$m_j \cdot \frac{v^2}{\rho} = F_{n,j}. \tag{14}$$

We consider a rotating structure revolving with angular velocity ω at fixed orbital radii. Hence, for the body $m_{j,1}$ the radius of curvature of its trajectory is $\rho = r_j$, and the velocity of the body is$v = \omega \cdot r_j$. After substitution of the above quantities and force (12) into Eq. (14), the differential motion equation of the body $m_{j,1}$ acquires the form

$$\omega^2 = \frac{G}{r_j^3} \left[m_0 + \sum_{i \neq j}^{N_2} \left[m_i \cdot \sum_{l=1}^{N_3} \frac{1 - r_{i,j} \cdot \cos \Delta\varphi_{j,1,i,l}}{\left(1 + r_{i,j}^2 - 2r_{i,j} \cdot \cos \Delta\varphi_{j,1,i,l}\right)^{3/2}} \right] + m_j \cdot f_{n3} \right], \tag{15}$$

where $j = 1, 2, \ldots N_2$.

Thus, the motion of the bodies making up a rotating structure is governed by N_2 Eq. (15). This system of equations is an algebraic system. As it was noted above, the unknown quantities here are the ring radii r_j and the body masses m_j. With the ring radii r_j specified, the body masses $m_j$are defined by Eq. (15). If necessary, it is the body masses m_j that can be specified; then, the ring radii r_j can be found from Eq. (15).

For generating a uniform algorithm for solving the

problem for multi-layer structures with and without a central body, we introduce an initial total mass m_{in} of the central body and all the peripheral bodies forming the first ring. We denote the mass fraction due to the central body in this structure as p_{m0}. Then, the mass of the central body is $m_0 = m_{in} \cdot p_{m0}$. A structure without a central body is defined by the equality $p_{m0} = 0$.

An analysis of potential applications of this problem shows that a good strategy towards solving the problem consists in specification of the geometry of the multi-layer axisymmetric structure, including the ring radii r_j, followed by the determination of the masses m_j (via solving system (15)). We therefore rewrite Eq. (15) in a different form. We isolate in the second term the non-dimensional acceleration of the body $m_{j,1}$ due to the action of a unit mass contained in the body$m_{i,1}$ of the i-th ring:

$$Q_{j,1,i,l} = \frac{1 - r_{i,j} \cdot \cos \Delta\varphi_{j,1,i,l}}{\left(1 + r_{i,j}^2 - 2r_{i,j} \cdot \cos \Delta\varphi_{j,1,i,l}\right)^{3/2}}, \tag{16}$$

where $r_{i,j} = r_i/r_j$ is the non-dimensional ratio of the radii.

Then, the non-dimensional acceleration of the body $m_{j,1}$ due to the interaction of that body with all the bodies in the i-th ring is

$$a_{j,i} = \sum_{l=1}^{N_3} Q_{j,1,i,l}. \tag{17}$$

Since the bodies forming the ring impart each body in the same ring with a non-dimensional acceleration f_{n3}, we introduce a designation

$$a_{j,j} = f_{n3}. \tag{18}$$

We normalize the masses of all the bodies by the initial mass m_{in}, and denote the non-dimensional masses of the bodies as

$$m_{ud,j} = m_j/m_{in}; \quad m_{ud,0} = m_0/m_{in}. \tag{19}$$

Then, using formulas (16)–(19), we can rewrite Eq. (15) as the following system of linear algebraic equations:

$$\sum_{i=1}^{N_2} a_{j,i} \cdot m_{ud,i} = b_j, \quad j = 1,2\ldots N_2, \tag{20}$$

where

$$b_j = c_i \cdot r_{ud,j}^3 - m_{ud,0}; \quad r_{ud,j} = r_j/r_1; \quad c_i = r_1^3\omega^2/(m_{in} \cdot G). \tag{21}$$

In the system of linear algebraic Eq. (20), the specified parameters are N_2, N_3, $\varphi_{j,1}$, m_{in}, p_{m0}, ω, and r_j, and the unknown quantities are the non-dimensional masses $m_{ud,i}$ of the peripheral bodies. The mass of the central body remains unchanged, i.e. $m_0 = \text{const}$. On the other hand, the non-dimensional masses of all the peripheral bodies $m_{ud,j}$ result from the solution of Eq. (20).

Solution of the equations

The solution of linear algebraic equation system (20) is given (Korn and Korn 1968) by

$$m_{ud,i} = D_i/D, \tag{22}$$

where D is the determinant of the matrix $a_{j,I,}$

$$D = \det\left[a_{j,i}\right] \tag{23}$$

and D_i is the determinant of the matrix $a_{j,i}$ in which the i-th column is replaced with the free-term column b_i.

In the case of two rings, we have

$$D = \begin{vmatrix} a_{11} & a_{12} \\ a_{21} & a_{22} \end{vmatrix} = a_{11}a_{22} - a_{21}a_{12}$$

and, also,

$$D_1 = \begin{vmatrix} b_1 & a_{12} \\ b_2 & a_{22} \end{vmatrix} = b_1a_{22} - b_2a_{12} \text{ and } D_2 = \begin{vmatrix} a_{11} & b_1 \\ a_{22} & b_2 \end{vmatrix} = a_{11}b_2 - a_{22}b_1$$

Then, in view of (22), we can write the solutions of Eq. (20) for the masses of one body in the first and second layer as

$$m_{ud,1} = (b_1a_{22} - b_2a_{12})/(a_{11}a_{22} - a_{21}a_{12}); \quad m_{ud,2} = (a_{11}b_2 - a_{21}b_1)/(a_{11}a_{22} - a_{21}a_{12}) \tag{24}$$

Solutions (24) define a rotating structure formed by two layers. In each layer, there are N_3bodies, where N_3 is an arbitrary integer number. The total number of the bodies in such a two-layer structure is $N = 2N_3 + 1$.

In the case of three or a greater number of layers, the solutions of system (20) emerge as too cumbersome expressions hard to be used for calculating the non-dimensional masses $m_{ud,i}$. Specification of a large number of initial parameters also becomes difficult. Moreover, the solution of the linear system of Eq. (20) may yield negative masses $m_{ud,i}$. That is why variation of initial parameters of the rotating system has to be applied to obtain positive values for all masses. For a greater number of rings N_2 and for a greater number of bodies in the rings N_3 the latter leads to a greater amount of computations. For executing the computations, several computers programs have been developed. It was found that the above problems can be solved most easily via numerical solution of the algebraic equation system (15) or (20). The computer programs were written in the FORTRAN language (Smulsky 2013a). In the first solution version of the problem in the program RtCrcStr.for, the equation system (15) was solved by the iteration method. In the second solution version of the problem in the program RtCrcSt2.for, for solving the equation system (20) the Gauss method was used. The latter version being simpler, below we consider this version. Nevertheless, it should be noted here that, at certain N_2-to-N_3 ratios, it is easier to get a solution of equations using RtCrcStr.for.

The initial data for a particular rotating structure are specified in the file RtCrcSt2.dat. Instead of the angular velocity ω, we deal with the structure's rotational period $P_{rd} = 2\pi/\omega$. In the program, the distances and the times are used as relative quantities. That is why the period $P_{rd}$is specified in sidereal years. The ring radii are specified in the data file RtCrcSt2.dat with the help of a new parameter o_{kr} defined by the relation

$$r_j = j \cdot o_{kr} \cdot r_1, \tag{25}$$

where r_1 is the radius of the first ring. The value of o_{kr} cannot be smaller than 0.5.

The radius of the first ring can be determined from the existence condition of this ring at some specified mass of the central body m_0 and at some initial body mass m_1 in the first ring (Smulsky 1999, 2004; Smul'skii 2003). The latter condition results from motion Eq. (15) on disregard of other rings:

$$r_1 = \left[\frac{G(m_0 + m_1 \cdot f_{n3})}{\omega^2}\right]^{1/3} \tag{26}$$

The angles that define the position of the first bodies $\varphi_{j,1}$ can be specified in two ways: (1) all angles $\varphi_{j,1} = 0$ and (2) the angle $\varphi_{j,1} = 0$ alternates with the angle $\varphi_{j,1} = 0.5 \cdot \Delta\varphi_0$. Other patterns of the radii r_j and angles $\varphi_{j,1}$ can be specified in additional files that can be included in the data file RtCrcSt2.dat.

The executable module of the program RtCrcSt2.exe generates several output data files, including a special file with specified values of all kinematic parameters of the structure. That file contains initial data and conditions for the program Galactica (Smulsky 2012a). In Galactica, a high-precision method for numerical integration of differential motion equations for material points interacting with one another by the Newton law of gravitation is implemented. The system Galactica is available in free access (Smulsky 2012b, c). This system computes the dynamics of the structures made up by interacting material points, and it allows one to investigate the evolution of such structures in time.

All structures obtained with the help of the program RtCrcSt2.exe were verified using the program Galactica. It should be noted here that Galactica integrates non-simplified equations, not Eqs. (15) or (20). The program Galactica contains an option allowing graphical representation of treated systems. Below, we reproduce the displayed images of several structures obtained after the first step in integration of their differential motion equations.

A description of the program RtCrcSt2.for and its functioning algorithm, and also the listing of the program, can be found elsewhere (Smulsky 2013a). The executable files are available athttp://www.ikz.ru/~smulski/Data/RtCrcStr/.

Examples of rotating structures

In the present study, multi-layer rotating structures involving 1, 2, 3, 4, 5, 15, 30, 100, 103 and 1,000 rings have been calculated. In the rings, the numbers of the bodies, 2, 5, 8, 10, 29, 30 and 999, were specified parameters. The total number of the bodies reached one million. Configurations with various values of the angle of the first body $\varphi_{j,1}$ in the rings, and also with various alternation patterns of the angles in neighbor rings, were treated. Structures with different starting masses were calculated. As the initial masses, the Sun's and Earth's masses were set. Structures with and without a central body were considered.

A structure involving five rings with each of the rings comprising eight bodies ($N_2 = 5$ and$N_3 = 8$) with rotational period $P_{rd} = 1$ year is shown in Figure 1. The numbers of the rings are counted from the center O, and the numbers of the bodies, from the x_O-axis. The angles $\varphi_{j,1}$ of the first bodies in the rings sequentially assume values 0, $\Delta\varphi_0/2$, 0, $\Delta\varphi_0/2$ and 0. The initial mass $m_{in} = 1.98912 \times 10^{30}$ kg was specified to be somewhat heavier than the Sun's mass. For the ring radii and for the masses of one body in the rings normalized, respectively by the radius of the first ring and by the mass of one body in the first ring, the values $r_{ud,j}$ = 0, 1, 2.005, 2.985, 4.082, and 4.980, and $m_{ud1,j} = m_j/m_1$ = 1.256, 1, 2.957, 2.250, 7.712 and 2.973 were obtained. Here, the ring with radius rud,0rud,0 = 0 represents the central body. In the calculated structure, the radius of the first ring and the mass of one body in this ring are respectively $r_1 = 1.481 \times 10^{11}$ m and $m_1 = 1.568 \times 10^{30}$ kg. The mass of the whole structure is $m_{SS} = 2.138 \times 10^{32}$ kg, this mass being 108 times greater than the Sun's mass.

Consider now three structures comprising fifteen rings ($N_2 = 15$) with thirty bodies contained in each of the rings ($N_3 = 30$); those structures have different alternation patterns of the angles $\varphi_{j,1}$of the first body in the rings. The rotating structure with zero angles of the first body in all rings,$\varphi_{j,1} = 0$, is shown in Figure 2. The line segments at the peripheral bodies indicate their velocity vectors.

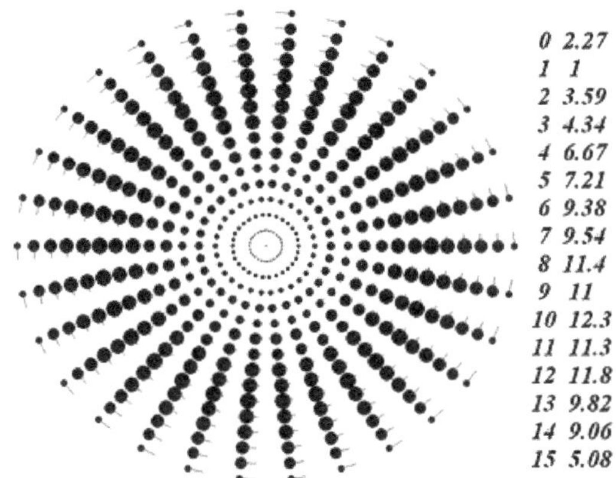

Figure 2: Image of a multi-layer axisymmetric rotating structure as seen on PC display. The characteristics of the structure were obtained by integration of differential motion equations performed by the program Galactica during one time step: $N_2 = 15$, $N_3 = 30$,$\varphi_{j,1} = 0$, $P_{rd} = 1$ year; the mass of the central body is equal to the Sun's mass. Additionally, the number columns in the figure show the ring radii and the masses of one body in the rings normalized, respectively by the radius of the first ring$r_1 = 1.493837 \times 10^{11}$ m and by the mass of one body in the first ring$m_1 = 8.684966 \times 10^{29}$ kg. The total mass of the whole structure is$m_{SS} = 3.218489 \times 10^{33}$ kg.

As it is seen from Figure 2, the structure possesses a radial-beam configuration. In homographic dynamics (Grebenikov 2010), that structure would be termed the configuration comprising 15 concentrically embedded equilateral 30-gons. From the second number column in Figure 2, we see that the least masses belong to bodies in the first ring. Providing that the mass of the central body m_0 is equal to the Sun's mass, the mass m_1 appears to be 2.27 times lighter than the massm_0. On going from the first to tenth ring, the masses of the bodies show almost a monotonic increase and, afterwards, they decrease in value. In that structure, the heaviest mass of the central body is $m_{10} = 12.3 \cdot m_1$.

In the structure shown in Figure 2 and in two subsequent structures, the ring radii increase in value due to the addition of the radius of the first ring r_1 (see Eq. (25)). In the structure shown in Figure 1, the ring radii show a different variation pattern. The latter can be attributed to the fact that, in the latter case, a proportional increase of ring radii results in an emergence of negative masses in the solution of Eq. (20). It should be noted that the structure shown in Figure 1 was obtained using the program RtCrcStr.for.

Figure 3 shows a multi-layer rotating structure with sequential alternation of the angles $\varphi_{j,1}$ of the first body in the rings. In homographic dynamics (Grebenikov 2010), this structure would be termed a configuration formed by 15 mutually embedded equilateral 30-gons rotated with respect to each other through the angle $\pi/15$. In comparison with the structure shown in Figure 2, the total mass of this structure has increased by a factor of 1.236. Here, the masses of the bodies in the rings also increase in value starting from the center. Nonetheless, the monotonic increase of the masses is violated after the tenth ring, and the heaviest bodies occur in the last ring.

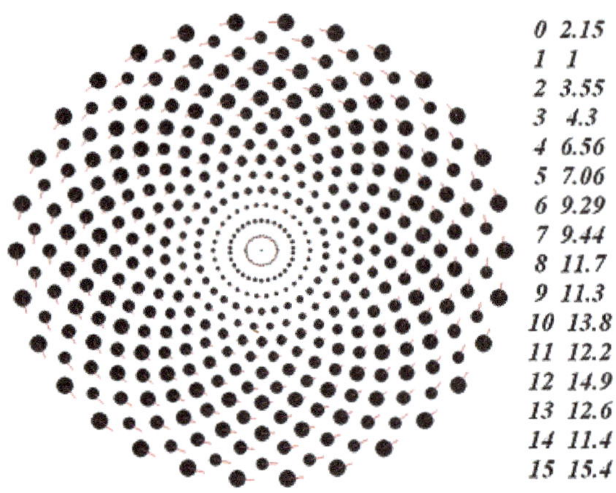

Figure 3: Multi-layer axisymmetric structure with $N_2 = 15$ and $N_3 = 30$; the angle $\varphi_{j,1} = 0$ alternates with the angle $\varphi_{j,1} = 0.5{\cdot}\Delta\varphi_0$; $P_{rd} = 1$ year; $r_1 = 1.493837 \times 10^{11}$ m;$m_1 = 9.172058 \times 10^{29}$ kg and $m_{SS} = 3.977302 \times 10^{33}$ kg. The rest designations are the same as in Figure 2.

In the structure shown in Figure 4, the angle $\varphi_{j,1} = \Delta\varphi_0/2$ recurs each three layers. Here, the total mass of the system has also increased in comparison with the radial-beam structure of Figure 2; yet, this mass proved to be lighter than the mass of the structure shown in Figure 3. The mass of the bodies in the rings also increases with increasing ring radius. In the rings with$\varphi_{j,1} = \Delta\varphi_0/2$, the body masses are normally heavier than in the neighbor rings. The heaviest mass is in the ring with number 12. Among the three examples considered above, this mass is the heaviest one, being twenty times greater than the mass of the bodies in the first ring.

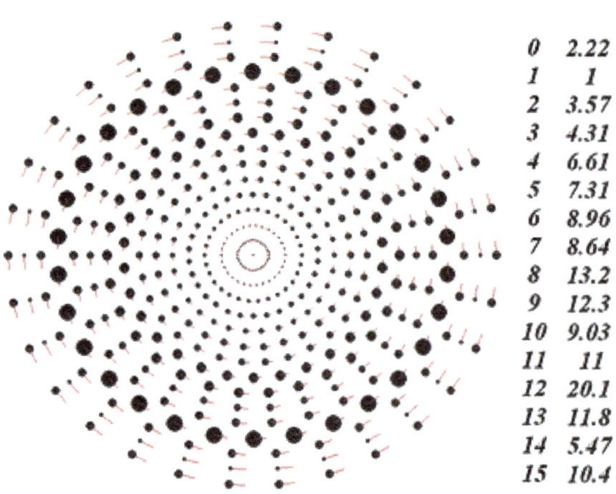

Figure 4: Multi-layer axisymmetric structure with $N_2 = 15$ and $N_3 = 30$; the angles $\varphi_{j,1} = 0$ and$\varphi_{j,1} = 0.5{\cdot}\Delta\varphi_0$ recur each three layers; $P_{rd} = 1$ year; $r_1 = 1.493837 \times 10^{11}$ m,$m_1 = 8.889051 \times 10^{29}$ kg and $m_{SS} = 3.570164 \times 10^{33}$ kg. For the rest designations, see Figure 2.

The majority of homographic problems are represented by mutually embedded regular polygons (Grebenikov 2010). On the other hand, irregular polygons such as, for instance, mutually embedded rhombs, are also met (Grebenikov 2010; Diarova and Zemtsova 2006). Such a rhombic multi-layer structure [see Figure 20 in (Grebenikov 2010)] can be constructed as a structure comprising several rings. In Figure 5a, this structure is shown as a structure comprising four rings with two bodies contained in each of the rings ($N_2 = 4$ and $N_3 = 2$). Here, the ring radii and the masses of one body in the rings normalized respectively by the radius of the first ring and by the mass of one body in this ring are $r_{ud,j} = 0$, 1, 1.697, 1.703 and 2.40, and$m_{ud1,j} = 1{,}161$, 1, 193.47, 490.54 and 23.76. Unlike in the previous structures, the angle of the first body in the first ring is $\varphi_{1,1} = 0.5{\cdot}\Delta\varphi_0 = \pi/2$ (see Figure 5a). That is why this body occupies position at the ordinate axis. That structure, comprising two rhombs, one rhomb being embedded into the other, is identical to the structure shown in Figure 20 of monograph (Grebenikov 2010).

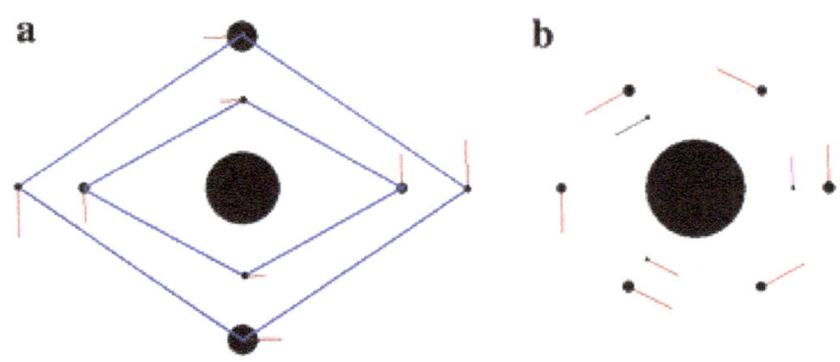

Figure 5: **a** The four-layer axisymmetric structure with $N_2 = 4$ and $N_3 = 2$; the angle $\varphi_{j,1} = 0.5{\cdot}\Delta\varphi_0$alternates with the angle $\varphi_{j,1} = 0$; $P_{rd} = 2$ year; $r_1 = 1.491651 \times 10^{11}$ m,$m_1 = 1.695854 \times 10^{27}$ kg and $m_{SS} = 4.373151 \times 10^{30}$ kg. **b** The three-layer axisymmetric structure with $N_2 = 3$ and $N_3 = 3$; $\varphi_{j,1} = 0$, $0.5{\cdot}\Delta\varphi_0$, and 0; $P_{rd} = 1.5$ year;$r_1 = 1.491990 \times 10^{11}$ m, $m_1 = 9.018554 \times 10^{26}$ kg and $m_{SS} = 2.971883 \times 10^{30}$ kg. The rest designations are the same as in Figure 2.

As it was shown above, in the four-layer ring structure of Figure 5a the dimensionless radii of the second and third rings are roughly identical, $r_{ud,3} = 1.697$ and $r_{ud,4} = 1.703$. Nonetheless, the angles of the first body in those rings are shifted with respect to each other by the angle $0.5{\cdot}\Delta\varphi_0$. It follows from here that axisymmetric rotating structures having $2{\cdot}N_3$ bodies in their separate rings can be constructed. For creating such structures, it is required that the radii of neighbor rings in them be identical while the angles of the first body in those rings, different. As an example, Figure 5b shows a three-layer structure with identical radii of the second and third layers. Here, the non-dimensional radii and the non-dimensional masses are respectively$r_{ud,j} = 0$, 1, 1.375 and 1.375, and $m_{ud1,j} = 21{,}844$, 1, 175.8 and 193.8. Evidently, here the radii of the second and third rings are identical while the masses of the bodies forming the rings are different. The latter structure is identical to the homographic configuration involving a regular hexagon which in turn contains a concentric equilateral triangle [see Figure 26 in monograph (Grebenikov 2010)].

The example in Figure 5b exhibits an interesting feature. Mathematically, this structure is defined as a structure that comprises three rings ($N_2 = 3$), whereas, physically, here we have two rings, or two layers. Yet, the bodies in the second layer exhibit two possible mass patterns. Thus, the presented examples prove that a wide set of homographic configurations (Grebenikov2010), or planar central configurations (Perko and Walter 1985), can be represented by multi-layer axisymmetric rotating structures.

As it was noted above, a structure comprising 103 rings and 29 bodies in each ring (N_2 = 103 and N_3 = 29) and a structure comprising 1,000 rings and 999 bodies in each ring (N_2 = 1,000 and N_3 = 999) were constructed. In the former structure, the total number of the bodies wasN = 2,988, and in the latter structure this number was N = 999,001. In addition, structures with the starting mass m_{in} equal to the Earth's mass were calculated. The total number of the rings in such structures was N_2 = 6, 15, and 103. The generated files with input data and initial conditions for the program Galactica proved to be identical to the files obtained for an initial mass m_{in} somewhat heavier than the Sun's mass. This means that the results obtained for the Sun's mass (those results are shown in Figures 1, 2, 3, 4, 5) simultaneously present results for the structures whose central mass is equal to the Earth's mass.

Stability problems for examined structures and their applications

Despite the fact that the characteristics of examined structures were obtained as an exact solution of the problem, the solution results are inevitably expressed with numbers having a finite length of their representation with digits. As a result, the obtained characteristics of a structure obviously differ from the exact characteristics. The mentioned difference presents one factor that makes the structure change its configuration during numerical integration of the equations. The second factor is the precision of the integration method. If the structure is a stable one, then, with running time, the bodies making up the structure start executing low-amplitude oscillations being, as a rule, azimuthal oscillations. If the structure is unstable, then this structure finally suffers disintegration.

In the above-mentioned homographic-dynamics studies, especially in publication (Grebenikov2010), problems on the stability of rotating structures were treated analytically within the framework of Hamiltonian dynamics. As it was noted in (Grebenikov 2010), there exist more than 100 definitions of the notion "stability", and in space dynamics one can encounter up to thirty such definitions. Unfortunately, such methods fail to produce direct predictions for the evolution of particular structures. That is why, for predicting the dynamic evolution of a structure, it is required to perform numerical integration of differential motion equations for involved bodies. Here, the integration accuracy should be high enough, and the inaccuracy, small, so that to not distort the predicted behavior of the structure. All those requirements are met in the program Galactica.

From the standpoint of the force interaction, all multi-layer rotating structures present unstable structures. Yet, the lifetime of such structures can vary over a wide range. Figure 6 shows the structure of Figure 1 after 1.6 revolutions. The changes in the structure have emerged because of the disintegration of the first, inner, ring. In this ring, the eight bodies have combined in pairs. Further motion of the paired bodies leads to a collapse of the whole structure.

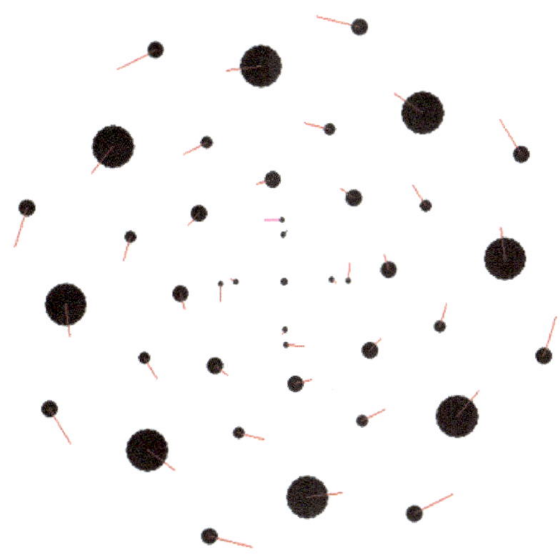

Figure 6: The onset of decomposition of the axisymmetric structure with $N_2 = 5$, $N_3 = 8$, and$P_{rd} = 1$ year in which the angle $\varphi_{j,1} = 0$ alternates with the angle $\varphi_{j,1} = 0.5 \cdot \Delta\varphi_0$. The structure has executed 1.6 revolutions; the starting configuration of the structure is shown in Figure 1.

The program Galactica was used to investigate the dynamics of axisymmetric structures in studies (Smulsky 2008, 2011, 2013b, 2014; Mel'nikov et al. 2008; Smul'skii 2011). Those studies have revealed a broad range of possible behaviors that could be displayed by axisymmetric structures. Of primary interest is the lifetime of investigated structures. This time is longer the lower is the non-dimensional rotational velocity of the system. If an axisymmetric structure presents a substructure of some greater structure, then the lifetime of the axisymmetric structure can increase in value. In the latter case, the bodies forming the structure execute oscillations as they move along their trajectories. Such oscillations act to prevent the bodies from approaching each other, and those oscillations may therefore infinitely prolong the lifetime of the structure (Mel'nikov et al. 2008).

Of special interest is using the program Galactica for performing studies of the dynamic properties of axisymmetric rotating structures. Such studies may lead to application of examined structures in various problems of celestial and space dynamics. For instance, single-layer axisymmetric structures were used in developing compound models for Earth and Sun's rotation in (Smulsky 2008, 2011; Mel'nikov et al. 2008). In such studies, a qualitative picture for the evolution of Earth's axis was revealed. In addition, the oscillations of peripheral bodies in such a model have suggested an idea of possible oscillating motion of Earth's continents in latitude direction (Mel'nikov et al. 2008). The compound model for Sun's rotation proved helpful in revealing the Mercury-perihelion excess rotation required for the development of a description of Solar-system dynamics based on the Newton law of gravitation.

Using the investigated structures for modeling planetary disks and disk galaxies also attracts interest. With body trajectories uniformly expanded in space, an axisymmetric structure becomes a spheroidal structure. Of special interest are the construction of such structures and the study of their dynamics. Very probably, new ideas in this field will prove helpful in gaining a better insight into the nature of globular star clusters and into the processes proceeding in such clusters.

CONCLUSIONS

An exact solution to the problem of N bodies arranged in an in-plane multi-layer axisymmetric rotating structure has been obtained. The axial symmetry of such a structure consists in that the geometric and dynamic characteristics of the structure remain unchanged on its rotation through an angle $\Delta\varphi_0$. Characteristics of particular structures with various numbers of involved layers and bodies (up to one million bodies) have been determined. The obtained characteristics were verified by numerical integration of the differential motion equations of the bodies forming the structures. Examination of the dynamic properties of the rotating structures with the help of the program Galactica will enable application of obtained data in various celestial- and space-mechanics models. The multi-layer rotating structures involve all the planar central configurations presently known in homographic dynamics.

ACKNOWLEDGEMENTS

In mid-March of 2010, Eugeny Aleksandrovich Grebenikov presented me a manuscript version of his monograph (Grebenikov 2010), which I have read with vivid interest. Later, from time to time I looked through the already published book. Gradually, polygonal configurations have transformed into ring body structures in which the interaction between bodies was expressed in terms of forces. As a result of my several-year activities in the field, the solution reported in the present publication was invented. Unfortunately, it was too late for me to make E. A. Grebenikov acquainted with the obtained solution. That is why I dedicate this solution to his memory. The calculations for many bodies with extended representation of numbers were performed on supercomputers of the Siberian Supercomputing Center, Siberian Branch of the Russian Academy of Science, Novosibirsk.

Dedicated to E. A. Grebenikov's memory.

REFERENCES

1. Albouy A (1996) The symmetric central configurations of four equal masses. American Mathematical Society, Providence
2. Albouy A, Fu Y, Sun SZ (2008) Symmetry of planar four-body convex central configurations. Proc R Soc A 464:1355–1365
3. Bang D, Elmabsout B (2001) Configurations polygonales en équilibre relative. C R Acad Sci Paris t Série II b 329:243–248
4. Diacu F (1990) The masses in a symmetric centered solution of the n-body problem. Proc AMS 109:1079–1085
5. Diarova D, Zemtsova NI (2006) The instability of the rhombus-like central configurations in Newtonian 9-body problem. In: Computer algebra in scientific computing (CASC 2006), LNCS 4194, Springer, Berlin Heidelberg, Chisinau, September 11–15, pp 141–148
6. Diarova DM, Zemtsova NI, Ikhsanov EV (2006) Existence of central configurations in a model for the Newtonian problem for eight bodies. In: Proc. of ISA RAN "Dynamics of Linear and non-Linear Systems", vol. 25(1), pp 64–71 (in Russian)
7. Grebenikov EA (2010) Mathematical problems in homographic dynamics. MAX Press, Moscow, p 256 (in Russian)
8. Grebenikov EA, Kozak-Skovorodkina D, Yuakubyak M (2002) Computer-algebra methods in the many-body problem. People's Friendship University of Russia, Moscow, p 212
9. Grebenikov EA, Diarova DM, Zemtsova NI (2006). Existence of stable rhombus-like central configurations in the Wintner sense for the Newtonian model of nine bodies. In: Theoretical and applied problems in non-linear analysis, collection of scientific papers, Computing Center, Russian Academy of Science, pp 65–76
10. Gutsu VD, Diarova DM, Zemtsova NI (2007) Stability study for stationary solutions to the rhombus-like bounded problem for ten bodies. In: Theoretical and applied problems in non-linear analysis, collection of scientific papers. Computing Center, Russian Academy of Science, pp 99–109
11. Hampton M, Moeckel R (2006) Finiteness of relative equilibria of the four-body problem. Invent Math 163(2):289–312
12. Korn G, Korn T (1968) Mathematical handbook for scientists and engineers: definitions, theorems, and formulas for reference and review. Nauka, Moscow, p 720
13. Mel'nikov VP, Smul'skii II, Smul'skii YI (2008) Compound modeling of earth rotation and possible implications for interaction of continents. Rus Geol Geophys 49:851–858. doi:10.1016/j.rgg.2008.04.003
14. Moeckel R (1990) On central configurations. Math Z 205:499–517
15. Perko LM, Walter EL (1985) Regular polygon solutions of N-body problem. Proc AMS 94:301–309
16. Saari DG (1980) On the role and properties of N body central configurations. Celest Mech 21:9–20
17. Shi J, Xie Z (2010) Classification of four-body central configurations with three equal masses. J Math Anal Appl 363:512–524
18. Silushik A (2003) Konieczne i wystarczajace warunki istnienia homograficznych rozwianzan w specjalnym zagadnieniu 7-u i 10-u cial. In: Proceeding of the international workshop. Byelorussian State University, Brest, pp 206–210
19. Smul'skii II (2003) The axisymmetric problem of gravitational interaction of N bodies. Matematicheskoe Modelirovanie 15(5):27–36 (in Russian).http://www.smul1.newmail.ru/Russian1/IntSunSyst/Osvnb4.doc. Accessed 19 Mar 2015
20. Smul'skii II (2011) Multilayer ring structures. Phys Part Nuclei Lett 8(5):436–440. doi:10.1134/S1547477111050189. http://www.ikz.ru/~smulski/Papers/PHPL436.pdf. Accessed 19 Mar 2015
21. Smulsky JJ (1999) The theory of interaction. Novosibirsk University, Scientific Publishing Center of UGG SB of RAS, Novosibirsk, p 294 (in Russian).http://www.ikz.ru/~smulski/TVfulA5_2.pdf. Accessed 19 Mar 2015

22. Smulsky JJ (2004) The theory of interaction. PH Cultural Information Bank, Ekaterinburg, p 302. http://www.ikz.ru/~smulski/TVEnA5_2.pdf. Accessed 19 Mar 2015

23. Smulsky JJ (2008) Numerical modeling of the evolution of the satellite of a rotating body. In Col. the theoretical and applied tasks of the nonlinear analysis. Russian Academy of Sciences: A. A. Dorodnicyn Computing Center. Moscow, pp 100–117 (In Russian).http://www.ikz.ru/~smulski/Papers/ModSun07c.pdf. Accessed 19 Mar 2015

24. Smulsky JJ (2011) New components of the Mercury's perihelion precession. Nat Sci 3(4):268–274. doi:10.4236/ns.2011.34034

25. Smulsky JJ (2012a) Galactica software for solving gravitational interaction problems. Appl Phys Res 4(2):110–123. doi:10.5539/apr.v4n2p110

26. Smulsky JJ (2012b) The system of free access Galactica to compute interactions of N bodies. I. J Mod Educ Comput Sci 11:1–20. doi:10.5815/ijmecs.2012.11.01. http://www.mecs-press.org/

27. Smulsky JJ (2012c) The Galactica Software. Institute of Earth Cryosphere, SB RAS. Tyumen.http://www.ikz.ru/~smulski/GalactcW/. Accessed 19 Mar 2015

28. Smulsky JJ (2013a) Multi-layer axisymmetric rotating structures. Institute of Earth Cryosphere SB RAS, Tyumen. Dep. in VINITI 28.10.2013, No. 303-B2013, p 27 (in Russian).http://www.ikz.ru/~smulski/Papers/OsMVStr.pdf. Accessed 19 Mar 2015

29. Smulsky JJ (2013b) The axisymmetric Coulomb interaction and orbital motion instability. Institute of Earth Cryosphere, SB RAS. Tyumen. Dep. in VINITI 28.10.2013, No. 304-B2013 2013, p 30 (in Russian). http://www.ikz.ru/~smulski/Papers/KulInt2.pdf. Accessed 19 Mar 2015

30. Smulsky JJ (2014) Axisymmetric Coulomb interaction and research of its stability by system Galactica. Open Access Libr J 1(e773):1–23. doi:10.4236/oalib.1100773

31. Xia Z (1991) Central configurations with many small masses. J Differ Equ 91:168–179

32. Yu X, Zhang S (2015) Central configurations formed by two twisted regular polygons. J Math Anal Appl 425:372–380

CHAPTER 8 STABILITY OF THE MOONS ORBITS IN SOLAR SYSTEM IN THE RESTRICTED THREE-BODY PROBLEM

Sergey V. Ershkov

INTRODUCTION

The stability of the motion of the moon is the ancient problem which leading scientists have been trying to solve during last 400 years. A new derivation to estimate such a problem from a point of view of relative motions in restricted three-body problem (R3BP) is proposed here.

Systematic approach to the problem above was suggested earlier in KAM- (Kolmogorov-Arnold-Moser-) theory [1] in which the central KAM-theorem is known to be applied for researches of stability of Solar System in terms of restricted three-body problem [2–5], especially if we consider photogravitational restricted three-body problem [6–8] with additional influence of Yarkovsky effect of nongravitational nature [9].

KAM is the theory of stability of dynamical systems [1] which should solve a very specific question in regard to the stability of orbits of so-called "small bodies" in Solar System, in terms of restricted three-body problem [3]: indeed, dynamics of all the planets is assumed to satisfy restrictions of restricted three-body problem (such as infinitesimal masses and negligible deviations of the main orbital elements).

Nevertheless, KAM also is known to assume the appropriate Hamilton formalism in proof of the central KAM-theorem [1]; the dynamical system is assumed to be Hamilton's system and all the mathematical operations over such a dynamical system are assumed to be associated with a proper Hamilton system.

According to the Bruns theorem [5], there are no other invariants except well-known 10 integrals for three-body problem (including integral of energy and momentum); this is a classical example of Hamilton's system. But in case of restricted three-body problem, there are no other invariants except only one, Jacobian-type integral of motion [3].

Such a contradiction is the main paradox of KAM-theory; it adopts all the restrictions of restricted three-body problem, but nevertheless it proves to use the Hamilton formalism, which assumes the conservation of all other invariants (the integral of energy, momentum, etc.).

To avoid ambiguity, let us consider a relative motion in three-body problem [2].

Equations of Motion

Let us consider the system of ODE for restricted three-body problem in barycentric Cartesian coordinate system, at given initial conditions [2, 3]:

$$m_1 \mathbf{q}_1'' = -\gamma \left\{ \frac{m_1 m_2 (\mathbf{q}_1 - \mathbf{q}_2)}{|\mathbf{q}_1 - \mathbf{q}_2|^3} + \frac{m_1 m_3 (\mathbf{q}_1 - \mathbf{q}_3)}{|\mathbf{q}_1 - \mathbf{q}_3|^3} \right\},$$

$$m_2 \mathbf{q}_2'' = -\gamma \left\{ \frac{m_2 m_1 (\mathbf{q}_2 - \mathbf{q}_1)}{|\mathbf{q}_2 - \mathbf{q}_1|^3} + \frac{m_2 m_3 (\mathbf{q}_2 - \mathbf{q}_3)}{|\mathbf{q}_2 - \mathbf{q}_3|^3} \right\},$$

$$m_3 \mathbf{q}_3'' = -\gamma \left\{ \frac{m_3 m_1 (\mathbf{q}_3 - \mathbf{q}_1)}{|\mathbf{q}_3 - \mathbf{q}_1|^3} + \frac{m_3 m_2 (\mathbf{q}_3 - \mathbf{q}_2)}{|\mathbf{q}_3 - \mathbf{q}_2|^3} \right\}, \quad (1)$$

where q_1, q2 and q_3 mean the radius vectors of bodies m_1, m_2, and m_3, respectively; γ is the gravitational constant.

Citation: Sergey V. Ershkov, "Stability of the Moons Orbits in Solar System in the Restricted Three-Body Problem,"Advances in Astronomy, vol. 2015, Article ID 615029, 7 pages, 2015. doi:10.1155/2015/615029

The system above could be represented for relative motion of three bodies as shown below (by the proper linear transformations):

$$(\mathbf{q}_1-\mathbf{q}_2)''+\gamma(m_1+m_2)\frac{(\mathbf{q}_1-\mathbf{q}_2)}{|\mathbf{q}_1-\mathbf{q}_2|^3}$$

$$=\gamma m_3\left\{\frac{(\mathbf{q}_3-\mathbf{q}_1)}{|\mathbf{q}_3-\mathbf{q}_1|^3}+\frac{(\mathbf{q}_2-\mathbf{q}_3)}{|\mathbf{q}_2-\mathbf{q}_3|^3}\right\},$$

$$(\mathbf{q}_2-\mathbf{q}_3)''+\gamma(m_2+m_3)\frac{(\mathbf{q}_2-\mathbf{q}_3)}{|\mathbf{q}_2-\mathbf{q}_3|^3}$$

$$=\gamma m_1\left\{\frac{(\mathbf{q}_3-\mathbf{q}_1)}{|\mathbf{q}_3-\mathbf{q}_1|^3}+\frac{(\mathbf{q}_1-\mathbf{q}_2)}{|\mathbf{q}_1-\mathbf{q}_2|^3}\right\},$$

$$(\mathbf{q}_3-\mathbf{q}_1)''+\gamma(m_1+m_3)\frac{(\mathbf{q}_3-\mathbf{q}_1)}{|\mathbf{q}_3-\mathbf{q}_1|^3}$$

$$=\gamma m_2\left\{\frac{(\mathbf{q}_1-\mathbf{q}_2)}{|\mathbf{q}_1-\mathbf{q}_2|^3}+\frac{(\mathbf{q}_2-\mathbf{q}_3)}{|\mathbf{q}_2-\mathbf{q}_3|^3}\right\}. \tag{2}$$

Let us designate the following:

$$\mathbf{R}_{1,2}=(\mathbf{q}_1-\mathbf{q}_2),$$

$$\mathbf{R}_{2,3}=(\mathbf{q}_2-\mathbf{q}_3),$$

$$\mathbf{R}_{3,1}=(\mathbf{q}_3-\mathbf{q}_1). \tag{*}$$

Using of (*) above, let us transform the previous system to another form:

$$\mathbf{R}''_{1,2}+\gamma(m_1+m_2)\frac{\mathbf{R}_{1,2}}{|\mathbf{R}_{1,2}|^3}=\gamma m_3\left\{\frac{\mathbf{R}_{3,1}}{|\mathbf{R}_{3,1}|^3}+\frac{\mathbf{R}_{2,3}}{|\mathbf{R}_{2,3}|^3}\right\},$$

$$\mathbf{R}''_{2,3}+\gamma(m_2+m_3)\frac{\mathbf{R}_{2,3}}{|\mathbf{R}_{2,3}|^3}=\gamma m_1\left\{\frac{\mathbf{R}_{1,2}}{|\mathbf{R}_{1,2}|^3}+\frac{\mathbf{R}_{3,1}}{|\mathbf{R}_{3,1}|^3}\right\},$$

$$\mathbf{R}''_{3,1}+\gamma(m_1+m_3)\frac{\mathbf{R}_{3,1}}{|\mathbf{R}_{3,1}|^3}=\gamma m_2\left\{\frac{\mathbf{R}_{2,3}}{|\mathbf{R}_{2,3}|^3}+\frac{\mathbf{R}_{1,2}}{|\mathbf{R}_{1,2}|^3}\right\}. \tag{3}$$

Analysing system (3) we should note that if we sum all the above equations one to each other it would lead us to the result below:

$$\mathbf{R}''_{1,2}+\mathbf{R}''_{2,3}+\mathbf{R}''_{3,1}=0. \tag{4}$$

If we also sum all the equalities (*) one to each other, we should obtain

$$\mathbf{R}_{1,2}+\mathbf{R}_{2,3}+\mathbf{R}_{3,1}=0. \tag{**}$$

Under assumption of restricted three-body problem, we assume that the mass of small 3rd body $m_3 \ll m_1, m_2$, respectively; besides, for the case of moving of small 3rd body $m3$ as a moon around the 2nd body $m2$, let us additionally assume $|\mathbf{R}_{2,3}| \ll |\mathbf{R}_{1,2}|$.

So taking into consideration (**), we obtain from system (3) the following:

$$\mathbf{R}''_{1,2}+\gamma(m_1+m_2)\frac{\mathbf{R}_{1,2}}{|\mathbf{R}_{1,2}|^3}=0,$$

$$\mathbf{R}''_{2,3}+\gamma(m_2+m_3)\frac{\mathbf{R}_{2,3}}{|\mathbf{R}_{2,3}|^3}$$

$$=\gamma m_1\left\{\frac{\mathbf{R}_{1,2}}{|\mathbf{R}_{1,2}|^3}-\frac{(\mathbf{R}_{1,2}+\mathbf{R}_{2,3})}{|\mathbf{R}_{1,2}+\mathbf{R}_{2,3}|^3}\right\}$$

$$\mathbf{R}_{1,2}+\mathbf{R}_{2,3}+\mathbf{R}_{3,1}=0, \tag{5}$$

where the 1st equation of(5) describes the relative motion of 2 massive bodies (which are rotating around their common centre of masses on Kepler's trajectories); the 2nd describes the orbit of small 3rd body m_3 (moon) relative to the 2nd body m_2 (planet), for which we could obtain according to the trigonometric "Law of Cosines" [10]:

$$\mathbf{R}''_{2,3}+\gamma(m_2+m_3)\frac{\mathbf{R}_{2,3}}{|\mathbf{R}_{2,3}|^3}$$

$$+\frac{\gamma m_1}{|\mathbf{R}_{1,2}|^3}\left(1+3\cos\alpha\frac{|\mathbf{R}_{2,3}|}{|\mathbf{R}_{1,2}|}\right)\mathbf{R}_{2,3}$$

$$\cong -3\cos\alpha\left(\frac{\gamma m_1}{|\mathbf{R}_{1,2}|^3}\mathbf{R}_{1,2}\right)\frac{|\mathbf{R}_{2,3}|}{|\mathbf{R}_{1,2}|}, \tag{6}$$

where α is the angle between the radius-vectors $R_{2,3}$ and $R_{1,2}$.

Equation (6) could be simplified under the additional assumption above $|\mathbf{R}_{2,3}| \ll |\mathbf{R}_{1,2}|$ for restricted mutual motions of bodies m_1 and m_2 in R3BP [3] as below:

$$\mathbf{R}''_{2,3}+\left(\frac{\gamma(m_2+m_3)}{|\mathbf{R}_{2,3}|^3}+\frac{\gamma m_1}{|\mathbf{R}_{1,2}|^3}\right)\cdot\mathbf{R}_{2,3}=0. \tag{7}$$

Moreover, if we present (7) in the form below

$$\mathbf{R}''_{2,3}+\gamma(1+\xi+\eta)\cdot m_2\cdot\frac{\mathbf{R}_{2,3}}{|\mathbf{R}_{2,3}|^3}=0,$$

$$\xi=\left(\frac{m_1}{m_2}\cdot\frac{|\mathbf{R}_{2,3}|^3}{|\mathbf{R}_{1,2}|^3}\right),\ \eta=\left(\frac{m_3}{m_2}\right), \tag{8}$$

then (8) describes the relative motion of the centre of mass of 2nd giant-body $_{m2}$ (planet) and the centre of mass of 3rd body (moon) with the effective mass $(\xi \cdot m_2 + m_3)$, which are rotating around their common centre of masses on the stable Kepler's elliptic trajectories. Besides, if the dimensionless parameters ξ, $\eta \to 0$, then (8) should describe a quasi-circle motion of 3rd body (moon) around the 2nd body m_2 (planet).

THE COMPARISON OF THE MOONS IN SOLAR SYSTEM

As we can see from (8), ξ is the key parameter which determines the character of moving of the small 3rd body m_3 (the moon) relative to the 2nd body m_2 (planet). Let us compare such a parameter for all considerable known cases of orbital moving of the moons in Solar System [11] (Table 1).

DISCUSSION

As we can see from Table 1, the dimensionless key parameter ξ, which determines the character of moving of the small 3rd body m_3 (moon) relative to the 2nd body m_2 (planet), is varying for all variety of the moons of the planets (in Solar System) from the meaning 0.0004 · 10−6 (for Proteus of Neptune) to the meaning 54.46 · 10−6 (for Iapetus of Saturn), but it still remains to be negligible enough for adopting the stable moving of the effective mass ($\xi \cdot m_2 + m_3$) on quasielliptic Kepler's orbit around their common centre of masses with the 2nd body m_2.

Equation (8) and the corresponding parameter ξ play a key role in this paper. As for the physical meaning of (8), it describes the relative motion of the centre of mass of 2nd giant-body m_2 (planet) and the centre of mass of 3rd body (moon) with the effective mass($\xi \cdot m_2 + m_3$), which are rotating around their common centre of masses on the stable Kepler's elliptic trajectories. In case the dimensionless parameters $\xi, \eta \to 0$ then (8) should describe a quasi-circle motion of 3rd body (moon) around the 2nd body m_2 (planet).

For example, (8) refers to the classical two-body problem if ξ is a constant; nevertheless, ξ is fluctuating with time during orbital motion in R3BP, and hence (8) actually describes a perturbed two-body problem and its solution is nonconstant elliptic instead of fixed elliptic. As for physical explanation on the effective mass($\xi \cdot m_2 + m_3$), it seems that $\xi \cdot m_2$ could be also considered as the secular part of the third-body perturbation.

As for the connection (similarities, differences, etc.) between equation of relative motion (8) and the classical perturbed two-body problem (with the main perturbation being third-body gravity), they are roughly equivalent, but the proposed ansatz is obviously an alternative approach, which could be more effective for the investigations of mutual relative motion and stability of the moons orbits in Solar System. If the total sum of dimensionless parameters ($\xi + \eta$) is negligible then (8) should describe a stable quasi-circle orbit of 3rd body (moon) around the 2nd body m_2 (planet). Let us consider the proper examples which deviate (differ) to some extent from the negligibility case $(\xi + \eta) \to 0$ above (Table 1) [11]:

(1) Nereid-Neptune:

$$(\xi+\eta) = (\mathbf{35.81} + 0{,}29)\cdot 10^{-6}, \quad \text{eccentricity } \mathbf{0.7507}. \tag{9}$$

(2) Triton-Neptune:

$$(\xi+\eta) = (0.01 + 210)\cdot 10^{-6}, \quad \text{eccentricity } 0.000016. \tag{10}$$

(3) Iapetus-Saturn:

$$(\xi+\eta) = (\mathbf{54.46} + 3.4)\cdot 10^{-6}, \quad \text{eccentricity } 0.0286. \tag{11}$$

(4) Titan-Saturn:

$$(\xi+\eta) = (2.2 + 240)\cdot 10^{-6}, \quad \text{eccentricity } 0.0288. \tag{12}$$

(5) Io-Jupiter:

$$(\xi+\eta) = (0.168 + 47)\cdot 10^{-6}, \quad \text{eccentricity } 0.0041. \tag{13}$$

(6) Callisto-Jupiter:

$$(\xi+\eta) = (14.89 + 58)\cdot 10^{-6}, \quad \text{eccentricity } 0.0074. \tag{14}$$

(7) Ganymede-Jupiter

$$(\xi+\eta) = (2.73 + 79)\cdot 10^{-6}, \quad \text{eccentricity } 0.0013 \tag{15}$$

(8) Phobos-Mars:

$$(\xi+\eta) = (0{,}217 + 0{,}02)\cdot 10^{-6}, \quad \text{eccentricity } \mathbf{0.0151}. \tag{16}$$

(9) Moon-Earth:

$$(\xi+\eta) = (5{,}532 + 12{,}300)\cdot 10^{-6}, \quad \text{eccentricity } 0.0549 \tag{17}$$

The obvious extreme exception is the Nereid (moon of Neptune) from this scheme; Nereid orbits Neptune in the prograde direction at an average distance of 5,513,400 km, but its high eccentricity of 0.7507 takes it as close as 1,372,000 km and as far as 9,655,000 km [11].

The unusual orbit suggests that it may be either a captured asteroid or Kuiper belt object or that it was an inner moon in the past and was perturbed during the capture of Neptune's largest moon Triton [11]. One could suppose that the orbit of Nereid should be derived preferably from the assumptions of R4BP (the case of restricted four-body problem) or more complicated cases.

Table 1: Comparison of the averaged parameters of the moons in Solar System

Masses of the planets (Solar System), kg	Ratio m_1 (Sun) to mass m_2 (planet)	Distance $\lvert \mathbf{R}_{1,2} \rvert$ (between Sun and planet), AU	Parameter η, ratio m_3 (moon) to mass m_2 (planet)	Distance $\lvert \mathbf{R}_{2,3} \rvert$ (between moon and planet) in 10^3 km	Parameter $\xi = ((m_1/m_2)\cdot(\lvert\mathbf{R}_{2,3}\rvert^3/\lvert\mathbf{R}_{1,2}\rvert^3))$
Mercury, $3.3 \cdot 10^{23}$	(332,946/0.055)	0.387 AU			
Venus, $4.87 \cdot 10^{24}$	(332,946/0.815)	0.723 AU			
Earth, $5.97 \cdot 10^{24}$	1 Earth = 332,946 kg	1 AU = 149,500,000 km	$\mathbf{12,300 \cdot 10^{-6}}$	383.4	Moon: $\mathbf{5,532 \cdot 10^{-6}}$
Mars, $6.42 \cdot 10^{23}$	(332,946/0.107)	1.524 AU	(1) Phobos: $0.02 \cdot 10^{-6}$ (2) Deimos: $0.003 \cdot 10^{-6}$	(1) Phobos: 9.38 (2) Deimos: 23.46	(1) Phobos: $0.22 \cdot 10^{-6}$ (2) Deimos: $3.4 \cdot 10^{-6}$
Jupiter, $1.9 \cdot 10^{27}$	(332,946/317.8)	5.2 AU	(1) Ganymede: $\mathbf{79 \cdot 10^{-6}}$ (2) Callisto: $\mathbf{58 \cdot 10^{-6}}$ (3) Io: $\mathbf{47 \cdot 10^{-6}}$ (4) Europa: $25 \cdot 10^{-6}$	(1) Ganymede: **1,070** (2) Callisto: **1,883** (3) Io: 422 (4) Europa: 671	(1) Ganymede: $2.73 \cdot 10^{-6}$ (2) Callisto: $14.89 \cdot 10^{-6}$ (3) Io: $0.17 \cdot 10^{-6}$ (4) Europa: $0.67 \cdot 10^{-6}$
Saturn, $5.69 \cdot 10^{26}$	(332,946/95.16)	9.54 AU	(1) Titan: $\mathbf{240 \cdot 10^{-6}}$ (2) Rhea: $4.1 \cdot 10^{-6}$ (3) Iapetus: $3.4 \cdot 10^{-6}$ (4) Dione: $1.9 \cdot 10^{-6}$ (5) Tethys: $1.09 \cdot 10^{-6}$ (6) Enceladus: $0.19 \cdot 10^{-6}$ (7) Mimas: $0.07 \cdot 10^{-6}$	(1) Titan: **1,222** (2) Rhea: 527 (3) Iapetus: **3,561** (4) Dione: 377 (5) Tethys: 294.6 (6) Enceladus: 238 (7) Mimas: 185.4	(1) Titan: $2.2 \cdot 10^{-6}$ (2) Rhea: $0.18 \cdot 10^{-6}$ (3) Iapetus: $\mathbf{54.46 \cdot 10^{-6}}$ (4) Dione: $0.07 \cdot 10^{-6}$ (5) Tethys: $0.03 \cdot 10^{-6}$ (6) Enceladus: $0.016 \cdot 10^{-6}$ (7) Mimas: $0.008 \cdot 10^{-6}$
Uranus, $8.69 \cdot 10^{25}$	(332,946/14.37)	19.19 AU	(1) Titania: $40 \cdot 10^{-6}$ (2) Oberon: $35 \cdot 10^{-6}$ (3) Ariel: $16 \cdot 10^{-6}$ (4) Umbriel: $13.49 \cdot 10^{-6}$ (5) Miranda: $0.75 \cdot 10^{-6}$	(1) Titania: 436 (2) Oberon: 584 (3) Ariel: 191 (4) Umbriel: 266.3 (5) Miranda: 129.4	(1) Titania: $0.08 \cdot 10^{-6}$ (2) Oberon: $0.2 \cdot 10^{-6}$ (3) Ariel: $0.01 \cdot 10^{-6}$ (4) Umbriel: $0.019 \cdot 10^{-6}$ (5) Miranda: $0.002 \cdot 10^{-6}$
Neptune, $1.02 \cdot 10^{26}$	(332,946/17.15)	30.07 AU	(1) Triton: $\mathbf{210 \cdot 10^{-6}}$ (2) Proteus: $0.48 \cdot 10^{-6}$ (3) Nereid: $0.29 \cdot 10^{-6}$	(1) Triton: 355 (2) Proteus: 118 (3) Nereid: **5,513**	(1) Triton: $0.01 \cdot 10^{-6}$ (2) Proteus: $0.0004 \cdot 10^{-6}$ (3) Nereid: $35.81 \cdot 10^{-6}$

As we can see from consideration above, in case of Earth's Moon, such dimensionless key parameters increase simultaneously to the crucial meanings $\xi = 0.0055$ and $\eta = 0.0123$, respectively, $(\xi + \eta) = 0.0178$. It means that the orbit for relative motion of the moon in regard to the Earth could not be considered as quasi-elliptic orbit and should be considered as nonconstant elliptic orbit with the effective mass $(\xi \cdot m_2 + m_3)$ placed in the centre of mass for the moon.

As we know [3, 4], the elements of that elliptic orbit depend on the position of the common centre of masses for 3rd small body (moon) and the planet (Earth). But such a position of their common centre of mass should obviously differ for the real mass $m3$ and the effective mass $(\xi \cdot m2 + m3)$ placed in the centre of mass of the 3rd body (moon). So, the elliptic orbit of motion of the moon derived from the assumptions of R3BP should differ from the elliptic orbit which could be obtained from the assumptions of R2BP (the case of restricted two-body problem: it means mutual moving of 2 gravitating masses without the influence of other central forces).

As for the meanings of the terms quasi-elliptic, quasicircle, and nonconstant elliptic, "quasi" means that the main orbital elements of the orbit of moon around the planet still remain approximately the same without essential alterations (due to negligible influence of moon's gravity in a frame of the R3BP), but the term "nonconstant elliptic" means that (8) describes actually a perturbed two-body problem and its solution is nonconstant elliptic instead of fixed elliptic.

REMARKS ABOUT THE ECCENTRICITIES OF THE ORBITS

According to the definition [11], the orbital eccentricity of an astronomical object is a parameter that determines the amount by which its orbit around another body deviates from a perfect orbit:

$$e = \sqrt{1 + \frac{2\varepsilon h^2}{\mu^2}}, \tag{18}$$

where ε is the specific orbital energy, h is the specific angular momentum, μ is the sum of the standard gravitational parameters of the bodies, and $\mu = \gamma \cdot m_2 \cdot (1 + \xi + \eta)$; see (8).

The specific orbital energy equals the constant sum of kinetic and potential energy in a 2-body ballistic trajectory [11]:

$$\varepsilon = -\frac{\mu}{2a} = \text{const}; \tag{19}$$

here a is the semimajor axis. For an elliptic orbit, the specific orbital energy is the negative of the additional energy required to accelerate a mass of one kilogram to escape velocity (parabolic orbit).

Thus, assuming $\xi = (t)$, we should obtain the following from the equality above:

$$-\frac{\gamma \cdot m_2 \left(1 + \xi(t) + \eta\right)}{2a(t)} = \text{const},$$

$$\Longrightarrow a(t) = a_0 \cdot \left(1 + \xi(t) + \eta\right), \tag{20}$$

where $\xi(t)$ is the periodic function depending on timeparameter t, which is slowly varying during all the timeperiods from the minimal meaning $\xi > 0$ to the maximal meaning ξ_{max}, preferably $(\xi_{max} - \xi_{min}) \rightarrow 0$.

Besides, we should note that, in an elliptical orbit, the specific angular momentum h is twice the area per unit time swept out by a chord of ellipse (i.e., the area which is totally covered by a chord of ellipse during its motion per unit time, multiplied by 2) from the primary to the secondary body [11], according to Kepler's 2nd law of planetary motion.

Since the area of the entire orbital ellipse is totally swept out in one orbital period, the specific angular momentum h is equal to twice the area of the ellipse divided by the orbital period, as represented by

$$h = b\sqrt{\frac{\gamma(1 + \xi + \eta) \cdot m_2}{a}}, \tag{21}$$

where b is the semiminor axis. So, from (20) we should obtain that, for the constant specific angular momentum h, the semiminor axis b should be constant also.

Thus, we could express the components of elliptic orbit as follows:

$$x(t) = a_0 \cdot \left(1 + \xi(t) + \eta\right) \cdot \cos t,$$
$$y(t) = b \cdot \sin t, \tag{22}$$

which could be schematically imagined as it is shown in Figures 1(a), 1(b), and 1(c).

As for the chosen parameters in Figures 1(a), 1(b), and 1(c), meanings of the parameter $a_0 \cdot \eta$ are varying in the range from 0.0123 (Figure 1(a)) to 10.123 (Figure 1(b)) and 49.5 (Figure 1(c)); parameter $a_0 \cdot \xi(t)$ is varying in the range from $0.0055 \cdot (0.9 + 0.1 \cdot \sin t)$(Figure 1(a), b=1) to $0.55 \cdot (0.9 + 0.1 \cdot \sin t)$ (Figure 1(b), $b = 10$) and $100 \cdot (1.0 + 0.5 \cdot \sin t)$ (Figure 1(c), b= 150). Obviously, we can see that the orbit of moon at the end of Figures 1(a), 1(b), and 1(c) quite differs from the elliptic one.

CONCLUSION

We have considered the equations of motion of three-body problem in a Lagrange form (for the relative motions of 3 bodies in regard to each other). Analyzing such a system of equations, we explore the case of moon's motion of negligible mass $m3$ around the 2nd of two giant-bodies m_1, m_2, the mass of which is assumed to be less than the mass of central body m_2. Besides, only the natural satellites which are massive enough to have achieved hydrostatic equilibrium have been considered. Twenty-two of such midsized natural satellites for planets of Solar System, including Earth's Moon, are known; see Table 1

The elegant derivation of a key parameter ξ that determines the character of the moving of the moon relative to the planet has been proposed. We also obtain that the equations of motion R3BP should describe the relative mutual motion of the centre of mass of 2nd giant-body$m2$ (planet) and the centre of mass of 3rd body (moon) with additional effective mass $\xi \cdot m_2$ placed in that centre of mass ($\xi \cdot m_2 + m_3$), where ξ is the dimensionless dynamical parameter (nonconstant, but negligible). Thus, they should be rotating around their common centre of masses on Kepler's elliptic orbits. So, the case R3BP of "3-body problem" for the moon's orbit was elegantly reduced to the case R2BP of "2-body problem" (the last one is known to be stable for the relative motion of "planet-satellite" pairs [3, 4]).

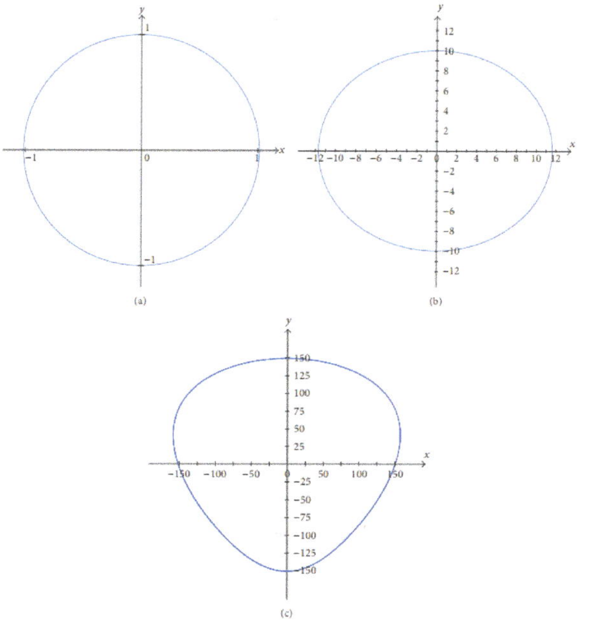

Figure 1: Orbits of the moon, schematically imagined.

For negligible effective mass ($\xi \cdot m_2 + m_3$) it gives equations of motion which should describe a quasi-elliptic orbit of 3rd body (moon) around the 2nd body m_2 (planet) for most of the moons of the planets in Solar System. But the orbit of Earth's Moon should be considered as nonconstant elliptic motion for the effective mass $0.0178 \cdot m_2$ placed in the centre of mass for the 3rd body (moon). The position of their common centre of masses should obviously differ for the real mass $m_3 = 0.0123 \cdot m_2$ and for the effective mass $(0.0055 + 0.0123) \cdot m_2$ placed in the centre of mass of the moon.

ACKNOWLEDGMENT

The author devotes this paper to his wife for her heart love which is the only source of his scientific spirit of creation.

REFERENCES

1. V. I. Arnold, Mathematical Methods of Classical Mechanics, Springer, New York, NY, USA, 1978.
2. J. Lagrange, OEuvres, vol. 6, Gautier-Villars, Paris, France, 1873, edited by M. J. A. Serret.
3. V. Szebehely, Theory of Orbits. The Restricted Problem of Three Bodies, Yale University, New Haven, Conn, USA; Academic Press, London, UK, 1967.
4. G. N. Duboshin, "Nebesnaja mehanika. Osnovnye zadachi i metody," in Handbook for Celestial Mechanics, Nauka, Moscow, Russia, 1968, (Russian).
5. H. Bruns, "Über die integrale des Vielkörper-problems," Acta Mathematica, vol. 11, no. 1–4, pp. 25–96, 1887.
6. Shankaran, J. P. Sharma, and B. Ishwar, "Equilibrium points in the generalised photogravitational non-planar restricted three body problem," International Journal of Engineering, Science and Technology, vol. 3, no. 2, pp. 63–67, 2011.
7. Y. A. Chernikov, "The photogravitational restricted three-body problem," Soviet Astronomy, vol. 14, p. 176, 1970.
8. J. Singh and O. Leke, "Stability of the photogravitational restricted three-body problem with variable masses," Astrophysics and Space Science, vol. 326, no. 2, pp. 305–314, 2010. · ·
9. S. V. Ershkov, "The Yarkovsky effect in generalized photogravitational 3-body problem," Planetary and Space Science, vol. 73, no. 1, pp. 221–223, 2012.
10. E. Kamke, Hand-Book for ODE, Science, Moscow, Russia, 1971.
11. S. S. Sheppard, The Giant Planet Satellite and Moon Page, Departament of Terrestrial Magnetism at Carniege Institution for Science, 2015, http://en.wikipedia.org/wiki/Solar_System.

CHAPTER 9

KEPLER ACCURACY MODEL FOR CO-PERIODIC SATELLITE SEPARATION EXTREMA

Robert D. Furber

Keywords: Proper relative extrema, Co-periodic orbits, Satellite relative motion, Configuration manifold, Estimated condition number

INTRODUCTION

A requirement to determine extrema in the separation between satellites and times of their occurrence arises in many contexts, from the avoidance of collisions to the establishment of communications or other types of cross-links. An important subset are cases of co-periodic orbits. For purely Kepler motion, when only a single epoch of input data is available, errors in the predicted extrema due to real world perturbations will grow with time from the epoch to the predicted times of the extrema. However, large numbers of satellites are in systems under centralized command and control. Two examples are the Global Positioning System (GPS) of thirty satellites in 12 h orbits, and the Tracking and Data Relay Satellite System (TDRSS) of geosynchronous satellites, some of which may be collocated. A specialized treatment of any system must take advantage of available resources, which may include tracking data collection hardware, orbit determination methodology, operational simulation software to generate filtered high fidelity estimates of future orbital states, station keeping strategies, and Monte Carlo representations of data collection errors. State variables estimated by numerical integration of state transition matrices are variance/covariance uncertainty measures, and several forms of six parameters describing the best estimate (having maximum likelihood) of the position and velocity, or their equivalents, of the center of mass of each satellite at any time. The simple Kepler model employed here can be added to the operational software (OS) of these systems. To limit the growth of errors in the predicted extrema due to perturbations, the OS states of motion would determine algorithm inputs for an early epoch, and the algorithm's initial estimates of the times of the extrema during the next period would be used by the OS to simulate states of motion in neighborhoods of those times. An iterative process, again using this algorithm, would lead to predictions of extrema and of the conditioning of the algorithm for the case considered, reflecting the fidelity of the OS.

In contrast the problem of non co-periodic two-satellite relative motion is of interest in studies of the increasing risk of collisions among debris and functional satellite assets (Klinkrad 2006). Several examples will illustrate the powerful methods that have been applied to this more general problem.

A two dimensional invariant manifold, embedded in three dimensional configuration space, to which any solution is confined for the general problem of two arbitrary Kepler elliptical orbits is determined by Gurfil and Kholshevnikov (2006). The manifold is invariant because any solution starting on it will remain and evolve on it. Five examples are explicated. The first illustrates the construction of a relative motion manifold, while others determine separation extrema. The square of the separation function is expressed in terms of the leader satellites's eccentric anomaly at which a relative extreme separation is reached. These anom- alies are determined from the roots of a trigonometric polynomial of degree 8. In each of these examples degenerate features allow reduction of the order of the polynomial, and solutions are straightforward. For general cases in which the full 8 degree polynomial must be solved the reader is referred to Vassiliev and Kholshevnikov (1999).

Citation: Robert D. Furber, Kepler accuracy model for co-periodic satellite separation extrema, DOI: 10.1007/s10569-014-9534-y.

A trigonometric polynomial representation of the square of the separation function, of order no less than 8 in the angular variable determining position on one ellipse, is derived by Vassiliev and Kholshevnikov (1999). The roots of this polynomial are the critical points (angular variable values). Reference is made to an earlier article by one of them (Vassiliev) in which a polynomial with this property was found of degree 16. Imposing the critical condition on the separation function is shown to be equivalent to the vanishing of three trigonometric polynomials in the sine and cosine of the critical angular variable. These polynomials are identified as elements of a ring of polynomials with real coefficients depending on the orbital elements. Using the MAPLE system of computer algebra a Gröbner basis is constructed for the ideal consisting of the polynomials of the ring which vanish. Of the 21 polynomials comprising this basis only one is independent of the follower's angular variable, which proves that it coincides with the candidate trigonometric polynomial. A property of the Gröbner basis shows that its leading term is of eighth order. This establishes a range for the number of real roots from 4 to 16. The authors experimented with different orbits, were not able to find a case in which the number of real roots is as high as 16, and assert that the upper bound is still to be determined. Some degenerate cases were considered and proved consistent with simpler interpretations admitted by the degeneracy. The authors claim that unlike traditional iterative schemes this method is insensitive to the occurrence of close multiple roots.

Maruskin (2010) derives a closed form for the distance function in a topology in which two spheres constitute the configuration manifold for the space of bounded orbits of constant energy. The topology is that of a cone passing through the direct product of the two spheres, and results in a five-dimensional manifold embedded in R6 . Analytic steps lead to an expres- sion for the separation function in terms of orbital elements. See also Kholshevnikov (2008) for development of the topology and properties of another abstract space representation of the general solution.

The methods of Gurfil and Kholshevnikov (2006), Vassiliev and Kholshevnikov (1999), Maruskin (2010) and Kholshevnikov (2008) are algebraically determinate, and involve no explicit dependence on iterative or sampling processes. A treatment of the general case of two arbitrary closed or open Keplerian orbits is that of Armellin et al. (2010). A global optimizer (COSY-GO) based on Taylor model methods results in a "rigorous enclosure of the global minimum of the objective function[s]", and is applied effectively to the Potentially Hazardous Asteroid (PHA) Apophis. A pair of nonlinear equations for the critical point anomalies is derived, and orbital parameters are treated as interval sets to allow consideration of orbit determination uncertainties and osculation of the elements. Future work planned by the authors will increase the rigor of the analysis by taking account of the interval part of the Taylor model, a further challenge, given the necessity of implementation in floating point arithmetic, which does not have uniform intervals. These authors provide a generous set of references on this subject.

This study is conventional, involves a sampling strategy to locate initial estimates of the critical points, and the Newton–Raphson method to converge on final estimates. Analytical forms for the first and second derivatives of the squared separation are required. The literature relating to this approach is too vast, long-established, and well-known to readers of this journal to describe.

The objective is a methodology to determine the number of digits justified for any co- periodic case. This generality bespeaks a treatment to determine the conditioning of the algorithm rather than a detailed treatment of an actual satellite system's uncertainties. The condition number estimate is a measure of the sensitivity of the solution to the data upon which the predictions depend. For example, an estimated condition number on the order of 109 would reduce by nine the number of significant digits of the solution compared to the maximum number expected in the absence of ill-conditioning. The effect of perturbations on the condition numbers is beyond the scope of this article. The Kepler motion model leads to ill-conditioning in those cases for which some Kepler elements become indeterminate, i.e., those in which the inclination and/ or eccentricity vanish. A possible similar algorithm based on equinoctial elements (Battin 1999) (p. 490) should achieve better conditioning in some of those cases, and may be the subject of a future study to complement this one.

Convenience will be served by reference to proper local extrema: "A function $f : \mathbf{R} \rightarrow \mathbf{R}$ is said to have a proper (or strict) local maximum at p in **R** provided there exists $E > 0$ such that if $0 <| x - p |< E$, then $f(x) < f(p)$** (Posey and Vaughan 1986). This definition differs from that of local extrema for which $f(x) \leq f(p)$. It is a functional definition because in a series of cases, as one of constant separation is approached the number of proper relative extrema vanishes instead of becoming infinite by definition. In what follows "proper relative extrema" often will be shortened to "extrema".

Section 2 summarizes Kepler solution formulae for the one fixed center problem, in which the mass of a satellite is negligible compared to a point mass at the origin of the inertially non- rotating reference frame (Battin 1999) (p. 109, footnote). In Sect. 3 an equation of constraint between two parameters is derived. Either of these may be considered among the eight defining a case and is included to facilitate applications to geosynchronous orbits. Section 4 derives formulae for the square of the separation function and its first two derivatives with respect to the time-like mean argument of latitude of satellite number 1. Section 5 expresses the form taken by the Newton–Raphson method. In Sect. 6 a baseline is established, the condition number estimation method is expressed, and test results are reported. Conclusions are in Sect. 7. Appendix A describes two arbitrarily chosen cases. Electronic Supplementary Material consisting of the following are linked to the online version of this article: The first file is a "pseudo-code", intended to show the architecture of the coded algorithm. The other files contain input files, the gfortran code source file, and an output file all suitable to be downloaded and merged with user codes.

Kepler's laws

Since the motion of a single satellite in the fixed center problem requires 6 classical orbital elements (or their equivalents), the relative motion of two satellites requires 12. Imposing the co-periodic condition reduces this to 10, first because the two semi major axes become one, and second, because the difference between the mean anomalies must be constant at all times, implying that only the single difference between the mean anomalies at the epoch is required. Furthermore, only the magnitude of the separation vector is needed. Since this is invariant under rotations of the reference frame, only the difference between the right ascensions of the nodes is required, which reduces the number of parameters to 9. While this is true when the separation distance is expressed in conventional units, the semi-major axis is a scaling factor, and later will be chosen as the unit of length. By dividing out this scaling factor and understanding that it is the unit of length, the algorithm will depend on eight independent parameters. Seven are straightforward orbital elements or differences between them. The eighth can be either of two parameters, subject to a constraint.

This algorithm will be tested by approaching one case known to cause certain classical elements to become indeterminate: that of co-orbital circular motion. This is a region of the domain of the algorithm' s inputs for which ill-conditioning should occur. Here no greater effort is made to characterize the behavior of this condition number estimator over the eight dimensional domain of inputs.

Because the motion is periodic, the number of proper relative extrema in one period will always be even, and in the form of minimum/maximum pairs.

Conventional symbols for the orbital elements are used, except that the symbol for the semi major axis will be r_c instead of a. The symbols for the right ascension of the ascend- ing node, inclination, argument of perigee, eccentricity and true anomaly are, respectively, Ω_j, i_j, ω_j, e_j, and ν_j, where the j (= 1, 2) index distinguishes satellites 1 and 2. The meanargument of latitude is the sum of the mean anomaly and the argument of perigee. For exam- ple, the mean argument of latitude of satellite 2 is $M_2 + \omega_2$, where M_2 is the mean anomaly of satellite 2. Summarizing: the derivation identifies seven of the independent parameters as:the difference between the right ascensions of the ascending nodes, $\delta\Omega = \Omega_2 - \Omega_1$, the twoinclinations, the two arguments of perigee, and the two eccentricities. The eighth parameteris either the difference between the mean arguments of latitude, called δ, or the difference between the mean equator crossing longitudes, which will be called $\delta\xi_0$.

The separation function and its derivatives are expressed in terms of well-known explicit functions of the true anomalies and arguments of latitude, $u_j = \nu_j + \omega_j$. Since the method requires the derivatives of the separation function with respect to time, and since the true anomalies and arguments of latitude are related to the time-like mean anomalies by Kepler's transcendental equation, this equation, the method for its solution and other necessary clas- sical element results will be stated next in terms of these symbols.

Kepler's second law expresses the orbit as an ellipse with one focus at the center of attraction, and polar coordinates r and ν. In this section the j subscripts will not be needed since only one satellite is involved. However, subscripts will return temporarily to indicate the order of corrections for the iterative process.

Standard texts such as (Moulton 1956) derive the following equation relating the true anomaly to E, the eccentric anomaly:

$$\tan \nu/2 = \sqrt{(1+e)/(1-e)}\tan E/2 \tag{1}$$

This determines the true anomaly once the eccentric anomaly is known. Given a time and corresponding mean anomaly, Kepler's equation,

$$M = E - e \sin E \tag{2}$$

allows the calculation of E. Of the many methods available for solving Kepler's equation, that of Lagrange (Battin 1999) (p. 202, equation (5.17)) is used to find an initial approximation to E in terms of M, which includes terms to the seventh power in the eccentricity. While for eccentricities greater than 0.6624.., the Lagrange series is known not to converge for some values of M, this is not a problem here since only an initial estimate is needed, only the first seven terms are retained, and the following iterative procedure will provide convergence.

To this first approximation to E, E0, there corresponds a value of M, called M0, which is determined by Kepler's equation as M0 = E0 − e sin E0. The remaining steps are describedin reference (Moulton 1956) (p. 162), where a Taylor series expansion of M about M0 in powers of E − E0 leads to the iterative procedure for improving the estimate of E until the correction satisfies a smallness tolerance criterion called τk at which the iterations stop.[1]

Two final standard results required in what follows in Sect. 4 are the second integral of the motion expressing the conservation of (specific) orbital angular momentum as

$$r^2 d\nu/dt = \sqrt{r_c \mu (1-e^2)}, \tag{3}$$

and Kepler's harmonic law expressing the mean orbital angular speed as

$$dM/dt = \sqrt{\mu/r_c^3}, \tag{4}$$

where μ is the product of the mass at the center of attraction and the universal gravitational constant.

EQUATION OF CONSTRAINT

An equation of constraint exists between the difference of the mean equator crossing longi- tudes and the phase angle, or difference of the mean arguments of latitude. It can be derived as follows. Let the Earth be the center of attraction and consider a right handed Earth centered reference frame called the C-frame, which rotates with respect to inertial space at the mean orbital rate about the Earth's polar axis. The ZC axis is toward the north celestial pole, and, at the epoch, the XC axis is toward the zero longitude point on the equator. For geosynchronous orbits this frame is identified with the conventional Earth-fixed frame in which the XC axis is toward the zero longitude point on the equator at all times. Let H be the angle between the XC axis and the direction of the vernal equinox. For geosynchronous orbits H is the Greenwich hour angle of the vernal equinox. Let the longitude-like coordinate be called ξ. For geosynchronous orbits ξ is the longitude. Let $\xi_0 j$ represent the mean equator crossing ξ coordinate of the jth satellite. That is, if $\xi_{\uparrow j}$ and $\xi_{\downarrow j}$ are, respectively, the ξ coordinates of satellite j when it crosses the equatorial plane from south to north and from north to south during one orbital period, then

$$\xi_{0j} = 1/2(\xi_{\uparrow_j} + \xi_{\downarrow_j}) \tag{5}$$

Given the definition of H, the ξ coordinates of the ascending and descending nodes can be written

$\xi ANj(t) = \Omega j - H(t)$,

$\xi DNj(t) = \pi + \Omega j - H(t)$. (6)

Let the time dependence of H and M_1 be expressed as $H(t) = H(0) + \dot{u}'t$ and $M_1(t) \quad M_1(0)+\dot{u}'t$, where $\dot{u}'$ where u˙ represents both the common mean orbital rate of the satellites (identified by Eq. 4 as $\sqrt{\mu/r_c^3}$), and the rate at which the C-frame rotates about the north polar axis. Solving the second equation for u˙ t and substituting into the first yield H(t) = H(0) + M1(t) − M1(0) which allows Eqs. 6 to become ξAN $H(t) = H(0) + M_1(t) - M_1(0)$

$$\xi_{ANj}(t) = \Omega_j - H(0) + M_1(0) - M_1(t),$$
$$\xi_{DNj}(t) = \pi + \Omega_j - H(0) + M_1(0) - M_1(t). \quad (7)$$

Representing the times of ascending and descending nodal passages by $t_{\uparrow j}$ and $t_{\downarrow j}$, Eqs. 7 imply that

$$\xi_{\uparrow_j} = \xi_{ANj}(t_{\uparrow_j}) = \Omega_j - H(0) + M_1(0) - M_1(t_{\uparrow_j}),$$
$$\xi_{\downarrow_j} = \xi_{DNj}(t_{\downarrow_j}) = \pi + \Omega_j - H(0) + M_1(0) - M_1(t_{\downarrow_j}) \quad (8)$$

By the definition of δ, $M_1(t_{\uparrow 2})$ and $M_1(t_{\downarrow 2})$ in Eq. 8 can be replaced by $M_2(t_{\uparrow_2})+\omega_2-\omega_1-\delta$ and $M_2(t_{\downarrow_2}) + \omega_2 - \omega_1 - \delta$, respectively, resulti

$$\xi_{\uparrow_2} = \Omega_2 - H(0) + M_1(0) - M_2(t_{\uparrow_2}) - \omega_2 + \omega_1 + \delta,$$
$$\xi_{\downarrow_2} = \pi + \Omega_2 - H(0) + M_1(0) - M_2(t_{\downarrow_2}) - \omega_2 + \omega_1 + \delta. \quad (9)$$

At $t_{\uparrow j}$ or $t_{\downarrow j}$ the arguments of latitude become 0 or π, and the true anomalies $-\omega_j$ or $\pi - \omega_j$. Equation 1 yields the corresponding eccentric anomalies, and Eq. 2 the mean anomalies on the right of the j = 1 case of Eqs. 8 and on the right of Eq. 9. Substituting from the j = 1 case of Eqs. 8 and from Eqs. 9 into Eq. 5 yields:

$$\xi_{01} = \pi/2 + \Omega_1 - H(0) + M_1(0) - [M_1(t_{\uparrow 1}) + M_1(t_{\downarrow 1})]/2.$$
$$\xi_{02} = \pi/2 + \Omega_2 - H(0) + M_1(0) - [M_2(t_{\uparrow 2}) + M_2(t_{\downarrow 2})]/2 - \omega_2 + \omega_1 + \delta.$$

Since $\delta\xi_0 = \xi_{02} - \xi_{01}$, the last two equations lead to the constraint:

$$\delta\xi_0 = \delta + \delta\Omega - \omega_2 + \omega_1 - [M_2(t_{\uparrow 2}) + M_2(t_{\downarrow 2}) - M_1(t_{\uparrow 1}) - M_1(t_{\downarrow 1})]/2. \quad (10)$$

For geosynchronous motion $\delta\xi_0$ can be chosen as the eighth given parameter, and Eq. 10 can be solved for δ. This constraint is not meaningful for vanishing eccentricities and/or inclinations.

Each satellite's equatorial plane crossings' ξ coordinates (longitudes) are required to be within that satellite's assigned "longitude box". Substituting from the j=1 case of Eq. 8 and from Eq. 9 the differences are c $\delta\xi_j = \xi_{\downarrow j} - \xi_{\uparrow j} = \pi - M_j(t_{\downarrow j}) + M_j(t_{\uparrow j})$ To second order in the eccentricities, this becomes $\delta\xi_j = 4e_j \sin\omega_j$.

SEPARATION FUNCTION AND DERIVATIVES

Let ρ be the separation distance between satellites 1 and 2. Then,

$$(\rho/r_c)^2 = (\mathbf{r}_2 - \mathbf{r}_1)\cdot(\mathbf{r}_2 - \mathbf{r}_1)/r_c^2$$

where r_j are the position vectors relative to the center of attraction. Let the unit of length be r_c by redefining the symbols for the magnitudes, r_j, to represent the ratio of the magnitudes to r_c. Then the last equation can be rewritten as $\rho^2 = r_2^2 + r_1^2 - r_2 r_1 2\hat{\mathbf{r}}_2 \cdot \hat{\mathbf{r}}_1$, where the hats signify unit vectors. To simplify the notation, let

$$A = r_2^2 + r_1^2,$$
$$B = r_2 r_1,$$
$$C = 2\hat{\mathbf{r}}_2 \cdot \hat{\mathbf{r}}_1, \quad (11)$$

so that ρ^2 becomes

$$\rho^2 = A - B{\cdot}C. \quad (12)$$

Recall that the independent variable is time, and that for any time the mean anomalies are known. Therefore, as shown in Sect. 2, the true anomalies and arguments of latitude are to be considered implicit functions of time. The radial coordinates are implicit functions of time by virtue of the equation of the ellipse expressed in units of r_c as

$$r_j = (1 - e_j^2)/(1 + e_j \cos \nu_j). \tag{13}$$

Both position unit vectors are expressed in the conventional right-handed Earth-centered inertial frame, the I-frame, in which the X_I axis is toward the vernal equinox, and the Z_I axis is toward the north celestial pole. The result is (Battin 1999 Problem 3-21 page 125):

$$\hat{\mathbf{r}}_j^{(I)} = \begin{pmatrix} \cos \Omega_j \cos u_j - \cos i_j \sin \Omega_j \sin u_j \\ \sin \Omega_j \cos u_j + \cos i_j \cos \Omega_j \sin u_j \\ \sin i_j \sin u_j \end{pmatrix} \tag{14}$$

The first step is to differentiate A, B and C as defined by Eq. 11 each, twice, with respect to time. Then the Newton–Raphson procedure is used to find the extrema of ρ as those values provided by Eq. 12 at the times when the first time derivative of ρ^2 vanishes.

Since the semi-major axes are the same, the mean anomalies, M_j , have the same proportionality to time, and can be treated as equivalent stand-ins for time. Specifically, given Eq. 4 for the mean orbital rate the time differentiation operator can be written:

$$d/dt = (dM_j/dt)(d/dM_j) = \sqrt{\mu/r_c^3}\, d/dM_j, \tag{15}$$

Therefore, common factors of d M_j /dt can be divided out of any result of differentiation with respect to time. Given the redefinition of r_j stated in the first paragraph of this section an expression for du_j /d M_j follows from Eq. 3. Dividing that expression throughout by r_c^2 and recalling Eq. 4 yield the following form in length units of r_c:

$$r_j^2 d\nu_j/dM_j = r_j^2 du_j/dM_j = \sqrt{1 - e_j^2} \tag{16}$$

It will be convenient to introduce a special symbol, u', for the mean argument of latitude of satellite 1:

$$u' = M_1 + \omega_1.$$

Since δ is the difference between the mean arguments of latitude $(\delta = (M_2 + \omega_2) - u')$

$$d/dM_1 = d/dM_2 = d/du'. \tag{17}$$

The choice of u'as the time-like independent variable implies that the true anomalies are determined by solving Kepler's equation. Consider A. By the first of Eqs. 11:

$$dA/du' = 2\left[r_2 dr_2/du' + r_1 dr_1/du'\right] \tag{18}$$

By Eqs. 13 and 16 the dr_j/du' factors in the last result become

$$dr_j/du' = e_j \sin \nu_j \Big/ \sqrt{1 - e_j^2}. \tag{19}$$

Substituting into Eq. 18 yields

$$dA/du' = 2\left[r_2 e_2 \sin \nu_2 \Big/ \sqrt{1 - e_2^2} + r_1 e_1 \sin \nu_1 \Big/ \sqrt{1 - e_1^2}\right] \tag{20}$$

The result for the first derivative of B is

$$dB/du' = r_2 e_1 \sin \nu_1 \Big/ \sqrt{1 - e_1^2} + r_1 e_2 \sin \nu_2 \Big/ \sqrt{1 - e_2^2}. \tag{21}$$

Differentiation of Eq. 20 yields (again using Eqs. 13 and 16):

$$d^2A/du'^2 = 2\left(\left[\frac{e_2 \sin \nu_2}{\sqrt{1 - e_2^2}}\right]^2 + \left[\frac{e_1 \sin \nu_1}{\sqrt{1 - e_1^2}}\right]^2 + \frac{e_2 \cos \nu_2}{r_2} + \frac{e_1 \cos \nu_1}{r_1}\right) \tag{22}$$

Similar use of Eqs. 16, 17 and 19 applied to the result of differentiating Eq. 21 yields:

$$d^2B/du'^2 = \frac{r_2}{r_1^2} e_1 \cos \nu_1 + \frac{r_1}{r_2^2} e_2 \cos \nu_2 + 2\frac{e_2 e_1 \sin \nu_2 \sin \nu_1}{\sqrt{1 - e_2^2}\sqrt{1 - e_1^2}} \tag{23}$$

Using Eq. 14 in $C = 2\hat{\mathbf{r}}_1 \cdot \hat{\mathbf{r}}_2$ yields a sum of four terms each of which contains one of the following four forms as a common factor: $\sin u_2 \sin u_1, \sin u_2 \cos u_1, \cos u_2 \sin u_1, \cos u_2$. The products of sines and cosines of the nodes appear in forms like $\sin \Omega_2 \sin \Omega_1 + \cos \Omega_2 \cos \Omega_1$, and by trigonometric identities become sines and cosines of $\delta\Omega = \Omega_2 - \Omega_1$. Trigonometric identities such as $u_2 \cos u_1 = [\cos(u_2 + u_1) + \cos(u_2 - u_1)]/2$ lead to a new expression with four new terms, each involving a factor of a sine or cosine of either the difference $u_2 - u_1$, or the sum $u_2 + u_1$. With the following definitions,

$$g = -(\cos i_2 + \cos i_1)\sin\delta\Omega,$$
$$h = (1 + \cos i_2 \cos i_1)\cos\delta\Omega + \sin i_2 \sin i_1,$$
$$m = -(\cos i_2 - \cos i_1)\sin\delta\Omega,$$
$$n = (1 - \cos i_2 \cos i_1)\cos\delta\Omega - \sin i_2 \sin i_1, \tag{24}$$

the result is

$$C = g\sin(u_2 - u_1) + h\cos(u_2 - u_1) + m\sin(u_2 + u_1) + n\cos(u_2 + u_1) \tag{25}$$

Note that C is form invariant for co-periodic and non co-periodic orbits. Differentiating Eq. 25 and using Eqs. 15 and 16 yield

$$dC/du' = (g\cos(u_2 - u_1) - h\sin(u_2 - u_1))\left[\sqrt{(1-e_2^2)}/r_2^2 - \sqrt{(1-e_1^2)}/r_1^2\right]$$
$$+(m\cos(u_2 + u_1) - n\sin(u_2 + u_1))\left[\sqrt{(1-e_2^2)}/r_2^2 + \sqrt{(1-e_1^2)}/r_1^2\right] \tag{26}$$

Differentiation of the last expression and substitution from Eqs. 16 and 19 yield

$$d^2C/du'^2 = -2(g\cos(u_2 - u_1) - h\sin(u_2 - u_1))\left[e_2\sin\nu_2/r_2^3 - e_1\sin\nu_1/r_1^3\right]$$
$$-2(m\cos(u_2 + u_1) - n\sin(u_2 + u_1))\left[e_2\sin\nu_2/r_2^3 + e_1\sin\nu_1/r_1^3\right]$$
$$-(g\sin(u_2 - u_1) + h\cos(u_2 - u_1))\left[\sqrt{1-e_2^2}/r_2^2 - \sqrt{1-e_1^2}/r_1^2\right]^2$$
$$-(m\sin(u_2 + u_1) + n\cos(u_2 + u_1))\left[\sqrt{1-e_2^2}/r_2^2 + \sqrt{1-e_1^2}/r_1^2\right]^2. \tag{27}$$

THE NEWTON–RAPHSON METHOD

The nine functions derived in the last section for A, B, and C and their first and second derivatives with respect to u', (the first two of Eq. 11, and Eqs. 20, 21, 22, 23, 25, 26 and 27) will be used in applying the classical Newton–Raphson method to locating the relative extrema of ρ^2. Equation 12 implies that

$$\rho^2 = A - B{\cdot}C,$$
$$d\rho^2/du' = dA/du' - C{\cdot}dB/du' - B{\cdot}dC/du',$$
$$d^2\rho^2/du'^2 = d^2A/du'^2 - 2dB/du'{\cdot}dC/du' - d^2B/du'^2{\cdot}C - B{\cdot}d^2C/du'^2. \tag{28}$$

The Newton–Raphson method requires initial approximations to the values of u' for which $\frac{d\rho^2}{du'}$ passes through zero. As noted in Sect. 1 branch-and-bound optimizers utilizing domain reduction methods (Armellin et al. 2010) and employing the techniques of interval analysis (Moore et al. 2009) also have been used to locate these critical points. The author has chosen instead to rely upon the conventional sampling method described in this section. It is simple and fails only if the limit of the arithmetic precision of the host software is approached either while locating the neighborhood of u in which the sign of $\frac{d\rho^2}{du'}$ changes or in estimating a proper relative extreme separation. Before describing this sampling method the limiting case of near circular motion will be mentioned. In that case the above formulae imply that a solution is

$$u'_{10} = 1/2\,(\arctan(m/n) - \delta)$$

where m and n are given by Eqs. 24. This results because for circular orbits $r_1 = r_2 = 1$ and C reduces to $g\sin\delta + h\cos\delta + m\sin(2u' + \delta) + n\cos(2u' + \delta)$. Equating the derivative of the resulting form for ρ^2 to zero implies that $m\cos(2u' + \delta) = n\sin(2u' + \delta)$ and leads to this estimate of one of the critical points. Since the orbits are circular the approximations to the other three critical points differ from this approximation by π/2, π, and 3π/2 radians.

$$m\cos(2u' + \delta) = n\sin(2u' + \delta)$$

Initial approximations are found by sampling the values of $\frac{d\rho^2}{du'}$, and identifying those u' at which this rate of change changes sign. The sampling is at intervals of 1/8 degree of the true anomaly of satellite 1 and over the 360 degree range of ν_1. To each such initial approximation to the true anomaly at which the sign of $\frac{d\rho^2}{du'}$ changes there corresponds a value of u'. Let the initial approximations to u' found in this way be called u'_{k0}, where k = 1, 2, 3, ...k_{max}.

Consider a Taylor series expansion of $d\rho^2/du'$ about any one of these points

$$d\rho^2/du'\big]_{u'=u'_k} = d\rho^2/du'\big]_{u'=u'_{k0}} + d^2\rho^2/du'^2\big]_{u'=u'_{k0}} (u'_k - u'_{k0}) + ...,$$

where u'_k represents the correct location of the kth zero of $\frac{d\rho^2}{du'}$. The classical Newton–Raphson method proceeds by dropping the terms of second and higher order in $u'_k - u'_{k0}$, setting the left side to zero, letting u'_{k1} be the new estimate of u'_k on the right, and solving for u'_{k1}, which yields: u k1

$$u'_{k1} = u'_{k0} - \frac{d\rho^2}{du'} \Big/ \frac{d^2\rho^2}{du'^2}\Big]_{u'=u'_{k0}} . \tag{29}$$

Equation 28 now can be reevaluated at $u' = u'_{k1}$, and a second correction calculated by Eq. 29, where u'_{k1} and u'_{k0} are replaced, respectively, by u'_{k2} and u'_{k1}. This is repeated until the absolute value of the correction becomes smaller than the tolerance, τ_n The choice of τ_n and τ_k (see footnote 1) will be discussed in the next section.

CONDITIONING ESTIMATES

Following a description of the condition number formulation, coded equations are derived, and a baseline and strategy are developed for estimating the accuracy of solutions. Two tests demonstrate the strategy: In the first the limit of a sequence of decreasingly eccentric cases is that of co-planar circular orbits with δ = 1 degree. In a second sequence δ = 0. The second test involves highly eccentric orbits.

Adopting the conventions of Higham, (cf. pp. 462 ff) (Higham 2002) the input data vector is called d_j, j = 1, ..8 , where

$$d^T = (d_1, d_2, ..., d_8)^T = (\delta, \delta\Omega, i_1, i_2, \omega_1, \omega_2, e_1, e_2)^T . \tag{30}$$

Because the conditioning of the u'_k and ρ_k $(k = 1, .., k_{max})$ differ two solution vectors are defined as

$$f^T_{u'}(d) = (f_{u'1}(d), ...f_{u'k_{max}}(d))^T = (u'_1, ..., u'_{k_{max}})^T$$
$$f^T_{\rho}(d) = (f_{\rho 1}(d), ...f_{\rho k_{max}}(d))^T = (\rho_1, ..., \rho_{k_{max}})^T . \tag{31}$$

In the spirit of Higham functions $F_{u'}(x_{u'}; d)$ and $F_\rho(x_\rho; d)$ are defined as

$$F_{u'}(x_{u'}; d) = x_{u'} - f_{u'}(d)$$
$$F_\rho(x_\rho; d) = x_\rho - f_\rho(d), \tag{32}$$

where $F_{u'}(x_{u'}; d) = 0$ and $F_\rho(x_\rho; d) = 0$ are satisfied by $x^*_{u'} = f_{u'}(d)$ and $x^*_\rho = f_\rho(d)$ respectively. Higham concludes that the condition numbers are [his equation (25.11)]:

$$cond(F_{u'}, x^*_{u'}; d) = ||F_{u'x}(x^*_{u'}; d)^{-1} F_{u'd}(x^*_{u'}; d)||(||d||/||x^*_{u'}||)$$
$$cond(F_\rho, x^*_\rho; d) = ||F_{\rho x}(x^*_\rho; d)^{-1} F_{\rho d}(x^*_\rho; d)||(||d||/||x^*_\rho||), \tag{33}$$

where $F_{u'x}$ and $F_{\rho x}$ are Jacobians of partial derivatives of components of $F_{u'}$ and F_ρ with respect to components of $x_{u'}$ and x_ρ, and $F_{u'd}$ and $F_{\rho d}$ are Jacobians of partial derivatives of components of $F_{u'}$ and F_ρ with respect to the components of d. Therefore, by Eqs. 32 each of $F_{u'x}$ and $F_{\rho x}$ is the $k_{max} \times k_{max}$ identity matrix. Let $J^{u'}$ and J^ρ represent $F_{\rho d}(x^*_\rho; d)$ so that the last two equations become

$$cond(F_{u'}, x^*_{u'}; d) = ||J^{u'}||(||d||/||x^*_{u'}||)$$
$$cond(F_\rho, x^*_\rho; d) = ||J^\rho||(||d||/||x^*_\rho||). \tag{34}$$

Explicitly, the elements of the Jacobians are

$$J^{u'}_{ij} = \partial F_{u'i}/\partial d_j\big]_{x_{u'}=x^*_{u'}} = -\,\partial f_{u'i}(d)/\partial d_j\big]_{x_{u'}=x^*_{u'}}$$
$$J^{\rho}_{ij} = \partial F_{\rho i}/\partial d_j\big]_{x_\rho=x^*_\rho} = -\,\partial f_{\rho i}(d)/\partial d_j\big]_{x_\rho=x^*_\rho} . \tag{35}$$

The"2-norm" or Euclidean length is chosen for the norms of d, $x^*_{u'}$ and x^*_ρ. Therefore,

$$||d|| = \left[\sum_{j=1}^{8} |d_j|^2\right]^{1/2},$$

$$||x_{u'}^*|| = \left[\sum_{i=1}^{k_{max}} |x_{u'i}^*|^2\right]^{1/2},$$

$$||x_\rho^*|| = \left[\sum_{i=1}^{k_{max}} |x_{\rho i}^*|^2\right]^{1/2}. \tag{36}$$

Using Frobenius norms, i.e.:

$||J^{u'}||_F = [trace(J^{u'*}J^{u'})]^{1/2}$ and $||J^{\rho}||_F = [trace(J^{\rho*}J^{\rho})]^{1/2}$ in Eq. 34 results in c

$$cond(F_{u'}, x_{u'}^*; d) = [trace(J^{u'*}J^{u'})]^{1/2}||d||/||x_{u'}^*||,$$
$$cond(F_\rho, x_\rho^*; d) = [trace(J^{\rho*}J^{\rho})]^{1/2}||d||/||x_\rho^*||. \tag{37}$$

The matrices $J^{u'}$ and $J^{\rho}$of Eqs. 35 are determined by differencing the results of executing the algorithm with sequential offsets of each element of the data vector, d. Centered finite differencing for the partial derivatives involves the following steps: Invoke the Kronecker delta symbol and define the jth element of the data vector for positive and negative offsets of the lth element of the data vectors as

$$pd_j^l = d_j + \epsilon_d \delta_j^l$$
$$md_j^l = d_j - \epsilon_d \delta_j^l, \tag{38}$$

Let the ith element of the resulting solution vectors be

$$pf_{u'i}^l = f_{u'i}(pd^l),$$
$$mf_{u'i}^l = f_{u'i}(md^l),$$
$$pf_{\rho i}^l = f_{\rho i}(pd^l),$$
$$mf_{\rho i}^l = f_{\rho i}(md^l). \tag{39}$$

By Eqs. 35 the centered finite difference approximations to the Jacobians become (Abramowitz and Stegun 1964) (p. 883, formula 25.3.21)

$$J_{ij}^{u'} = -\partial f_{u'i}/\partial d_j = -\left(pf_{u'i}^j - mf_{u'i}^j\right)/2\epsilon_d,$$
$$J_{ij}^{\rho} = -\partial f_{\rho i}/\partial d_j = -\left(pf_{\rho i}^j - mf_{\rho i}^j\right)/2\epsilon_d. \tag{40}$$

These Jacobians have k_{max} rows and 8 columns.

The choice of d is half the square root of the Fortran numeric inquiry function called EPSILON. On an IEEE compliant iMac in double-precision that is about 7.45 × 10−9. This is a good strategy in this case because differences are required of almost equal quantities. The estimates of the condition numbers should be of the correct order of magnitude and can afford to lose many significant digits. Substitutions from Eqs. 40 for the Jacobians, and from Eqs. 36 for $||d||$, $||x_{u'}^*||$ and $||x_\rho^*||$ into Eqs. 37 yield the condition numbers of $f_{u'}$ and f_ρ. H

Higham (2002) states that "Accuracy refers to the absolute or relative error of an approximate quantity. Precision is the accuracy with which the basic arithmetic operations +,-,*,/ are performed, and for floating point arithmetic is measured by the unit roundoff u". Since the unit roundoff is half the EPSILON intrinsic function introduced in the last paragraph, about 1.1×10^{-16} on the iMac in double precision, the number of significant digits in the data may be taken as 16, and neither tolerance can be chosen smaller than 10−16. Because there is no reason to impose a smaller tolerance on the stopping of the Kepler equation solver than on the stopping of the Newton–Raphson iterator the same choice is made for τ_k and τ_n. The solutions will have fewer than sixteen significant digits because they are generated by the algorithm, and it is characterized by less than perfect conditioning. (Perfection would mean the condition number estimates would be one.) A baseline case for identifying the highest accuracy possible is necessary. It involves the important parts of the code, presents no special difficulty in converging on solutions, and is determined by the data in Table 1, which also includes the tolerances, "earth rate factor", initially sampled satellite 1 true anomaly and sampling interval. The rate factor, per f, multiplies the earth rate physical constant, $\omega_\oplus$, and determines the scale factor r_c. For example setting per f to unity results in

$$r_c = \left[\mu/(perf \cdot \omega_\oplus)^2\right]^{1/3} = 42164.174420503 \text{ km},$$

where $\omega_\oplus$ and μ are:

$$\omega_\oplus = 7.292115145999999 \times 10^{-05} \text{ rad/s},$$
$$\mu = 398600.5 \text{ km}^3/\text{s}^2.$$

These physical parameters only determine the scale factor. The accuracy estimates are based on the normalized units in which the angles and separations are respectively in radians and r_c units. In this sense the accuracy estimates do not depend on per f, $\omega_\oplus$, or μ.

Table 1: Baseline case data vector, tolerances, rate factor, sat 1 initial true anomaly, and sampling interval

$\delta = 30$ deg	$\delta\Omega = 70$ deg
$i_1 = 5$ deg	$i_2 = 25$ deg
$\omega_1 = 100$ deg	$\omega_2 = 300$ deg
$e_1 = .5$	$e_2 = .1$
$\tau_n = 10^{-14}$	$\tau_k = 10^{-14}$
$perf = 1$, $\nu_1(0) = 0$ deg, $\Delta\nu = 1/8$ deg	

Only one pair of proper relative extrema were found for the baseline ($k_{max} = 2$). A second run exchanged the roles of the satellites and, therefore, required changes of the signs of δ and δΩ. The results of the two runs were compared, and the smallest number of common most significant digits was confirmed to be 14 for both solutions. The small condition numbers were 2.13 and 7.24 respectively. Let. $N_{u'}]_{d.p.}$ and $N_\rho]_{d.p.}$ represent double precision accuracies based on condition numbers. The baseline then implies that in general since both formulae provide the 14 significant digits in the baseline case, when the logarithms are rounded to nearest integers.

$$N_{u'}]_{d.p.} = 14 - \log[cond(F_{u'}, x^*_{u'}; d)],$$
$$N_{\rho}]_{d.p.} = 15 - \log[cond(F_{\rho}, x^*_{\rho}; d)], \quad (41)$$

A second test was performed. Creating single precision codes allowed another two solutions to be found and compared by interchanging satellite labels. This showed that the smallest number of most significant digits held in common by both solutions was six. Using logic similar to that just applied to the double precision baseline results in the following formulae for single precision condition number accuracy estimates:

$$N_{u'}]_{s.p.} = 6 - \log[cond(F_{u'}, x^*_{u'}; d)]$$
$$N_{\rho}]_{s.p.} = 7 - \log[cond(F_{\rho}, x^*_{\rho}; d)], \quad (42)$$

Between the single- and double-precision results the condition numbers themselves were in agreement to 5 places for $cond^{(F_{u'}, x^*_{u'}; d)}$ and 3 for $cond^{(F_{\rho}, x^*_{\rho}; d)}$. In the tests reported later in this section the last four formulae will allow accuracy estimate comparisons for stressing cases.

The number of data vectors involved in a case totals seventeen. One of these determines the nominal solution, eight correspond to offsets by $+\epsilon_d$ to each data vector element, and eight to offsets by $-\epsilon_d$. Each calculation first finds the k_{max} initial approximations to the mean arguments of latitude u'_{k0}, in a sequence monotone increasing in the angles over the (0,360) degree range. Each code resorts the extrema found by the filter and leaves them monotone increasing in the u'_{k0} and starting with the smallest.

With these conventions established the first sequence of cases for the first stress test will be considered. Six data elements are common to all cases in the sequence and are given in Table 2, where it should be understood that for runs involving the interchange of identities the sign of δ must be reversed

In each precision and for each case, a straight run, with the larger eccentricity considered e1, is followed by a second with the identities interchanged. As in the baseline case, visual inspection of the solutions allows identification of the number of left-most digits held in common. These estimates are called $N^{vis}_{u'}$ and N^{vis}_{ρ}. The condition number-based estimates of accuracy are as defined by Eqs. 41 and 42, where the subscripts distinguish the two precision models. The subscripts will be dropped unless needed. For each $x^*_{u'}$ solution some correspondence is to be expected among the $-\log(\tau_k)$, $N_{u'}$, and $N^{vis}_{u'}$ measures of accuracy, and for each x^*_{ρ} solution some correspondence is expected between $-\log(\tau_k)$ and N_{ρ}. In this sequence the visual inspections show very high agreement in the N^{vis}_{ρ} measure.

Table 2: Vector elements held constant for the first stress test sequence 1a

$\delta = 1$ deg	$\delta\Omega = 0$ deg
$i_1 = 5$ deg	$i_2 = 5$ deg
$\omega_1 = 330$ deg	$\omega_2 = 330$ deg

The co-planar and near circular nature of the motion renders the separation extremely close to constant, and this agreement is not a meaningful measure of accuracy for these cases. In the double precision estimates as the first derivative of the square of the separation is sampled through zero, the calculated separation is unchanged to 17 significant digits over a range of u'on the order of 1/8th degree, and results from taking the square root of ρ^2, which cannot be accurate to 17 significant digits.

In both sequences of the first test the smaller of the two eccentricities is set to d . This ensures that negative eccentricities will not occur in constructing the Jacobians. Specifically, $md^7_7 = e_1 - \epsilon_d$ and $md^8_8 = e_2 - \epsilon_d$ (cf. Eqs. 38) are prevented from becoming negative. Under an interchange of satellite identities the $N_{u'}$ will not be in exact agreement because the u'_k solutions are then those of the other satellite. They are in agreement to within three significant digits. The N_ρ are in excellent agreement as expected.

Based upon the data in Table 2, Fig. 1 shows accuracy estimates for which the larger eccentricity, elarger, ranges from 5×10^{-8} to 10^{-1} for sequences 1a (left) and 1b (right). Attempts to drop below the smallest value considered for elarger caused the filter to fail. The rate became so negligible that its absolute value approached the unit roundoff, and the filter was unable to detect changes of sign in $\frac{d\rho^2}{du'}$. For $\delta = 1$ degree (left), in both precisions, Nu drops sharply at a largest eccentricity of 1.19×10^{-2}, where the number of extrema changes from 2 to 4 as the log of the largest eccentricity increases. In this same neighborhood of the abscissa the double precision N_ρ starts improving from about 12–13 and the single precision from 4 to 5 at $e_{larger} = 10^{-1}$. For $\delta = 0$ (right) in both precisions the condition number estimates (in dot-dashes for Nu and dots for Nρ) are superimposed. Also, in the double precision results the tolerance-based estimates (in solid) and the straight and interchanged identity estimates (in dashes) are consistently higher than the condition number estimates by about one significant digit, while that is not true of the single-precision results. In sequence 1b k_{max} remains unchanged at 4. I

In each sequence the coplanar decreasingly eccentric orbits' condition results in the separation of the limiting case (of smallest larger eccentricity) becoming $2r_c \sin \delta/2$. In units of r_c this is 0.017 for sequence 1a, and near vanishing for sequence 1b.

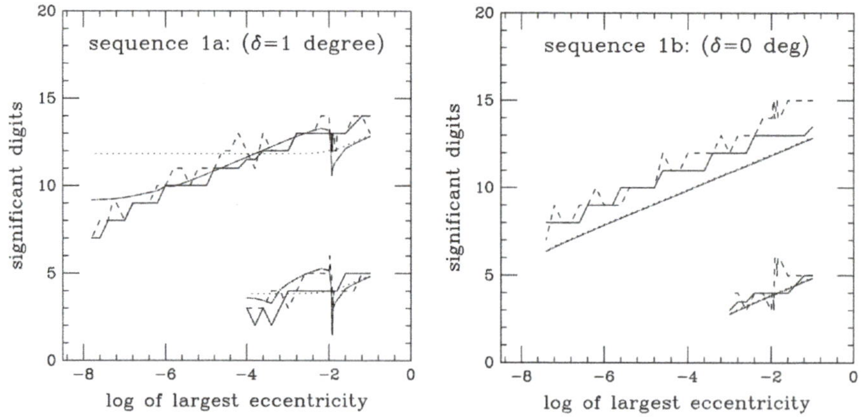

Figure 1: Accuracy estimates are shown as a non-zero constant separation is approached [δ = 1 degree (left)] and as the separation vanishes [δ = 0 (right)]. In both sequences the upper curves are double- and the lower single-precision estimates. Solid are $-log(\tau_k)$, *dot-dashed* $N_{u'}$, *dots* N_ρ, and *dashes* are $N_{u'}^{vis}$. The single precision codes could not converge on solutions for abscissae less than −4

In sequence 1b, for the smallest value of elarger, 5 × 10^{-8}, the code calculates the function $\rho^2(u')$ displayed in left panel of Fig. 2. This results from the classical cancellation effect of subtracting almost equal numbers and losing significance. Because the first and second derivatives of this function are much smoother and do not suffer from cancellation the Newton–Raphson method is successful in taming this so well that the proper extrema estimates are correct to 6 significant digits. The double precision results for that case are in Table 3.

High Eccentricity Test: Again six elements of the data vector are unchanged and are as described in Table 2, except that δ is again 0 degree. Table 4 shows the accuracy estimates and solution elements in conventional units. The right panel of Fig. 2 displays the behavior of the rate of change of ρ^2 for this test.

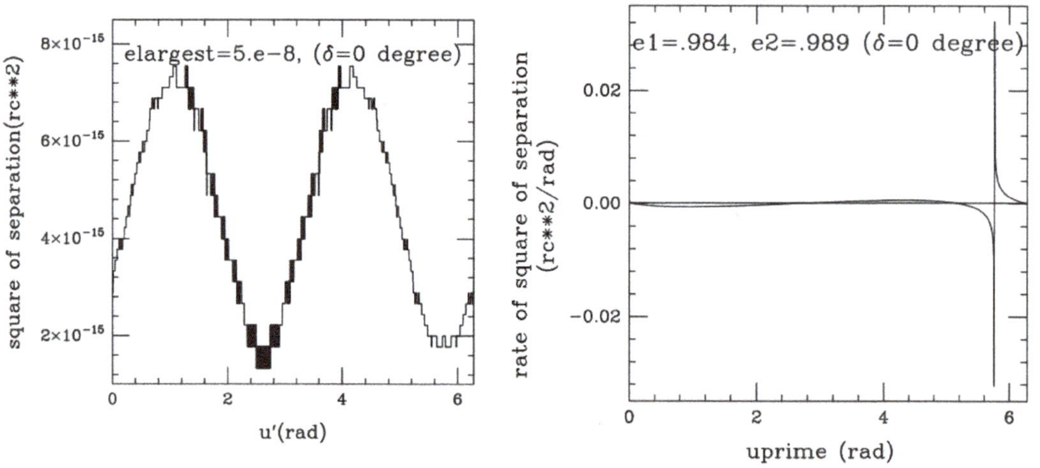

Figure 2: Separation squared versus u' is shown for the limiting case of sequence 1b (left). Note the effects of approaching the limit in the number of significant digits. The rate of change of the separation squared versus u' is shown for the high eccentricity test (right)

Table 3: Solutions and accuracies for $e_{larger} = 5 \times 10^{-8}$ and δ = 0 deg from sequence 1b

$u'_1 = 1.04163$ rad	$\rho_1 = 8.56007 \times 10^{-08}$ rc
$u'_2 = 2.61243$ rad	$\rho_2 = 3.65002 \times 10^{-08}$ rc
$u'_3 = 4.18322$ rad	$\rho_3 = 8.68880 \times 10^{-08}$ rc
$u'_4 = 5.75402$ rad	$\rho_4 = 4.47035 \times 10^{-08}$ rc
$cond(F_{u'}, x^*_{u'}; d) = 3.36 \times 10^7$	$cond(F_\rho, x^*_\rho; d) = 3.06 \times 10^8$
$N_{u'} = 6.47$	$N_\rho = 6.51$

Table 4: High eccentricity test: input tolerance(s) and eccentricities, and resulting accuracies and extrema for geosynchronous orbits

τ	e_{larger}	$e_{smaller}$	$N_{u'}$	$N_{u'}^{vis}$	N_ρ	N_ρ^{vis}
10^{-13}	0.989	0.984	11.63	12	11.62	12
$u'_1 = 1.99685639500$ deg			$\rho_1 = 1343.89338779$ km			
$u'_2 = 150.000000000$ deg			$\rho_2 = 210.820872102$ km			
$u'_3 = 298.003143605$ deg			$\rho_3 = 1343.89338779$ km			
$u'_4 = 330.000000000$ deg			$\rho_4 = 210.820872103$ km			

CONCLUSIONS

Using an unperturbed Kepler model of relative motion for co-periodic satellites, three methods were compared to determine the number of significant digits in estimating separation function extrema. One is based on the minimum tolerances for the numerical processes that allow convergence, a second on the symmetry expected in the separation function when satellite labels are interchanged, and the third on the formalism of condition number estimation. Two tests' results suggest that the latter is a reliable measure of conditioning. More extensive testing of the algorithm is beyond the scope of this article because the domain of the eight input parameters is too large. The range of the output function is a case-dependent even number, k_{max}, of timed relative extrema, where a small change in an input parameter can cause a change in k_{max}. It is likely that uncertainties caused by ill-conditioning will be gratifyingly negligible compared to those caused by real world input uncertainties and modeling errors in physically meaningful cases.

The ideal case of Kepler motion can be represented to any accuracy by this algorithm provided the precision of the arithmetic is high enough.

This study is most applicable to common period systems of satellites. However, similar algebra and logic are easily applied to the more general ten parameter problem of not-necessarily-equal period relative motion extrema, the subject of a future study.

REFERENCES

1. Abramowitz, M., Stegun, L.A.: Handbook of Mathematical Functions With Formulas, Graphs, and Mathematical Tables. U.S. Government Printing Office, Washington, DC, 20402 (1964)
2. Armellin, R., Di Lizia, P., Berz, M., Makino, K.: Computing the critical points of the distance function between two Keplerian orbits via rigorous global optimization. Celest. Mech. Dyn. Astron. 107, 377–395 (2010)
3. Battin, R.H.: An introduction to the mathematics and methods of astrodynamics, Revised edn. American Institute of Aeronautics and Astronautics Inc, Reston, Virginia (1999)
4. Gurfil, P., Kholshevnikov, K.V.: Manifolds and metrics in the relative spacecraft motion problem. J Guid. Control Dyn. 29, 1004–1010 (2006)
5. Higham, N.J.: Accuracy and Stability of Numerical Algorithms. Society for Industrial and Applied Mathematics, Philadelphia (2002)
6. Kholshevnikov, K.V.: Metric spaces of Keplerian orbits. Celest. Mech. Dyn. Astron. 100, 169–179 (2008)
7. Klinkrad, H.: Space Debris, Models and Risk Analysis. Springer, Berlin (2006)
8. Maruskin, J.M.: Distance in the space of energetically bounded Keplerian orbits. Celest. Mech. Dyn. Astron. 108, 265–274 (2010)
9. Moore, R.E., Kearfott, R.B., Cloud, M.J.: Introduction to Interval Analysis. Society for Industrial and Applied Mathematics, Philadelphia (2009)
10. Moulton, F.R.: An Introduction to Celestial Mechanics. The Macmillan Co., NY (1956)
11. Posey, E.E., Vaughan, J.E.: Extrema and nowhere differentiable functions. Rocky Mt. J. Math. 16(4) (Fall 1986)
12. Vassiliev, N.N., Kholshevnikov, K.V.: On the distance function between two Keplerian elliptic orbits. Celest. Mech. Dyn. Astron. 75, 75–83 (1999)

CHAPTER 10

A HAMILTONIAN APPROACH TO THE PLANAR OPTIMIZATION OF MID-COURSE CORRECTIONS

E. Iorfida, P. L. Palmer, M. Roberts

Keywords: Primer vector, Trajectory optimization, Indirect method, Mid-course corrections, Deep space maneuvers

INTRODUCTION

The addition of mid-course corrections (or deep space maneuvers, DSM) along an interplanetary transfer has the advantage of decreasing the overall cost of a trajectory. In fact, applying a DSM will modify not only the trajectory but also the boundary conditions (BC) of the orbit (Bernelli-Zazzera et al. 2007). In an impulsive thrust case, this implies modifying the departure and arrival ΔVs since the cost of a mission can be defined as sum of impulses (magnitudes of the ΔVs). Generally, this process is done in the scenario of the patched conics method, which means considering a sequence of Keplerian arcs linked together (Battin 1999). The inclusion of a DSM along a transfer orbit has the main drawback of increasing the search space complexity of an optimization problem (Vasile and De Pascale 2006). For this reason, it can be useful to have a method that gives a first guess on where to perturb the orbit for the possible addition of a mid-course correction.

In the early 1960s, Lawden introduced the concept of the primer vector in Lawden (1963). Derived with an indirect optimal control method, this variable is associated with the equation of motion and it has to satisfy a set of necessary conditions (NC) for a transfer to be optimal. In particular if the NCs are not fulfilled, it is energetically favourable to add a DSM, see Conway (2010). For a linear system, Prussing (1995) proved that these conditions are also sufficient.

Later studies such as Lion and Handelsman (1968), Jezewski and Rozendaal (1968), Vinh (1972) and Jezewski (1975) extensively showed where and how to perturb the trajectory for the addition of the correction (with gradient based optimization methods). At present, the primer vector theory is mainly exploited in global optimization scenario as a first estimation of the position of the DSM and therefore it allows mission planners to reduce the size of the search space (Olympio et al. 2006). However, due to its computational complexity, its usage is narrowed also to simple three-body problem cases as in Luo et al. (2010).

The main disadvantage of Lawden's theory is that it is based on solving a boundary value problem (BVP) that involves the state transition matrix (STM) of the motion. Currently, there are no known analytic solutions to the problem and, subsequently, no insights into what determines whether a given transfer will be optimal or not. In papers like Goodyear (1965), a formal solution to the differential equations used to derive the STM is provided. Even if the method presented in Goodyear (1965) is valid for every type of Keplerian transfer, it does not give a complete understanding of the structure of the problem. In fact, it is defined in a general inertial frame and therefore no simplifications in the equations (36 terms expressed as sums and multiplications of series) are possible.

This paper presents a novel approach to the theory which simplifies the structure of the problem. In particular, the case where departure, arrival and transfer orbits are co-planar is analyzed.

Citation: E. Iorfida, P. L. Palmer, M. Roberts, A Hamiltonian approach to the planar optimization of mid-course corrections, doi:10.1007/s10569-015-9666-8.

PRIMER VECTOR THEORY PROCEDURE

The equations of motion can be written as

$$\dot{\mathbf{x}}(t) = \begin{bmatrix} \dot{\mathbf{r}}(t) \\ \dot{\mathbf{v}}(t) \end{bmatrix} = \begin{bmatrix} \mathbf{v}(t) \\ \mathbf{g}[\mathbf{r}(t)] + \Gamma \mathbf{u}(t) \end{bmatrix} \tag{1}$$

where x(t) is the state, r(t) is the position vector, v(t) is the velocity vector, g(r) is the gravitational acceleration, u(t) represents the unit vector in the direction of the thrust and Γ is the impulsive thrust acceleration (Conway 2010).

As described in Bryson and Ho (1975), there exists a Hamiltonian that defines the system given in Eq. 1:

$$H = \Gamma + \boldsymbol{\lambda}_r{}^T \mathbf{v}(t) + \boldsymbol{\lambda}_v{}^T [\mathbf{g}(\mathbf{r}) + \Gamma \mathbf{u}], \tag{2}$$

where $\lambda_{r/v}$ are the co-state variables associated, respectively, to the position and the velocity of Eq. 1. In a classic optimization problem, the cost function can be defined as the total propellant consumption. For an impulsive thrust case, the cost is simply the sum of the magnitudes of the N impulses fired along the transfer arc (Conway 2010)

$$J = \sum_{i=1}^{N} \Delta V_i. \tag{3}$$

The adjoint equations referred to Eq. 2 which minimize the cost are

$$\dot{\boldsymbol{\lambda}}_r{}^T = -\frac{\partial H}{\partial \mathbf{r}} = -\boldsymbol{\lambda}_v \mathbf{G}(\mathbf{r}) \tag{4}$$

$$\dot{\boldsymbol{\lambda}}_v{}^T = -\frac{\partial H}{\partial \mathbf{v}} = -\boldsymbol{\lambda}_r{}^T \tag{5}$$

where $\mathbf{G}(\mathbf{r}) \triangleq \frac{\partial \mathbf{g}(\mathbf{r})}{\partial \mathbf{r}}$ is the 3×3 gravity gradient matrix.

The primer vector p(t) and its derivative p˙(t) have been defined in Lawden (1963) as

$$\mathbf{p}(t) \triangleq -\boldsymbol{\lambda}_v(t) \tag{6}$$

$$\dot{\mathbf{p}}(t) = \boldsymbol{\lambda}_r(t). \tag{7}$$

Equations 4, 5, 6 and 7 combined together, give the 'primer vector equation':

$$\ddot{\mathbf{p}} = \mathbf{G}(\mathbf{r})\mathbf{p}. \tag{8}$$

Furthermore, Lawden characterized the theory with a set of NC for the impulsive thrust case, when the cost is simply expressed as a sum of the magnitudes of ΔVs (as in Eq. 3). The satisfaction of the NCs defines the optimality of a trajectory [extensive explanation in Jezewski (1975)]:

The primer vector and its derivative are continuous everywhere.

The magnitude of the primer vector is always below 1, aside from the instants where the impulse occurs, where p=1.

At the impulse times, the primer vector is a unit vector in the direction of the thrust.

As a consequence of the above conditions, $dp/dt = \dot{p} = \dot{\mathbf{p}}^T \mathbf{p} = 0$ at an intermediate impulse.

Equation 7 needs to be integrated to obtain the primer vector profile along the transfer orbit and then to verify if the NC are satisfied. Its solution can be expressed in the form of

$$\begin{bmatrix} \mathbf{p}(t) \\ \dot{\mathbf{p}}(t) \end{bmatrix} = \boldsymbol{\Phi}(t, t_0) \begin{bmatrix} \mathbf{p}(t_0) \\ \dot{\mathbf{p}}(t_0) \end{bmatrix} \tag{9}$$

where Φ is the STM associated to the state (Jezewski 1975). The third NC gives the BC on the primer vector, defining it as unit vector in the direction of the thrust:

$$\mathbf{p}_0 \equiv \mathbf{p}(t_0) = \frac{\Delta \mathbf{V}_0}{\|\Delta \mathbf{V}_0\|}$$
$$\mathbf{p}_f \equiv \mathbf{p}(t_f) = \frac{\Delta \mathbf{V}_f}{\|\Delta \mathbf{V}_f\|} \tag{10}$$

However, both initial conditions on $\mathbf{p}(t)$ and $\dot{\mathbf{p}}(t)$ are required for the propagation of Eq. 9. p_0 is known from Eq. 10 and the initial condition of $\dot{p}$ comes inverting Eq. 9 as:

$$\dot{\mathbf{p}}(t_0) = \boldsymbol{\Phi}_{12}{}^{-1}(t_f, t_0) \cdot [\mathbf{p}(t_f) - \boldsymbol{\Phi}_{11}(t_f, t_0) \cdot \mathbf{p}(t_0)] \tag{11}$$

where $\boldsymbol{\Phi}_{11}$ and $\boldsymbol{\Phi}_{12}$ represent partition matrices of Φ and they are the only unknowns of Eq. 11. As a consequence, this procedure implies solving a BVP where firstly the STM is integrated along the arc to solve Eq. 11 for p˙(t0) and, afterwards, Eq. 9 is propagated to verify Lawden's NCs.

POLAR COORDINATES TRANSFORMATION

This paper presents a novel procedure framed in a local polar coordinate system defined by $\hat{\mathbf{e}}_r, \hat{\mathbf{e}}_\vartheta$ and $\hat{\mathbf{e}}_h$. They are, respectively, unit vectors in the direction of the local position vector, of the derivative of $\hat{\mathbf{e}}_r$ with respect to the true anomaly (ν) and of the orbital momentum vector (h):

$$\hat{\mathbf{e}}_r = \frac{\mathbf{r}}{\|\mathbf{r}\|}; \quad \hat{\mathbf{e}}_\vartheta = -\hat{\mathbf{e}}_r \times \hat{\mathbf{e}}_h; \quad \hat{\mathbf{e}}_h = \frac{\mathbf{h}}{\|\mathbf{h}\|}. \tag{12}$$

It can be demonstrated that this coordinate system is an eigenvector basis of the gravity gradient matrix G (Eq. 8), defined to first order approximation as

$$\mathbf{G} = -\frac{\mu}{r^3}\mathbf{I}_3 + \frac{3\mu}{r^5}\mathbf{r} \cdot \mathbf{r}^T. \tag{13}$$

The corresponding eigenvalues of G, with respect to the unit vectors of Eq. 12, are $2\mu/r^3$, of algebraic multiplicity 1, and $-\mu/r^3$, of algebraic multiplicity 2.

Combining together Eq. 8 and the eigen-decomposition of G, the primer vector equation can be expressed in polar form as

$$\ddot{\mathbf{p}} = \mathbf{G}\mathbf{p} = \frac{\mu}{r^3}(2p_r\hat{\mathbf{e}}_r - p_\vartheta\hat{\mathbf{e}}_\vartheta - p_h\hat{\mathbf{e}}_h). \tag{14}$$

Finally the first derivative of the primer vector with respect to time in polar coordinates can be derived:

$$\frac{d\mathbf{p}}{dt} = \underbrace{\left(\dot{p}_r - \frac{h}{r^2}p_\vartheta\right)\hat{\mathbf{e}}_r + \left(\dot{p}_\vartheta + \frac{h}{r^2}p_r\right)\hat{\mathbf{e}}_\vartheta}_{\text{IN PLANE}} + \underbrace{\dot{p}_h\hat{\mathbf{e}}_h}_{\text{OUT OF PLANE}}. \tag{15}$$

From Eq. 15, it is clear that the out-of-plane component is decoupled from the in-plane components.

The analysis related to the out-of-plane component is treated in a complementary paper from the authors, now in press.

In-plane components

Concerning the in-plane part of Eq. 15, p_r and p_ϑ do not have formal analytic solutions. Even so, it is possible to find an analytic approximation for which the time-variant variable is considered to be constant for a small integration step size.

From Eq. 15, two new variables can be defined:

$$\begin{cases} P_r \triangleq \dot{p}_r - \dfrac{h}{r^2} p_\vartheta \\ P_\vartheta \triangleq \dot{p}_\vartheta + \dfrac{h}{r^2} p_r \end{cases} . \tag{16}$$

P_r and P_ϑ are the components of the first derivative of the primer vector with respect to time in the local orbital reference frame. Therefore, the in-plane part of Eq. 15 becomes

$$\dot{\mathbf{p}} = P_r \hat{\mathbf{e}}_r + P_\vartheta \hat{\mathbf{e}}_\vartheta . \tag{17}$$

Taking into account the derivatives with respect to time of the eigenvectors:

$$\begin{cases} \dot{\hat{\mathbf{e}}}_r = \dot{\nu} \hat{\mathbf{e}}_\vartheta \\ \dot{\hat{\mathbf{e}}}_\vartheta = -\dot{\nu} \hat{\mathbf{e}}_r \\ \dot{\hat{\mathbf{e}}}_h = 0 \end{cases} \tag{18}$$

the second derivative of the primer vector can be easily derived as

$$\ddot{\mathbf{p}} = \left(\dot{P}_r - \frac{h}{r^2} P_\vartheta\right)\hat{\mathbf{e}}_r + \left(\dot{P}_\vartheta + \frac{h}{r^2} P_r\right)\hat{\mathbf{e}}_\vartheta \tag{19}$$

The primer vector equation in polar form (Eq. 14) combined with Eq. 16 and Eq. 19 gives the derivatives of P_r and P_ϑ

$$\begin{cases} \dot{P}_r = \dfrac{2\mu}{r^3} p_r + \dfrac{h}{r^2} P_\vartheta \\ \dot{P}_\vartheta = -\dfrac{\mu}{r^3} p_\vartheta - \dfrac{h}{r^2} P_r \end{cases} \tag{20}$$

To simplify the structure of the equations, two new variables can be defined:

$$\sigma \triangleq \frac{r}{l} \equiv \frac{1}{1 + e\cos\nu}$$
$$\omega \triangleq \frac{h}{r^2} = \frac{h}{(l\sigma)^2} \tag{21}$$

which satisfy also the following equivalence

$$\sigma \cdot \omega^2 = \frac{r}{l}\frac{h^2}{r^4} = \frac{\mu}{r^3}\frac{r}{r}\frac{l}{l} = \frac{\mu}{r^3} \tag{22}$$

where $l = a(1 - e^2)$ is the semi-latus rectum of the orbit (Battin 1999).

The new variables, linked together, give a new first order ODE set in $[p_r, p_\vartheta, P_r, P_\vartheta]$:

$$\begin{bmatrix} \dot{p}_r \\ \dot{p}_\vartheta \\ \dot{P}_r \\ \dot{P}_\vartheta \end{bmatrix} = \begin{bmatrix} 0 & \omega & 1 & 0 \\ -\omega & 0 & 0 & 1 \\ 2\sigma\omega^2 & 0 & 0 & \omega \\ 0 & -\sigma\omega^2 & -\omega & 0 \end{bmatrix} \begin{bmatrix} p_r \\ p_\vartheta \\ P_r \\ P_\vartheta \end{bmatrix} \tag{23}$$

Equation 23 can be represented in compact matrix format as

$$\dot{\mathbf{p}}_{\text{polar}}(t) = \mathbf{M}(t)\mathbf{p}_{\text{polar}}(t). \tag{24}$$

Palmer Coordinates Conversion and Integration Scheme

The system of Eq. 23 can be written in terms of the Hamiltonian:

$$H_{\text{PV}} = \frac{1}{2}({P_r}^2 + {P_\vartheta}^2) + \omega(p_\vartheta P_r - p_r P_\vartheta) - \frac{1}{2}\sigma\omega^2(2{p_r}^2 - {p_\vartheta}^2) \tag{25}$$

If H_{PV} is compared to the one of Hill's problem, H_{Hill} of Quinn et a_l. (2010), there is a noticeable similarity:

$$H_{\text{Hill}} = \frac{1}{2}({p_x}^2 + {p_y}^2) + \Omega(y p_x - x p_y) - \frac{1}{2}\Omega^2(2x^2 - y^2) \tag{26}$$

The substantial difference between the two Hamiltonian forms is that, while in H_{Hill} the coefficient Ω is constant during the motion, in H_{PV} both σ and ω vary in time. However, for the circular orbit case (Hill's problem), σ=1 and ω=Ω which implies a one-to-one correspondence between the two Hamiltonians. This fact validates a fundamental result: for a circular orbit transfer there is an analytic solution to the planar primer vector problem, which is exactly the Hill's solution.

For Hill's problem, the motion of the in-plane state vector of Eq. 26, $\mathbf{z} = (\Omega x, \dot{x}, \Omega y, \dot{y})^T$, can be separated into two parts: an oscillation and a linear drift. This result is presented in Palmer (2007), where the transformation of coordinates $\tilde{\mathbf{p}} = \mathcal{P}\mathbf{z} = (p_1, p_2, p_3, p_4)^T$ on z allows this separation to be explicit (as a note, the original notation of Palmer (2007) has been changed to avoid confusion).

The combination of Eqs. 16, 25 and 26 results in the relationship between the state vector zz and ppolarppolar:

$$\begin{cases} x = p_r \\ y = p_\vartheta \end{cases} ; \begin{cases} p_x = P_r = \dot{x} - \Omega y \\ p_y = P_\vartheta = \dot{y} + \Omega x \end{cases} \Rightarrow \begin{cases} \dot{x} = p_x + \Omega y = P_r + \omega p_\vartheta \\ \dot{y} = p_y - \Omega x = P_\vartheta - \omega p_r \end{cases} \tag{27}$$

where the frequency associated with z is Ω, while the one of p_{polar} is ω.

If the transformation of coordinates presented in Palmer (2007) is applied to the primer vector problem (Eq. 27), it becomes

$$\begin{bmatrix} p_1 \\ p_2 \\ p_3 \\ p_4 \end{bmatrix} = \begin{bmatrix} -3 & 0 & 0 & -2 \\ 0 & 1 & 0 & 0 \\ 0 & 2 & -1 & 0 \\ 2 & 0 & 0 & 1 \end{bmatrix} \begin{bmatrix} \omega p_r \\ P_r + \omega p_\vartheta \\ \omega p_\vartheta \\ P_\vartheta - \omega p_r \end{bmatrix} \tag{28}$$

Integration of the Palmer Coordinates

The coordinates $\tilde{\mathbf{p}}$ (also called *Palmer coordinates*) can be expressed as functions of $(p_r, p_\vartheta, P_r, P_\vartheta)$ as shown in Eq. 28, where two couplings are evident: one between p_1 and p_4 and the other between p_2 and p_3.

If $\tilde{\mathbf{P}}$ is first combined with the system of Eq. 23 and then differentiated with respect to time (while σ, and therefore ω, are kept constant for small intervals of time), it yields

$$\begin{bmatrix} \dot{p}_1 \\ \dot{p}_2 \\ \dot{p}_3 \\ \dot{p}_4 \end{bmatrix} = \omega \begin{bmatrix} 0 & 4\sigma - 3 & -2(\sigma - 1) & 0 \\ 2\sigma - 3 & 0 & 0 & 4(\sigma - 1) \\ 4(\sigma - 1) & 0 & 0 & 8\sigma - 5 \\ 0 & -2(\sigma - 1) & \sigma - 1 & 0 \end{bmatrix} \begin{bmatrix} p_1 \\ p_2 \\ p_3 \\ p_4 \end{bmatrix} \tag{29}$$

From this system of equations, it is then possible to differentiate even further to a second order system as

$$\begin{bmatrix} \ddot{p}_1 \\ \ddot{p}_2 \\ \ddot{p}_3 \\ \ddot{p}_4 \end{bmatrix} = \omega^2 \begin{bmatrix} -(2\sigma - 1) & 0 & 0 & -2(\sigma - 1) \\ 0 & -(2\sigma - 1) & 2(\sigma - 1) & 0 \\ 0 & -2(\sigma - 1) & 3(\sigma - 1) & 0 \\ 2(\sigma - 1) & 0 & 0 & 3(\sigma - 1) \end{bmatrix} \begin{bmatrix} p_1 \\ p_2 \\ p_3 \\ p_4 \end{bmatrix} \tag{30}$$

In Eq. 30, the variables are coupled again $(p_1$ with p_4 and p_2 with $p_3)$.

The structure of the problem allows to have a 4th order ODE in p_1 in the form of

$$p_1{}^{IV} - \omega^2(\sigma - 2)p_1{}^{II} - \omega^4(\sigma - 1)(2\sigma + 1) = 0. \tag{31}$$

Assuming that the coefficients of Eq. 31 are constant for a small interval of time, the solution will be $p_1 \approx e^{\omega\lambda t}$.

The characteristic equation in λ of Eq. 31 is

$$\lambda^4 - \omega^2(\sigma - 2)\lambda^2 - \omega^4(\sigma - 1)(2\sigma + 1) = 0, \tag{32}$$

whose solutions are

$$\lambda_1^{\pm} = \frac{\sqrt{2}}{2\omega}\sqrt{(\sigma - 2) \pm \sqrt{9\sigma^2 - 8\sigma}}$$
$$\lambda_2^{\pm} = -\frac{\sqrt{2}}{2\omega}\sqrt{(\sigma - 2) \pm \sqrt{9\sigma^2 - 8\sigma}}. \tag{33}$$

The nature of $\lambda_{1,2}{}^{\pm}$± depends upon the value of σ. If $\Delta_\sigma \triangleq 9\sigma^2 - 8\sigma$, there are three cases (the interval σ<0 is not taken into consideration because it does not have any physical meaning):

- $\boldsymbol{\Delta_\sigma > 0}$: $\sigma > 8/9$. $\lambda_{1,2}{}^{\pm}$ can be either real (if $\text{sign}((\sigma - 2) \pm \Delta_\sigma) > 0$) or imaginary (if $\text{sign}((\sigma - 2) \pm \Delta_\sigma) < 0$);
- $\boldsymbol{\Delta_\sigma = 0}$: σ can be either 0 or 8/9 and $\lambda_{1,2}{}^{\pm}$ are always imaginary;
- $\boldsymbol{\Delta_\sigma < 0}$: $0 < \sigma < 8/9$ and $\lambda_{1,2}{}^{\pm}$ are generally complex.

In the case of an elliptic orbit the nature of the eigenvalues can be summarized, as in Fig. 1, into three distinguished areas:

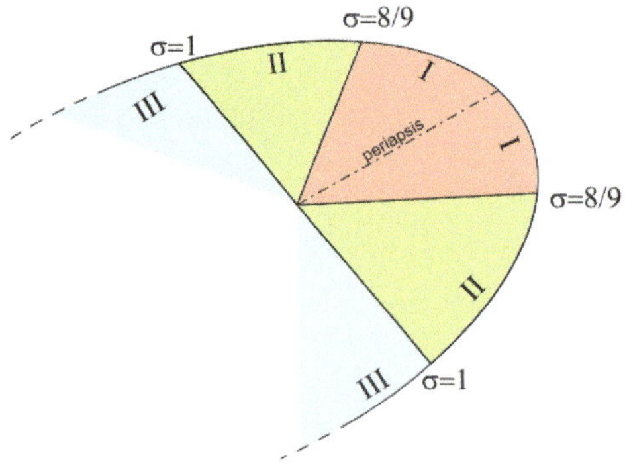

Figure 1: Elliptical transfer orbit divided into three regions (I, II, III) according to the nature of the eigenvalues $\lambda_{1,2}{}^{\pm}$

- Area (I) $0 < \sigma < 8/9$: corresponds to the case of $\Delta_\sigma < 0$, and includes the periapsis of the transfer orbit, which is the point where σ has its minimum value, $\sigma_{\min} = (1+e)^{-1}$. In this interval the eigenvalues are complex.
- Area (II) $8\,8/9 \leq \sigma < 1$: corresponds to the case of $\Delta_\sigma > 0$ with $\text{sign}((\sigma - 2) \pm \Delta_\sigma) < 0$, and the eigenvalues are imaginary. For $\sigma = 1$, that is $r = l, \lambda_{1,2}{}^{+} = 0$ while $\lambda_{1,2}{}^{-}$ are imaginary.
- Area (III) σ>1: corresponds to the case of $\Delta_\sigma > 0$; $\lambda_{1,2}{}^{+}$are real, while $\lambda_{1,2}{}^{-}$ are imaginary. At the apoapsis (included in this area) σ reaches its maximum value $\sigma_{\max} = (1 - e)^{-1}$.

For values of the eccentricity smaller than $1/8$, $\Delta \geq 0$ and $\sigma \geq 8/9$, therefore there is no Area (I). In other words, the eigenvalues are either purely imaginary or real (Area (II) and Area (III)).

After the aforementioned considerations, the complete solution of Eq. 31 can be derived, and it simply is

$$p_1 = Ae^{\omega\lambda_1^+ t} + Be^{\omega\lambda_2^+ t} + Ce^{\omega\lambda_1^- t} + De^{\omega\lambda_2^- t} \tag{34}$$

where $\mathbf{A} = (A, B, C, D)^T$ is a vector of coefficients given by the initial conditions of p_1.

The other components of $\tilde{\mathbf{p}}$ can be evaluated as for p_1, resulting in a formal matrix expression

$$\tilde{\mathbf{p}} = \boldsymbol{\Sigma}(\sigma)^{-1}\tilde{\mathbf{E}}(t,\sigma)^{-1}\mathbf{A} \tag{35}$$

$\tilde{\mathbf{E}}(t,\sigma)$ is a diagonal matrix with independent exponential functions as elements of the diagonal, functions of time t and $\omega(\sigma)$:

$$\tilde{\mathbf{E}} = \begin{bmatrix} e^{\omega\lambda_2^+ t} & 0 & 0 & 0 \\ 0 & e^{\omega\lambda_1^+ t} & 0 & 0 \\ 0 & 0 & e^{\omega\lambda_2^- t} & 0 \\ 0 & 0 & 0 & e^{\omega\lambda_1^- t} \end{bmatrix} \tag{36}$$

while $\boldsymbol{\Sigma}(\sigma)$ is a polynomial matrix in σσ, written as

$$\boldsymbol{\Sigma} = \frac{(\sigma - 1)}{(2\sigma + 1)\sqrt{\Delta_\sigma}} \begin{bmatrix} (2\sigma + 1)\xi^- & -\lambda_1^+\gamma^- & \lambda_1^+\delta^- & -(2\sigma + 1) \\ (2\sigma + 1)\xi^- & \lambda_1^+\gamma^- & -\lambda_1^+\delta^- & -(2\sigma + 1) \\ -(2\sigma + 1)\xi^+ & \lambda_1^-\gamma^+ & -\lambda_1^-\delta^+ & (2\sigma + 1) \\ -(2\sigma + 1)\xi^+ & -\lambda_1^-\gamma^+ & \lambda_1^-\delta^+ & (2\sigma + 1) \end{bmatrix} \tag{37}$$

The parameters $\xi^\pm$, $\gamma^\pm$ and $\delta^\pm$ are also only functions of σ and they are defined as

$$\begin{cases} \delta^\pm \triangleq [(-3\sigma + 1) \mp \sqrt{\Delta_\sigma}] \\ \gamma^\pm \triangleq \dfrac{1}{4(1 - \sigma)}[(24\sigma^2 - 25\sigma + 4) \pm (8\sigma - 5)\sqrt{\Delta_\sigma}] \\ \xi^\pm \triangleq \dfrac{(5\sigma - 4) \pm \sqrt{\Delta_\sigma})}{4(1 - \sigma)} \end{cases} \tag{38}$$

Conversion From Time to True Anomaly

The propagation process presented so far, is based on time as the integration variable. In orbital dynamics it is sometimes more convenient to use true anomaly rather than time; therefore, the variable conversion is

$$\frac{d\nu}{dt} = \omega \Rightarrow \frac{d\nu}{\omega} = dt \Rightarrow \frac{\nu}{\omega} = t \tag{39}$$

Since both σ and ω have been considered constant along an integration step, the conversion from time to true anomaly does not affect the solutions of Eq. 35, a part from a factor ω.

The ODE of Eq. 29 can be converted into true anomaly as

$$\tilde{\mathbf{p}}' = \begin{bmatrix} 0 & 4\sigma - 3 & -2(\sigma - 1) & 0 \\ 2\sigma - 3 & 0 & 0 & 4(\sigma - 1) \\ 4(\sigma - 1) & 0 & 0 & 8\sigma - 5 \\ 0 & -2(\sigma - 1) & \sigma - 1 & 0 \end{bmatrix} \tilde{\mathbf{p}}, \tag{40}$$

where the symbol (′) represents the derivative with respect to ν, and in compact form it is equivalent to

$$\tilde{\mathbf{p}}' = \boldsymbol{\Lambda}\tilde{\mathbf{p}}. \tag{41}$$

This implies that the solution shown in Eq. 35 remains in the same form:

$$\mathbf{A} = \mathbf{E}(\nu, \sigma)\boldsymbol{\Sigma}(\sigma)\tilde{\mathbf{p}}. \tag{42}$$

The only difference appears in E, as

$$\mathbf{E}(\nu,\sigma) = \begin{bmatrix} e^{\lambda_2^+ \nu} & 0 & 0 & 0 \\ 0 & e^{\lambda_1^+ \nu} & 0 & 0 \\ 0 & 0 & e^{\lambda_2^- \nu} & 0 \\ 0 & 0 & 0 & e^{\lambda_1^- \nu} \end{bmatrix} \tag{43}$$

where the exponents are explicit functions in ν. Through the transformation from time to true anomaly, the importance of the Palmer coordinates is even more clear. In fact, using $\tilde{\mathbf{p}}$ allows a solution which is a linear combination of the eigenvalues of the matrix M (Eq. 24). This solution depends only on σ through the polynomial matrix Σ and the diagonal matrix E, with an explicit propagation variable ν.

Matrices Propagation Scheme

Since Eq. 42 assumes that σ is constant for small interval of ν, it is required to apply a 'double' propagation for the Palmer coordinates. This means that every time that the true anomaly is propagated to the next step, the corresponding value of σ must be updated, as shown in the scheme of Fig. 2.

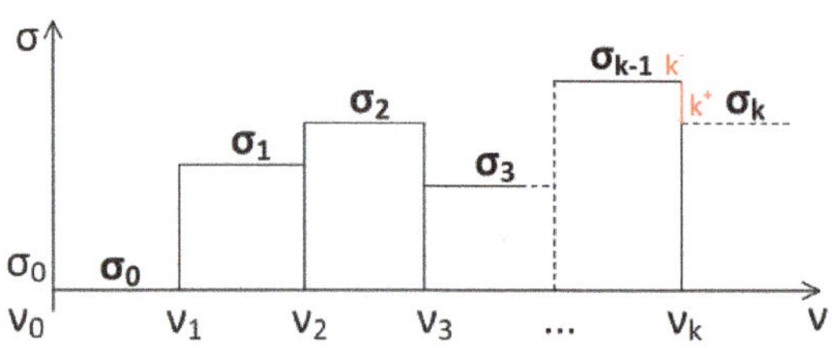

Figure 2: Step Propagation scheme of ν and σ, with σ=const along every interval of ν. A general k_{th} state is highlighted with the two limiting values ($k^{\pm}$) in *red*.

At the initial time the coefficients depend upon the initial value of the Palmer coordinates as

$$\mathbf{A}_0 = \mathbf{E}(\nu_0,\sigma_0)\boldsymbol{\Sigma}(\sigma_0)\tilde{\mathbf{p}}_0 \tag{44}$$

Knowing $\mathbf{A}_0, \tilde{\mathbf{p}}$ can be propagated to the step '1', but only just before the update of σ. In fact the value of $\tilde{\mathbf{p}}$ after the update will be different. To have a formal expression, this condition can be considered at a generic *k*th propagation state (see Fig. 2), where the values of the Palmer coordinates just before $(\tilde{\mathbf{p}}_{k^-})$ and after $(\tilde{\mathbf{p}}_{k^+})$ the variation of σσ are

$$\tilde{\mathbf{p}}_{k^-} = \boldsymbol{\Sigma}^{-1}(\sigma_{k-1})\mathbf{E}^{-1}(\nu_k,\sigma_{k-1})\mathbf{A}_{k-1} \tag{45}$$

$$\tilde{\mathbf{p}}_{k^+} = \boldsymbol{\Sigma}^{-1}(\sigma_k)\mathbf{E}^{-1}(\nu_k,\sigma_k)\mathbf{A}_k \tag{46}$$

In the limit towards $k, \tilde{\mathbf{p}}_{k^\pm}$ are not the same because they depend on two different σs. However, for the primer vector the continuity must be guaranteed along the whole transfer arc. For this reason, a simple propagation scheme (Fig. 2) has to be used, instead of a more complex one (e.g. leapfrog method). In fact, it guarantees that a 'continuity check' on the primer vector is executed at every update of σ.

From Eq. 28 a relationship between $\mathbf{p}_{polar}$ and $\tilde{\mathbf{p}}$ can be derived

$$\tilde{\mathbf{p}} = \begin{bmatrix} -\omega & 0 & 0 & -2 \\ 0 & \omega & 1 & 0 \\ 0 & \omega & 2 & 0 \\ \omega & 0 & 0 & 1 \end{bmatrix}\mathbf{p}_{polar} \equiv \mathbf{K}(\omega)\mathbf{p}_{polar}. \tag{47}$$

The matrix K(ω), depending on ω, will be different before and after the update. Subsequently, the relations between the Palmer coordinate and the primer vector at a generic *k*th step are

$$\mathbf{p}_{polar\,k} = \mathbf{K}^{-1}(\omega_{k-1})\tilde{\mathbf{p}}_{k^-} \tag{48}$$

$$\tilde{\mathbf{p}}_{k^+} = \mathbf{K}(\omega_k)\mathbf{p}_{polar\,k} \tag{49}$$

Combining Eqs. 48 and 49 gives the matrix which guarantees the continuity of the primer vector, defined as $\mathcal{I}_{\omega_k}$:

$$\tilde{\mathbf{p}}_{k^+} = \mathbf{K}(\omega_k)\mathbf{K}^{-1}(\omega_{k-1})\tilde{\mathbf{p}}_{k^-} \triangleq \mathcal{I}_{\omega_k}\tilde{\mathbf{p}}_{k^-} \tag{50}$$

The procedure for the propagation of the Palmer coordinates from its initial conditions $\tilde{\mathbf{p}}_0$ to a generic *k*th state can be then formalized as

$$\tilde{\mathbf{p}}_k = \left(\prod_{i=0}^{k-1}\mathcal{I}_{\omega_i}\boldsymbol{\Sigma}^{-1}(\sigma_i)\mathbf{E}^{-1}(\nu_{i+1},\sigma_i)\mathbf{E}(\nu_i,\sigma_i)\boldsymbol{\Sigma}(\sigma_i)\right)\tilde{\mathbf{p}}_0. \tag{51}$$

In summary, Eq. 51 can be written in matrix form as

$$\tilde{\mathbf{p}}_k = \boldsymbol{\Psi}(\nu_k,\nu_0)\tilde{\mathbf{p}}_0 \tag{52}$$

where $\boldsymbol{\Psi}(\nu_k,\nu_0)$ represents the propagation matrix of the Palmer coordinates.

The final step of the scheme is to convert the Palmer coordinates back into primer vector at the BC. Combining the definitions of P_r and P_ϑ (Eq. 16) and Eq. 28, there is (where ptotptot clearly refers to the the primer vector and its derivative with respect to time):

$$\tilde{\mathbf{p}} = \begin{bmatrix} -3\omega & 0 & 0 & -2 \\ 0 & 0 & 1 & 0 \\ 0 & -\omega & 2 & 0 \\ 2\omega & 0 & 0 & 1 \end{bmatrix}\begin{bmatrix} p_r \\ p_\vartheta \\ \dot{p}_r \\ \dot{p}_\vartheta \end{bmatrix} \equiv \mathbf{B}(\omega)\mathbf{p}_{tot}. \tag{53}$$

Equations 52 and 53 together yield to

$$\mathbf{p}_{tot\,f} = \mathbf{B}^{-1}(\omega_f)\boldsymbol{\Psi}(\nu_f,\nu_0)\mathbf{B}(\omega_0)\mathbf{p}_{tot\,0} \triangleq \tilde{\boldsymbol{\Phi}}_{(f,0)}\mathbf{p}_{tot\,0} \tag{54}$$

where $\tilde{\boldsymbol{\Phi}}_{(f,0)}$ has been defined as the total propagation matrix of the primer vector.

Equation 54 summarizes the whole propagation scheme for the primer vector.

Independence From the Semi-Major Axis of the Transfer Orbit

The value of the 'continuity' matrix, II_ω, defined in Eq. 50, is actually independent from the explicit value of ω (and therefore from *a*) and it is equal to

$$\mathcal{I}_{\omega_k} = \begin{bmatrix} (\omega_k/\omega_{k-1}+2) & 0 & 0 & (-2\omega_k/\omega_{k-1}+2) \\ 0 & (2\omega_k/\omega_{k-1}-1) & (-\omega_k/\omega_{k-1}+1) & 0 \\ 0 & (2\omega_k/\omega_{k-1}-2) & (-\omega_k/\omega_{k-1}+2) & 0 \\ (\omega_k/\omega_{k-1}-1) & 0 & 0 & (2\omega_k/\omega_{k-1}-1) \end{bmatrix} \tag{55}$$

where the ratio ω_k/ω_{k-1} can be found to be dependent only on the true anomalies before and after the update (see definitions of ω and σ from Eq. 21)

$$\frac{\omega_k}{\omega_{k-1}} = \left(\frac{l \cdot \sigma_{k-1}}{l \cdot \sigma_k}\right)^2 = \left(\frac{1 + e\cos(\nu_k)}{1 + e\cos(\nu_{k-1})}\right)^2 \tag{56}$$

Therefore, since l does not appear in any other terms of the matrices of Eq. 51, Ψ is not a function of the semi-major axis of the transfer orbit.

On the contrary, by the definition of Eq. 53, B is linearly dependent on ωω. This makes $\tilde{\Phi}_{(f,0)}$ to be a function of the initial and final values of ω in a 'block-form' as

$$\tilde{\Phi}_{(f,0)} = \mathbf{B}^{-1}(\omega_f)\Psi(\nu_f, \nu_0)\mathbf{B}(\omega_0) \propto \begin{bmatrix} 1 & 1 & (1/\omega_f) & (1/\omega_f) \\ 1 & 1 & (1/\omega_f) & (1/\omega_f) \\ \omega_0 & \omega_0 & 1 & 1 \\ \omega_0 & \omega_0 & 1 & 1 \end{bmatrix} \tag{57}$$

For the final solution of Eq. 54, it is necessary to have the complete initial state $\mathbf{p}0_{tot}$, but at this stage of the procedure, the part concerning the derivative is not yet known.

A transformation of the primer vector components from the derivative with respect to time, to the derivative with respect to true anomaly (as in Eq. 39) gives

$$\frac{dp}{dt} = \frac{dp}{d\nu}\frac{d\nu}{dt} = \frac{dp}{d\nu}\omega. \tag{58}$$

The product between the primer vector in true anomaly, which

$$\tilde{\Phi}_{(f,0)} \cdot \mathbf{p}_{tot,\nu_0} = \mathbf{p}_{tot,\nu_f} \propto \begin{bmatrix} 1 & 1 & (\omega_0/\omega_f) & (\omega_0/\omega_f) \\ 1 & 1 & (\omega_0/\omega_f) & (\omega_0/\omega_f) \\ \omega_0 & \omega_0 & \omega_0 & \omega_0 \\ \omega_0 & \omega_0 & \omega_0 & \omega_0 \end{bmatrix} \tag{59}$$

Having $\mathbf{p}_{tot,\nu_f} = (p_{rf}, p_{\vartheta f}, \omega_f p'_{rf}, \omega_f p'_{\vartheta f})^T$, from Eq. 59 clearly p_{rf} and $p_{\vartheta f}$ are independent from the semi-major axis, since the ratio between the initial and final ωs of Eq. 59 is of the same form of Eq. 56. The last two rows of $\tilde{\Phi}_{(f,0)} \cdot \mathbf{p}_{tot,\nu_0}$ of Eq. 59 are linear in ω_0. Since they respectively correspond to $\omega_f p'_{rf}$ and $\omega_f p'_{\vartheta f}$, this implies that the derivatives in true anomaly of the primer vector are also both independent from a.

This is an important conclusion because it states that the primer vector and its derivative are not influenced by the value of the semi-major axis of the transfer orbit. As a result, the size of the search space is reduced and the optimization problem is simplified.

FIXED TRANSFER ORBIT AND VARIABLE DEPARTURE/ARRIVAL ORBITS

From the definition at the BC of the primer vector (Eq. 10), p_0 and p_f depend only on the directions of the initial and final ΔVs, but they are independent from their magnitudes. Therefore, a 'natural' application is to consider a problem where the ΔVs are parametrized by their directions. In the case where the transfer, departure and arrival orbits are coplanar, the in-plane components of the primer vector can be defined with respect to the orientation of the two local polar reference frames at initial and final position, expressed through two angles α and β:

$$\mathbf{p}_0 = \begin{bmatrix} \cos\alpha \\ \sin\alpha \end{bmatrix}$$

$$\mathbf{p}_f = \begin{bmatrix} \cos\beta \\ \sin\beta \end{bmatrix} \tag{60}$$

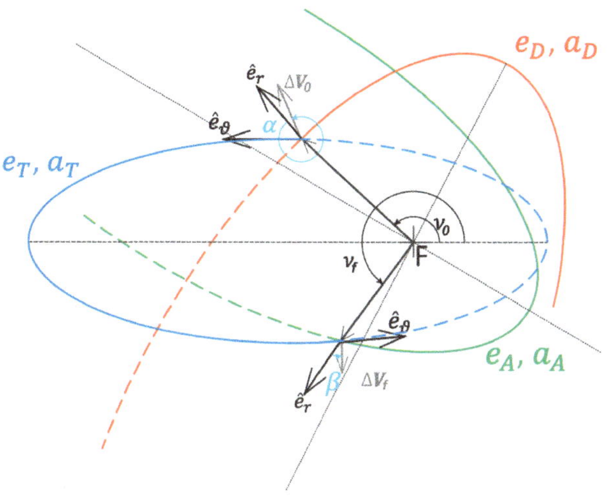

Figure 3: Geometric representation of a co-planar departure (e_D, a_D), transfer (e_T, a_T) and arrival (e_A, a_A) orbits case.

If the transfer orbit is fixed (therefore also the velocities v_0 and v_f), the departure and arrival orbits vary according to the orientation (and magnitude) of the ΔVs (see Fig. 3).

In the local boundary polar reference frames there is

$$\begin{aligned} \mathbf{v}_0 &= v_{r0}\hat{e}_{r0} + v_{\vartheta 0}\hat{e}_{\vartheta 0} \\ \mathbf{v}_f &= v_{rf}\hat{e}_{rf} + v_{\vartheta f}\hat{e}_{\vartheta f} \end{aligned}, \tag{61}$$

while the ΔVs are

$$\begin{aligned} \Delta\mathbf{V}_0 &= \Delta V_0(\cos\alpha\hat{e}_{r0} + \sin\alpha\hat{e}_{\vartheta 0}) \\ \Delta\mathbf{V}_f &= \Delta V_f(\cos\beta\hat{e}_{rf} + \sin\beta\hat{e}_{\vartheta f}) \end{aligned}. \tag{62}$$

Defining v_D and v_A respectively as the velocity at r_0 on the departure orbit and the velocity at r_f on the arrival orbit, they will be

$$\begin{aligned} \mathbf{v}_D &= \mathbf{v}_0 - \Delta\mathbf{V}_0 = (v_{r0} - \Delta V_0\cos\alpha)\hat{e}_{r0} + (v_{\vartheta 0} - \Delta V_0\sin\alpha)\hat{e}_{\vartheta 0} \\ \mathbf{v}_A &= \mathbf{v}_f + \Delta\mathbf{V}_f = (v_{rf} + \Delta V_f\cos\beta)\hat{e}_{rf} + (v_{\vartheta f} + \Delta V_f\sin\beta)\hat{e}_{\vartheta f} \end{aligned} \tag{63}$$

With a fixed transfer orbit, the total state transition matrix $\tilde{\Phi}_{(f,0)}$ of Eq. 54 is fixed as well. In addition, fixing the transfer orbit means fixing its eccentricity and its semi-major axis. This implies that for this scenario the optimality can vary only if the directions of the ΔVs, α and β, change. In an optimization process, such a problem has a 5 dimensional search space: e,v_0 and v_f, which define the transfer orbit, and α and β, which define the BC on the ΔVs. Hence, the analysis can be approached differently, having restricted the problem to those 5 dimensions.

Lawden's NCs state that the primer vector is equal to 1 at the BC and at each intermediate impulse (Lawden 1963). In other words, for a trajectory to be optimal, p must be always smaller than 1 along the arc and equal to 1 at the boundaries, where the ΔVs are fired. Within this scenario, a set of transfer orbits with the same eccentricity can be parametrized by, for example, fixing the initial true anomaly and varying v_f.

Figure 4 illustrates the case where a fixed transfer orbit and fixed v_0 are considered. The maps show the maximum value of the primer vector according to gradual increasing values of v_f, for fixed e=0.8 and v_0= 5 deg. They are plotted in a complete interval (0–360 deg) of α and βwith a computational step of 5 deg.

As already explained, an optimal transfer will have an overall maximum value of the primer vector equal to 1 (dark blues area in Fig. 4). Non-optimal trajectories have p>1 and therefore in Fig. 4 they are represented in colours ('non-optimality islands').

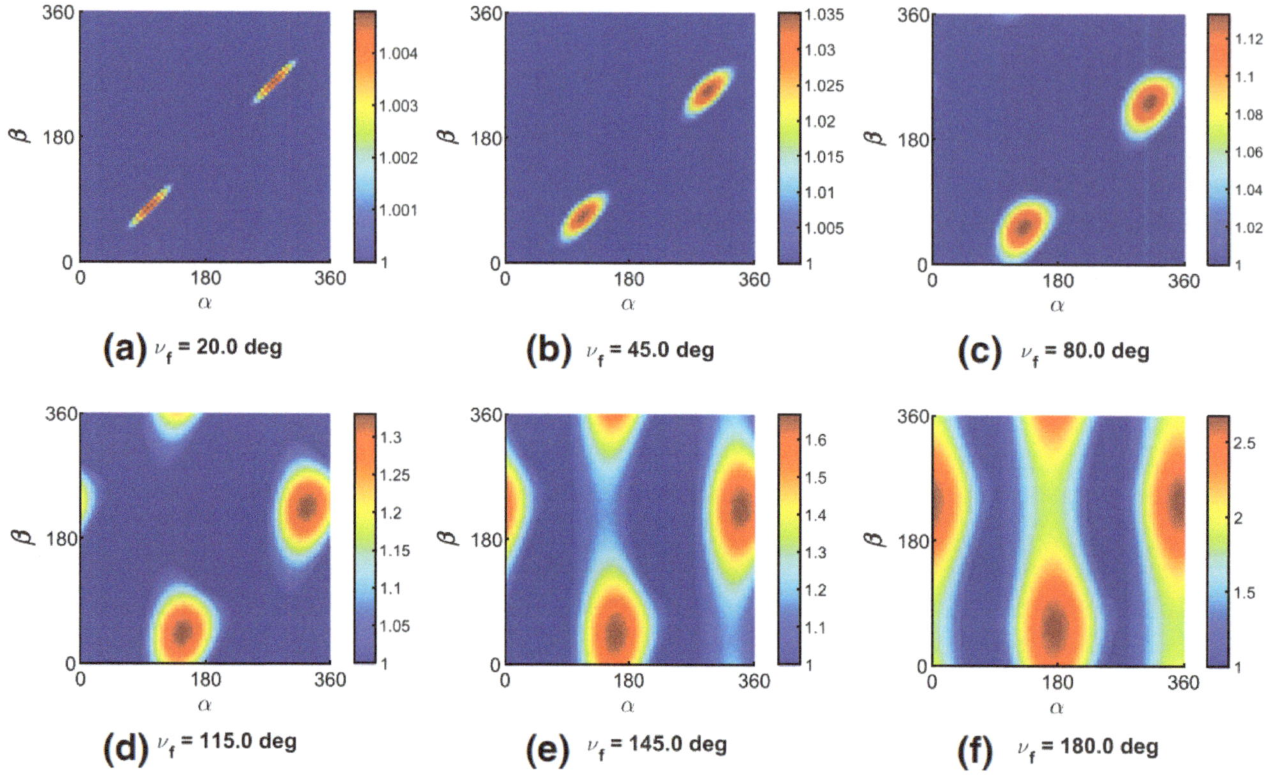

Figure 4: Sequence of maximum values of primer vector maps for a fixed transfer orbit and increasing value of v_f, versus α and β. The fixed parameters of the transfer orbit are $e = 0.8$, $a = 0.95$ DU, $\nu_0 = 5$deg.

The problem is very well structured with a precise geometry: for every specific v_f case, the 'non-optimal' islands have always the same dimension, shape and orientation. In particular for small span of true anomalies the 'non-optimal' islands are very stretched but small in size, therefore almost the whole map of Fig. 4a represents an optimal case. There are two main motivations behind this result, one is that for small Δv, the modulus of the primer vector does not change significantly. In fact, it starts and ends in 1 but its variation along the interval is minimal due to the small span between the initial and final true anomaly. The other reason can be found in the accuracy of the numerical computation (10^{-16}). Therefore, in Fig. 4a it can be seen that even 'non-optimal' islands have very low peak values.

Along the different plots shown in Fig. 4, the islands rotate and grow in size and eventually merge for increasingly higher spans of Δv. The maximum value of the primer vector is always located in the central area of the islands and it increases with an increasing interval of true anomalies. In fact, while in the case where Δv= 15 deg there is $p_{max} \approx 1$, in the last case ($\Delta\nu = 175$ deg), $p_{max} \approx 3$. These kind of maps allow mission planners not only to visualize the possibility to be in an optimal region but also to change the BC, represented either by the direction of the thrusts (α and β) or by the true anomalies, to obtain an optimal transfer.

With the same kind of approach, a different scenario can be to exploit the boundary derivatives of the primer vector. It is clear that for an optimal transfer, since the profile of the primer vector has to be smaller than 1, the initial derivative must be negative while the final derivative must be positive. This condition does not ensure that overall the primer vector will be optimal, but it gives information on the possibility of having an optimal trajectory and on the BC of the problem. On the other hand, if $\dot{p}_0$ is positive and $\dot{p}_f$ is negative they definitely define non-optimal trajectories. The expansions at first order of the initial and final derivatives are

$$\begin{cases} \dot{p}_0 \approx \dot{p}_{r0} p_{r0} + \dot{p}_{\vartheta 0} p_{\vartheta 0} \\ \dot{p}_f \approx \dot{p}_{rf} p_{rf} + \dot{p}_{\vartheta f} p_{\vartheta f} \end{cases} \tag{64}$$

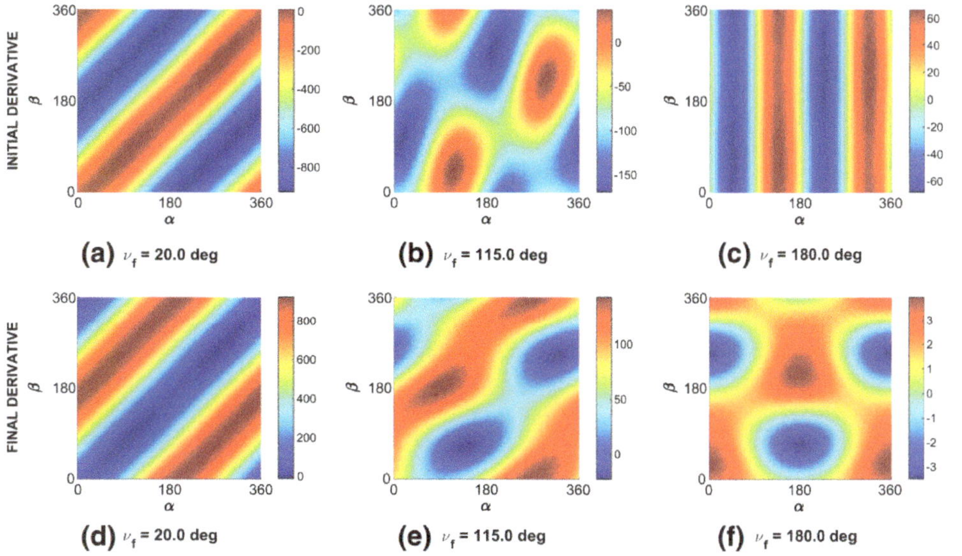

Figure 5: Sequence of initial and final derivatives maps for a fixed transfer orbit and increasing value of ν_f, versus α and β. The characteristics of the transfer orbit are e = 0.8, a = 0.95 DU, $\nu_0 = 5$ deg.

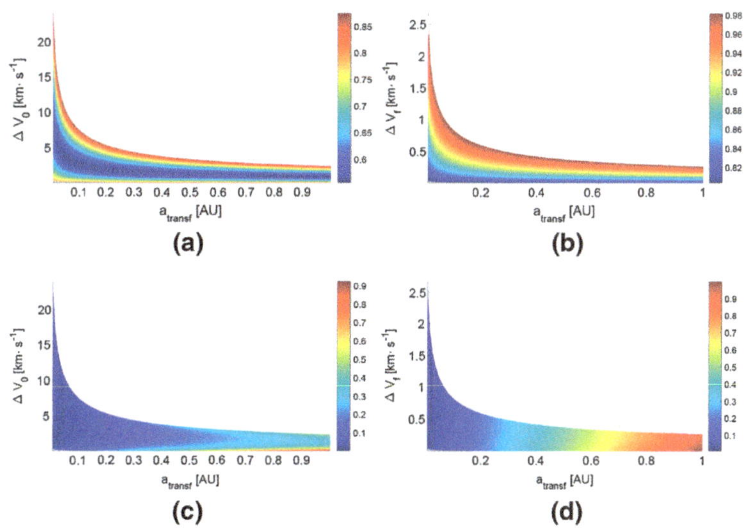

Figuer 6: Departure (*left-side*) and arrival (*right-side*) orbits Keplerian elements (eccentricity and semi-major axis) versus magnitudes of initial and final ΔVs and semi-major axis of the transfer orbits.

and they are shown in Fig. 5; specifically, $\dot{p}_0$ is represented in Fig. 5a–c, whereas $\dot{p}_f$ in Fig. 5d–f. As in Fig. 4, the structure of the maps of Fig. 5 is very well defined. These plots are evaluated (as in the previous case) for fixed values of $e = 0.8$ and $\nu_0 = 5$ deg, while ν_f is increasing and αand β are defined in the interval from 0 deg to 360 deg.

For small true anomaly spans($\nu_f = 20.0$ deg), both the derivatives have a sort of 'stripe' configuration, with opposite values. In fact, it is clear from Fig. 5a, d that $\dot{p}_0 > 0$ and $\dot{p}_f < 0$ and they reach a minimum value of ≈ -800 and $\approx +800$ respectively. Therefore, the two derivatives define the same boundary optimality areas. On the other hand, for larger $\Delta\nu$, $\dot{p}_0$ and $\dot{p}_f$ tend to smaller absolute values and they both have positive and negative values. Furthermore, as ν_f increases, the two structures are no longer comparable: they move and stretch tending, in the end, into two completely different scenario. At $\nu_f = 180$ deg, in fact, the initial derivative has a vertical stripes structure (Fig. 5c), while the final derivative has well-defined rounded non-optimal areas (negative value, represented by dark blue and green regions) which have the same size (Fig. 5f).

For both cases (Figs. 4, 5), the most interesting area of the map is represented when the variables cross their critical boundary value (that is 1 for the primer vector and 0 for its derivative). This aspect represents, in fact, the marginal conditions from which a trajectory changes its optimality properties.

The important characteristic of the results presented so far is that they encompass an entire set of orbits with same primer vector profile but different transfer orbits and BC. Specifically, what is varying are the semi-major axis of the transfer orbit (atransfatransf) and the magnitudes of the ΔVs at the beginning and at the end (therefore the departure and arrival orbits themselves).

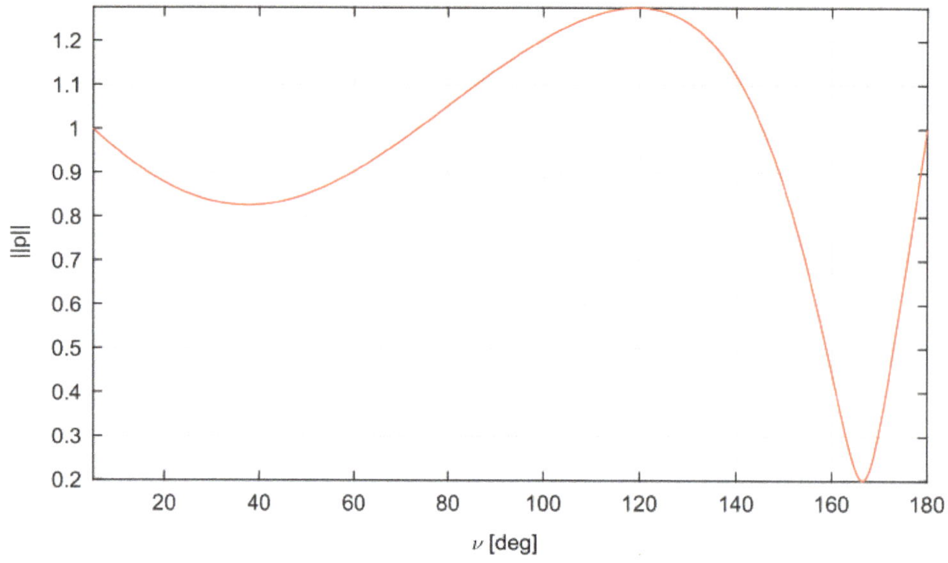

Figure 7: Primer vector profile, in common to all departure and arrival orbits of Fig. 6.

An interesting complete scenario is represented in Fig. 6, where it is shown how *e* and *a* of the departure and arrival orbits vary according to different magnitudes of $\mathbf{v}_{0,f}$ and atransfatransf. The fixed parameters are the 5 aforementioned dimensions of the search space: BC in direction of ΔVs (α and β), eccentricity of the transfer orbit and BC in position (v_0 and v_f). In this specific case the values are $\alpha = 30$ deg, $\beta = 60$ deg, $e = 0.5$, $\nu_0 = 5$ deg and $\nu_f =$ 180 deg. The simulations have been run with the Sun as central body, a set of semi-major axes of the transfer orbit that goes from 0.01 AU to 1.0 AU and a set of magnitudes of $\mathbf{v}_{0,f}$,f that go from 1 to 80 % of the magnitudes of $\mathbf{v}_{0,f}$ f (see Eq. 63).

On the left-hand side of Fig. 7 are the Keplerian elements of the possible departure orbits, while on the right-hand side are the equivalent parameters for the arrival orbits. With the chosen intervals of atransfatransf and $\Delta\mathbf{v}_{0,f}$, the eccentricities of the departure orbits result in a possible variation from roughly 0.5 to 0.87, while the ones of the arrival orbits have higher values (0.8–0.96). On the other hand the values of the semi-major axes of the two orbits are roughly comparable, but their variation is different with respect to atransfatransf and $\mathbf{v}_{0,f}$.

The main conclusion of this result is that all kinds of departure and arrival orbits summarized in Fig. 6 will meet the same objective, in fact they all correspond to the same primer vector profile (shown in Fig. 7), that in this case is non-optimal.

In summary, this procedure gives a different approach to the primer vector theory and allows one to pursue a complete analysis on entire classes of departure and arrival orbits with common characteristics for specific fixed parameters of the transfer trajectory.

CONCLUSIONS

This paper shows how applying a reference system transformation to the primer vector equation causes the decoupling between the in-plane and out-of-plane coordinates. The case when the transfer, departure and arrival orbits are coplanar is analyzed. Furthermore, a coordinates transformation is applied with the support of the Hamiltonian of the equations. The Palmer coordinates allow the simplification of the structure of the analysis and give a complete analytical approximation which, given the boundary conditions of the problem, can propagate the primer vector along the arc. For the circular transfer case, Hill's solution is also a solution for the primer vector problem. The primer vector is shown to be independent from the semi-major axis of the transfer orbit and the mathematical complexity and computational time of the problem are reduced. Furthermore, fixing the transfer orbit, it is established that the dimensions of the search space can be reduced to 5. Two approaches to understand the likelihood of the optimality of a specific set of transfers are examined in the paper, with related optimality maps. Finally it is shown that a single profile of primer vector is related to a set of departure and arrival orbits through the directions of the initial and final ΔVs and the semi-major axis of the transfer trajectory. The simplicity of the new procedure allows the exploitation of many other properties of the problem, not presented in this paper.

ACKNOWLEDGMENTS

This research has been funded by the European Commission through the Marie Curie Initial Training Network PITN-GA-2011-289240, AstroNet-II The Astrodynamics Network.

REFERENCES

1. Battin, R.H.: An Introduction to the Mathematics and Methods of Astrodynamics, Revised edn. AIAA Education Series (1999)

2. Bernelli-Zazzera, F., Lavagna, M., Armellin, R., Di Lizia, P., Topputo, F., Berz, M., et al.: Global Trajectory Optimization: Can We Prune the Solution Space When Considering Deep Space Maneuvers?. In: Technical report, ESA (2007)

3. Bryson, A.E., Ho, Y.: Applied Optimal Control, Optimization, Estimation, and Control. Taylor & Francis, London (1975)

4. Conway, B.A. (ed.): Spacecraft Trajectory Optimization. Cambridge University Press, Cambridge (2010)

5. Goodyear, W.H.: Completely general closed-form solution for coordinates and partial derivatives of the two-body problem. Astron. J. **70**(3), 189–192 (1965)

6. Jezewski, D.J.: Primer Vector Theory And Applications. In: Technical report, Lyndon B. Johnson Space Center, NASA (1975)

7. Jezewski, D.J., Rozendaal, H.L.: An efficient method for calculating optimal free-space n-impulse trajectories. AIAA J. **6**(11), 2160–2165 (1968)

8. Lawden, D.F.: Optimal Trajectories for Space Navigation. Butterworths Mathematical Texts, London (1963)

9. Lion, P.M., Handelsman, M.: Primer vector on fixed-time impulsive trajectories. AIAA J. **6**(1), 127–132 (1968)

10. Luo, Y., Zhang, J., Li, H., Tang, G.: Interactive optimization approach for optimal impulsive rendezvous using primer vector and evolutionary algorithms. Acta Astronaut. **67**, 396–405 (2010)

11. Olympio, J.T., Marmorat, J.P., Izzo, D.: Global Trajectory Optimization: Can We Prune the Solution Space When Considering Deep Space Maneuvers?. In: Technical report, ESA (2006)

12. Palmer, P.L.: Reachability and optimal phasing for reconfiguration in near circular orbit formations. J. Guid. Control Dyn. **30**(5), 1542–1546 (2007)

13. Prussing, J.E.: Optimal impulsive linear systems: sufficient conditions and maximum number of impulses. J. Astronaut. Sci. **48**(2), 195–206 (1995)

14. Quinn, T., Perrine, R.P., Richardson, D.C., Barnes, R.: A symplectic integrator for Hill's equations. Astron. J. **139**, 803–807 (2010)

15. Vasile, M., De Pascale, P.: Preliminary design of multiple gravity-assist trajectories. J. Spacecr. Rockets **43**(4), 794–805 (2006)

16. Vinh, N.X.: Integration of the primer vector in a central force field. J. Optim. Theory Appl. **9**(1), 51–58 (1972)

CHAPTER 11

ANALYTICAL AND SEMI-ANALYTICAL TREATMENT OF THE SATELLITE MOTION IN A RESISTING MEDIUM

S. E. Abd El-Bar and F. A. Abd El-Salam

INTRODUCTION

The problem of the orbital motion of celestial bodies in a resisting medium can be traced back to Newton's Principle [1] and Laplace's analysis of a possibly finite velocity of gravitation [2]. However, with the advent of the space age there arose the need of precisely predicting the orbital behavior of the newly launched bodies.

When the artificial satellites move in Low Earth Orbits (in brief LEO) their orbital motion is perturbed by the resisting force. Therefore, an adequate analytical description of the drag force is required to predict precisely the orbital perturbations due to atmospheric drag. These perturbations due to the atmospheric drag are more complicated and difficult to deal with than orbits perturbed by gravity because of the nonconservative property of the drag force. The adequate allowance for atmospheric drag turned out to be a problem for unforeseen complexity [3–8]. In order to be able to solve the equations of motion including the drag force analytically, one is forced to do many approximations. On the other hand, from the computational point of view, having an analytical solution at hand, we can jump from the initial conditions to the new state, which makes the computation extremely fast.

The order of magnitude of the atmospheric drag acting on a satellite depends on the altitude of the satellite. The orbital behavior of an LEO satellite could be strongly influenced by atmospheric drag [4, 9, 10], and therefore affects the quality of the remote sensing of the satellite [11, 12]. Traditionally, the problem of orbit disturbance by atmospheric drag is solved by numerical integration [13–18]. An analytical solution gives the theoretical integrals and shows the physical effects with very clear spectral properties [19–24]. This may give a direct insight into the physical phenomenon of the disturbance [25–27]. Bezdeka and Vokrouhlicky [28] developed a semianalytic theory of motion for close-Earth spherical satellites including drag and gravitational perturbations. Xu et al. [29] derived an analytical solution of a satellite orbit disturbed by atmospheric drag. They first transformed the disturbance force vector and rotated it to the orbital frame so that it can be used in the simplified Gaussian equations of satellite motion. The disturbances are separated into three parts: shortperiodic terms with triangular functions of the mean anomaly$M$, longperiodic terms with triangular functions of the argument of perigee and inclination (ω,), and secular terms [nonperiodic functions of semimajor axis and eccentricity (a,e)].

The main effects of the force are to bring out large secular decreases in the elements a, the semimajor axis, and e, the eccentricity, that cause the satellite plunges toward Earth. This perturbing influence is one of the most important perturbing forces in the altitude regime from 150 to 600 km which is called the drag force. The drag force, $\vec{F}_D$, per unit mass of the satellite can be represented by

$$\vec{F}_D = -\frac{1}{2}C_D\left(\frac{A}{m}\right)\rho\left\|\vec{V}\right\|\vec{V}, \tag{1.1}$$

Citation: S. E. Abd El-Bar and F. A. Abd El-Salam, "Analytical and Semi-Analytical Treatment of the Satellite Motion in a Resisting Medium," Journal of Applied Mathematics, vol. 2012, Article ID 489849, 18 pages, 2012. doi:10.1155/2012/489849.

where C_D is a dimensionless drag coefficient, empirically determined to have a value close to 2; it can be taken as 2.2 with an error (standard deviation) which should not exceed 5% unless the assumption about the accommodation coefficient is greatly in error. But in fact the drag coefficient depends on the mean free path of the atmospheric molecules with respect to the linear dimensions of the satellite; therefore, the orbital theory should be applied with caution for large satellites in their last revolutions [4]. After that Sehnal [30] derived a theoretical expression for the dimensionless drag coefficient (see (3.1) there). $\vec{V}$ is the velocity of the satellite relative to the atmosphere (called the ambient velocity), A represents the effective cross-sectional area of the satellite, which is to be found by averaging all possible projected areas of the satellite onto a plane perpendicular to $\vec{V}$, m is the mass of the satellite, and ρ is the density of the atmosphere. The density ρ depends on position and time in a very complicated manner. Its variation with height depends on the form of the atmosphere, since it decreases rapidly with slight growing in the altitude; the effect of the flattening of the atmosphere is rather significant.

ROTATION OF THE ATMOSPHERE

This effect is so small, so a simple approximation is justified. The differential rotation of the atmosphere complicates the numerical calculation and it gives no sense in this very small effect. So we will assume that the atmosphere rotates with the same angular velocity of the Earth $\vec{\sigma}$, then

$$\vec{V} = \vec{v} - \vec{\sigma} \times \vec{r}, \tag{2.1}$$

where $\vec{v}$ is the orbital velocity of the satellite, and $\vec{r}$ its radius vector. Thus

$$V^2 = v^2 - 2\vec{v} \cdot (\vec{\sigma} \times \vec{r}) + (\vec{\sigma} \times \vec{r})^2. \tag{2.2}$$

If $(\hat{i}, \hat{j}, \hat{k})$ are the base vectors of the rectangular geocentric equatorial frame associated with the spherical coordinates (r, λ, δ), then $\vec{\sigma} = \sigma\hat{k}$ with $\sigma = 7.27205217 \times 10^{-5}$ rad/s, and

$$V^2 = v^2 - 2\sigma h \cos I + \sigma^2 r^2 \cos^2\delta, \tag{2.3}$$

where I is the orbital inclination, δ is the satellite's declination, and h is the integral of areas. Now the ambient velocity can be written in the form:

$$V^2 = v^2 k_R, \tag{2.4}$$

Where

$$k_R = 1 - \frac{2\sigma h \cos I}{v^2} + \frac{\sigma^2 r^2 \cos^2\delta}{v^2} \tag{2.5}$$

The third term in k_R may usually be neglected, being less than about 1/250, and the second term which is of order of 1/15 may be evaluated at perigee, where the density is much greater than elsewhere on the orbit; thus we may effectively regard k_R as a constant.

THE ATMOSPHERIC DRAG FORCE MODEL

We have assumed that the atmospheric density is approximated as an exponential function which is given by

$$\rho = \rho_{p_o} \exp[\beta(r_{p_o} - r)] \tag{3.1}$$

where ρ_{p_o} is the atmospheric density at the initial perigee point r_{p_o}, β is taken as a constant, $1/\beta$ is about 50 kilometers for most satellite orbits, and r is given by relation:

$$r = a(1 - e\cos E) \tag{3.2}$$

where a, e are the orbital semimajor axis and eccentricity and E is the eccentric anomaly of the satellite. Now

$$\rho = \rho_{p_o} \exp[\beta(a_o - a - x_o) + \beta x \cos E] \tag{3.3}$$

where $x = ae$, $x_o = a_o e_o$, $\vec{V}$ is the velocity of satellite relative to ambient atmosphere and is generally through

$$\vec{V} = \vec{v} - \vec{\omega}_a \times \vec{r}, \tag{3.4}$$

where $\vec{v}$ is the velocity of satellite with respect to the Earth's center, $\vec{\omega}_a$ is the west-to-east angular velocity of the atmosphere and $\vec{r}$ is the position vector of the satellite.

The radial V_s, transverse V_t, and orthogonal V_w components of the velocity are given by

$$\begin{aligned} V_s &= \sqrt{\frac{\mu}{a(1-e^2)}}\, e \sin f, \\ V_t &= \sqrt{\frac{\mu}{a(1-e^2)}}(1 + e\cos f) - r\omega_a \cos I, \\ V_w &= \omega_a r \sin I \cos(f + \omega), \end{aligned} \tag{3.5}$$

where $\mu = G(M_\oplus + m) \approx GM_\oplus$, the Earth's gravitational parameter, is the product of the universal constant of gravity G and the total mass of the Earth $M_\oplus$, the mass of the satellite m, and f, ω are the true anomaly and the argument of the perigee of the satellite. Then, the radial S_D, transverse T_D, and orthogonal W_D components of the drag force are given by

$$\begin{aligned} S_D &= -\frac{1}{2} C_D \frac{A}{m} \rho \left\|\vec{V}\right\| \sqrt{\frac{\mu}{a(1-e^2)}}\, e \sin f, \\ T_D &= -\frac{1}{2} C_D \frac{A}{m} \rho \left\|\vec{V}\right\| \left[\sqrt{\frac{\mu}{a(1-e^2)}}(1 + e\cos f) - r\omega_a \cos I\right] \\ W_D &= -\frac{1}{2} C_D \frac{A}{m} \rho \left\|\vec{V}\right\| r\omega_a \sin I \cos(f + \omega), \end{aligned} \tag{3.6}$$

Where

$$\left\|\vec{V}\right\| = \left[\frac{\mu}{a(1-e^2)}\left(e^2 + 2e\cos f + 1\right) + r^2\omega_a^2\left(\cos^2 I + \sin^2 I \cos^2(f + \omega)\right) - 2r\omega_a \cos I \sqrt{\frac{\mu}{a(1-e^2)}}(1 + e\cos f)\right]^{1/2}. \tag{3.7}$$

Using the relations

$$\cos f = \frac{\cos E - e}{1 - e\cos E}, \tag{3.8}$$

$$\sin f = \frac{\sqrt{1 - e^2}\sin E}{1 - e\cos E}. \tag{3.9}$$

Equations (3.6), can be written in the form

$$S_D = -b\frac{n}{1-\cos E}ae\rho\left\|\vec{V}\right\|\sin E,$$

$$T_D = -b\frac{n(1-e^2)}{1-\cos E}a\rho\left\|\vec{V}\right\|\left[1-d\frac{(1-ed\cos E)^2}{1-e^2}\right],$$

$$W_D = -b\omega_a a\rho\left\|\vec{V}\right\|(1-e\cos E)\sin I\cos(f+\omega), \tag{3.10}$$

Where

$$\left\|\vec{V}\right\| = \sqrt{\frac{\mu}{a}}\sqrt{\frac{1+e\cos E}{1-e\cos E}}\left\{1-d\frac{1+e\cos E}{1-e\cos E}+\frac{\omega_a^2}{2n^2}\left[\frac{1+e\cos E}{1-e\cos E}\right]^2\right.$$
$$\left.+\left[\left(1-e^2\cos^2E\right)\sin^2I\cos^2(f+\omega)+e^2\cos^2I\sin^2I\right]\right\}+\cdots,$$

$$d = \frac{\omega_a}{n}\sqrt{1-e^2}\cos I, \tag{3.11}$$

where $n = \sqrt{\mu/a^3}$ is the mean motion.

For a close Earth's satellite the ratio ω_a/n is about 1/15; the third term in the expansion is thus only about 1/500of the leading term and will be ignored. For satellites that are not close, the ratio will be larger but the atmospheric density that will multiply this expression will then be so small that the product will still be unimportant. we thus take

$$\left\|\vec{V}\right\| = \sqrt{\frac{\mu}{a}}\sqrt{\frac{1+e\cos E}{1-e\cos E}}\left[1-d\frac{1+e\cos E}{1-e\cos E}\right]$$

$$b = \frac{1}{2}C_D\left(\frac{A}{m}\right). \tag{3.12}$$

LAGRANGE'S PLANETARY EQUATIONS

The orbit is uniquely determined by 6 elements, namely, the orbital elements. The Lagrange planetary equations are a set of 6 differential equations which measure the rate change of these orbital elements a,,i,ω,Ω and any time element, for example, the mean anomaly

$$\dot{I} = \frac{r\cos(f+\omega)}{na^2\sqrt{1-e^2}}W,$$

$$\dot{\Omega} = \frac{r\sin(f+\omega)}{na^2\sqrt{1-e^2}\sin I}W,$$

$$\dot{a} = \frac{2e\sin f}{n\sqrt{1-e^2}}S+\frac{2a\sqrt{1-e^2}}{nr}T,$$

$$\dot{\omega} = -\dot{\Omega}\cos I-\frac{\sqrt{1-e^2}}{nae}\left[S\cos f-T\left(1+\frac{r}{a(1-e^2)}\right)\sin f\right],$$

$$\dot{e} = \frac{\sqrt{1-e^2}\sin f}{na}S+\frac{\sqrt{1-e^2}}{nae}\left[\frac{a(1-e^2)}{r}-\frac{r}{a}\right]T,$$

$$\dot{M} = n-\frac{2}{na^2}Sr-\dot{\omega}\sqrt{1-e^2}-\dot{\Omega}\sqrt{1-e^2}\cos I. \tag{4.1}$$

There are many forms of the Lagrange planetary equations. In many cases, some of orbital elements are undefined, and hence, we have to replace this set by a new set of elements valid to define the orbit, and the planetary equations will take a new form in terms of the new element. For the equatorial orbits I=0 or I=180o or even when I, e have small values we have near equatorial near circular orbits and the periapsis is undefined and hence the time of pericentric passage,T and ω are undefined or at least meaningless and should be replaced by other elements as the longitude of periapsis, ϖ, defined as $\varpi=\omega+\Omega$.

THE EARTH'S GRAVITY FORCE MODEL

The gravitational field of the Earth is quite irregular. However, it can be represented by an infinite series of spherical harmonics, choosing the principal axes with origin of coordinates at centre of mass in addition assuming also that the gravitational field of the Earth is axially symmetric; then the form adopted is, [31]

$$U_\oplus = -\frac{\mu}{r}\left[1-\sum_{n=2}^{\infty}\left(\frac{R_\oplus}{r}\right)^n J_n P_n(\sin\delta)\right] \tag{5.1}$$

where $R_\oplus$ is the equatorial radius of the Earth, μ is the Earth's gravitational parameter, (r,δ) are the appeared geocentric coordinates of the satellite, J_n are zonal harmonic coefficients, and $P_n(\sin\delta)$ the Legendre Polynomials are givenby

$$P_n(\sin\delta) = \frac{1}{2^n}\sum_{j=0}^{N}(-1)^j\frac{(2n-2)!(\sin\delta)^{n-2j}}{j!(n-j)!(n-2j)!},\quad (n=0,1,2,3,\ldots), \tag{5.2}$$

Where

$$N = \begin{cases} \frac{n}{2} & n \text{ even,} \\ \frac{n-1}{2} & n \text{ odd.} \end{cases} \tag{53}$$

After some lengthy straightforward algebraic calculation we can evaluate the force components in the radial S_G, transverse T_G, and normal W_G components as

$$\begin{aligned} S_G = &-\frac{3}{4}J_2\frac{\mu R_\oplus^2}{r^4}\left[2-3\sin^2 I(1-\cos 2(f+\omega))\right] \\ &+J_3\frac{R_\oplus^3\mu}{2r^5}\sin I\left[\left(15\sin^2 I-12\right)\sin(f+\omega)-5\sin^2 I\sin 3(f+\omega)\right] \\ &+J_4\frac{5R_\oplus^4\mu}{64r^6}\left[24-120\sin^2 I+105\sin^4 I+20\left(6\sin^2 I-7\sin^4 I\right)\right. \\ &\qquad\left.\times\cos 2(f+\omega)+35\sin^4 I\cos 4(f+\omega)\right], \\ T_G = &-3J_2\frac{\mu R_\oplus^2}{r^4}\sin I\cos I\sin F_{1,1}\cos F_{1,1}-J_3\frac{3R_\oplus^3\mu}{4r^5}\sin I\cos F_{1,1}\left[-2+5\sin^2 I-\sin^2 I\cos F_{2,2}\right] \\ &-J_4\frac{5R_\oplus^4\mu}{8r^6}\sin^2 I\cos F_{1,1}\left[\left(-12+21\sin^2 I\right)\sin F_{1,1}-7\sin^2 I\sin F_{3,3}\right], \\ W_G = &-3J_2\frac{\mu R_\oplus^2}{r^4}\sin I\cos I\sin(f+\omega) \\ &-J_3\frac{3R_\oplus^3\mu}{4r^5}\cos I\left[-2+5\sin^2 I-\sin^2 I\cos 2(f+\omega)\right] \\ &-J_4\frac{5R_\oplus^4\mu}{8r^6}\sin I\cos I\left[\left(-12+21\sin^2 I\right)\sin(f+\omega)-7\sin^2 I\sin 3(f+\omega)\right]. \end{aligned} \tag{5.4}$$

THE ANALYTICAL SOLUTION

In this section, we proceed to calculate the first-order approximation of the orbital elements caused by the drag and the gravity forces during revolution by integrating Lagrange's planetary equations (4.1) from $E=-\pi$ to $E=\pi$, and retain only terms to order $O(e^2)$, after substitution of $\{S = S_D + S_G,\ T = T_D + T_G, W = W_D + W_G\}$ into the Lagrange planetary equations we have the perturbations in the elements as

$$\Delta\sigma = \Delta\sigma_D + \Delta\sigma_G, \tag{6.1}$$

where $\sigma\equiv(a,e,I,\omega,\Omega,M)$ denotes any of the orbital elements, $\Delta\sigma_D$ the perturbations due to the drag force, and $\Delta\sigma_G$ the perturbations due to the gravity force is given by

$$\begin{aligned} \Delta a_D = -4\pi a^2 H_o\Bigg\{&\left[\left(1-2\frac{\omega_a}{n}\cos I\right)+\frac{3}{2}\left(1+\frac{\omega_a}{n}\cos I\right)e^2\right]I_o \\ &+2eI_1+\frac{1}{2}\left(\frac{3}{2}+\frac{\omega_a}{n}\cos I\right)I_2e^2\Bigg\}, \end{aligned}$$

$$\Delta a_G = J_2\sum_{i=1}^{5}\sum_{j=-2}^{2}\alpha''_{ij}\cos F_{i,j}+J_3\sum_{k=1}^{7}\sum_{l=-3}^{3}\beta''_{kl}\sin F_{k,l}+J_4\sum_{m=1}^{9}\sum_{n=-4}^{4}\gamma''_{mn}\cos F_{m,n},$$

$$\Delta e_D = -4\pi abH_0\left\{\frac{e}{2}\left(1+\frac{5\omega_a}{2n}\cos I\right)I_0+\left[\left(1-\frac{2\omega_a}{n}\cos I\right)-e^2\left(\frac{5}{8}-\frac{2\omega_a}{n}\cos I\right)\right]I_1\right.$$

$$\left.+\frac{e}{2}\left(1+\frac{3\omega_a}{2n}\cos I\right)I_2+\frac{e^2}{8}I_3\right\},$$

$$\Delta e_G = J_2\sum_{i=1}^{5}\sum_{j=-2}^{2}\overline{\delta}_{ij}\cos F_{i,j}+J_3\left\{X_1 f+\sum_{m=1}^{7}\sum_{n=-3}^{3}\overline{\sigma}_{mn}\sin F_{m,n}\right\}+J_4\left\{X_2 f+\sum_{k=1}^{9}\sum_{l=-4}^{4}\overline{Q}_{kl}\cos F_{k,l}\right\},$$

$$\Delta I_D = -\pi ab\omega_a H_o\frac{\sin I}{n}\left\{\left[\left(1-\frac{\omega_a}{n}\cos I\right)\right.\right.$$

$$\left.+\frac{e^2}{2}\left(\frac{9}{4}-\frac{3\omega_a}{n}\cos I+\frac{17}{8}\cos 2\omega\right)+\frac{55\omega_a}{8n}\cos I\cos 2\omega\right]I_o$$

$$+e\left[-1+\frac{3\omega_a}{n}\cos I+\frac{1}{2}\left(\frac{5\omega_a}{n}\cos I-9\right)\cos 2\omega\right]I_1$$

$$+\left[\left(1-\frac{\omega_a}{n}\cos I\right)\right.$$

$$\left.+\frac{e^2}{2}\left(5-\frac{17\omega_a}{n}\cos I\right)+\left(1-\frac{21\omega_a}{n}\cos I\right)\cos 2\omega\right]I_2$$

$$\left.+\frac{e}{2}\left(\frac{\omega_a}{n}\cos I-1\right)\cos 2\omega I_3+\frac{e^2}{16}\left(9-\frac{11\omega_a}{n}\cos I\right)\cos 2\omega I_4\right\},$$

$$\Delta I_G = J_2\sum_{i=1}^{3}Z_i\cos F_{i,2}+J_3\left\{X_3 f+\sum_{k=1}^{5}\sum_{l=-1}^{3}\overline{\overline{C}}_{kl}\sin F_{k,l}\right\}+J_4\left\{X_4 f+\sum_{p=1}^{7}\sum_{q=-2}^{4}\overline{\overline{S}}_{pq}\cos F_{p,q}\right\},$$

$$\Delta\Omega_D = -ab\omega_a H_0\frac{\omega_a}{n}\left\{\left(\left[\frac{3e}{4}\left(1-\frac{\omega_a}{n}\cos I\right)+\frac{e^2}{8}\left(41-\frac{64\omega_a}{n}\cos I\right)\right]\sin 2\omega\right.\right.$$

$$\left.+\frac{e^2}{2}\left(1-\frac{\omega_a}{n}\cos I\right)\cos 2\omega\right)I_o$$

$$+\left[\frac{e}{2}\left(9-\frac{7\omega_a}{n}\cos I\right)-\frac{e^2}{2}\left(6-\frac{11\omega_a}{n}\cos I\right)\right]\sin 2\omega I_1$$

$$+\left(\left[\left(1-\frac{\omega_a}{n}\cos I\right)+e\left(1-\frac{\omega_a}{n}\cos I\right)\right.\right.$$

$$\left.\left.+\frac{e^2}{4}\left(9-\frac{33\omega_a}{n}\cos I\right)\sin 2\omega+\frac{e^2}{2}\left(1-\frac{\omega_a}{n}\cos I\right)\cos 2\omega\right]\right)I_2$$

$$+\left[\frac{e}{2}\left(\frac{3\omega_a}{n}\cos I-1\right)-\frac{e^2}{4}\left(4-\frac{9\omega_a}{n}\cos I\right)\right]\times\sin 2\omega I_3$$

$$+\frac{e}{8}\left[\left(2-\frac{2\omega_a}{n}\cos I\right)+\left(5-\frac{13\omega_a}{n}\cos I\right)\right]\sin 2\omega I_4$$

$$\left.+\frac{e^2}{4}\frac{\omega_a}{n}\cos I\sin 2\omega I_5\right\},$$

$$\Delta\Omega_G = J_2\left\{X_8 f+\sum_{b=1}^{3}\sum_{c=0}^{2}\overline{p}_{bc}\sin F_{b,c}\right\}+J_3\left\{X_9 f+\sum_{\alpha=1}^{5}\sum_{\beta=-1}^{3}\overline{q}_{\alpha\beta}\cos F_{\alpha,\beta}\right\}$$

$$+J_4\left\{X_{10} f+\sum_{i=1}^{7}\sum_{j=-2}^{4}\overline{p}_{ij}\sin F_{i,j}\right\},$$

$$\Delta\omega_D = \Delta\Omega_D\cos I,$$

$$\Delta\omega_G = J_2\left\{X_5 f+\sum_{p=1}^{5}\sum_{q=-2}^{2}\overline{U}_{pq}\sin F_{p,q}\right\}+J_3\left\{X_6 f+\sum_{k=1}^{7}\sum_{l=-3}^{3}\overline{V}_{kl}\cos F_{k,l}\right\}$$

$$+J_4\left\{X_7 f+\sum_{m=1}^{9}\sum_{n=-4}^{4}\overline{Z}_{mn}\sin F_{m,n}\right\},$$

$$\Delta M = -2\pi n, \tag{6.2}$$

Where

$$I_0(x) = \frac{1}{\pi}\int_0^{\pi} \cos nE \exp(x\cos E)dE \tag{6.3}$$

is the first kind modified Bessel function of order n, and by using the values of $I_0(c)$ and $I_1(c)$ one finds that

$$\begin{aligned} I_2 &= I_0 - \frac{2}{c}I_1, \\ I_3 &= -\frac{4}{c}I_0 + \left(1+\frac{8}{c^2}\right)I_1, \\ I_4 &= \left(1+\frac{24}{c^2}\right)I_0 - \left(1-\frac{6}{c^2}\right)I_1, \\ I_5 &= -\frac{12}{c}\left(1+\frac{16}{c^2}\right)I_0 + \left(1+\frac{72}{c^2}+\frac{384}{c^4}\right)I_1, \end{aligned} \tag{6.4}$$

where H_o is the scale height. It is given by

$$H_o = \rho_{po}\mathrm{exe}[\beta(a_o - a - x_o)] \tag{6.5}$$

Finally the nonvanishing coefficients included in the gravity perturbations are previously computed by the authors [32].

THE NUMERICAL SOLUTION

It is well known that the numerical solution methods provide accurate ephemeris of a satellite with respect to a force of a complex mathematical formulation. So in this section we will solve numerically the equations of motion after substituting the drag force components S_D, T_D, W_D, given by (3.10) together with the Earth's gravity force components S_G, T_G, W_G, given by (5.4) (we will use Earth's oblate model and retaining terms up to fourth zonal harmonics $J4$).

There are many methods to get the numerical solutions of any first-order differential equations. We will use one of the best methods of the numerical integration schemes, namely, Runge-Kutta fourth-order method.

If the Earth was assumed spherical, had no atmosphere, and the satellite is isolated from the other bodies in the solar system, the orbit of a satellite would be an ellipse of constant size and shape in a plane whose direction remains fixed relative to the stars. The orbital elements a,,i,ω,Ω would remain unchanged. This simple thought rather insipid situation is upset by the effects of the considered perturbing forces, the Earth's oblateness, and the atmospheric drag.

The Earth's oblateness leads to two major gravitational perturbations on the satellite orbits, the first is the rotation of the orbital plane about the Earth's axis in the direction opposite to the satellite's motion so that (if$I$<90o) the angle steadily decreases while I remains constant. The second effect is the rotation of the major axis in the orbital plane so that the argument of perigee ω increases. The chief effects of the atmospheric drag on a satellite orbit are fortunately quite different from those of the gravitational field. Since the atmospheric density decreases rapidly as height above the Earth increases, a satellite in an orbit of appreciable eccentricity is affected mostly by drag within a small section of the orbit where it is closest to the Earth. To first-order approximation, the effect of atmospheric drag is to retard the satellite as it passes perigee, with the result that it does not swing out so far from the Earth at the subsequent apogee. The apogee height is reduced while the perigee height remains almost constant. The orbit contracts and becomes more nearly circular; the orbital elements (a,) decrease steadily [4].

Now, a computer program of fourth-order Runge-Kutta method is applied for the solution of this system using the initial values and a fixed step size. Also, we will choose one of the numerical examples to show the decay in two orbital elements, namely, (a,). Also we will give a numerical example to the analytical solution given by (6.1) in order to compare the results of the numerical and analytical solutions.

NUMERICAL EXAMPLE

In this paragraph, we give a numerical example concerning the artificial satellite ROHINI (1980 62A) [33]. The data are as follows.

The satellite mass and its cross-sectional areas are, respectively, (m)=35.443kg, (A)=0.319019×10^{-6}. The initial orbital elements (semimajor axis, eccentricity, and the inclination) of the satellite are a_0=6989.2057km, e_0=0.04367712°,$_0$=44.67198° ω_0=239.3378° km,Ω_0=174.1602°, and M_0=25.63974°. And one orbital revolution was elapsed in 1.58835208 hours. The required physical and dynamical constants are

$$\begin{gathered} R_{\text{Å}} = 6378.135\,\text{km}, \quad \mu = 398600.8\,\text{km}^3/\text{sec}^{-2}, \quad J_2 = 1.08263\times10^{-3}, \\ J_3 = -2.53648\times10^{-6}, \quad J_4 = -1.6233\times10^{-6}, \quad \omega_a = 7.27205217\times10^{-6}\,\text{rad/sec}, \\ \rho_0\,(300\,\text{km}) = 0.0221\,\text{kg}\,\text{km}^{-3}, \quad \rho_0\,(250\,\text{km}) = 0.0681\,\text{kg}\,\text{km}^{-3}, \\ \rho_0\,(200\,\text{km}) = 0.271\,\text{kg}\,\text{km}^{-3}, \quad C_D = 2.2. \end{gathered} \tag{8.1}$$

The results of the numerical example of the analytical solution as well as the numerical solution are represented on Figures 1(a), 2(a), 3(a), 4(a), 5(a), 6(a), 7(a), 8(a), and 9(a) to show the decay of the semimajor axis, and Figures 1(b)–9(b) to show the change in the eccentricity. To read the figures properly we used the letters B, C, and D to denote to the numerical example of the analytical solutions while we used the letters E, F, and G to denote for the numerical solutions. The description of these letters is as follows.B: with drag and rotation, C: with drag without rotation, D: without drag, E: with drag and rotation, F: with drag without rotation, and G: without drag.

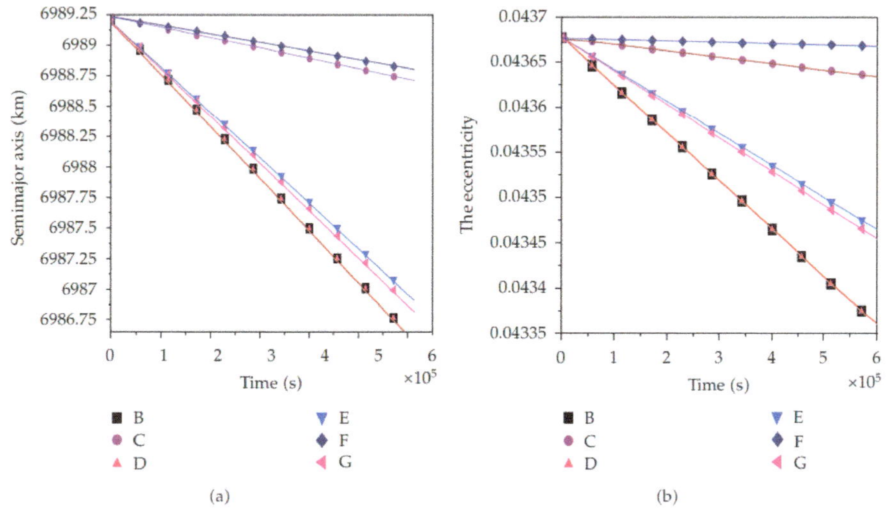

Figure 1: (a) For one hundred revolutions at 300 km height above the Earth's surface. (b) For one hundred revolutions at 300 km height above the Earth's surface.

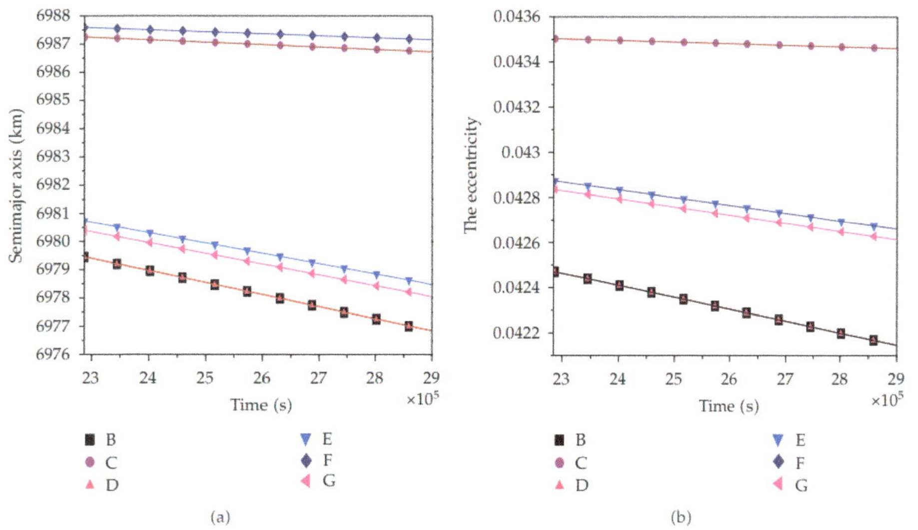

Figure 2: (a) For five hundred revolutions at 300 km height above the Earth's surface. (b) For five hundred revolutions at 300 km height above the Earth's surface.

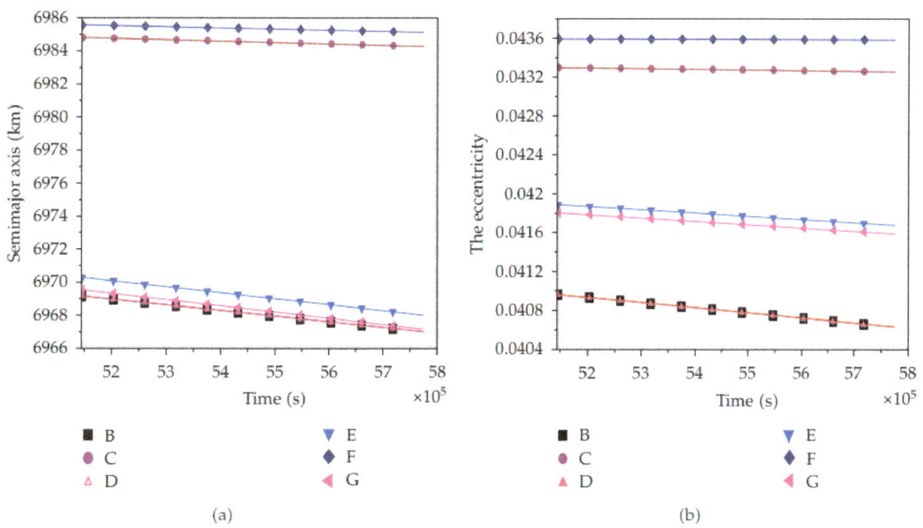

Figure 3: (a) For one thousand revolutions at 300 km height above the Earth's surface. (b) For one thousand revolutions at 300 km height above the Earth's surface.

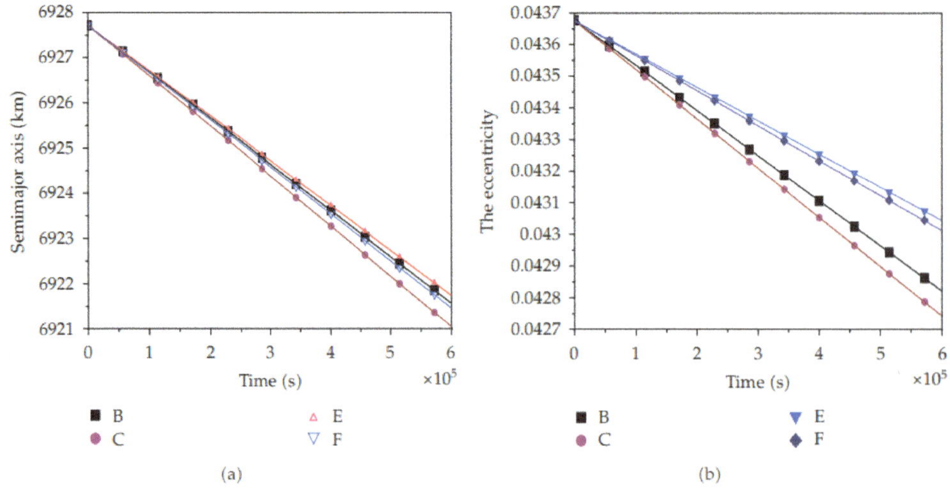

Figure 4: (a) For one hundred revolutions at 250 km height above the Earth's surface. (b) For one hundred revolutions at 250 km height above the Earth's surface.

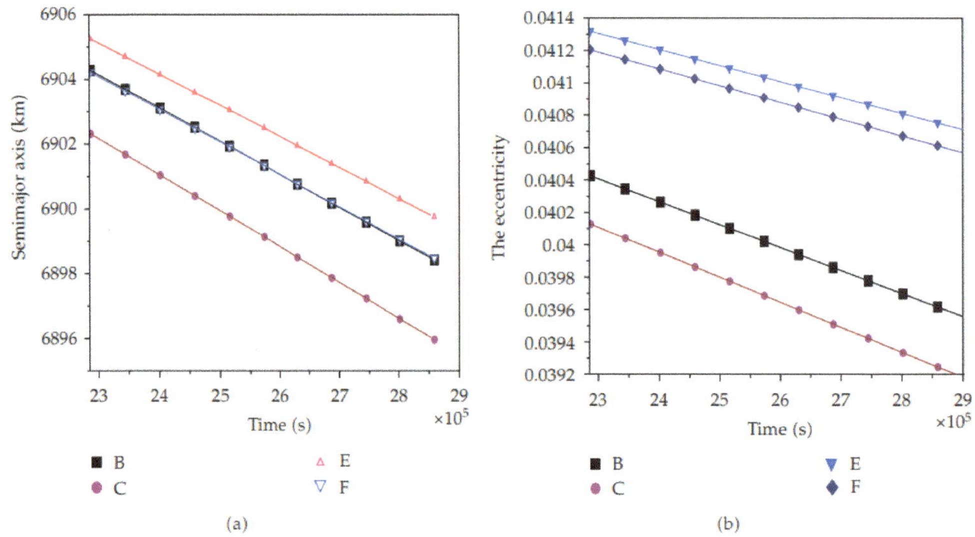

Figure 5: (a) For five hundred revolutions at 250 km height above the Earth's surface. (b) For five hundred revolutions at 250 km height above the Earth's surface.

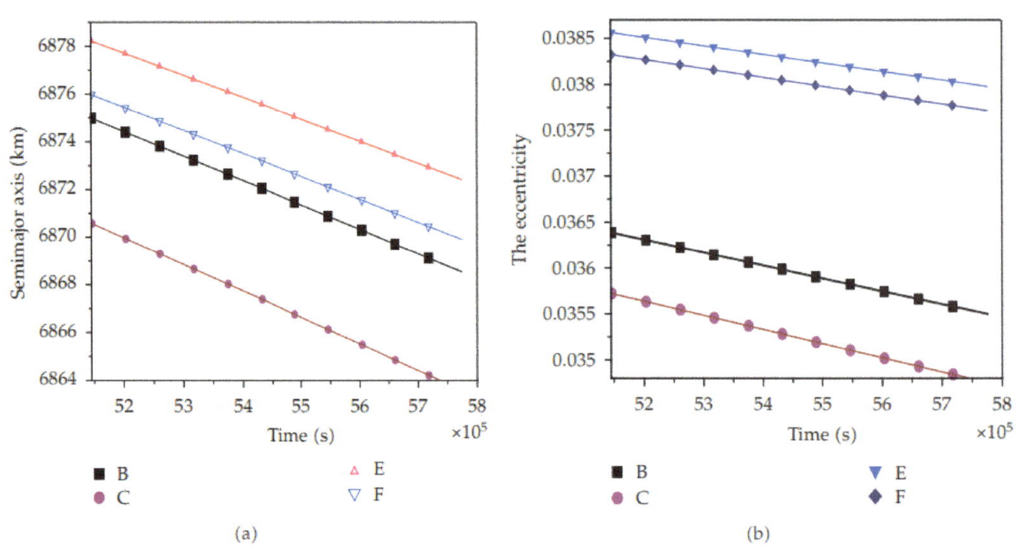

Figure 6: (a) For one thousand revolutions at 250 km height above the Earth's surface. (b) For one thousand revolutions at 250 km height above the Earth's surface.

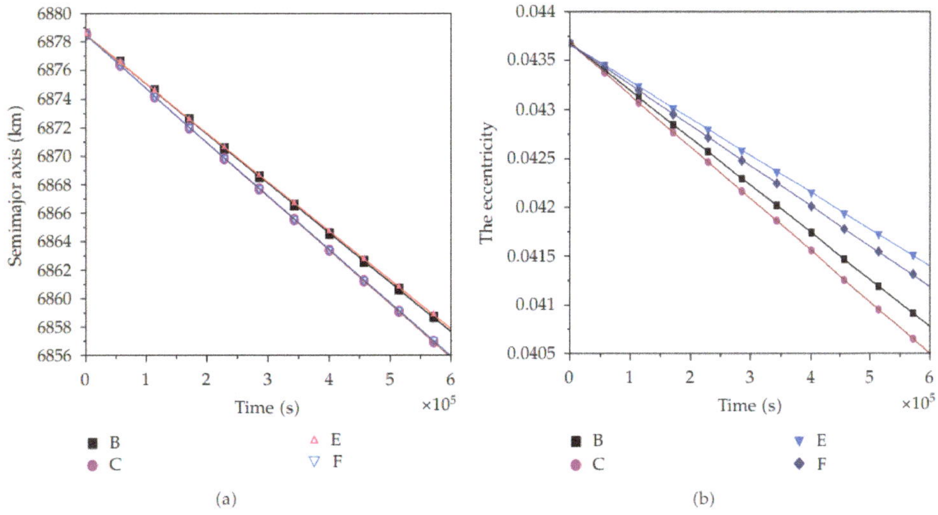

Figure 7: (a) For one hundred revolutions at 200 km height above the Earth's surface. (b) For one hundred revolutions at 200 km height above the Earth's surface.

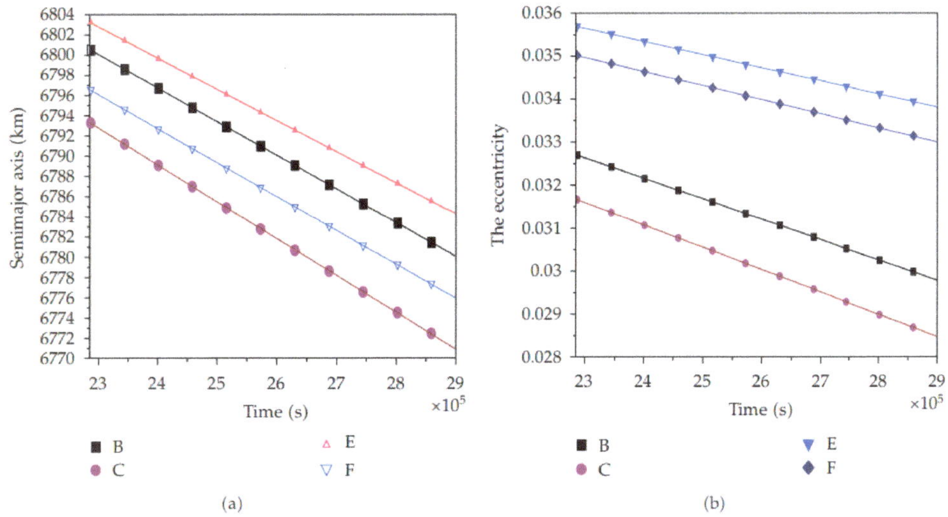

Figure 8: (a) For five hundred revolutions at 200 km height above the Earth's surface. (b) For five hundred revolutions at 200 km height above the Earth's surface.

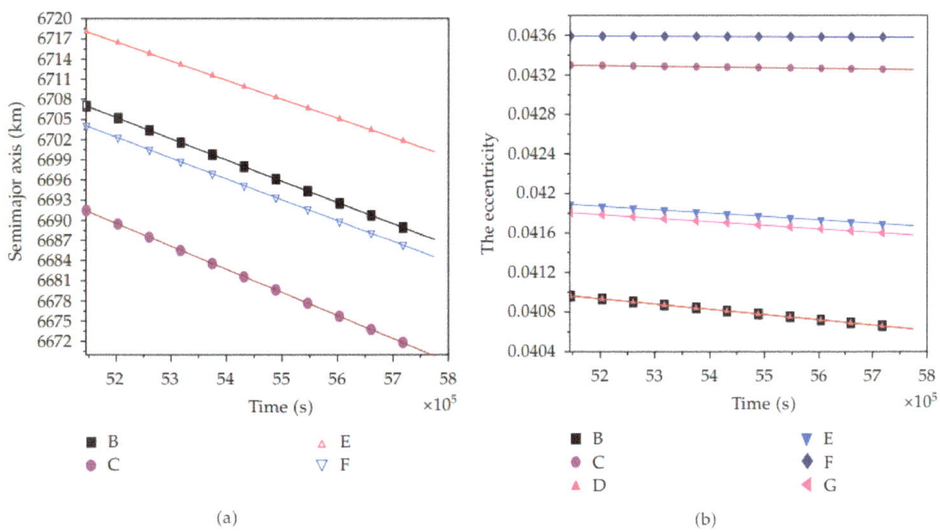

Figure 9: (a) For five thousand revolutions at 200 km height above the Earth's surface. (b) For five thousand revolutions at 200 km height above the Earth's surface.

All figures show the decrease in the semimajor axis and in the eccentricity for the rotating and nonrotating atmosphere at different altitudes.

ANALYSIS OF THE SOLUTIONS

The following dynamical effects are very clear. The decay of the semimajor axis is shown on Figures 1(a) to 9(a). We depicted the analytical and numerical solutions on one figure at different altitudes. Similar situation is done for the eccentricity on Figures 1(b)–9(b).(1)The decrease in the semimajor and in the eccentricity is significant (much bigger) at the high altitudes at the beginning of the motion but it oscillates afterwards, that is, the slope of the rate of change is sometimes high, sometimes moderate, and sometimes low. Therefore, we can conclude that the perturbation on the long range is nonlinear (see Figures 10(a) and 10(b)).(2)The effect of the atmospheric rotation is so small.(a)The atmospheric rotation appeared clearly on the figures of numerical solution because we have used the well-known fourth-order Runge-Kutta method.(b)The atmospheric rotation disappeared on the figures of the analytical solution since the approximation made in the analytical solution is much bigger than this effect. (3)Disregarding the drag force makes the rate of change of the semimajor axis and the rate of change of the eccentricity very little as seen clearly from the slopes of the figures of the solutions with drag and those without drag. This justifies the fact that at Low Earth Orbits (in brief LEO) the dominant secular perturbation is due to the drag force.(4)The difference between the solutions with drag and those without drag is sometimes big, sometimes low, and sometimes moderate. This signifies that the perturbation is nonlinear.

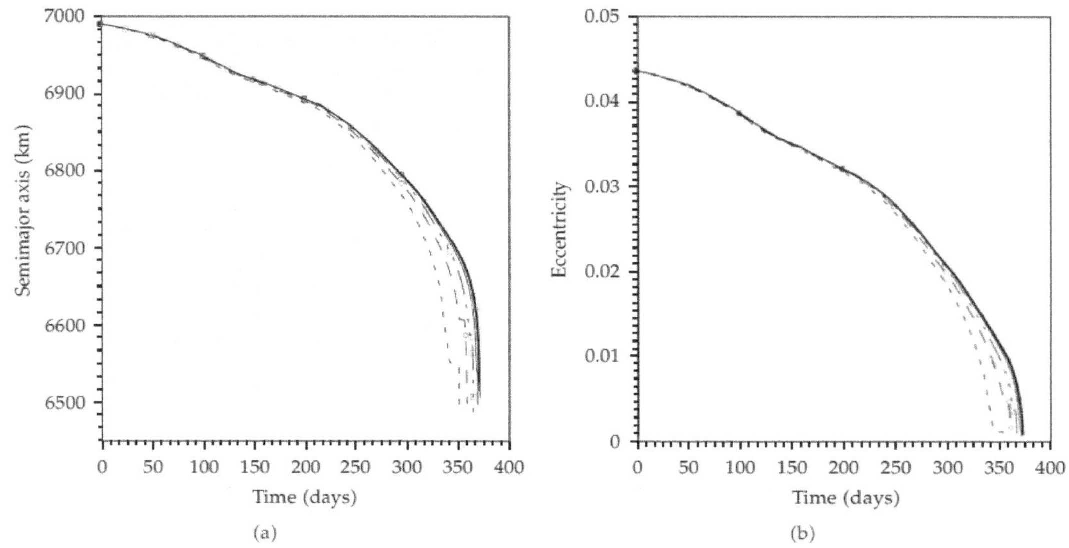

Figure 10: (a) Long range perturbation in the semimajor axis. (b) Long range perturbation in the eccentricity.

CONCLUSION

In this paper, we studied the effect of the atmospheric drag and the asphericity of the Earth on the orbital dynamics of an artificial satellite in a low Earth orbit. We approximated the density of the atmosphere to an exponential function including the rotation of the atmosphere. We obtained a first-order analytic atmospheric drag perturbations. We evaluated the short periodic perturbations of all orbital elements due to the geopotential. To obtain a more precise treatment, we integrated the equations of motion numerically using the Runge-Kutta method of fourth order. The validity of the theory is checked using the already decayed Indian satellite ROHINI where its data are available. A good agreement between the observed line elements of the satellite and our predicted solutions encourage us in future to apply our theory on any similar future missions.

ACKNOWLEDGMENTS

The authors are deeply indebted and thankful to the deanship of the scientific research and his helpful and distinct team of employees at Taibah University, Al-Madinah Al-Munawwarah, Saudia Arabia. This research work was supported by a Grant no. (617/1431).

REFERENCES

1. I. Newton, Philosophiae Naturalis Principia Mathematica, Book 2, Section 4, University of California Press, Berkeley, Calif, USA, 1934, English translation by F. Cajori, Newton.s Principia.
2. P. S. Laplace, Traité de Mécanique Céleste, Tom 4, Par 2 Courcier, 1805.
3. D. Brouwer, "Review of celestial mechanics," Annual Review of Astronomy and Astrophysics, vol. 1, pp. 219–234, 1963.
4. D. G. King-Hele, Theory of Satellite Orbits in an Atmosphere, Butterworth, London, UK, 1964.
5. D. G. King-Hele, A Tapestry of Orbits, Cambridge University Press, Cambridge, UK, 1992.
6. D. Brouwer and G. M. Clemence, Methods of Celestial Mechanics, Academic Press, New York, NY, USA, 1961.
7. V. A. Chobotov, Ed., Orbital Mechanics, AIAA, Washington, DC, USA, 1991.
8. D. Boccaletti and G. Pucacco, Theory of Orbits. Vol. 1, Springer, Berlin, Germany, 1996.

9. P. D. Van Kamp, Principles of Astronomy, Freemann & Co., San Francisco, Calif, USA, 1967.
10. R. H. Battin, An Introduction to the Mathematics and Methods of Astrodynamics, AIAA, Reston, Va, usa, 1999.
11. S. Herrick, Astrodynamics. Vol. 2, Van Nostrand Reinhold, London, uk, 1972.
12. D. A. Vallado, Fundamentals of Astrodynamics and Applications, Springer, Berlin, Germany, 3rd edition, 2007.
13. J. Meeus, Astronomical Algorithms, Willmann-Bell, Richmond, Va, USA, 1991.
14. O. Montenbruck and E. Gill, Satellite Orbits: Models, Methods and Applications, Springer, Heidelberg, Germany, 2000.
15. G. Seeber, Satellite Geodesy: Foundations, Methods, and Applications, Walter de Gruyter, Berlin, Germany, 2003.
16. G. Xu, GPS—Theory, Algorithms and Applications, Springer, Heidelberg, Germany, 2nd edition, 2007.
17. G. Xu, Orbits, Springer, Heidelberg, Germany, 2008.
18. G. Xu, T. H. Xu, T. K. Yeh, and W. Chen, "Analytic solution of satellite orbit disturbed by atmospheric drag," Monthly Notices of the Royal Astronomical Society, vol. 410, no. 1, pp. 654–662, 2010.
19. C. Cui, Deutsche Geodätische Kommission bei der Bayerischen Akademie der Wissenschaften, Reihe C, 357, 1990.
20. M. Schneider and C. F. Cui, Deutsche Geodätische Kommission bei der Bayerischen Akademie der Wissenschaften, Reihe A: Theoretische Geodäsie, 121, 2005.
21. S. M. Kudryavtsev, "Long-term harmonic development of lunar ephemeris," Astronomy & Astrophysics, vol. 471, pp. 1069–1075, 2007.
22. J. Licandro, A. Alvarez-Candal, J. De León, N. Pinilla-Alonso, D. Lazzaro, and H. Campins, "Spectral properties of asteroids in cometary orbits," Astronomy and Astrophysics, vol. 481, no. 3, pp. 861–877, 2008. ·
23. A. Pál, "An analytical solution for Kepler's problem," Monthly Notices of the Royal Astronomical Society, vol. 396, no. 3, pp. 1737–1742, 2009. ·
24. D. Lynden-Bell, "Analytic orbits in any central potential," Monthly Notices of the Royal Astronomical Society, vol. 402, no. 3, pp. 1937–1941, 2010. ·
25. W. Torge, Geodesy, Walter de Gruyter, Berlin, Germany, 1991.
26. J. Desmars, S. Arlot, J.-E. Arlot, V. Lainey, and A. Vienne, "Estimating the accuracy of satellite ephemerides using the bootstrap method," Astronomy and Astrophysics, vol. 499, no. 1, pp. 321–330, 2009. ·
27. J. R. Touma, S. Tremaine, and M. V. Kazandjian, "Gauss's method for secular dynamics, softened,"Monthly Notices of the Royal Astronomical Society, vol. 394, no. 2, pp. 1085–1108, 2009.
28. A. Bezdeka and D. Vokrouhlicky, "Semianalytic theory of motion for close-Earth spherical satellites includingdragand gravitational perturbations," Planetary and Space Science, vol. 52, pp. 1233–1249, 2004.
29. G. Xu, X. Tianhe, W. Chen, and T.-K. Yeh, "Analytical solution of a satellite orbit disturbed by atmospheric drag," Monthly Notices of the Royal Astronomical Society, vol. 410, no. 1, pp. 654–662, 2011. ·
30. L. Sehnal, "Dynamics and Astrometry of natural and artificial celestial bodies," in Proceedings of the Conference on Astrometry and Celestial Mechanics, pp. 325–332, Poznan, Poland, 1994.
31. P. M. Fitzpatrick, Principles Of Celestial Mechancics, Academic Press, London, UK, 1970.
32. F.A. Abd El- Salam and S. Abd El-Bar, "Computation of the different errors in the ballistic missiles range," Journal of the Applied Mathematics, vol. 2011, Article ID 349737, 16 pages, 2011. ·
33. R. K. Sharma and L. Mani, "Study of RS-1 orbital decay with KS differential equations," Indian Journal of Pure and Applied Mathematics, vol. 6, pp. 833–842, 1985.

CHAPTER 12

EARTH'S ROTATION: A CHALLENGING PROBLEM IN MATHEMATICS AND PHYSICS

José M. Ferrándiz, Juan F. Navarro, Alberto Escapa and Juan Getino

Keywords: Earth rotation nutation precession polar motion UT_1

RELEVANCE AND FEATURES OF THE EARTH ROTATION PROBLEM

The accurate determination and prediction of the orientation and the motion of the Earth in the space is needed in various fields, especially since the advent of the space age. Direct examples in which that knowledge is essential are: carrying out astronomical observations from an observatory located on the Earth's surface, making observations of spacecrafts from ground-located tracking stations, observing the Earth from the space, determination and prediction of satellite orbits, etc.

A good knowledge of the Earth's orientation is necessary for any applications related to pinpointing of points or objects with respect to the Earth at a global scale. There is a very broad set of such applications, ranging from popular handy simple navigation devices to the most sophisticated investigations of Space Geodesy that address the quantification of the physical effects of climate change. The most popular issue is the determination of sea level variation, whose magnitude is typically of a few millimetres per year. Besides, there is a variety of geodetic research aimed at finding the fingerprints of different geophysical processes: mass movements in oceans, ice sheets, terrestrial water storages, displacement fields associated with earthquakes, etc. (Plag*et al.*2009a, 2010) All of those geodetic studies have very demanding requirements of accuracy. The GGOS (Global Geodetic Observing System) initiative developed by the International Association of Geodesy (IAG) targeted the requirements of accuracy on the level of 1 mm in position and 1 mm/year in stability (Plag*et al.*2009b).

The IAG considers Earth rotation as one of the three pillars of Geodesy, because of the relevant role it plays in Space Geodesy, the other two being Earth geometry and gravity. Those "three pillars" provide the basis for the realization of the reference systems required to assign time-dependent coordinates to points and objects, and to describe the Earth's motion in space. This is not at all new: a quick look to the table of contents of some classic treatises like Tisserand(1891) would suffice to appreciate how the interaction of those pillars have fostered theoretical advances in many fields, such as Mathematics, Physics, Astronomy, Geodesy or Geophysics.

The solution to the Earth rotation problem consists mainly in the determination of the rotation matrix linking the celestial and the terrestrial reference frames. Nowadays, one of its most remarkable features is the extreme requirements of accuracy that must be fulfilled in the near future, at the level of a millimetre on the tangent plane to the planet surface, which corresponds to an angle about 30 μμ as from the Earth centre, roughly speaking. Due to its relevance and the broad range of its applications, there is an international service in charge of monitoring and predicting the Earth rotation, the International Earth Rotation and Reference Systems Service (IERS). It was established in 1987 by the International Astronomical Union (IAU) and the International Union of Geodesy and Geophysics (IUGG). IERS is also responsible of the realization and maintenance of the celestial and terrestrial reference frames associated with Earth rotation, namely the International

Citation: José M. Ferrándiz, Juan F. Navarro, Alberto Escapa, Juan Getino, Earth's Rotation: A Challenging Problem in Mathematics and Physics, doi:10.1007/s00024-014-0879-7

Celestial Reference Frame (ICRF) (Fey*et al.*2004) and the International Terrestrial Reference Frame (ITRF) (Altamimi*et al.*2011). More information appears in the Annual Reports yearly published by IERS (Dick2011). This service provides the international community with combined solutions for the EOP (Earth Orientation Parameters) (Bizouard and Gambis2009) and also publishes the IERS Conventions, which are widely used not only in the field of Earth rotation, but in satellite orbit determination and many other geodetic or geophysical applications (Petit and Luzum2010).

The Earth rotation is affected by many factors that must be accounted for to obtain solutions suitable to meet the present needs. Apart from the mathematical methods used to derive solutions, the main physical influences come from:

- Lunisolar gravitational attraction and planetary attraction.
- Earth figure and tensor of inertia (really not constant but time-varying).
- Earth internal structure: fluid outer core (FOC), solid inner core (SIC), etc.
- Effects at the boundaries of the inner layers, with dissipations and topography.
- Deformations (which produce geometric and dynamical effects).
- Tides: solid earth tides, ocean tides.
- Many other geophysical influences: redistribution of ice–water-vapor masses, currents, winds, hydrology, magnetism, post-glacial rebound, earthquakes, etc.

THE RIGID-EARTH MODEL: A FIRST STEP TOWARDS THE SOLUTION

Assuming the Earth is a rigid body is a logic first step, which has fulfilled the practical needs of accuracy for centuries. The solutions for nutations are close enough to the actual non-rigid Earth nutations, since the maximum differences between them (for each frequency) are below 30 mas (milliarcseconds), about 1 m on the Earth surface, and therefore irrelevant in ancient observations. The accuracy was thus satisfactory for applications until the development of highly accurate space geodetic techniques. Besides, the rigid model allows a great simplification for several reasons:

- In this case, perturbations only arise from the gravitational attraction of celestial bodies on an Earth with a constant tensor of inertia.
- The theoretical definition of the terrestrial frame is trivial, since there are no intricacies associated to deformations.
- Rigid body rotations have been widely studied for centuries, and there are lots of well-known topics easily found in the literature: Euler equations for rigid body rotation, systems of variables, integrability issues, etc.
- There are several well-established approaches at hand: Newtonian, Eulerian, Lagrangian, Hamiltonian.
- The unperturbed motion is essentially the Euler-Poinsot problem; therefore, it is integrable (in the Liouville sense) and convenient to derive asymptotic solutions by means of perturbation methods.

Eulerian formulation The formulation using Euler equations (1749) is the most extended and well-known in rotational dynamics. In the Newtonian framework, we consider an inertial reference system F centered at the barycenter O of the body and a system B attached to the body, moving with angular velocity ω. The velocity of a body particle P with respect to both frames holds $v_{\mathcal{F}} = v_B + \omega \times r = \omega \times r$, since P is in rest in B. Therefore, the absolute angular momentum M can be expressed in the body frame as $M = \Pi\omega$, Π being the matrix of inertia, constant in this case. If L stands for the external torque, the basic Newtonian equation in the inertial frame, $\mathrm{d}M_{\mathcal{F}}/\mathrm{d}t = L_{\mathcal{F}}$, writes in the body frame as

$$\frac{\mathrm{d}\boldsymbol{M}}{\mathrm{d}t} + \boldsymbol{\omega} \times \boldsymbol{M} = \boldsymbol{L}, \quad \text{or} \quad \Pi\frac{\mathrm{d}\omega}{\mathrm{d}t} + \boldsymbol{\omega} \times \Pi\boldsymbol{\omega} = \boldsymbol{L}. \tag{1}$$

It is usual to choose the body axes of BB aligned with the principal axes of inertia of the body, and then Euler equation (1) reduces to the most familiar form

$$A\dot{\omega}_1 + (C - B)\omega_2\omega_3 = L_1,$$
$$B\dot{\omega}_2 + (A - C)\omega_1\omega_3 = L_2,$$
$$C\dot{\omega}_3 + (B - A)\omega_1\omega_2 = L_3.$$

Let us notice that those equations provide the derivatives of the angular velocities, but not the angles specifying the orientation of the body system B relative to the inertial one FF. This information is necessary to determine the rotational motion, since usually the torque L depend on the attitude of the body. Therefore, Eq. (1) must be complemented with other equations. One choice in Astronomy is to describe the attitude by means of the Euler angles ψ, θ, φ, by performing a sequence of three consecutive rotations with respect to the 3, 1, and 3 axes, respectively (Fig. 1, left). The time derivatives of the Euler angles are related to the components of the angular velocity vector ω by the kinematical equations (Woolard1953a; Leimanis1965)

$$\frac{\mathrm{d}\varphi}{\mathrm{d}t} = \omega_1\frac{\sin\psi}{\sin\theta} + \omega_2\frac{\cos\psi}{\sin\theta},$$
$$\frac{\mathrm{d}\theta}{\mathrm{d}t} = \omega_1\cos\psi - \omega_2\sin\psi,$$
$$\frac{\mathrm{d}\psi}{\mathrm{d}t} = \omega_3 - \omega_1\cot\theta\sin\psi - \omega_2\cot\theta\cos\psi. \tag{2}$$

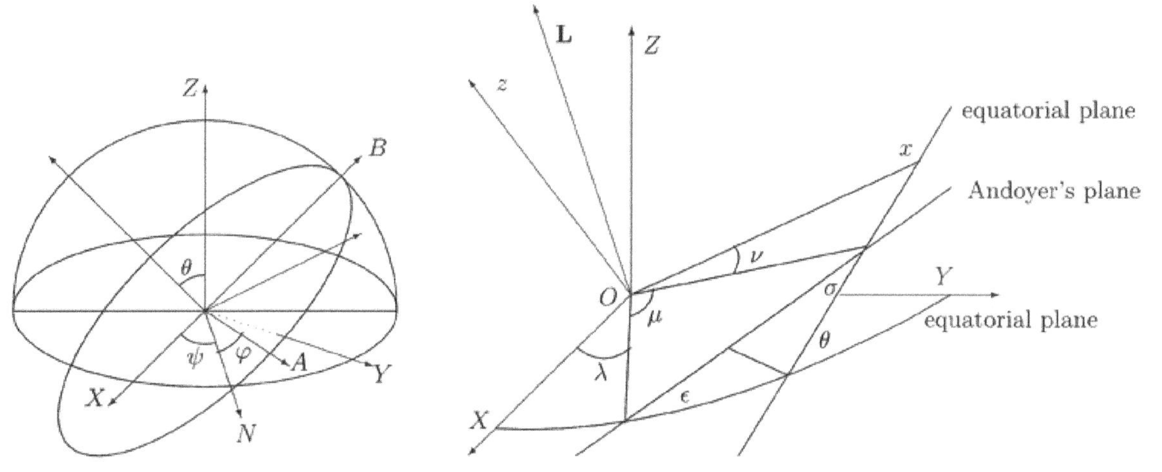

Figure 1: Euler (*left*) and Andoyer (*right*) variables.

Variational formulations Eulerian formulations are seldom used nowadays in Earth rotation studies. An exception is the solution by Roosbeek and Dehant (1998), that followed the Eulerian method, with direct computation of torques. Woolard (1953a) used a Lagrangian approach to derive the most accurate solution of his epoch, which was adopted by the IAU for a period.Bretagnon*et al.* (1988) later computed a highly accurate solution with that formalism. Nevertheless, the most successful variational approach has been the Hamiltonian one. It allowed systematic derivations of accurate solutions with the concourse of perturbation methods based on the Lie series. Kinoshita (1977) established such an approach to rigid Earth rotation that became a model to follow. His solution was the best of its class and was the base of the non-rigid solution by Wahr, adopted by the IAU in 1980. Accuracy was improved later byKinoshita and Souchay (1990) and Souchay and Kinoshita (1996, 1997). The fundamentals of this method are presented in the next section.

THE HAMILTONIAN TREATMENT OF RIGID EARTH ROTATION

Variables. A key of the success of the Hamiltonian method application to rotation problems is the use ofAndoyer variables, also named after Serret (Tisserand1891; Deprit and Elipe1993). The fixed frame F of coordinates (*OXYZ*) is transformed into the moving frame F (*Oxyz*) by means of five rotations, since the plane orthogonal to the angular momentum vector (often called Andoyer plane) is used as an intermediate step to go from the equinoctial plane *OXY* to the equator *Oxy*. The scheme of rotations is 3–1–3–1–3, and the corresponding angles are (λ,I,μ,σ,ν). They are shown in Fig. 1, to the right.

Three of those five angles (λ,μ,ν) are canonical coordinates and the other two (I,σ) are auxiliary angles related to the conjugate momenta through $\cos I = \Lambda/M, \cos\sigma = N/M$. The canonical momenta are denoted as Λ, M, N, and they have a clear dynamical meaning: M is the modulus of the angular momentum vector M, N its component along the polar or figure axis O_z of the body, and Λ its component along the *OZ* axis of the fixed frame. Let us point that the notation (λ,μ,ν;Λ,M,N) corresponds to (h,g,l;H,G,L) used by Kinoshita; the auxiliary angle I (obliquity of the Andoyer plane) has the same notation and σσ stands for Kinoshita's *J*, the angle between M and the body figure axis. Further analyses of the Andoyer set appear inEfroimsky and Escapa (2009).

The components of M in the body system have the expressions

$$M_1 = M\sin\sigma\sin\nu, \quad M_2 = M\sin\sigma\cos\nu, \quad M_3 = M\cos\sigma. \tag{3}$$

Unperturbed problem In Andoyer variables, the kinetic energy is

$$T = \frac{1}{2}\left(M^2 - N^2\right)\left(\frac{\sin^2\nu}{A} + \frac{\cos^2\nu}{B}\right) + \frac{1}{2}\frac{N^2}{C}. \tag{4}$$

It is especially simple in the case of axial symmetry, since all the coordinates are cyclic:

$$T = \frac{1}{2}\frac{M^2 - N^2}{A} + \frac{1}{2}\frac{N^2}{C}. \tag{5}$$

The free motion of a symmetric body is easily described in Andoyer variables. As the potential V=0, the Hamiltonian $\mathcal{H} = T + V = T$. Therefore, the moments M, N, Λ, are constant and angles II, σ are as well. Coordinate λ (representing an ecliptic longitude) is also constant. The remaining two angles, μ,ν, are linear functions of time. The sum μ+ν corresponds to the diurnal rotation with angular velocity ΩEΩE and νν provides the rotation of the angular momentum around the figure (or polar) axis. Its frequency is proportional to the dynamical ellipticity H=(C−A)/A. That is the simplest case of the free polar motion, known as Euler free oscillation.

Let us remark that the solution to the angular variables is computed directly and no previous solution to the angular velocity is needed, unlike in the approaches based on Euler–Liouville equations. The longitude and obliquity of the Earth equator, λ_f, I_f can be computed from the approximate relationships

$$\lambda_f - \lambda = \sigma\frac{\sin\mu}{\sin I}, \quad I_f - I = \sigma\cos\mu \tag{6}$$

which hold up to first order in σσ, of the order of 10−610−6 rad. Those differences are known as Oppolzer terms. The classic precession-nutation angles used in Astronomy are $\psi = -\lambda_f$ and $\epsilon = -I_f$, since they are reckoned in the opposite direction.

Perturbing potential The free rotation of a rigid earth is perturbed by the gravitational attraction of the Sun, Moon and planets. The gravitational potential due to a body of mass m*m*and coordinates (r^*, α, δ) in the fixed frame has a known expansion as a series of spherical harmonics (SH) multiplied by the corresponding Stokes coefficients. In the case of the Earth, the zonal term due to the Earth oblateness (associated to P_{20} and also named MacCullagh's term) is at least 1,000 times larger than the others, and thus provides a good approximation to the potential

$$V \simeq V_0 = \frac{Gm^*}{a^{*3}}(C - A)\left(\frac{a^*}{r^*}\right)^3 P_2(\sin\delta), \tag{7}$$

It must be expressed in terms of the canonical variables and the coordinates of the disturbing body, assumed to be known functions of time provided by some ephemeris, usually referred either to an inertial (or quasi-inertial) system or to a non-inertial system with known motion (e.g., ecliptic of date). Nutation theories have relied on analytical or semi-analytical ephemeris since a long time ago (Newcombe1898; Woolard1953b, etc.), and they still use such ephemeris, instead of numerical ones as in the JPL series (e.g., DE432, Folkner*et al.*1994, 2014). The main reason is that nutation theories intend to derive semi-analytical solutions in which the constituent frequencies are explicit, and resorting to numerical ephemeris would introduce additional complications. The ephemeris commonly used are ephemeris VSOP87 (Bretagnon1982) and ELP2000 (Chapront-Touzé1980), respectively. The transformation of the standard expansion in SH is a difficult task, since it requires performing five rotations. Kinoshita successively applied Wigner's theorem of transformation of SH under rotation and calculated the second degree SH of Moon and Sun to obtain up to the order of σ

$$V = \sum_{p=S,M} k'_p \sum_i \left[\frac{1}{2}(3\cos^2\sigma - 1)B_i \cos\Theta_i - \frac{1}{2}\sin 2\sigma \sum_{\tau=\pm 1} C_{i,\tau}\cos(\mu - \tau\Theta_i)\right] \tag{8}$$

where subindex p stands for the perturbing body (S = Sun, M = Moon), the parameter $k'_p = 3Gm_p(C - A)/a_p^3$ factorises the main terms of the potential generated by body p and the coefficients B_i, $C_{i,\tau}$, depending on the variable I, are

$$B_i = -\frac{1}{6}(3\cos^2 I - 1)A_i^{0)} - \frac{1}{2}\sin 2IA_i^{1)} - \frac{1}{4}\sin^2 IA_i^{2)},$$

$$C_{i,\tau} = -\frac{1}{4}\sin 2IA_i^{0)} + \frac{\tau}{4}\sin I(1 + \tau\cos I)A_i^{1)} + \frac{1}{2}(1 + \tau\cos I)(-1 + 2\tau\cos I)A_i^{2)}. \tag{9}$$

As for arguments Θ_i, we have that $\Theta_i = m_{1i}l_M + m_{2i}l_S + m_{3i}F + m_{4i}D + m_{5i}\Omega$, where l_M, l_S, F, D, and Ω are the *Delaunay* arguments of the Moon and the Sun. Within our level of approximation, we can assume that $d\Theta_i/dt = n_i$, the mean motion nini being constant. Let us stress that the canonical variable λ is implicitly contained in Ω through $\Omega = \Omega_0 - \lambda$, where Ω_0 is the mean longitude of the Moon referred to the origin of longitude on the ecliptic of date. The numerical values of coefficients $A_i^{j)}$, Θ_i and n_i, as well as of the list of the five integer numbers $(m_{1i}, m_{2i}, m_{3i}, m_{4i}, m_{5i})$ associated to each value of the index ii, depend on the orbital theories of the Moon and the Sun. They were first computed by Kinoshita (1977) and updated by Kinoshita and Souchay (1990) and Navarro (2002).

Note on the Efficient Expansion of the Potential

A main difficulty of the Hamiltonian theory in the rigid case is the handling of terms of the gravitational potential due to lunisolar attraction. The expansion of the potential contains spherical harmonics (SH) of the perturbing bodies (Sun, Moon and Planets) beyond McCullagh's approximation. Their spherical coordinates are given by numerical or semi-analytical ephemeris, which provide them as multiple Fourier series whose arguments are linear combinations of the orbital variables of the relevant body. The semi-analytical expansion of those SH is a difficult task because of the large number of terms (thousands), and the help of computer algebra is essential.

The best option is designing and using special purpose symbolic manipulators to handle the so-called Poisson series. More advanced processors exist, capable of manipulating the full expansions of the potential, including canonical variables (Kinoshita series) and performing transformations of SH, even rotations applying Wigner's Theorem (Navarro andFerrándiz2002).

Since the early 1960s, investigators have used computers to generate analytical expressions. The first symbolic processors were developed to work with Poisson series, that is, multivariate Fourier series whose coefficients are multivariate Laurent series,

$$\sum_{i_1,\ldots,i_n}\sum_{j_1,\ldots,j_m} C_{i_1,\ldots,i_n}^{j_1,\ldots,j_m} x_1^{i_1}\ldots x_n^{i_n} \frac{\cos}{\sin}(j_1\phi_1 + \cdots + j_m\phi_m)$$

where $C_{i_1,\ldots,i_n}^{j_1,\ldots,j_m} \in \mathbb{R}, \quad i_1,\ldots,i_n, j_1,\ldots,j_m \in \mathbb{Z}$, and $x_1,\ldots,x_n$ and $\phi_1,\ldots,\phi_m$ are called polynomial and angular variables, respectively. These processors were applied to problems in non-linear mechanics or non-linear differential equations in the field of celestial mechanics. One of their first applications was concerned with the orbital motion of the Moon. Delaunay devised his perturbation method to treat the lunar problem and spent 20 years doing algebraic calculations by hand to solve it. Deprit*et al.* (1971) extended the solution of Delaunay's work with the help of a special purpose symbolic processor, and Henrard (1979) pushed it to order 25. This solution was improved by iteration by Chapront-Touzé (1980), and planetary perturbations were also introduced by Chapront-Touzé (1980). Later, analytical theories for the rotation of the Earth (Kinoshita1977) were treated with the help of symbolic computation packages. Nowadays, there are many open problems that require massive symbolic computation. To cite one example, we will refer to the analytical theory of the resonant motion of Mercury. Motivated by the projects of space missions like BepiColombo and MESSENGER,D'Hoedt and Lemaitre (2004) developed a spin–orbit model for the rotation of Mercury. The computation of the spherical harmonics of Mercury are performed with the use of the planetary theory "Variations Séculaires des Orbites Plantaires" (VSOP) (Bretagnon1988). The VSOP87 analytical solution of the motion of Mercury contains trigonometric series that represent the coordinates of the body (elliptic, rectangular or spherical coordinates according to the version).

For instance, the solution for the distance Sun–Mercury (r) is given as a Poisson series containing 2,371 terms, and so, the calculus of 1/r1/r through a Taylor expansion requires high accuracy symbolic computation with Poisson series containing hundreds of thousands of terms.

Many Poisson series processors have been developed until now, as PSP (Broucke1970), mechanised algebraic operations (MAO) (Rom1969), Trigonometric Manipulator (TRIGMAN) (Jefferys1970), MSNam (Henrard1986), PARSEC (Richardson1989), and others.

Analytical Solutions up to the Second Order

The Hamiltonian method allows the derivation of highly accurate asymptotic solutions, depending analytically on the canonical variables and the arguments of lunisolar and planetary orbits, considered as known functions of time. Accurate solutions need to derive perturbations up to the second order, using, e.g., Hori's perturbation method (1966). The Hamiltonian can be cast in the form

$$\mathcal{H}(p,q)=\mathcal{H}_0(p,q)+\mathcal{H}_1(p,q)+\mathcal{H}_2(p,q), \qquad (10)$$

H_i being of order $O(\varepsilon^i)$, where i is a nonnegative integer and ε is a small parameter measuring the perturbation. We will sketch this procedure at the second order in ε following a similar method to that of Kinoshita (1977).

The algorithm consists in performing a canonical transformation from the actual canonical set (p,q) to a new one (p^*,q^*). This transformation is given at the second order by the generating function $\mathcal{W}=\mathcal{W}_1+\mathcal{W}_2$, with $\mathcal{W}_i=O(\varepsilon^i)$, which depends on the transformed set (p^*,q^*) of canonical variables. The transformed Hamiltonian at the second order has a similar form

$$\mathcal{H}^*(p^*,q^*)=\mathcal{H}_0^*(p^*,q^*)+\mathcal{H}_1^*(p^*,q^*)+\mathcal{H}_2^*(p^*,q^*) \qquad (11)$$

with $\mathcal{H}_i^*(p^*,q^*)=O(\varepsilon^i)$. In addition, some extra conditions are imposed on $\mathcal{H}_i^*$ in order to ensure that H* is easier to integrate than HH. In particular, we force H* to be free from periodic terms, that is to say, we combine the Lie transformation with an averaging method. By so doing, the transformed Hamiltonian H* and the generating function W are determined by the so–called equations of the method (Hori1966), which can be written up to the second order as

$$\begin{aligned}
&\mathcal{H}_0^*=\mathcal{H}_0,\quad \mathcal{H}_1^*=\mathcal{H}_{1\mathrm{sec}},\\
&\mathcal{H}_2^*=\mathcal{H}_{2\mathrm{sec}}+\frac{1}{2}\{\mathcal{H}_1+\mathcal{H}_{1\mathrm{sec}};\mathcal{W}_1\}_{\mathrm{sec}},\\
&\mathcal{W}_1=\int_{\mathrm{UP}}\mathcal{H}_{1\mathrm{per}}\,dt,\quad \mathcal{W}_2=\int_{\mathrm{UP}}\mathcal{H}_{2\mathrm{per}}\,dt\\
&\qquad+\frac{1}{2}\{\mathcal{H}_1+\mathcal{H}_{1\mathrm{sec}};\mathcal{W}_1\}_{\mathrm{per}},
\end{aligned} \qquad (12)$$

where the subscripts per and sec denote the periodic or secular part of the corresponding function, and the Poisson brackets are computed in the (p*,q*) canonical set. The integrals are evaluated along the solutions to the unperturbed problem generated by the Hamiltonian $\mathcal{H}_0^*$, obtained by literal substitution of the variables (p,q) by the variables (p*,q*) in H_0. The time evolution of the transformed canonical variables (p*,q*) is determined by solving the Hamiltonian equations

$$\frac{dp^*}{dt}=-\frac{\partial\mathcal{H}^*}{\partial q^*},\quad \frac{dq^*}{dt}=\frac{\partial\mathcal{H}^*}{\partial p^*} \qquad (13)$$

The variation of a function f(p,q)f(p,q) of the canonical variables can be computed at the second order by the expression $f(p,q)=f^*(p^*,q^*)+\Delta f(p^*,q^*)$ with

$$f^*(p^*,q^*)=f(p^*,q^*);\ \Delta f=\Delta_1 f+\Delta_2 f+\Delta_3 f$$

$$\rightarrow\begin{cases}\Delta_1 f=\{f^*;\mathcal{W}_1\}\\ \Delta_2 f=\{f^*;\mathcal{W}_2\}\\ \Delta_3 f=\frac{1}{2}\{\{f^*;\mathcal{W}_1\};\mathcal{W}_1\}.\end{cases}$$

The determination of the transformed Hamiltonian H* and the generating function W allows to describe the time evolution of any variable of the Earth rotation up to the second order in the perturbation parameter ε. For the sake of brevity, only a few expressions corresponding to a first order integration are displayed, following Getino and Ferrándiz (1995). The first order generating function W_1 is

$$\mathcal{W}_1=K_0'\left\{\frac{1}{2}(3\cos^2\sigma-1)W_a-\frac{1}{2}\sin 2\sigma W_b\right\} \qquad (14)$$

with

$$W_a=\sum_i\frac{B_i}{n_i}\sin\Theta_i,\quad W_b=\sum_{\tau=\pm1}\sum_i\frac{C_i(\tau)}{n_\mu-n_i}\sin(\mu-\tau\Theta_i), \qquad (15)$$

n_μ, n_i being the mean motions of μ and τ_i, i.e., $\left(\frac{d\mu}{dt},\frac{d\Theta_i}{dt}\right)$ respectively.

The perturbations of all the canonical variables can be obtained in a straight forward manner by taking derivates before being simplified doing σ=0, in short as

$$\Delta(\Lambda,M,N)=-\frac{\partial W}{\partial(\lambda,\mu,\nu)},\quad \Delta(\lambda,\mu,\nu)=\frac{\partial W}{\partial(\Lambda,M,N)} \qquad (16)$$

The first order nutations of the angular momentum axis (or Andoyer plane), are

$$\begin{aligned}
\Delta\lambda&=-K_0\frac{1}{\sin I}\frac{\partial W_a}{\partial I}=-\frac{K_0}{\sin I}\sum_i\frac{\partial}{\partial I}\left(\frac{B_i}{n_i}\right)\sin\Theta_i,\\
\Delta I&=K_0\frac{1}{\sin I}\frac{\partial W_a}{\partial\lambda}=\frac{K_0}{\sin I}\sum_i(-m_5)\frac{B_i}{n_i}\cos\Theta_i,
\end{aligned} \qquad (17)$$

with $K_0=K_0'/M$. The nutations of the figure axis are obtained by adding the *Oppolzer* terms

$$\begin{aligned}
\Delta(\lambda_f-\lambda)&=\frac{K_0}{\sin I}\sum_{\tau=\pm1}\sum_i\frac{\tau C_i(\tau)}{n_\mu-\tau n_i}\sin\Theta_i,\\
\Delta(I_f-I)&=K_0\sum_{\tau=\pm1}\sum_i\frac{C_i(\tau)}{n_\mu-\tau n_i}\cos\Theta_i.
\end{aligned} \qquad (18)$$

No other approach but the Hamiltonian succeeded in computing the nutations up to the second order. Solutions provide the longitude of the equinox and the obliquity of the equator as Poisson series of the arguments Θ_i, with coefficients depending analytically of Andoyer variables and of numbers A_i). Final series result after numerical evaluation. Solutions can be computed for any of the three axes of interest: axis of figure, angular momentum and angular velocity. The number of accounted terms is very high:

- REN 2000 solution (Souchayet *al.*1999) contains several 1,000 terms of lunisolar and planetary origin; the truncation level is approximately 0.1 μμas.
- FGN 2000 solution (Ferrándiz, Navarro and Getino) fully derived by computer algebra has a similar number of terms. A detailed second order solution showing the origin of the various terms was published by Getino*et al.* (2010).

Solutions only include the perturbations due to the external potential ("forced nutations" and precession). The secular part of the solution, arising from Eq. (13) provides the precession. The non-rigidity effects on precession are so small (Williams1994) that is not difficult to read that the precession is independent of the considered Earth model, which is not really true (Ferrándiz*et al.*2004, 2007).

Effect of the Liquid Core

Poincaré equations In (1891), Chandler detected variations of latitude in astrometric observations that pointed to a pole wobble, with a period of about 430 days, far from the Euler period for a rigid earth of about 305 days. The discovery of the so-called Chandler wobble (CW) stimulated the research on the effects of elasticity and the potential existence of a liquid core on Earth rotation. Poincaré (1901, 1910) developed the first satisfactory model for an Earth model consisting of a rigid mantle and a liquid core undergoing certain simple motion, often denoted as Poincaré model. He used two differentiated approaches to derive a set of equations quite similar to those of Euler,

$$\dot{M} + \omega \times M = L,\ \dot{M}_c - \delta\omega \times M_c = 0 \tag{19}$$

Here, M and L are the total angular momentum and torque acting on the whole earth, ω the angular velocity of the frame linked to the mantle and δω the relative angular velocity of the fluid core with respect to the mantle. Mc is the total angular momentum of the core, given by $M_c = \Pi_c(\omega + \delta\omega)$, Π_c being the tensor of inertia of the core in the mantle frame. Assuming axial symmetry and after neglecting the second and higher order terms in $\omega_1, \omega_2, \tilde{\omega}_3 = \omega_3 - \Omega_E$ and $\delta\omega$, the equations for $\tilde{\omega}_3$ and $\delta\omega$ happen to be uncoupled and the problem reduces to four linear equations. In this approximation, the complex variables $u = \omega_1 + i\omega_2$, and $v = \delta\omega_1 + i\delta\omega_2$ oscillate with two free frequencies

$$\sigma_1 = \frac{C - A}{A_m}\Omega, \quad \sigma_2 = -\Omega\left(1 + \frac{A}{A_m}\frac{C_c - A_c}{A_c}\right) \tag{20}$$

whilst the solution of the free polar motion (PM) is a linear combination of $e^{i\sigma_1 t}$, $e^{i\sigma_2 t}$The frequency σ_1 corresponds to CW, which replaces the Euler free oscillation of the rigid case. As A_m<A, frequency σ_1 is larger for a Poincaré Earth than for a rigid Earth, so that the period of the polar oscillation is shortened by the fluid core. The lengthening of the period is mainly due to the elastic yielding of the earth, as already explained by Newcombe (1892)—see also Getinoand Ferrándiz (1995) who performed more detailed calculations.

Besides, a second new free frequency σ_2 emerges due to the presence of the liquid core, that is known as NDFW (nearly diurnal free wobble) or RFCN (retrograde free core nutation), because it does not contribute solely to PM, but also gives rise to an observable nutation, named free core nutation (FCN). It was predicted by theory in early times (Vicente and Jeffreys1964), but its observation remained elusive for a long period and only could be evidenced after some years of very long baseline interferometry (VLBI) observations.

Poincaré (1910) also found the ratio of the amplitudes of the nutations of a rigid planet with and without a liquid core (in the linear approximation). That established the basis of the transfer function approach, which has been followed in most of the research on non-rigid earth nutations. Besides, he included in that paper a section treating a body with a fluid core contained in an elastic shell.

Hamiltonian approach to Poincaré's earth model Nevertheless, Poincaré did not perform any numerical evaluation of his solution to obtain values of the main nutations. Accurate solutions for a Poincaré model were computed much later by Getino (1995) and Getino and Ferrándiz(1997), but using their Hamiltonian method. Let the tensors of inertia be Π_m for the mantle, Π_c for the core and $\Pi = \Pi_m + \Pi_c$ for the whole Earth, which are assumed constant in a frame attached to the (rigid) mantle. If $\mathbf{M}, \mathbf{M}_m$ and M_c are, respectively, the angular momenta of the total Earth, the mantle and the core, they satisfy:

$$\begin{aligned}\mathbf{M} &= \mathbf{M}_m + \mathbf{M}_c = \Pi_m\,\omega + \Pi_\mathbf{c}\,(\omega + \delta\omega)\\ &= \Pi\,\omega + \Pi_c\,\delta\omega\,.\end{aligned} \tag{21}$$

Notice that setting $\mathbf{M}_c = \Pi_c(\omega + \delta\omega)$ means that an appropriate definition of the core rotation (Moritz1982) has been made, so that it is referred to a Tisserand frame (Moritz1982), as detailed in Getino (1995). The kinetic energy is thus

$$T = \frac{1}{2}\,(\mathbf{M} - \mathbf{M_c})^\mathbf{t}\,\Pi_m^{-1}(\mathbf{M} - \mathbf{M}_c) + \frac{\mathbf{1}}{\mathbf{2}}\,\mathbf{M}_c^\mathbf{t}\,\Pi_c^{-1}\,\mathbf{M}_c \tag{22}$$

This expression is canonically formulated by means of a set of canonical variables, λ, μ, ν, Λ, M, N for the whole Earth, and $\lambda_c, \mu_c, \nu_c, \Lambda_c, M_c, N_c$ for the core, with the help of the auxiliary angles σ, I, σ_c, I_c described by Getino (1995). The angular momenta M and $\mathbf{M}_c$ are given by

$$\mathbf{M} = \begin{pmatrix}\mathbf{K}\sin\nu\\ \mathbf{K}\cos\nu\\ \mathbf{N} = \mathbf{M}\cos\sigma\end{pmatrix},\quad \mathbf{M}_c = \begin{pmatrix}\mathbf{K}_c\sin\nu_c\\ -\mathbf{K}_c\cos\nu_c\\ \mathbf{N}_c = \mathbf{M}_c\cos\sigma_c\end{pmatrix} \tag{23}$$

where $K = M\sin\sigma,\ K_c = M_c\sin\sigma_c$. Note that σσ and σcσc are small quantities, of the order of 10^{-6} rad. The kinetic energy can be written as

$$\begin{aligned}T_0 &= \frac{1}{2A_m}\left[K^2 + \frac{A}{A_c}K_c^2\right] + \frac{K\,K_c}{A_m}\cos(\nu + \nu_c)\\ &+ \frac{1}{2\,C_m}\left[N^2 - 2\,N\,N_c + \frac{C}{C_c}\,N_c^2\right],\end{aligned}$$

$A = A_m + A_c,\ C = C_m + C_c$, being the principal moments of the total Earth. Let us remark that these hypotheses pose no problem related to the terrestrial frame, since the mantle is rigid and its principal axes are well defined. However, the number of canonical variables has been doubled, which increases the difficulty of the treatments. But there is an additional, essential difference with respect to the rigid case: resonance phenomena occur, which amplify the amplitudes of some Oppolzer terms, hence of nutations. That fact helps to constrain the values of certain geophysical parameters. Besides, the unperturbed problem is not integrable any more, irrespective of the axial symmetry of the body. The integrability issues in this model were studied by Ferrándiz and Barkin (2001).

THEORIES OF NON-RIGID EARTH NUTATIONS

Two-Layer Earth Models

As we pointed out above, explaining the observed CW period requires taking into account elasticity besides the liquid core. A number of solutions were developed between about 1950 and 1990 by considering Earth models composed of an elastic mantle and a liquid core, the standard two-layer model. They made use of the theory of elasticity, developed by Cauchy, Green, Poisson, Stokes, Lord Kelvin, etc. These approaches are very different, but they share some features:

- Kelvin solutions of the Laplace equations in terms of SH are used, as well as generalizations, including the assumption of variability for some parameters, such as density or Lamé parameters.
- Some simplifying hypotheses are usually made, as radial dependence of parameters or certain equilibrium conditions.
- In general, this procedure allows the reduction of the original continuous problem of elasticity to a discrete one, with the relevant parameters determined by quadratures assuming certain rheological models.

Jeffreys and Vicente (1957) proposed a variational formulation of Lagrangian type; therefore, the computation of the internal dissipative moments is avoided. In Molodenski's model (1961), the elastic equations for the mantle are approximated by spherical functions, and the fluid core is treated using hydrodynamical equations. SOS equations (Sasao*et al.*1980) had deep impact on later research. They are a simple generalization of Poincaré's, including elasticity and dissipations at the core–mantle boundary (CMB) due to friction and electromagnetic coupling. The original derivation was carried out by direct methods (Euler–Liouville). Using variational methods allowed a drastic simplification (Moritz).

The IAU 1980 nutation theory Wahr's solution Wahr (1981) was obtained by applying a certain transfer function to the rigid earth solution by Kinoshita (1977). An IAU Working Group proposed its adoption (Seidelmann1982) and the theory was endorsed by IAU as its first non-rigid Earth nutation theory in 1981. This solution follows the method of Smith and Whar: The partial differential equations of the elastic problem are transformed into an infinite system of ordinary differential equations through a series expansion of spheroidal and toroidal harmonics. A drastic truncation produces a finite system. The resulting equations are integrated numerically over the Earth volume, assuming a certain rheological model. This solution gives the nutations of an oceanless, elastic solid Earth with a fluid core. In the framework of IAU 1980, other effects not addressed in the official theory (oceanic, atmospheric, anelastic, etc.) are treated in the moving (terrestrial) reference frame, mainly using some versions of the Euler–Liouville equations, and are usually classified as "polar motion" terms (see Sect. 6).

Earth elasticity in the Hamiltonian method The Hamiltonian method contributed with a series of papers by Getino and Ferrándiz who introduced the Hamiltonian formalism to study an elastic Earth (1990, 1991, 1995). Let us notice that the definition of the body frame has no special difficulties under the assumption of linear elasticity, since the deformations have known expressions depending on constant Love numbers, and the variations of the principal axes and moments of inertia can be derived analytically (Barkin and Ferrándiz2000). More properties of the rotation of weakly deformable bodies are given by Barkin (1998, 2000a, b).

Getino and Ferrándiz (2000, 2001) also combined their previous results to derive an accurate Hamiltonian solution for a two-layered earth made of a liquid core and an inelastic mantle, and accounted for dissipation at CMB. That way of proceeding guarantees consistency of the new considered effects with the former pieces of theory. For instance, the main change when elasticity is put into the Poincaré model is the addition of a new term T_t to the Hamiltonian, which represents the increment of the kinetic energy due to the tidal deformation and is given by

$$\begin{aligned} T_t = & \frac{N - N_c}{A_m C_m} D_m^t \left[K_c \left(t_{13} \sin v_c - t_{23} \cos v_c\right)\right. \\ & - .-K \left(t_{13} \sin v + t_{23} \cos v\right)] \\ & - \frac{N_c}{A_c C_c} D_c^t K_c \left(t_{13} \sin v_c - t_{23} \cos v_c\right), \end{aligned} \tag{24}$$

where $D^t_{m,c}$ are constants related to the Love number k_2 and functions $t_{i,j}$ have expansions similar to the components of the potential.

Three-Layer Earth Models

The improvements of the space geodetic observation techniques since the late 1980s revealed that IAU1980 was not accurate enough. Besides, the launch and operation of new geodetic satellites improved the observational possibilities and contributed to obtaining more insight into matters such as bodily tides and other geophysical properties of the Earth. New investigations aimed at explaining the new results, among them Dehant*et al.* (1999). Mathews*et al.* (1991a, b) introduced a solid inner core in the basic Earth structure. An empirical nutation model was adopted in the IERS Conventions 1996 (McCarthy1996). In this context, an IAU Working Group on non-rigid Earth nutation theory started in 1994 and recommended that theories be based on geophysical models closer to the actual Earth (Dehant*et al.*1999).

Most of the theories developed in that epoch assumed a three-layered Earth made of elastic mantle, fluid outer core (FOC) and solid inner core (SIC). They had to rely upon a pre-existent rigid Earth solution, since they used a transfer function approach. Depending on theories, elasticity might be extended to deal with in-elastic or an-elastic assumptions, include dissipations in the inner layers boundaries or consider oceanic and atmospheric effects to some extent. Among those theories, we can cite first DD97 (Defraigne and Dehant1998), Sch97 (Schastok1997), Hg2000 (Huang*et al.*2001) among the main differentiated approaches.

In March 2000, three theories were selected as candidates to become the IAU 2000 nutation model (Dehant2002). They were:

- MHB2000 (Mathews*et al.*2002): a transfer function derived from a generalization of Poincaré–SOS equations was applied to REN2000. It was complemented with the Kinoshita–Souchay–Folgueira (1999) planetary perturbations for the rigid Earth.
- SF2000 (Shirai and Fukushima2000): applied a numerical convolution in the time domain to adjust parameters of Herrings transfer function.
- GF2000 (Getino and Ferrándiz2000): Hamiltonian, analytical theory for the Earth rotation, extending Kinoshita and Souchay's rigid Earth theory. It was complemented with the planetary non–rigid perturbations by Ferrándiz–Navarro–Getino and Huang et al. oceanic corrections, the final series being named FGHN.

All of them fit a low number of basic Earth parameters to observations and got similar accuracy, about 150 μμas in terms of wrms (weighted root mean squared) observations-model differences (if an empirical model for FCN is used). The accuracy of IAU 1980 was thus improved in more than one order of magnitude. MHB2000 was preferred and selected as IAU2000, and it is in force since 2003. In that year Fukushima (2003) published a new precession theory. The IAU1976 model of the precession (Lieske*et al.*1977) was changed 6 years later and the P03 model by Capitaine*et al.* (2003) was adopted as the IAU2006 precession model (Hilton*et al.*2006).

The Hamiltonian Method

Main features The Hamiltonian or global approach is the only one that allows the direct derivation of a non-rigid solution up to the second order of perturbation, in a fully consistent manner, since it is independent of any previous rigid Earth solution. That is because transfer function approaches are intrinsically linear. The rigid solution can be recovered when some parameters vanish. The calculation of some poorly known internal torques is avoided, since the approach is variational. The effect of geophysical Earth models is concentrated in a reduced set of parameters. Analytical solutions are convenient for several reasons, like fitting parameters, allowing the identification of resonances and providing more insight into the Earths interior and geophysical properties. Besides numerical methods have failed to provide good solutions in the non-rigid case so far: the attempts which have been successful within a fitting time interval (Krasinsky2006) showed a quick degradation when extrapolated beyond that interval (Capitaine*et al.*2009).

Free motion of a three layers Earth in the Hamiltonian approach The definition of the Andoyer variables for FOC and SIC takes into account the relations among a frame fixed to the mantle, OxmymzmOxmymzm, the Andoyer planes defined by the angular momentums of FOC and SIC, and frames "attached" to the FOC or SIC, $Ox_my_mz_m$. It originates a set of 18 canonical variables, λ, μ, ν, Λ, M, N for the total Earth, $\lambda_f, \mu_f, \nu_f, \Lambda_f, M_f, N_f$ for the fluid outer core, and $\lambda_s, \mu_s, \nu_s, \Lambda_s, M_s, N_s$ for the solid inner core, with auxiliary angles σ, I, σ_f, I_f, σsσs and I_s. Their geometrical meaning is displayed in Fig. 2. Denoting by M, M_f and M_s the absolute angular momenta of the whole earth, FOC and SIC, respectively; the canonical moments satisfy

$$
\begin{aligned}
M &= |\mathbf{M}|, & M_f &= |\mathbf{M_f}| & M_s &= |\mathbf{M_f}|,\\
N &= M\cos\sigma, & N_f &= M_f\cos\sigma_f, & N_s &= M_s\cos\sigma_s,\\
\Lambda &= M\cos I, & \Lambda_f &= M_f\cos I_f, & \Lambda_s &= M_s\cos I_s.
\end{aligned}
$$

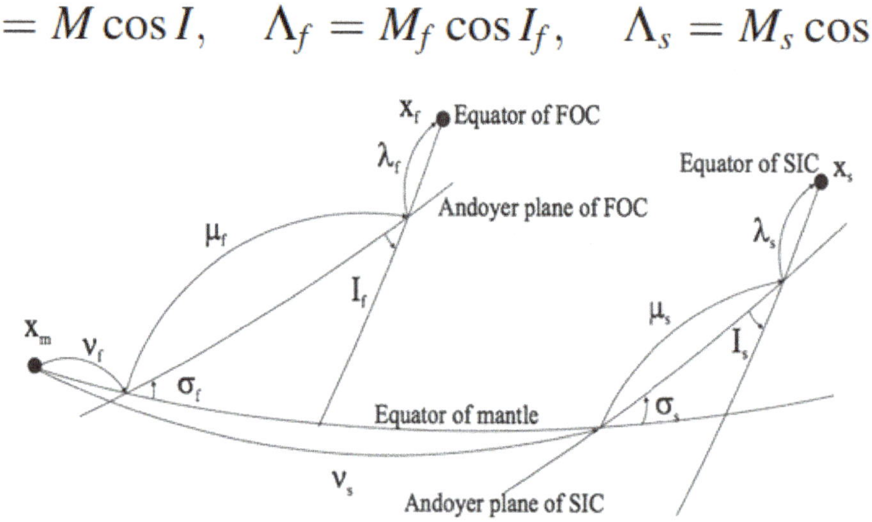

Figure 2: The Andoyer variables for FOC and SIC relate a fixed frame of mantle, $Ox_my_mz_m$, the plane defined by the angular momentum of FOC and SIC, Andoyer plane of FOC and SIC, and a fixed frame of FOC or SIC, $Ox_fy_fz_f$ or $Ox_sy_sz_s$.

The three layers Earth kinetic energy T is written as

$$T = \frac{1}{2}(\mathbf{M} - \mathbf{M_f} - \mathbf{M}_s)^t \Pi_m^{-1}(\mathbf{M} - \mathbf{M_f} - \mathbf{M}_s) + \frac{1}{2}\mathbf{M}_f{}^t \Pi_f^{-1}\mathbf{M_f} + \frac{1}{2}\mathbf{M}_s{}^t \Pi_s^{-1}\mathbf{M}_s,$$

Π, Π_f and Π_s being the respective inertia matrices. The angular momenta hold

$$\mathbf{M} = \begin{pmatrix} \mathbf{K}\sin\nu \\ \mathbf{K}\cos\nu \\ \mathbf{N} = \mathbf{M}\cos\sigma \end{pmatrix}, \quad \mathbf{M_f} = \begin{pmatrix} \mathbf{K}_f \sin\nu_f \\ -\mathbf{K}_f\cos\nu_f \\ \mathbf{N}_f = \mathbf{M}_f\cos\sigma_f \end{pmatrix}$$

$$\mathbf{M}_s = \begin{pmatrix} \mathbf{K}_s \sin\nu_f \\ -\mathbf{K}_s\cos\nu_f \\ \mathbf{N}_s = \mathbf{M}_s\cos\sigma_s \end{pmatrix},$$

with $K = M \sin\sigma, K_f = M_f \sin\sigma_f, K_s = M_s \sin\sigma_s$.

The explicit expression of the Hamiltonian is involved even for the unperturbed motion, especially if no restrictive hypothesis on the SIC attitude is made. Escapa*et al.* (2001) derived a solution to the linearised equations, which gives the frequencies of the four normal modes or free harmonic oscillations of the rotation pole in terms of the ellipticities and an additional small parameter δ:

$$m_1 = \frac{A}{A_m}e \quad \rightarrow \quad \text{CW or Chandler wobble,}$$

$$m_2 = -1 - \frac{A_f + A_m}{A_m}e_f \quad \rightarrow \quad \text{RFCN or retrograde free core nutation,}$$

$$m_3 = -1 + \delta \quad \rightarrow \quad \text{PFCN or prograde free core nutation,}$$

$$m_4 = e_s - \delta \quad \rightarrow \quad \text{ICW or inner core wobble.}$$

Recent progress in the Hamiltonian theory of non-rigid Earth Since the year 2000, many effects have been investigated by different authors and under various approaches. Several effects have been found to contribute to the nutations with direct or indirect terms reaching the magnitude of some tens of μμas. Most of them are not included in the current IAU or IERS models. Those terms are often referred to as of second order, although they can be cast in at least two distinct groups. The first group is made of second order terms in the sense of perturbation theory (crossing of the ordinary first order precession–nutation terms with themselves). They are part of a solution that is non-linear with respect to the dynamical ellipticity HH. The other group gathers small terms of various physical origins but sharing some properties: arising from unaccounted terms of the potential, like high frequency nutations (Escapa*et al.*2002) and indirect effect of sectorial and tesseral third order harmonics (Ferrándiz*et al.*2003), effect of fluid core on the precession (Ferrándiz*et al.*2004, 2007), direct effects of the actual rotation of the inner core (Escapa*et al.*2012), effects on nutations (Ferrándiz*et al.*2011) of the observed J_2 variation (Cheng and Tapley2004; Cheng*et al.*2011; Coxand Chao2002), and other time variations of the geopotential as unaccounted effects of tidal models (Ferrándiz*et al.*2011), etc. The Hamiltonian method provided a systematic, consistent procedure to approach all of them in the non-steady, non-rigid case.

THE SOLUTION FOR POLAR MOTION

Theories of earth rotation usually devote a part to calculate the frequencies of the unperturbed or free polar motion (PM), corresponding to the oscillations or wobbles of the axis of angular velocity or angular momentum around the figure axis or vice versa. Let us note that the differences between free periods in the rigid and non-rigid cases are more marked than those in the corresponding forced motions (nutation amplitudes), whose main components are the so-called Poisson terms, practically independent of the Earth model. Conversely, the amplitudes and phases of the polar wobbles are highly dependent on the Earth physics. Woolard already mentioned the relevance of geophysical effects on nutations and what he called diurnal nutations, although his terminology differs from the currently used.

The main components of nutations have long periods in the "inertial" frame. However, the main components of PM have long periods in the terrestrial or body-fixed frame, therefore they are in the diurnal band when seen from the inertial frame. However, the terrestrial frame is more convenient for their study as well as their determination since the advent of radio-interferometric techniques like VLBI.

In fact, the actual motion of the Earths pole, displayed in Fig. 3, has not been fully explained by any theory yet. It includes noticeable changes of amplitudes and phases of its main components (the Chandler wobble with an amplitude usually ranging from 100 to 200 μμas and the annual term with an amplitude nearby 100 μμas) as well as a long-term drift (Barkin2000a, b; Schuh*et al.*2001). A thorough review can be found in, e.g., Gross (2007).

Whilst nutations arise from a mainly astronomical forcing, the free wobbles of the pole are excited mainly by geophysical processes, and are difficult to predict (Chao and Gross1987;Dickey*et al.*2002, Gross*et al.*2005). For that reason, the solution for the forced PM is not derived analytically along with the nutations, but is computed from different equations using empirical time series providing the relevant excitation functions (Gross1992; Brzeziński1992).

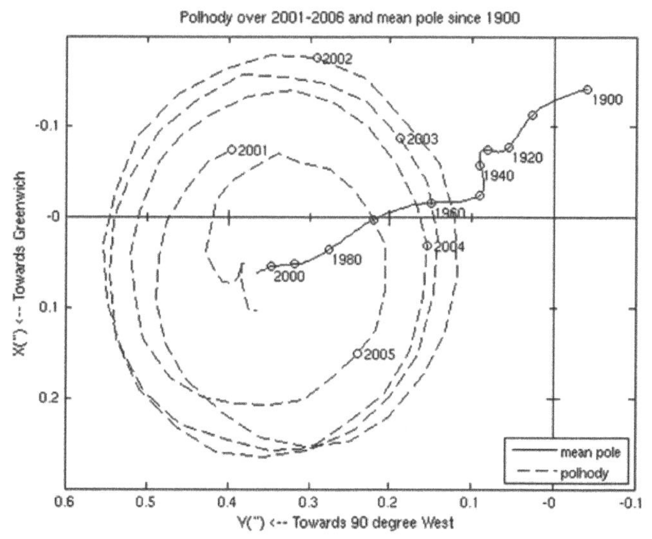

Figure 3: Motion of the Earth pole. Source: IERS.

That behaviour was essential for the definition of the set of Earth orientation parameters (EOP) currently in use. In 1982, an IAU Working Group on Nutation (Seidelmann1982) recommended the adoption of five EOP, namely: the precession/nutation angles ϵ, ψ, referred to the equinox and equator; UT1 (universal time 1), corresponding to the sidereal diurnal revolution and GMST or GAST (Greenwich Mean Sidereal Time or Greenwich Apparent Sidereal Time); the polar motion angles x, and y.

This set provides the transformation of coordinates from the celestial to the terrestrial frame (or vice-versa) by performing five rotations. The transformation is mathematically redundant, since the relative orientation of two reference systems can be specified by only three independent parameters. Nevertheless, that redundancy was convenient for the analysis of the VLBI observations of Earth rotation, which started in the early 1980s. It proceeded by fitting one set of five EOP to each observation session spanning a whole day.

PRESENT STATE OF THE EARTH ROTATION MODELLING AND OUTLOOK

Since the IERS establishment, EOP solutions are provided by IERS along with several Analysis Centres. Besides VLBI, other techniques contribute to determine a subset of EOP (UT1 and PM), namely satellite laser ranging (SLR) and GNSS (Global Navigation Satellite Systems). Time series of daily EOP values are produced by IERS (Bizouard and Gambis2009), by a combination of individual solutions computed by various associated Analysis Centres for each technique. Nowadays, IERS releases two sets of EOP related by a known transformation, since the nutation offsets dX,dYdX,dY and the Earth rotation angle (ERA) were recommended to replace the former three equinox-based EOP after a new paradigm was adopted by IAU in 2000, based in the use of the celestial intermediate origin and pole (CIO and CIP, respectively). Precise definitions of the main and auxiliary parameters and frames can be found in, e.g., the IERS Conventions 2010, Supplement to the Nautical Almanac (Urban and Seidelmann2013) or standards of fundamental astronomy (SOFA) documentation (Hohenkerk2010).

Current accuracy of EOP series is difficult to assess. Comparisons between combined solutions and individual solutions corresponding to different techniques and analysis centres provide some insight into their accuracy or uncertainty. Following the IERS Annual Report 2011 (Dick2011), the uncertainty of VLBI solutions may be near 90 μμas for nutations in average and in about 170 μμas for PM. The accuracy of precession/nutation models, when used to make forward predictions, is stabilised at around 150 μμas, in terms of wrms of the observation-model differences, and the figures are larger for PM prediction. The remarkable efforts made in the last years provided a better insight into the problem and unveiled new potential sources of error, but have not been compensated yet by a significant reduction of the residuals. Let us recall that the IAG's Global Geodetic Observing System (GGOS) initiative demands an accuracy of 1 mm to the systems of reference, besides a stability in time of 0.1 mm/year. That corresponds roughly to a value of 30 μμas for angular EOP (2 μμs for time).

From the observational side, the accuracy and performance of the major techniques is increasing. Therefore, series of more accurate EOP will be available in a few years. Besides, higher time resolution is expected. There are still many difficult open problems, such as magnetic effects (Huang*et al.*2011), motions of inner layers (Barkin and Vilke2004), relativistic effects (Klioner*et al.*2009), consistent and comprehensive treatment of a more realistic time-varying earth model, etc. Clearer separation of nutations and polar motion is also sought as we approach the EOP determination at a sub-diurnal rate (Nilsson*et al.*2010) and non-predictable constituents are accounted in the nutation angles like the free core nutation (FCN) (Lambert2007; Krasna*et al.*2013), whereas some short periodic predictable astronomical effects are included into PM (Getino*et al.*2001; Escapa*et al.*2002, Brzezinski2001).

In this context, the International Association of Geodesy (IAG) and the International Astronomical Union (IAU) set up a new Joint Working Group on Theory of Earth Rotation (or JWG ThER) in April 2013. The purpose of the new JWG is: "To promote the development of theories of Earth rotation that are fully consistent and that agree with observations and provide predictions of the Earth rotation parameters (ERP) with the accuracy required to meet the needs of the near future as recommended by, e.g., GGOS, the Global Geodetic Observing System of the IAG. Its structure is more complex than usual and adapts to the characteristics of the current EOP, as well as the specialised fields of research. The people in charge are:

- Chair: José M. Ferrándiz (IAU)
- Vice-Chair: Richard S. Gross (IAG)

The JWG is composed of three Sub-Working Groups (SWG):

- Precession/Nutation (Chair: Juan Getino)
- Polar Motion and UT1 (Chair: Aleksander Brzezinski)
- Numerical Solutions and Validation (Chair: Robert Heinkelmann)

These SWG should work independently but in parallel for the sake of efficiency, and they must be linked together as closely as the needs of consistency demand. More information is available in Ferrándiz and Gross (2014) and on the JWG website: http://web.ua.es/en/wgther/.

Future Prospects of the Hamiltonian Method

Meeting the stringent GGOS accuracy and stability goals is a challenging task, whose fulfilment requires a joint cooperative effort of the scientific community involved in the determination, modelling and prediction of Earth rotation. In the authors' opinion, the Hamiltonian approach can provide a valuable contribution to the theoretical modelling because of some of its features. First, the treatment addresses the Earth rotation globally, as a whole problem, and its previous results, described in former sections, show that the theory can incorporate any kind of geophysical models or effects that have been considered up to date, like the Earth division in solid and fluid layers, the various assumptions on its elastic behaviour, the dissipations at the layers boundaries, the time variation of the geopotential, etc. It also allows the incorporation of small corrections obtained independently by other theories. The inclusion of all the components in a sole Hamiltonian function (or more precisely formalism, to distinguish the generalised forces) helps to assess the magnitude of any neglected effect and ensures the self-consistency of the developments, so that there is no need to introduce corrections aimed at restoring consistency when some background models are updated. But the essential characteristic of the method is its capability to derive solutions with a prescribed level of accuracy in a systematic way, by calculating the approximate solution up to the suitable order of perturbation (usually first or second, depending on the magnitude of each group of terms), as well as to identify the contributions of the different effects included in the chosen geophysical model. This last property is not shared by any solution derived by numerical integration that also can reach high accuracy, but cannot separate the free motion component of the solution to the Earth attitude from the forced one, which is a difficulty according to the current conventions and EOP definitions.

ACKNOWLEDGMENTS

The authors acknowledge the valuable suggestions of the anonymous referees. This work has been partially supported by the Spanish government under Grants AYA2010-22039-C02-01 and AYA2010-22039-C02-02 from Ministerio de Economía y Competitividad (MINECO), the University of Alicante under Grant GRE11-08 and the Generalitat Valenciana, Grant GV/2014/072.

REFERENCES

1. Altamimi, Z., Collilieux, X. and Metivier, L. (2011), ITRF2008, an improved solution of the International Terrestrial Reference Frame, J. Geod. 85(8), 457–473.
2. Barkin, T. V. (1998), Unperturbed chandler motion and perturbation theory of the rotation motion of deformable celestial bodies, Astron. Astrophys. Trans. 17(3), 179–219.
3. Barkin, Y. V. (2000a), Towards on explanation of the secular motion of the earth's rotation axis pole, Astron. Astrophys. Trans. 19(1), 13–18.
4. Barkin, Y. V. (2000b), Perturbated rotational motion of weakly deformable celestial bodies, Astron. Astrophys. Trans. 19(1), 19–65.
5. Barkin, Y. V. and Ferrándiz, J. M. (2000), The motion of the Earth's principal axes of inertia caused by tidal and rotational deformations, Astron. Astrophys. Trans. 18, 605–620.
6. Barkin, Y. V. and Vilke, V. G. (2004), Celestial mechanics of planet shells, Astron. Astrophys. Trans. 23(6), 533–553.
7. Bizouard, C. and Gambis, D. (2009), The Combined Solution C04 for Earth Orientation Parameters consistent with International Terrestrial Reference Frame 2005, IAG Symp 134, 265–270.
8. Bretagnon, P. (1982), Theory for the motion of all the planets—The VSOP82 solution, Astron. Astrophys. 114, 278.
9. Bretagnon, P. (1988), Planetary theories in rectangular and spherical variables. VSOP 87 solution, Astron. Astrophys. 202, 304–315.
10. Bretagnon, P., Rocher, P., and Simon, J.-L. (1997), Theory of the rotation of the rigid Earth, Astron. Astrophys. 319, 305–317.
11. Broucke, R. (1970), How to assemble a Keplerian processor, Celest. Mech. 2, 9–20.
12. Brzeziński, A. (1992), Polar motion excitation by variations of the effective angular momentum function: considerations concerning deconvolution problem. Manuscr. Geod. 17, 3–20.
13. Brzeziński, A. (2001), Diurnal and sub-diurnal terms of nutation: a simple theoretical model for a nonrigid Earth, In N. Capitaine (ed.), Proc. of the Journées 2000—Systèmes de Référence Spatio-temporels, Observatoire de Paris, pp. 243–251.
14. Capitaine, N., Wallace, P. T. and Chapront, J. (2003), Expressions for IAU 2000 precession quantities, Astron. Astrophys. 412, 567–586.
15. Capitaine, N., Mathews, P. M., Dehant, V., Wallace, P. T. and Lambert, S. B. (2009), On the IAU 2000/2006 precession nutation and comparison with other models and VLBI observations, Celest. Mech. Dyn. Astron. 103, 179–190.
16. Chao, B. F. and R. S. Gross (1987), Changes in the Earths rotation and low-degree gravitational field induced by earthquakes, Geophys. J. Roy. Astr. Soc. 91, 569–596.
17. Chandler, S.C. (1891) On the variation of latitude. Astron. J. 11, 59–61.
18. Chapront-Touzé, M. (1980), La solution ELP du problème central de la Lune, Astron. Astrophys. 83–86.
19. Chapront-Touzé, M. (1982), Progress in the analytical theories for the orbital motion of the Moon, Celest. Mech. 26, 53–62.
20. Cheng, M. and Tapley, B. D. (2004), Variations in the Earth's oblateness during the past 28 years, J. Geophys. Res. 109, B09402.
21. Cheng, M. K., Ries, J. C. and Tapley, B. D. (2011), Variations of the Earth's Figure Axis from Satellite Laser Ranging and GRACE, J. Geophys. Res. 116, B01409.
22. Cox, C. M. and Chao, B. F. (2002), Detection of a large-scale mass redistribution in the terrestrial system since 1998, Science 297, 831–833.
23. Defraigne, P. and Dehant, V. (1998), New theoretical model for nutations and comparison with VLBI observations. In: Capitaine, N. (ed) Proc. Journées 1997—Systèmes de Référence Spatio-Temporels, Observatoire de Paris, pp 69–72.
24. Dehant, V., Defraigne, P. and Wahr, J. M. (1999a), Tides for a convective Earth, J. Geophys. Res. 104, 1035–1058.
25. Dehant V. et al. (1999b), Considerations concerning the non-rigid Earth nutation theory, Celest. Mech. Dyn. Astron. 72, 245–310.
26. Dehant, V. (2002), Report of IAU Working Group on 'Non-rigid Earth rotation theory', Highlights of Astronomy 12, 117–119.
27. Deprit, A., Henrard, J. and Rom, A. (1971), Analytical Lunar Ephemeris: Delaunay's Theory, Astron. J. 76, 269–272.
28. Deprit, A. and Elipe, A. (1993), Complete reduction of the Euler-Poinsot problem, J. Astronaut. Sci. 41, 603–628.
29. Dickey, J. O. et al. (2002), Recent Earth Oblateness Variations: Unraveling Climate and Postglacial Rebound Effects, Science, 298, 1975–1977.
30. D'Hoedt, S. and Lemaitre, A. (2004), The spin-orbit resonant rotation of Mercury: a two degree of freedom

Hamiltonian model, Celest. Mech. Dyn. Astron. 89, 267–283.

31. Dick, W. R. (ed) (2011), IERS Annual Report 2011. Verlag des Bundesamts fr Kartographie und Geodsie, Frankfurt AM.

32. Efroimsky, M. and Escapa, A. (2009), The theory of canonical perturbations applied to attitude dynamics and to the Earth rotation. Osculating and nonosculating Andoyer variables, Celest. Mech. Dyn. Astron. 98, Issue 4, 251–283.

33. Escapa, A., Getino, J. and Ferrándiz, J. M. (2001), Canonical approach to the free nutations of a three-layer Earth model, J. Geophys. Res. 106, 11387–11397.

34. Escapa, A., Getino, J. and Ferrándiz, J. M. (2002), Indirect effect of the triaxiality in the Hamiltonian theory for the rigid Earth nutations, Astron. Astrophys. 389, 1047–1054.

35. Escapa, A., Ferrándiz, J. M. and Getino, J. (2012), Influence of the inner core on the rotation of the Earth revisited, IAU Joint Discussion 7 "Space-time reference systems for future research", XXVIIIth General Assembly of the International Astronomical Union.

36. Ferrándiz, J. and Barkin, Y. (2001), On integrable cases of the Poincaré problem, Astron. Astrophys. Trans. 19, 769–780.

37. Ferrándiz, J. M., Escapa, A., Navarro, J. F., and Getino, J. (2003), Recent work on theoretical modelling of nutation. In: Richter, B., Schwegmann, W. and Dick, W.R. (eds) Proceedings of the IERS Workshop on Combination Research and Global Geophysical Fluids, IERS Technical Note 30, pp 163–167.

38. Ferrándiz, J. M., Navarro, J. F., Escapa, A. and Getino, J. (2004), Precession of the Nonrigid Earth: Effect of the Fluid Outer Core, Astron. J. 128, 1407–1411.

39. Ferrándiz, J. M., Navarro, J. F., Escapa, A., Getino, J. and Baenas, T. (2007), Influence of the mantle elasticity on the precessional motion of a two-layer Earth model, In: Lemaître, A. (ed) The rotation of celestial bodies, Press. Universitaires de Namur, pp 9–14.

40. Ferrándiz, J. M., Martínez-Ortiz, P. A. and García, D. (2011), *Effects of time gravity changes on the Earth* nutations, Geophysical Research Abstracts 13, EGU2011-4981.

41. Ferrándiz, J. M., Baenas, T. and Escapa, A. (2012), Effect of the potential due to lunisolar deformations on the Earth precession, Geophysical Research Abstracts 14, EGU2012-6175.

42. Ferrándiz, J. M. and Gross, R. S. (2014), The New IAU/IAG Joint Working Group on Theory of Earth Rotation, IAG Symp 143 (to appear).

43. Fey, A. L., Arias, E. F., Charlot, P., Feissel-Vernier, M., Gontier, A. M., Jacobs, C. S., Li, J. andMacmillan, D. S. (2004), The second extension of the International Celestial Reference Frame: ICRF-EXT. 1, Astron. J. 127, 3587–3608.

44. Folkner, W. M., Charlot, P., Finger, M. H., Williams, J. G., Sovers, O. J., Newhall, X., Standish, E. M. Jr. (1994), Determination of the extragalactic-planetary frame tie from joint analysis of radio interferometric and lunar laser ranging measurements, Astron. Astroph. 287, 279–289.

45. Folkner, W. M et al. (2014), JPL Interplanetary Network Progress Report 42–196, (2014) Available at http://ipnpr.jpl.nasa.gov/progress_report/42-196/196C.

46. Fukushima, T. (2003) A new precession formula, Astron. J. 126, 494–534.

47. Getino, J. and Ferrándiz, J. M. (1990), A Hamiltonian theory for an elastic earth: Canonical variables and kinetic energy, Celest. Mech. Dyn. Astron. 49, 303–326.

48. Getino, J. and Ferrándiz, J. M. (1991), A Hamiltonian Theory for an Elastic Earth—First Order Analytical Integration, Celest. Mech. Dyn. Astron. 51, 35–65.

49. Getino, J. and Ferrándiz, J. M. (1995), On the effect of the mantle elasticity on the Earth's rotation, Celest. Mech. Dyn. Astron. 61, 117–180.

50. Getino, J. and Ferrándiz, J. M. (1997), A Hamiltonian approach to dissipative phenomena between the Earth's mantle and core, and effects on free nutations, Geophys. J. Int. 130, 326–334.

51. Getino, J. and Ferrándiz, J. M. (2000), Effects of dissipation and a liquid core on forced nutations in Hamiltonian theory, Geophys. J. Int. 142, 703–715.

52. Getino, J. and Ferrándiz, J. M. (2000b), Advances in the Unified Theory of the Rotation of the Nonrigid Earth. In: Jhonston, T. et al. (ed) Towards models and constants for sub-microarcsecond astrometry, Proc. IAU Col. 180, pp 236–241 Geophys. J. Int. 142, 703–715.

53. Getino, J. and Ferrándiz, J. M. (2001), Forced nutations of a two-layer Earth model, Mon. Not. R. Astron. Soc. 322, 785–799.

54. Getino, J., Ferrándiz, J. M. and Escapa, A. (2001), Hamiltonian theory for the non-rigid Earth: semidiurnal terms, Astron. Astroph. 370, 330–341

55. Getino, J., Escapa, A. and Miguel, D. (2010), General theory of the rotation of the non-rigid Earth at the second order. I. The rigid model in Andoyer variables, Astron. J. 139, 1916–1934.

56. Gross, R. S. (1992), Correspondence between theory and observations of polar motion, Geophys. J. Int. 109, 162–170.

57. Gross, R. S., Fukumori, I. and Menemenlis, D. (2005), Atmospheric and oceanic excitation of decadal-scale Earth orientation variations, J. Geophys. Res. 110, B09405.

58. Gross, R. S. (2007), Earth rotation variations long period, In: Herring TA (ed) Physical Geodesy. Treatise on Geophysics vol 3, Elsevier, Oxford, 239–294.

59. Henrard, J. (1979), A New Solution to the Main Problem of Lunar Theory, Celest. Mech. 19, 337–355.

60. Henrard, J. (1986), Algebraic manipulation on computers for lunar and planetary theories. In:Kovalevsky, J. and Brumberg, V. (eds.) Proceedings IAU Symposium, 114, Reidel , pp 59–62.

61. Hilton, J. L., Capitaine, N., Chapront, J., Ferrndiz, J. M., Fienga, A., Fukushima, T., Getino, J.,Mathews,

P., Simon, J. L., Soffel, M., Vondrak, J., Wallace, P. and Williams, J. (2006), Report of the Internacional Astronomical Union Division I Working Group on precession and the ecliptic, Celest. Mech. Dyn. Astron. 94, 351–367.

62. Hohenkerk, C., and the IAU SOFA Board (2010), SOFA Tools for Earth Attitude. IAU. Available at http://www.iausofa.org
63. Hori, G. (1966), Theory of General Perturbation with Unspecified Canonical Variable, Publ. Astron. Soc. Jpn. 18, 287–296.
64. Huang, C. L., Jin, W. J. and Liao, X. H. (2001), A new nutation model of a non-rigid earth with ocean and atmosphere, Geophys. J. Int. 146, 126–133.
65. Huang, C. L., Dehant, V., Liao, X. H., Van Hoolst, T. and Rochester, M. G. (2011), On the coupling between magnetic field and nutation in a numerical integration approach, J. Geophys. Res. 116, B03403, doi:10.1029/2010JB007713.
66. Jefferys, W. H. (1970), A Fortran-based list processor for Poisson series. Celest. Mech. 2, 474–480.
67. Jeffreys, H. and Vicente, RO. (1957), The theory of nutation and the variation of latitude: the Roche model core, Month. Not. Roy. Astron. Soc. 117, 162–173.
68. Kinoshita, H. (1977), Theory of the rotation of the rigid Earth, Celest. Mech. Dyn. Astron. 15, 277–326.
69. Kinoshita, H. and Souchay, J. (1990), The theory of the nutation for the rigid earth model at the second order, Celest. Mech. Dyn. Astron. 48, 187–265.
70. Klioner, S. A., Gerlach, E., and Soffel, M. (2009), Relativistic aspects of rotational motion of celestial bodies, In: S. Klioner, K. Seidelmann, M. Soffel (eds.) Relativity in Fundamental Astronomy, Proc. of the IAU Symposium 261, Cambridge University Press, Cambridge, pp 112–123.
71. Krasinski, G.A. (2006), Numerical theory of rotation of the deformable Earth with the two-layer fluid core. Part 1: Mathematical model, Celest. Mech. Dyn. Astron. 96, 169–217.
72. Krásná, H., Böhm, K. and Schuh, H. (2013), Free core nutation observed by VLBI, Astron. Astrophys. 555, A29.
73. Lambert, S. (2007), Empirical model of the Free Core Nutation, Technical note, available athttp://syrte.obspm.fr/lambert/fcn/
74. Leimanis, E. (1965), The general problem of the motion of coupled rigid bodies about a fixed point. Springer-Verlag, Berlin-Heidelberg-New York.
75. Lieske, J. H., Lederle, T., Fricke, W., and Morando, B. (1977), Expressions for the Precession Quantities Based upon the IAU (1976) System of Astronomical Constants, Astron. Astrophys.58, 1–16.
76. Mathews, P. M., Herring, T. A., Buffet, B. A. and Shapiro, I. I. (1991a), Forced nutations of the Earth: Influence of inner core dynamics 1. Theory, J. Geophys. Res. 96, 8291–8242.
77. Mathews, P. M., Herring, T. A., Buffet, B. A. and Shapiro, I. I. (1991b), Forced nutations of the Earth: Influence of inner core dynamics 2. Numerical results and comparisons, J. Geophys. Res. 96, 8243–8257.
78. Mathews, P. M., Herring, T. A. and Buffet, B. A. (2002), Modelling of nutation and precession: New nutation series for nonrigid Earth and insights into the Earth's interior, J. Geophys. Res.107 B4, 2068–2094.
79. McCarthy, D. D. (ed) (1996), IERS Conventions, IERS Technical Note, 21, Observatoire de Paris, Paris, available at http://www.iers.org/TN21
80. Moritz, H. (1982), A variational principle for Moledensky's liquid-core problem, Bull. Geod.56, 381–400.
81. Navarro, J. F. and Ferrándiz, J. M. (2002), A new symbolic processor for the Earth rotation theory, Celest. Mech. Dyn. Astron. 82, 243–263.
82. Navarro, J. F. (2002), Teoría analítica de la rotación de la tierra rígida mediante manipulación simbólica, Doctoral Dissertation.
83. Newcombe, S. (1892), On the dynamics of the Earth's rotation, with respect to the periodic variations of latitude, Mon. Not. R. Astron. Soc. 248–249, 336–341.
84. Newcombe, S. (1898), Sur les formules de nutation basées sur les décisions de la conférence de 1896, Bull. Astron. 15, 241–246.
85. Nilsson, T., Böhm, J. and Schuh, H. (2010), Sub-diurnal Earth rotation variations observed by VLBI, Artificial Satellites, 45, No. 2.
86. Petit, G. and Luzum, B., IERS Conventions (2010), IERS Technical Note 36, Verlag des Bundesamtes für Kartographie und Geodäsie, Frankfurt am Main.
87. Plag, H. P., Gross, R. S. and Rotahacher, M. (2009a), Global geodetic observing system for geohazards and global change, Geosciences, BRGM's Journal for a Sustainable Earth 9, 96–103.
88. Plag, H. P. and Pearlman, M. (eds), Global Geodetic Observing System: Meeting the Requirements of a Global Society on a Changing Planet in 2020, Springer-Verlag, Berlin-Heidelberg, (2009b).
89. Plag, H. P., Rizos, C., Rothacher, M. and Neilan, R., The Global Geodetic Observing System (GGOS): Detecting the Fingerprints of Global Change in Geodetic Quantities. In: Advances in Earth Observation of Global Change, Chuvieco, E., Li, J., Yang, X. (eds.), Springer (2010).
90. Poincaré, H. (1901), Sur une forme nouvelle des *équations* de la mécanique, C.R. Acad. Sci. Paris, 132, 369–371.
91. Poincaré, H. (1910), Sur la précesion des corps déformables, Bull. Astronom. 27, 321–356.
92. Richardson, D. L. (1989), PARSEC: An interactive Poisson series processor for personal computing systems, Celest. Mech. Dyn. Astron. 45, 267–274.
93. Rom, A. (1969), Mechanized algebraic operations (MAO), Celest. Mech. 1, 301–319.
94. Roosbeek, F. and Dehant, V. (1998), RDAN97: An analytical development of rigid Earth nutations series using the torque approach, Celest. Mech. Dyn. Astron. 70, 215–253.

95. Sasao T., Okubo S., Saito M. (1980), A simple theory on the dynamical effects of a stratified fluid core upon nutational motion of the Earth In Fedorov, E. P., Smith, M. L., Bender, P. L. (eds) Nutation and the Earth's rotation, Proc. IAU Symp. 78, pp 165–183.

96. Schastok J. (1997), A new nutation series for a more realistic model Earth, Geophys. J. Int.,130, 137–150.

97. Schuh, H., Nagel, S. and Seitz, T. (2001), Linear drift and periodic variations observed in long time series of polar motion, J. Geod. 74: 701–710.

98. Seidelmann, P. K. (1982), 1980 IAU theory of nutation—the final report of the IAU Working Group on Nutation, Celest. Mech. 27, 79–106.

99. Shirai, T. and Fukushima, T. (2000), Numerical Convolution in the Time Domain and Its Application to the Nonrigid-Earth Nutation Theory, Astron. J., 119, 2475–2480.

100. Souchay, J. and Kinoshita, H. (1996), Corrections and new developments in rigid earth nutation theory. I. Lunisolar influence including indirect planetary effects, Astron. Astrophys.312, 1017–1030.

101. Souchay, J. and Kinoshita, H. (1997), Corrections and new developments in rigid-Earth nutation theory. II. Influence of second-order geopotential and direct planetary effect, Astron. Astrophys. 318, 639–652.

102. Souchay, J., Losley, B., Kinoshita, H. and Folgueira, M. (1999), Corrections and new developments in rigid Earth nutation theory III. Final tables REN-2000 including crossed-nutation and spin-orbit coupling effects, Astron. Astrophys. Suppl. Ser. 135, 111–131.

103. Tisserand, F.F. (1891), Traité de Mécanique Céleste, T. II Théorie de la figure des corps célestes et de leur mouvement de rotation. Gauthier Villars, Paris. Reprinted by Jacques Gabay, Paris, 1990.

104. Urban, S. E. and Seidelmann P. K. (eds) (2013), The Explanatory Supplement to the Astronomical Almanac. University Science Books, Mill Valley.

105. Vicente, R. O. and Jeffreys, H. (1964), Nearly diurnal nutation of the Earth, Nature 204, 120–121.

106. Wahr, J. M. (1981), The forced nutations of an elliptical, rotating, elastic and oceanless Earth. Geophys. J. Roy. Astron. Soc. 64, 705–727.

107. Williams, J. G. (1994), Contributions to the Earth's obliquity rate, precession, and nutation, Astron. J. 108, 711–724.

108. Woolard, E. W. (1953a) Theory of the rotation of the Earth around its center of mass, Goddard Space Flight Center.

109. Woolard, E. W. (1953b) A revedelopment of the theory of nutation, Astron. J. 58, 1–3.

CHAPTER 13

ON THE NEWTONIAN ANISOTROPIC CONFIGURATIONS

F. Shojai, M. R. Fazel, A. Stepanian and M. Kohandel

INTRODUCTION

The Newtonian theory of stellar structure is usually used to describe the stars with not extremely high densities [1, 2]. Even for high density compact objects like white dwarfs and neutron stars, applying Newtonian gravity leads to acceptable results, comparable to the results of the relativistic models [3].

The contents of the star is often modeled by a perfect fluid with an equation of state relating the mass density ρ and pressure P. Among all possible choices of this equation, the polytrope equation of state, $P = K\rho^{\gamma}$ is of great importance for many astrophysical situations [1, 2, 4] where K and $\gamma = 1 + \frac{1}{n}$ are both constants called the polytropic constant and polytropic exponent, respectively (n is called the polytropic index). The polytrope equation is the simplest equation of state useful for a wide range of fluid densities. Moreover, it leads to the scale symmetry and therefore enables introducing homology invariants [2].

On the other hand there are some phenomena leading to the local anisotropy of pressure inside a self gravitating system. For low density objects, the Newtonian approximation is valid. Some of the physical mechanisms for anisotropy in this regime are [5]:

- When there is an anisotropic velocity distribution in a collisionless gas [6, 7], the radial pressure satisfies the Jeans equation:

$$\frac{dP_r}{dr} = \rho\frac{d\phi}{dr} + \frac{2}{r}\Delta(r) \tag{1}$$

Citation: F. Shojai, M. R. Fazel, A. Stepanian, M. Kohandel , On the Newtonian anisotropic configurations, DOI: 10.1140/epjc/s10052-015-3478-6

where ϕϕ is the Newtonian gravitational potential of the fluid and we have considered a static and spherically symmetric distribution of matter. $\Delta = P_t - P_r$, where P_r and P_t are the radial and tangential pressure, respectively. In this case Δ measures the anisotropy of the velocity distributions:

$$\Delta = \rho(\langle V_r^2\rangle - \langle V_\phi^2\rangle) \tag{2}$$

where $\langle V_r^2\rangle$ and $\langle V_\phi^2\rangle$ are the radial and azimuthal velocity dispersions. (Note that according to the spherical symmetry assumption we have $\langle V_\phi^2\rangle = \langle V_\theta^2\rangle$). Equation (1) describes the hydrostatic equilibrium of a self gravitating object. To produce an anisotropic velocity distribution, consider a galactic halo of fermionic dark matter. The conservation of angular momentum of neutrinos streaming into the halo leads to an anisotropic pressure [8].

- For a slowly rotating system [2, 5], the equation of hydrostatic equilibrium in the first order is given by Eq. (1) in which $\Delta = -\frac{1}{3}\rho\omega^2 r^2$ where ω is the angular velocity.
- For a mixture of two non-interacting perfect fluids, the energy-momentum tensor is that of an anisotropic fluid in which ρ, P_r, and P_t are some functions of the mass densities and pressure of each component [9].
- For a low mass charged white dwarf, the repulsive electrostatic force can be regarded as a source of anisotropy [10].

According to the above cases, even in the Newtonian regime, generally we are dealing with two components of pressure.

Here we shall focus on the Newtonian anisotropic polytropes [4, 11]. In the next section, first we derive the anisotropic version of dimensionless Newtonian hydrostatic equation, Lane–Emden equation and its boundary conditions.

It is written in terms of physical quantities like the anisotropy function, in contrast to the form used in [4]. We extend the homology theorem for anisotropic polytropes and then in Sect. 3 we derive a new set of analytical solutions. There are some exact solutions of the Lane–Emden equation for special values of barotropic index in arbitrary spatial dimension [2].

In this paper, we use the ansatz that the form of the Lane–Emden equation does not change after introducing the anisotropy factor. First, a numerical analysis of the mass–radius relation of an anisotropic star is presented. We find the anisotropic modified form of the existing isotropic solutions by the above ansatz in Sect. 3. The authors of [12, 13], propose another ansatz on Δ in the modeling of relativistic stars. This is used in [4] to generate some anisotropic Newtonian polytrope solutions from isotropic solutions. Other various assumptions on Δ can be found in [11]. In Sect. 4, we focus on some integral theorems on the physical quantities of a star in the hydrostatic equilibrium obtained by Chandrasekhar in his book [1]. These are derived from Newtonian equations directly without any special assumption on the internal structure of the star or the equation of state. These theorems contain some inequalities containing the central pressure, the gravitational potential and the mean gravitational acceleration of a star. Here we extend these theorems to the general anisotropic equilibrium configuration. At the end, we present a perturbative treatment to study the effects of anisotropy in a star.

ANISOTROPIC POLYTROPES

The hydrostatic equilibrium of a star is governed by Eq. (1). The Newtonian potential is related to the density of the fluid by Poisson's equation:

$$\nabla^2\phi = -4\pi G\rho. \tag{3}$$

Considering the fluid obeys the polytropic equation of state along with Eq. (1), one gets

$$\phi(r) - \phi_0 = K(n+1)\left(\rho^{\frac{1}{n}} - \rho_0^{\frac{1}{n}}\right) - \int_0^r \frac{2\Delta(x)}{x\rho(x)}\mathrm{d}x \tag{4}$$

where ρ_0 and ϕ_0 denote the central density and gravitational potential. We take $\phi_0 = 0$. Substituting Eq. (4) into Eq. (3) leads to the following expression which is the fundamental equation of equilibrium:

$$K(n+1)\nabla^2\rho^{\frac{1}{n}} - \frac{1}{r^{N-1}}\frac{\mathrm{d}}{\mathrm{d}r}\left(r^{N-1}\frac{2\Delta(r)}{r\rho(r)}\right) = -4\pi G\rho \tag{5}$$

where N is the dimension of space.

Introducing the new dimensionless variable $\left(\pm\frac{(n+1)K}{4\pi G\rho_0^{1-\frac{1}{n}}}\right)^{-\frac{1}{2}} r,$, and with the following definitions:

$$\rho = \rho_0\theta^n, \qquad P_r = P_0\theta^{n+1}, \tag{6}$$

Equation (5) takes the following form:

$$\theta'' + \frac{N-1}{\xi}\theta' - \frac{2}{P_0(n+1)\xi\theta^n}\left[\Delta' + \frac{N-2}{\xi}\Delta - n\frac{\theta'}{\theta}\Delta\right] = \pm\theta^n \tag{7}$$

where a prime denotes differentiation with respect to the new radial coordinate, ξ, and the plus and minus signs correspond to $-\infty < n < -1, -1 < n < +\infty$, respectively. Note that $\theta = \theta(r)$is a dimensionless function. Equation (7) is the anisotropic version of the well-known Lane–Emden equation.

For n=−1, the polytrope equation of state gives $P_r = K$ if ρ≠0. Thus according to Eq. (4), the term involving Δ only contributes to the gravitational potential and the Poisson equation gives

$$\phi - \phi_0 = K\ln\left(\frac{\rho}{\rho_0}\right) - \int_0^r \frac{2\Delta(x)}{\rho(x)x}\mathrm{d}x. \tag{8}$$

This shows that the case with n=−1 is a special case, which must be treated separately. Whereas for the isotropic case (Δ=0), one gets ρ=0 and therefore, n=−1 is excluded from the further study for isotropic fluids.

The other important case arises when n=±∞ and therefore the polytrope equation of state reduces to

$$P_r = K\rho. \tag{9}$$

Replacing this in (1), we obtain

$$\phi - \phi_0 = K\ln\left(\frac{\rho}{\rho_0}\right) - \int_0^r \frac{2\Delta(x)}{\rho(x)x}\mathrm{d}x \tag{10}$$

The combination of the above expression when $\phi_0=0$ and (3) gives

$$\theta'' + \frac{N-1}{\xi}\theta' + \frac{2}{P_0\xi}e^{\theta}\left(\Delta' + \frac{N-2}{\xi}\Delta + \Delta\theta'\right) = e^{-\theta} \tag{11}$$

where $\rho = \rho_0 e^{-\theta},\ P_r = P_0 e^{-\theta}, \text{ and } \xi = \left(\frac{K}{4\pi G\rho_0}\right)^{-\frac{1}{2}} r$.

Either the Lane–Emden equation (7) (for n≠−1,±∞) or (11) (for n=±∞), contains two unknown functions, θ(ξ) and Δ(ξ). Thus we need another equation to obtain a closed system of equations.

Before describing our procedure to introduce the other equation, let us assume that Δ is a given function. To obtain a unique solution for θ one has to specify two boundary conditions. The first is $\theta(0) = 1$ (for $n \neq -1, \pm\infty$) and $\theta(0)$ (for n=±∞). In order to obtain the other condition, let us integrate (3):

$$\frac{\mathrm{d}\Phi}{\mathrm{d}r} = -r^{1-N}\int_0^r 4\pi G\rho x^{N-1}\mathrm{d}x. \tag{12}$$

Combination of the above relation with (1) and (6) produces

$$\frac{\mathrm{d}\theta}{\mathrm{d}\xi}|_{\xi\to 0} = -\frac{\xi}{N}|_{\xi\to 0} + \frac{2}{P_0(n+1)}\lim_{\xi\to 0}\left(\frac{\Delta(\xi)}{\xi}\right) \tag{13}$$

A similar calculation for n=±∞ leads to

$$\frac{\mathrm{d}\theta}{\mathrm{d}\xi}|_{\xi\to 0} = \frac{\xi}{N}|_{\xi\to 0} - \frac{2}{P_0}\lim_{\xi\to 0}\frac{\Delta(\xi)}{\xi}. \tag{14}$$

Equations (13) and (14) lead to the extensions of the Chandrasekhar's theorem [1] for the anisotropic polytrope case.

Theorem 1

The finite solutions of anisotropic Lane–Emden equation at the origin have to satisfy $\frac{\mathrm{d}\theta}{\mathrm{d}\xi}|_{\xi=0} = \frac{2}{P_0(n+1)}$

$\lim_{\xi\to 0}\left(\frac{\Delta(\xi)}{\xi}\right)$ *if* $n \neq -1, \pm\infty$ *and* $\frac{d\theta}{d\xi}|_{\xi=0} = \frac{2}{P_0}\lim_{\xi\to 0}$

$\frac{\Delta(\xi)}{\xi}$ *if* $n = \pm\infty$

Another important theorem is a generalization of the homology theorem [2].

Theorem 2

If the anisotropic Lane–Emden equation is satisfied by $\theta(\xi)$ *and* $\Delta(\xi)$ *then* $A^{\frac{2}{n-1}}\theta(A\xi)$ *and* $A^{\frac{2(n+1)}{1-n}}\Delta(A\xi)$ satisfy Eq. (7), and $\theta(A\xi) - \ln A^2$ *and* $\frac{1}{A^2}\Delta(A\xi)$ satisfy Eq. (11), where A is a constant.

The proof is straightforward. It can easily be obtained by substituting these expressions in (7) and (11). Thus we obtain a set of solutions parametrized by A.

One usual approach to solve (7) (or 11) with boundary conditions obtained in theorem I, is to suppose a special form for Δ. Then putting it into (7) (or 11) yields a differential equation for θ. If this procedure leads to smooth and non-negative pressures and density for the fluid, then they can be regarded as the physical solutions. However, knowing the function Δ, Eq. (7) (or 11) is a non-linear differential equation which in general one does not expect to obtain an analytical solution. Here we introduce a heuristic approach to get some analytical solutions. In order to do this, we assume that the anisotropy factor has no influence on the form of isotropic Lane–Emden equation. This means that Δ modifies only the coefficients of isotropic Lane–Emden equation. Hereafter, we adopt the minus sign on the right-hand side of Lane–Emden equation (7) (or 11), since the plus sign leads to an imaginary radial coordinate [2]. In the following we shall consider two cases.

Case 1

Δ modifies the coefficient of $\theta'\theta'$ in the isotropic Lane–Emden equation. In this case Eq. (7) can be considered as two separated equations as below

$$\frac{-2}{P_0(n+1)\theta^n}\left[\Delta' + \frac{N-2}{\xi}\Delta - n\frac{\theta'}{\theta}\Delta\right] = N_1\theta' \tag{15}$$

$$\theta'' + \frac{N_1 + N - 1}{\xi}\theta' = -\theta^n \tag{16}$$

where N_1 is an arbitrary constant. Equation (15) is a first order differential equation with the solution

$$\Delta(\xi) = -\frac{(n+1)P_0 N_1\theta^n}{2\xi^{N-2}}\int \theta'\xi^{N-2}d\xi. \tag{17}$$

Equations (16) and (17) with the boundary conditions (13) and $\theta(0)=1$ give the solutions of anisotropic polytropes.

Case 2

Δ modifies the coefficient of θ_n in the right-hand side of the isotropic Lane–Emden equation,

$$\frac{-2}{P_0(n+1)\theta^n\xi}\left[\Delta' + \frac{N-2}{\xi}\Delta - n\frac{\theta'}{\theta}\Delta\right] = N_2\theta^n \tag{18}$$

$$\frac{d^2\theta}{d\eta^2} + \frac{N-1}{\eta}\frac{d\theta}{d\eta} = -\theta^n \tag{19}$$

where $\eta = \sqrt{(N_2+1)}\ \xi,\ N_2 > -1$. The solution of Eq. (18) is

$$\Delta(\xi) = \frac{-(n+1)P_0 N_2\theta^n}{2(N_2+1)\xi^{N-2}}\int \theta^n\xi^{N-1}d\xi. \tag{20}$$

Now let us to interpret physically Eq. (7). The quantities of physical interest, for definite values of the polytropic index, are the stellar radius and mass. These are

$$R = \left(\frac{K(n+1)}{4\pi G}\right)^{\frac{1}{2}}\rho_0^{\frac{1-n}{2n}}\xi_1, \tag{21}$$

$$M = \int_0^R 4\pi r^2\rho dr = 4\pi\left(\frac{K(n+1)}{4\pi G}\right)^{\frac{3}{2}}\rho_0\int_0^{\xi_1}\theta^n\xi^2 d\xi \tag{22}$$

where ξ_1 is the first root of θ function and thus this point defines the surface of the star. Using the anisotropic Lane–Emden equation (7), we get

$$M = -4\pi\left(\frac{K(n+1)}{4\pi G}\right)^{\frac{3}{2}}\rho_0\int_0^{\xi_1}\frac{d}{d\xi}\times\left(\xi^2\frac{d\theta}{d\xi} - \frac{2}{P_0(n+1)}\Delta\xi\theta^{-n}\right)d\xi, \tag{23}$$

which with the boundary conditions $\theta_0=1$, Eq. (13), and the definition of surface ξ_1 gives

$$M = 4\pi\left(\frac{K(n+1)}{4\pi G}\right)^{\frac{3}{2}}\rho_0\left|\xi^2\frac{d\theta}{d\xi}\right|_{\xi_1} \tag{24}$$

This has the well-known form of the mass relation for isotropic star. Eliminating the central mass density between (21) and (24) we have

$$MR^{\frac{3-n}{n-1}} = 4\pi\left(\frac{K(n+1)}{4\pi G}\right)^{\frac{n}{n-1}}\xi_1^{\frac{n+1}{n-1}}\left|\frac{d\theta}{d\xi}\right|_{\xi_1} \tag{25}$$

knowing the polytropic index, $\left|\frac{d\theta}{d\xi}\right|_{\xi_1}$ can be obtained from (16) or (19) numerically and thus the constant on the right-hand side of Eq. (25) is found. Therefore the anisotropy factor does not appear explicitly in the mass or the mass–radius relation of a star. The effect of anisotropy is included in the θ function through its equation (parameter N_1 in (16) and variable η in (19)). Figures 1 and 2 show the mass–radius diagrams for low density white dwarfs with n=32 obtained from Eqs. (16) and (19), respectively.

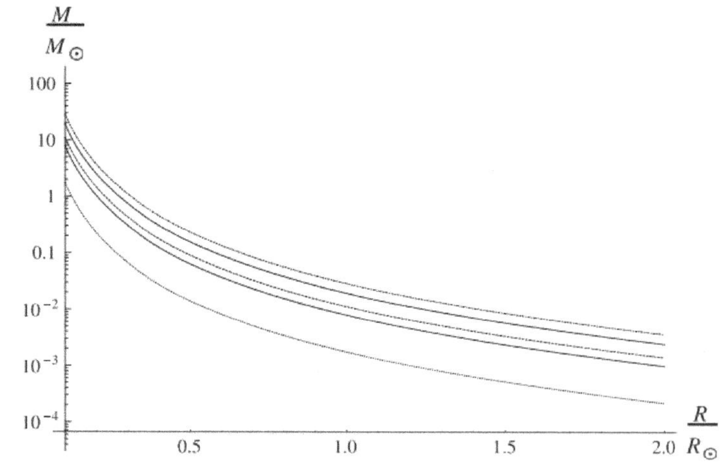

Figure 1: Plot of $\frac{M}{M_\odot}$ as a function of $\frac{R}{R_\odot}$ in the Case 1, for $n = \frac{3}{2}$ with N1=−1.5 (*dotted line*), N_1=0 (*thin line*, the isotropic case), $N_1 = -1.5$ (*dotted line*), $N_1 = 0$ (*thick line*), and N_1=2.5 (*dot-dashed line*).

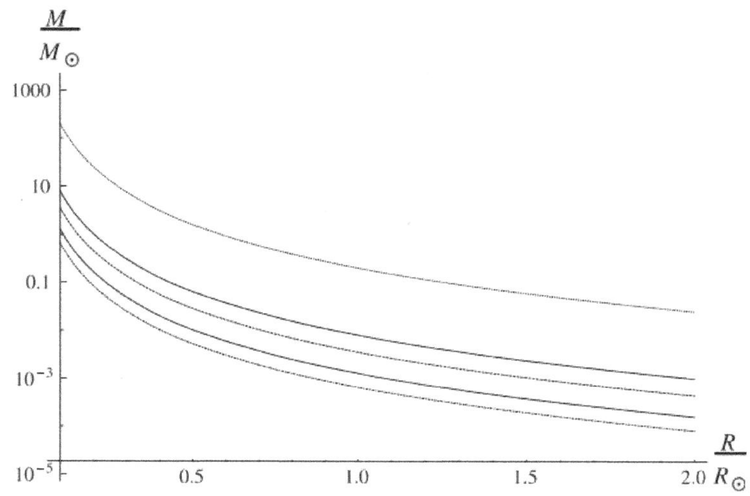

Figure 2: Plot of $\frac{M}{M_\odot}$ as a function of $\frac{R}{R_\odot}$ in the Case 2, for $n = \frac{3}{2}$ with N_2=−0.8 (*dotted line*), N_2=0 (*thin line*, the isotropic case), N_2=0.5 (*dashed line*), N_2=1.5 (*thick line*), and N_2=2.5 (*dot-dashed line*).

The influence of anisotropy on the mass ratio for a fix star radius is numerically plotted in Figs.3 and 4. Note that in the Case 1 the parameter N_1 scales the anisotropy, while for the Case 2, the parameter N_2 plays that role.

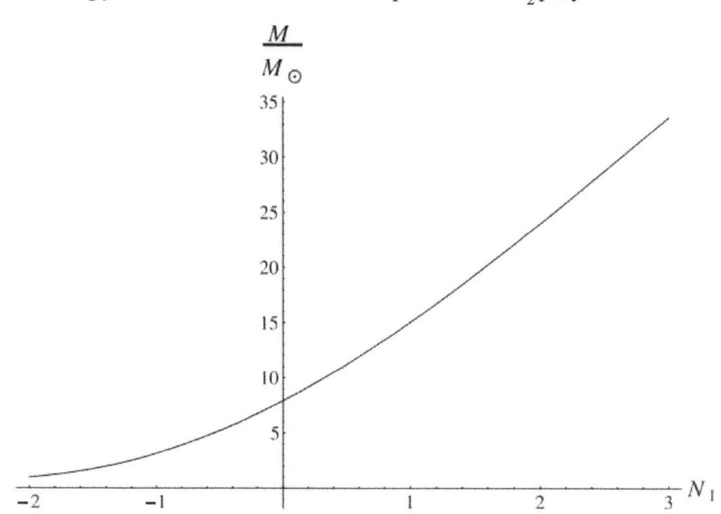

Figure 3: Plot of $\frac{M}{M_\odot}$ as a function of N_1 in the Case 1, for $n = \frac{3}{2}$ and $\frac{R}{R_\odot}$=10.

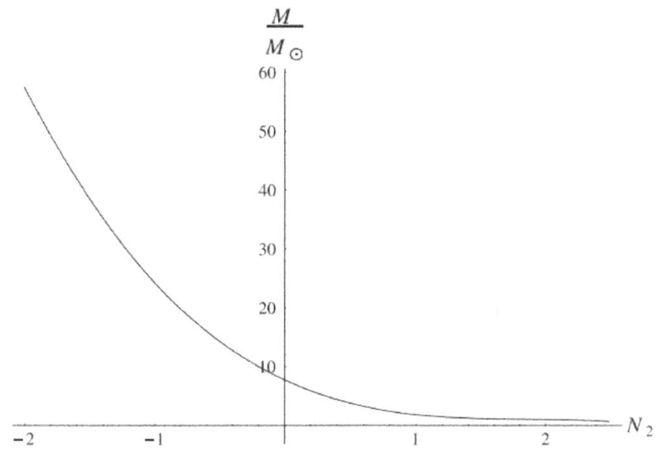

Figure 4: Plot of $\frac{M}{M_\odot}$ as a function of N_2 in the Case 2, for $n = \frac{3}{2}$ and $\frac{R}{R_\odot}=10$

EXACT ANALYTICAL SOLUTIONS OF ANISOTROPIC LANE–EMDEN EQUATION

Case I

n=0

With n=0, the density is constant, $\rho=\rho_0$, and Eq. (16) takes the following form:

$$\theta'' + \frac{S-1}{\xi}\theta' + 1 = 0 \tag{26}$$

in which $S=N_1+N$, where solution is

$$\theta = \begin{cases} -\frac{\xi^2}{2S} + 1, & S > 1, \\ -\frac{\xi^2}{2S} + A\frac{\xi^{2-S}}{2-S} + 1, & S \leq 1, S \neq 0 \end{cases} \tag{27}$$

in which A is a constant. In the above solution we have also implemented the boundary condition $\theta(0)=1$ and only those terms that lead to the non-singular solutions have been written.

We can now read ΔΔ from (17) considering the boundary condition (13):

$$\Delta = \begin{cases} \frac{(S-N)P_0}{2SN}\xi^2 & S > 1, \\ -\frac{(S-N)P_0}{2}\left(-\frac{\xi^2}{NS} + \frac{A}{N-S}\xi^{2-S}\right) & S \leq 1, S \neq 0, S \neq N \\ 0 & N = S = 1. \end{cases} \tag{28}$$

We see that, to satisfy the boundary conditions, we have to have $\Delta(0)=0$. The resulting ΔΔfunction is plotted in Fig. 5.

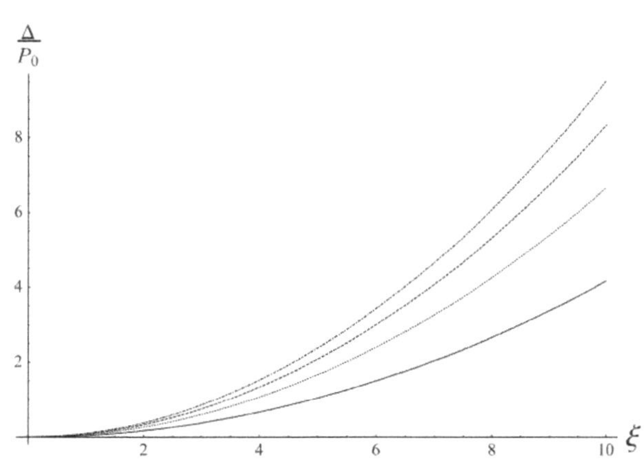

Figure 5: Plot of Δ for S>1 in three dimensions with S=4 (*thick line*), S=5 (*dotted line*), S=6 (*dashed line*), S=7 (*dot-dashed line*)

Also the radial and tangential pressures are determined from the expressions (6), (27), and (28).

n=1

In this case, the differential equation (16) reduces to

$$\theta'' + \frac{(S-1)}{\xi^2}\theta' + \theta = 0. \tag{29}$$

The solutions of (29) are given in terms of Bessel functions as follows:

$$\theta(\xi) = \xi^{-\frac{S-2}{2}}\left[C_1 J_{\frac{S-2}{2}}(\xi) + C_2 Y_{\frac{S-2}{2}}(\xi)\right] \tag{30}$$

where C_1 and C_2 are integration constants. Imposing the condition $\theta(0) = 1$, C_2 vanishes and we get

$$\theta(\xi) = \Gamma\left(\frac{S}{2}\right)\left(\frac{\xi}{2}\right)^{-\frac{S-2}{2}} J_{\frac{S-2}{2}}(\xi), \qquad S > 2, \tag{31}$$

in which the constant C_1 is expressed in terms of the Gamma function, $\Gamma(S_2)$. Substituting (31) into (17) leads to

$$\Delta(\xi) = \frac{2^{\frac{S}{2}}}{8}(S-N)P_0\xi^{3-\frac{S}{2}} J_{\frac{S-2}{2}}(\xi)\Gamma^2\left(\frac{S}{2}\right)\Gamma\left(\frac{N}{2}\right) F \times \left(\frac{N}{2}; 1+\frac{S}{2}, 1+\frac{N}{2}; -\frac{\xi^2}{4}\right) \tag{32}$$

where $F(a, b, c, x)$ is the hypergeometric function. These results also satisfy the boundary condition (13) with $\theta'(0) = \Delta(0) = 0$. The function (32) is depicted in Fig. 6.

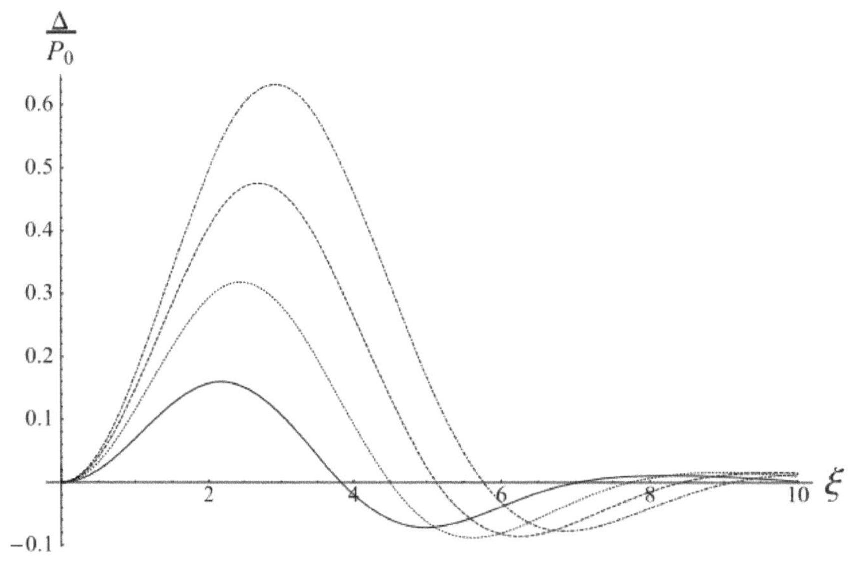

Figure 6: Plot of Δ for n=1 in three dimensions with S=4 (*thick line*), S=5 (*dotted line*), S=6 (*dashed line*), S=7 (*dot-dashed line*)

$n = \frac{S+2}{S-2}$

In this case, if one takes

$$\theta(\xi) = \left(\frac{4}{(n-1)^2}\right)^{n-1} \xi^{\frac{2}{1-n}} z(\xi), \qquad \xi = e^{-t}, \tag{33}$$

then Eq. (16) reduces to

$$\frac{d^2 z}{dt^2} = 4\frac{z \mp z^n}{(n-1)^2} \tag{34}$$

The above equation can be integrated as follows:

$$\frac{1}{2}\left(\frac{dz}{dt}\right)^2 = \frac{4}{(n-1)^2}\left[\frac{z^2}{2} \mp \frac{z^{n+1}}{n+1}\right] + C. \tag{35}$$

Considering $n > 1\ (S > 2)$, turning back to the original variables and using Eqs. (33) and (35), a simple calculation shows that the initial conditions on θ leads to z=0 and $dz/dt = 0$ at $\xi = 0$. This gives $C = 0$. Writing the above equation in terms of the original variables, we have

$$\frac{2\theta\theta'}{n-1} + \xi\frac{\theta'^2}{2} + \xi\frac{\theta^{n+1}}{n+1} = 0. \tag{36}$$

We proceed with integrating the above equation

$$\xi\frac{\theta^{\frac{n+1}{2}}}{\theta'} = D. \tag{37}$$

The constant D might be determined taking into account (13) and θ(0)=1

$$\frac{1}{D} = \lim_{\xi\to 0}\theta^{-\frac{S}{S-2}}\frac{\theta'}{\xi} = -\frac{1}{N} + \frac{2}{P_0(n+1)}\lim_{\xi\to 0}\frac{\Delta(\xi)}{\xi^2}. \tag{38}$$

As we see, the constant D depends on the Δ function near the origin. The differential equation (37) has the following solution:

$$\theta = \left[1 - \frac{\xi^2}{D(S-2)}\right]^{-\frac{S-2}{2}} \tag{39}$$

Employing (17), we get the following expression for the difference between the radial and tangential pressures:

$$\Delta(\xi) = -\frac{SP_0(S-N)}{ND(S-2)}\xi^2\left[1 - \frac{\xi^2}{D(S-2)}\right]^{-\frac{S+2}{2}} \times F\left(\frac{N}{2}, \frac{S}{2}, 1+\frac{N}{2}, \frac{\xi^2}{D(S-2)}\right) \tag{40}$$

Now, we are in the position to fix the constant D. Combining (38) and (40), one gets D=−S. The resulting Δ is shown in Fig. 7.

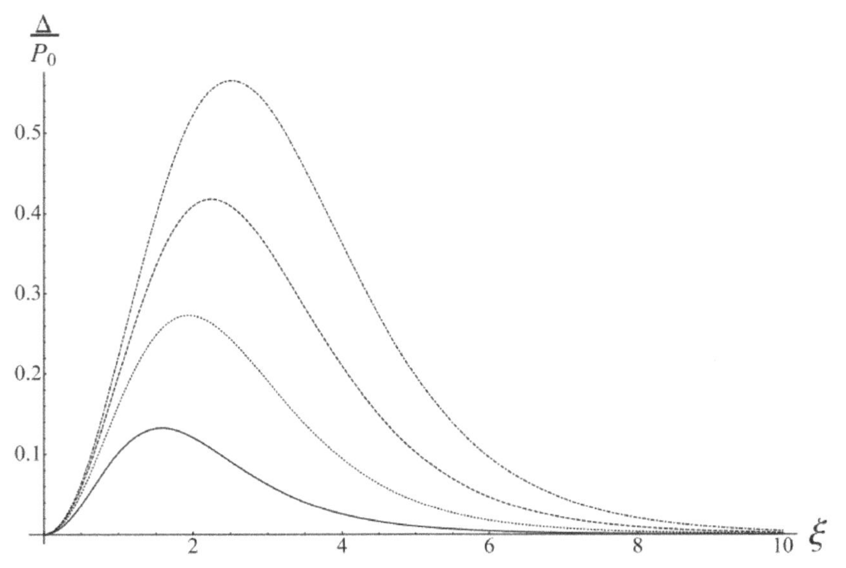

Figure 7: Plot of Δ for $n = \frac{S+2}{S-2}$ in three dimensions with S=4 (*thick line*), S=5 (*dotted line*), S=6 (*dashed line*), S=7 (*dot-dashed line*)

Case II

n=0

The solutions of differential equations (19) and (20) satisfying the boundary conditions are

$$\theta = \begin{cases} -\frac{N_2+1}{2N}\xi^2 + A\sqrt{N_2+1}\xi + 1, & N = 1, \\ -\frac{N_2+1}{2N}\xi^2 + 1, & N > 1, \end{cases}$$

(41)

$$\Delta = \begin{cases} -\frac{P_0 N_2}{2(N_2+1)N}\xi^2, & N \geq 2, \\ -\frac{P_0 N_2}{2(N_2+1)}(\xi^2 - \frac{\sqrt{(N_2+1)^3}}{N_2}A\xi), & N = 1. \end{cases}$$

(42)

The behavior of Δ is shown in Fig. 8.

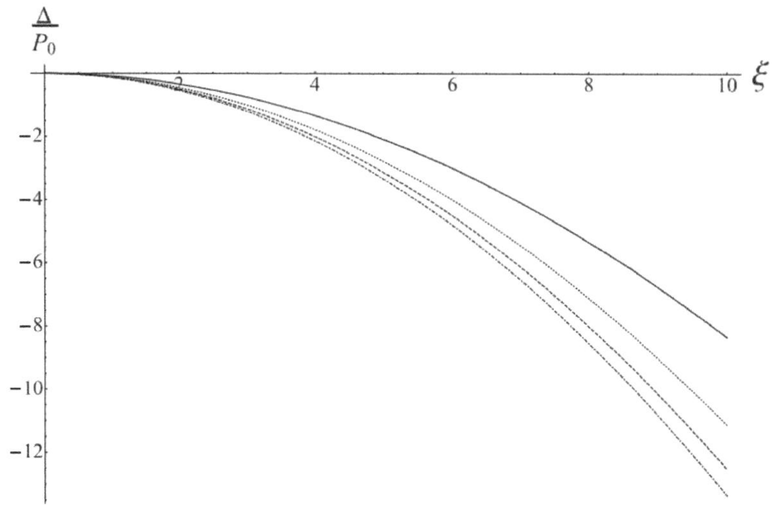

Figure 8: Plot of Δ for n=0 in three dimensions with N_2=1 (*thick line*), N_2=2 (*dotted line*), N_2=3 (*dashed line*), N_2=4 (*dot-dashed line*)

n=1

The situation for this case is also similar to the previous one and the substitution of S and ξ by N and η in (31) produces the solution of (19) with n=1. This outcome with the integration (20) gives the differences between the radial and tangential pressures:

$$\Delta(\xi) = -2^{N-2}\frac{P_0 N_2}{(N_2+1)^{\frac{1}{2}(N+1)}}\Gamma^2\left(\frac{N}{2}\right)\xi^{3-N}J_{\frac{N-2}{2}} \times(\sqrt{N_2+1}\xi)J_{\frac{N}{2}}(\sqrt{N_2+1}\xi),$$

(43)

which are depicted in Fig. 9.

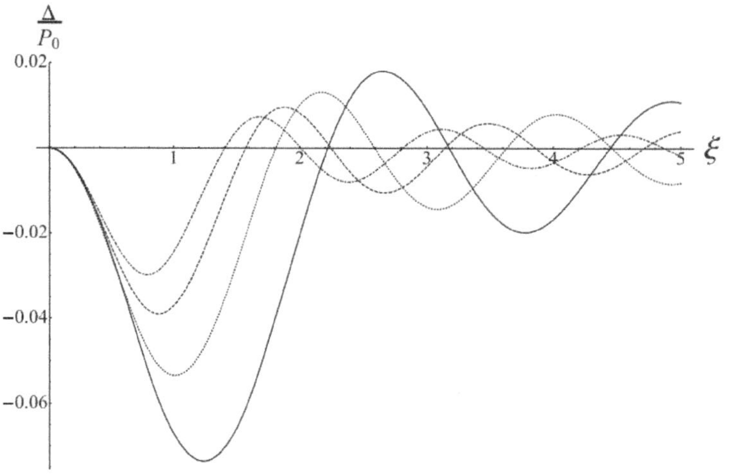

Figure 9: Plot of Δ for n=1 in three dimensions with N_2=11 (*thick line*), N_2=2 (*dotted line*), N_2=3 (*dashed line*), N_2=4 (*dot-dashed line*)

$n = \frac{N+2}{N-2}$

Similar to the two former cases, in this situation the solution to the (19) with $n = \frac{N+2}{N-2}$ might be obtained by changing S and ξ to N and η in (39),

$$\theta = \left[1 - \frac{N_2+1}{D(N-2)}\xi^2\right]^{-\frac{N-2}{2}}, \qquad N > 2. \tag{44}$$

Putting the above result in (20) gives

$$\Delta = -\frac{N P_0 N_2}{(N_2+1)(N-2)N}\left[1 - \frac{1+N_2}{D(N-2)}\xi^2\right]^{-(N+1)}\xi^2. \tag{45}$$

Combining (38) and (45), one gets $D = -\frac{N(N_2+1)}{2N_2-1}$. The graph of (45) is plotted in Fig. 10 for various values of N_2.

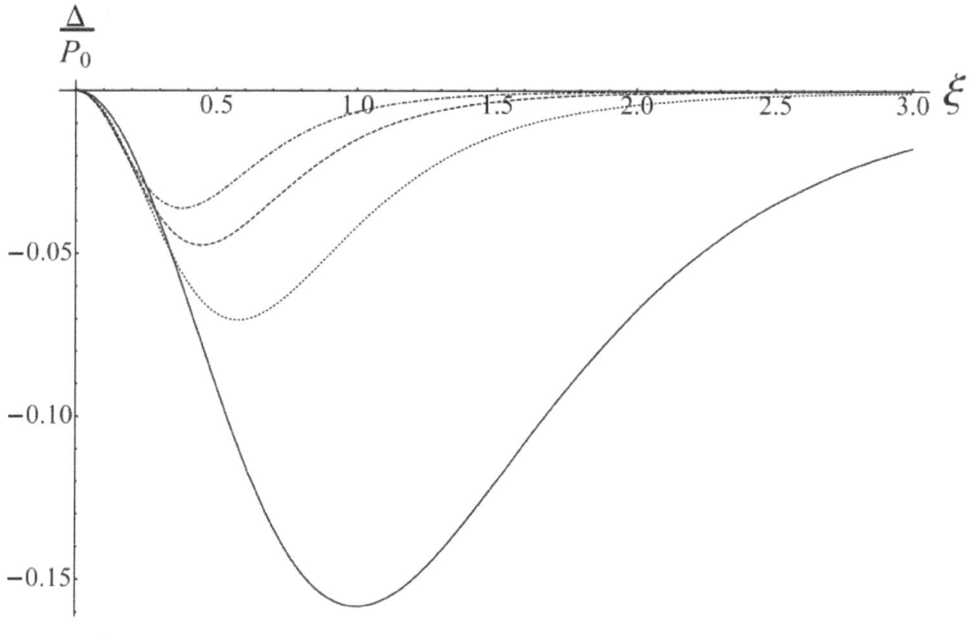

Figure 10: Plot of Δ for $n = \frac{N+2}{N-2}$ in three dimensions with N_2=1 (*thick line*), N_2=2 (*dotted line*), N_2=3 (*dashed line*), N_2=4 (*dot-dashed line*)

INTEGRAL THEOREMS

Chandrasekhar in his famous book [1] has discussed some inequalities for the physical quantities describing a star in the Newtonian gravitational equilibrium. In this section we will extend some of his results for anisotropic stars. The results are general and are not restricted to the polytropic case.

Let $m(r)$ be the mass contained inside the radius r, then

$$m' = 4\pi\rho r^2 \tag{46}$$

Equation (1) can now be written as

$$P_r' = -\frac{Gm\rho}{r^2} + \frac{2\Delta}{r} \tag{47}$$

where we have used the definition of gravitational potential. From (46) and (47) one can get the following equation:

$$\frac{1}{r^2}\left[\frac{r^2}{\rho}\left(P_r' - \frac{2\Delta}{r}\right)\right]' = -4\pi G\rho \tag{48}$$

These equations can be used to prove the following theorems.

Theorem 3

For any equilibrium configuration the function

$$I = P_{\mathrm{r}} + \frac{\mathrm{Gm}^2}{8\pi r^4} - 2\int_0^r \mathrm{d}\tilde{r}\,\frac{\Delta}{\tilde{r}} \tag{49}$$

does not increase outward.

Proof

Let us calculate

$$I' = P_{\mathrm{r}}' + \frac{\mathrm{Gmm}'}{4\pi r^4} - \frac{\mathrm{Gm}^2}{2\pi r^5} - \frac{2\Delta}{r} \tag{50}$$

Then by (47)

$$I' = -\frac{\mathrm{Gm}^2}{2\pi r^5} \le 0. \tag{51}$$

Corollary 1

For the central pressure, we have

$$P_0 > P_{\mathrm{r}} + \frac{\mathrm{Gm}^2}{8\pi r^4} - 2\int_0^r \mathrm{d}\tilde{r}\,\frac{\Delta}{\tilde{r}} > \frac{\mathrm{Gm}^2}{8\pi R^4} - 2\int_0^R \mathrm{d}\tilde{r}\,\frac{\Delta}{\tilde{r}} \tag{52}$$

in which $M = m(R)$ is the mass of the star. The last term in the right-hand side shows a lower bound on P_0. Moreover, for any arbitrary radius, the above relation leads to

$$P_{\mathrm{r}} > \frac{G}{8\pi}\left(\frac{M^2}{R^4} - \frac{m^2}{r^4}\right) - 2\int_r^R \mathrm{d}\tilde{r}\,\frac{\Delta}{\tilde{r}} \tag{53}$$

Theorem 4

For any equilibrium configuration

$$I_\nu \equiv \int_0^R \mathrm{d}m\,\frac{\mathrm{Gm}}{\tilde{r}^\nu} = 4\pi \int_0^R \mathrm{d}\tilde{r}\tilde{r}^{3-\nu}[(4-\nu)P_{\mathrm{r}} + 2\Delta] \tag{54}$$

if v<4.

Proof

From (47) we have

$$\frac{Gmm'}{4\pi r^4} = \frac{2\Delta}{r} - P_{\mathrm{r}}'. \tag{55}$$

Multiplying this by $r^{4-\nu}$ and then integrating from r to R gives

$$I_\nu = 8\pi \int_0^R \mathrm{d}\tilde{r}\tilde{r}^{\nu-3}\Delta - 4\pi \int_0^R \mathrm{d}P_{\mathrm{r}}\tilde{r}^{4-\nu}. \tag{56}$$

Integrating by parts the first integral, leads to (54).

For v=4, Eq. (56) gives

$$I_4 = 8\pi \int_0^R \mathrm{d}\tilde{r}\,\frac{\Delta}{\tilde{r}} + 4\pi P_0 \tag{57}$$

which also has a lower bound equal to $\frac{\mathrm{Gm}^2}{2R^4}$ according to (52).

The gravitational potential energy Ω of the configuration is $-I_1$. According to (56) it is given by

$$\Omega = -\int_0^R dV(3P_r + 2\Delta) \tag{58}$$

Denoting the mean value of gravitational acceleration by $\bar{g}$, we have

$$M\bar{g} = \int_0^R dm \frac{Gm}{r^2} = I_2 \tag{59}$$

and hence by (56)

$$M\bar{g} = 8\pi \int_0^R d\tilde{r}\tilde{r}\,(P_r + \Delta) = 8\pi \int_0^R d\tilde{r}\tilde{r} P_t. \tag{60}$$

This shows that for an anisotropic star only the tangential pressure contributes to finding the mean value of acceleration.

Theorem 5

For any equilibrium configuration

$$\begin{aligned}&\pi \nu P_0 R^{4-\nu} + \frac{(4-\nu)Gm^2}{8R^\nu}\\ &+2\pi\nu(4-\nu)\int_0^R d\tilde{r}\tilde{r}^{3-\nu}\int_0^R d\tilde{r}\frac{\Delta}{\tilde{r}}\\ &> 2\pi\nu\int_0^R d\tilde{r}\frac{\Delta}{\tilde{r}^{\nu-3}} + I_\nu > \frac{Gm^2}{2R^\nu}.\end{aligned} \tag{61}$$

Proof

By Theorem 3

$$\begin{aligned}&\frac{Gm^2}{8\pi R^4} - \frac{Gm^2}{8\pi r^4} - 2\int_0^R d\tilde{r}\frac{\Delta}{\tilde{r}} + 2\int_0^r d\tilde{r}\frac{\Delta}{\tilde{r}} < P_r < P_0\\ &-\frac{Gm^2}{8\pi r^4} + 2\int_0^r d\tilde{r}\frac{\Delta}{\tilde{r}}\end{aligned} \tag{62}$$

and by Theorem 4

$$\begin{aligned}&4\pi(4-\nu)\int_0^R d\tilde{r}\tilde{r}^{3-\nu}\left[\frac{Gm^2}{8\pi R^4} - \frac{Gm^2}{8\pi r^4} - 2\int_0^R d\tilde{r}\frac{\Delta}{\tilde{r}}\right.\\ &\left.+2\int_0^r d\tilde{r}\frac{\Delta}{\tilde{r}}\right]\\ &< I_\nu + 8\pi\int_0^R d\tilde{r}\tilde{r}^{3-\nu}\Delta < 4\pi(4-\nu)\int_0^R d\tilde{r}\tilde{r}^{3-\nu}\\ &\times\left[P_0 - \frac{Gm^2}{8\pi r^4} + 2\int_0^r d\tilde{r}\frac{\Delta}{\tilde{r}}\right].\end{aligned} \tag{63}$$

This inequality can be written as

$$\begin{aligned}&4\pi P_0 R^{4-\nu} + 8\pi\nu(4-\nu)\int_0^R d\tilde{r}\tilde{r}^{3-\nu}\int_0^r d\tilde{r}\frac{\Delta}{\tilde{r}}\\ &> I_\nu + 8\pi\int_0^R d\tilde{r}\tilde{r}^{3-\nu}\Delta + \frac{(4-\nu)}{2}\int_0^R d\tilde{r}\frac{Gm^2}{\tilde{r}^{\nu+1}}\\ &> \frac{Gm^2}{2R^\nu}.\end{aligned} \tag{64}$$

Inserting

$$\int_0^R d\tilde{r}\frac{\mathrm{Gm}^2}{\tilde{r}^{\nu+1}} = \frac{\mathrm{Gm}^2}{R^\nu} - 2I_\nu \tag{65}$$

in (64) and simplifying it, we get (61). □

Corollary 2

Setting v=1, we have

$$\left(\pi P_0 R^3 + \frac{3\mathrm{Gm}^2}{8R} + 6\pi \int_0^R d\tilde{r}\tilde{r}^2 \int_0^r d\tilde{r}\frac{\Delta}{\tilde{r}}\right) > 2\pi \int_0^R d\tilde{r}\tilde{r}^2 \Delta - \Omega > \frac{\mathrm{Gm}^2}{2R}, \tag{66}$$

which shows the upper and lower bounds of the potential energy, −Ω.

CONCLUDING REMARKS

In this paper we discussed how the anisotropy factor modifies the Lane–Emden equation and homology theorem. We obtained some theorems governing the characteristic functions of anisotropic star. These are the extension of Chandrasekhar's theorems. We performed a procedure to find the anisotropy factor, the radial pressure and the density functions exactly satisfying the Lane–Emden equation. We had two cases. In the first case the effect of anisotropy is to change the dimension of space from N to $N+N_1$ (which may be a non-integer number). For the second case, anisotropy shows itself as a rescaling of the radial coordinate in the function θ.

Moreover, it is straightforward to find some approximate analytical solutions of the polytropic stars with anisotropic pressure. For example, let us assume that the including anisotropy factor is equivalent to a slight modification of the polytropic index from its value in the absence of anisotropy, n0n0. This means that if an exact solution θ_0 of the isotropic Lane–Emden equation is known:

$$\theta_0'' + \frac{N-1}{\xi}\theta_0' = -\theta_0^{n_0} \tag{67}$$

Inclusion of the anisotropy factor leads to

$$\theta'' + \frac{N-1}{\xi}\theta' - \frac{2}{P_0(n_0+1)\xi\theta^{n_0}} \times \left[\Delta' + \frac{1}{\xi}\Delta - n_0\frac{\theta'}{\theta}\Delta\right] = -\theta^{n_0}, \tag{68}$$

which is equivalent to perturbing θ_0 and n_0 in Eq. (67) in such a way that $\theta = \theta_0 + \epsilon\theta_*$ and $n = n_0 + \epsilon$ satisfy (67)

$$\theta'' + \frac{N-1}{\xi}\theta' = -\theta^n, \tag{69}$$

where $|\epsilon| \ll 1$. Thus we have the following equation for θ^* and Δ:

$$\theta_*'' + \frac{N-1}{\xi}\theta_*' = -\theta_0^{n_0}\left(\ln\theta_0 + \frac{n_0\theta_*}{\theta_0}\right) \tag{70}$$

$$\frac{2}{P_0(n_0+1)\xi\theta^{n_0}}\left[\Delta' + \frac{1}{\xi}\Delta - n_0\frac{\theta'}{\theta}\Delta\right] = -\theta^n + \theta^{n_0}. \tag{71}$$

Expanding the right-hand side of (71) and assuming Δ= $\epsilon\Delta_*$, the following equation for Δ_* is obtained:

$$\frac{2}{P_0(n_0+1)\xi\theta_0^{n_0}}\left[\Delta_*' + \frac{1}{\xi}\Delta_* - n_0\frac{\theta_0'}{\theta_0}\Delta_*\right]$$
$$= -\theta_0^{n_0}(\ln\theta_0 + \frac{n_0\theta_*}{\theta_0}). \qquad (72)$$

Having n0n0, the functions θ_0 and thus θ_* can be determined from (67) and (70) and then Δ_* is obtained from (72). With the initial conditions $\theta_0(0) = 1, \theta_0'(0) = \theta_*(0) = \theta_*'(0) = 0$ and for the case $n_0 = 0$ and $N = 3$, we get $\theta_0 = 1 - \xi^2/6$. Inserting θ_0 into (69), we find that [14]

$$\theta_* = (3 - \frac{\xi^2}{6})\ln(1 - \frac{\xi^2}{6}) + \frac{2\sqrt{6}}{\xi}\ln\frac{\sqrt{6}+\xi}{\sqrt{6}-\xi} + \frac{5}{18}\xi^2 - 4 \qquad (73)$$

where $\xi \leq \sqrt{6}$. Thus:

$$\Delta = -\frac{P_0\epsilon}{2}\left[\frac{\xi^2}{9}[3\ln(1 - \frac{\xi^2}{6}) - 2] + \frac{4\sqrt{6}}{\xi}\arctan\frac{\xi}{\sqrt{6}} - 4\right]. \qquad (74)$$

The plot of Δ is shown in Fig. 11.

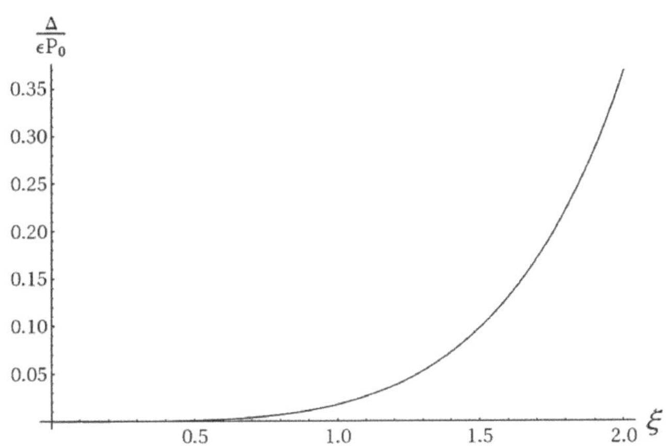

Figure 11: Plot of Δ for the analytical solution close to the exact solution ($n_0 = 0, N = 3$) as explained in the text.

One can find other solutions of (67), (70), and (72) numerically for arbitrary values of the barotropic index. For example, for $n_0=1$, the graph of Δ is plotted in Fig. 12.

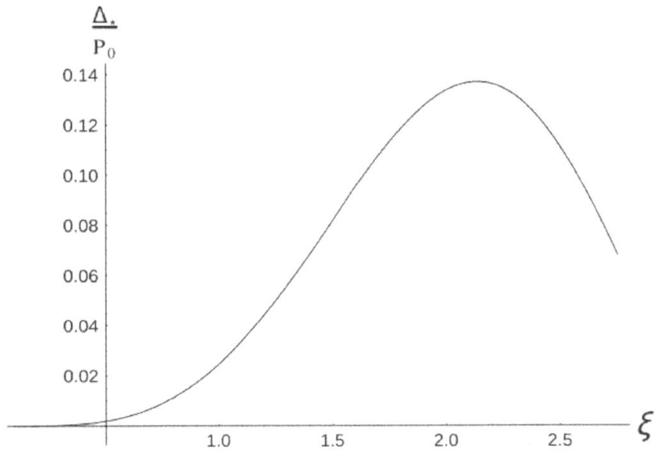

Figure 12: Plot of Δ for the numerical solution close to the exact solution (n_0=1 N=3) as explained in the text.

ACKNOWLEDGMENTS

This work was supported by a grant from University of Tehran.

REFERENCES

1. S. Chandrasekhar, *An Introduction to the Study of Stellar Structure* (The University of Chicago Press, Chicago, 1939)
2. G.P. Horedt, *Polytropes: Applications in Astrophysics and Related Fields* (Kluwer Academic Publishers, Amsterdam, 2004)
3. H. Rodrigues, Eur. J. Phys. 34, 667 (2013)
4. L. Herrera, W. Barreto, Phys. Rev. D 87, 087303 (2013)
5. L. Herrera, N.O. Santos, Phys. Rep. 286, 53 (1997)
6. J. Binney, S. Tremaine, *Galactic Dynamics* (Princeton University Press, Princeton, 2008)
7. P. Cuddeford, Mon. Not. R. Astron. Sot. 253, 414 (1991)
8. J. Ralston, L. Smith, Astrophys. J. 367, 54 (1991)
9. P. Alencar, P. Letelier, Phys. Rev. D 34, 343 (1986)
10. H. Liu, X. Zhang, D. Wen, Phys. Rev. D 89, 104043 (2014)
11. K. Dev, M. Gleiser, Gen. Relativ. Grav. 35, 8 (2003)
12. M. Cosensa, L. Herrera, M. Esculpi, L. Witten, Phys. Rev. D 25, 2527 (1982)
13. M. Cosensa, L. Herrera, M. Esculpi, L. Witten, J. Math. Phys. 22, 118 (1981)
14. Z.F. Seidov, R.K.H. Kuzakhmedov, Astron. Zh. 55, 1250 (1978)

CHAPTER 14

NEWTONIAN VIEW OF GENERAL RELATIVISTIC STARS

A. M. Oliveira, H. E. S. Velten, J. C. Fabris and I. G. Salako

INTRODUCTION

General relativity is the usual theory for dealing with gravitational phenomena [1]. Its building blocks, like the equivalence principle and predictions for the trajectories of planets and light in the solar system, have passed the most different tests. So far, we still do not have any clear evidence against GR though there are, of course, many alternative theories, like Brans–Dicke [2], f(R) [3] and others [4].

The application of GR to expanding backgrounds gives rise to cosmological solutions. For expanding, homogeneous and isotropic configurations we obtain from GR the well know Friedmann equations for the evolution of the cosmological scalar factor a(t)a(t). These cosmological solutions were obtained in 1922, a few years after Einstein presented the GR field equations.

A quite interesting fact is that in the 1930s Milne and McCrea started to develop a Newtonian description for cosmology [5, 6]. From their work, the so called Newtonian cosmology has appeared, from which one obtains the same relativistic solutions for matter dominated universes, i.e., Newtonian cosmology matches GR for pressureless fluids.

With the works of Milne in the 1950s and Harrisson in 1965, pressure effects were correctly incorporated into Newtonian cosmological solutions. This new approach became famous as the neo-Newtonian cosmology and all the GR Friedmann solutions could now be obtained [7, 8]. Therefore, via the neo-Newtonian formulation, cosmology can be studied with a much simpler formalism than the full GR theory.

Cosmology represents only one type of solutions which can be obtained from gravitational relativistic theories. Black holes configurations and gravitational waves are also examples of solutions. For static configurations GR is able to describe the structure of objects like stars. The GR solutions for isotropic stars are known as the Tolman–Oppeinheimer–Volkoff (TOV) equations from which we can solve the equilibrium configuration of the stellar interior [9, 10].

In general, nuclear reactions within the stellar interior induce energy flow via radiative convection. However, since for compact stars the nuclear timescale is much larger than the thermal and dynamical timescales one can assume the hydrostatic equilibrium for most of the star lifetime. In a first approximation, hydrostatic equilibrium in stars can be studied with Newtonian mechanics. From this approach one obtains the Lane–Emden equation [11] which is basically the Newtonian limit of the TOV system when pressure does not source gravitational effects in the stellar interior. Of course, since the pressure effects, mainly in extreme relativistic stars, are very important for the stellar properties, many systems in nature cannot be described via the Newtonian equations. One important example is a neutron star. Hence, one may wonder: Is it possible to reproduce the full TOV solutions using neo-Newtonian hydrodynamics? In other words: Is there the same equivalence between GR and the neo-Newtonian framework for a static configuration? This is the main goal of this paper. See also Refs. [12, 13] for interesting related investigations.

In the next section we briefly review the neo-Newtonian cosmology. This review will be useful to present the neo-Newtonian hydrodynamics which will be important for the rest of the paper. A crucial aspect of this discussion concerns the equivalence principle which will also be addressed here through the different possible definitions of mass. In Sect. 3 we briefly review the Newtonian hydrostatic equilibrium equation and the Tolman–Oppenheimer–Volkoff (full relativistic)

Citation: A. M. Oliveira, H. E. S. Velten, J. C. Fabris, I. G. Salako , Newtonian view of general relativistic stars, doi:10.1140/epjc/s10052-014-3170-2.

equation for star. Then we propose the implementation of the neo-Newtonian hydrodynamics for stellar configurations. We show in Sect. 4 numerical results for the mass–radius diagram for neutron stars adopting well-known equations of state (EoS) given in the literature. We conclude in the final section.

NEWTONIAN HYDRODYNAMICS IN AN EXPANDING BACKGROUND: COSMOLOGY

The purpose of this section is to present the neo-Newtonian hydrodynamics applied to cosmology. However, let us start with the standard case of Newtonian equations. The basic equations of Newtonian hydrodynamics for an inviscid perfect fluid are the following:

$$\dot{\rho} + \nabla \cdot (\rho \vec{v}) = 0, \tag{1}$$

$$\rho \frac{d\vec{v}}{dt} \equiv \rho[\dot{\vec{v}} + (\vec{v} \cdot \nabla)\vec{v}] = -\nabla p, \tag{2}$$

where ρρ is the fluid density, pp its pressure, and v⃗ v→ its the velocity field. The dot means derivative with respect to the cosmic time tt, i.e., a˙=∂taa˙=∂ta.

The above system of equations becomes suitable to study cosmology adopting the velocity field $\vec{v} = H(t)\vec{r}$ (Hubble's law) where $H(t) = \frac{\dot{a}(t)}{a(t)}$, being $a(t)$ the scale factor. It is worth to note the trivial solution for the continuity equation (1) $\rho(a) = \rho_0/a^3$, where today's scale factor a0=1a0=1gives today's density of the fluid ρ_0.

Gravitational interaction is coupled to Euler's equation (2) as

$$\dot{\vec{v}} + (\vec{v} \cdot \nabla)\vec{v} = -\frac{\nabla p}{\rho} - \nabla \Psi, \tag{3}$$

where the gravitational potential ΨΨ obeys the Poisson equation

$$\nabla^2 \Psi = 4\pi G \rho. \tag{4}$$

Equations (1) and (3) provide a fluid picture of the cosmic medium which is gravitationally self-interacting via the Poisson equation (4). In the Newtonian cosmology the Friedmann equations read

$$\frac{\dot{a}^2}{a^2} + \frac{(-2E)}{a^2} = \frac{8\pi G}{3}\rho \quad \text{and} \quad \dot{H} + H^2 = -\frac{4\pi G}{3}\rho, \tag{5}$$

where E is a constant of integration associated to the energy of system. The pressure is not dynamically relevant for the homogeneous and isotropic background. With Newtonian cosmology it is not possible to model a radiation dominated phase or even to study a late time dark energy dominated epoch. This approach is restricted to a description of the Einstein–de Sitter universe.

Including Pressure: the neo-Newtonian Cosmology

A simple way to include the pressure effects and, at the same time, keep the simplicity of the Newtonian physics is the use of the neo-Newtonian equations developed during the 1950s by McCrea [7] and by Harrison in the 1960s [8]. Later, during the 1990s, an important analysis concerning the perturbative behavior of the neo-Newtonian equations has helped to set the final form for the fluid equations in this approach [14] (see also [15]). This set of equations re

$$\dot{\rho} + \nabla \cdot (\rho \vec{v}) + p\nabla \cdot \vec{v} = 0, \tag{6}$$

$$\dot{\vec{v}} + (\vec{v} \cdot \nabla)\vec{v} = -\nabla \Psi - \frac{\nabla p}{\rho + p} \tag{7}$$

$$\nabla^2 \Psi = 4\pi G(\rho + 3p) \tag{8}$$

Combining Eqs. (6), (7), and (8) one obtains equations for the expansion of the homogeneous and isotropic background that are exactly the same as the relativistic Friedmann equations,

$$\frac{\dot{a}^2}{a^2} + \frac{(-2E)}{a^2} = \frac{8\pi G}{3}\rho, \tag{9}$$

$$\dot{H} + H^2 = -\frac{4\pi G}{3}(\rho + 3p) \tag{10}$$

The main idea behind the neo-Newtonian formalism relies on the following substitutions: firstly, it is necessary to redefine the concept of inertial and passive-gravitational mass density. With the redefinition

$$\rho_i \rightarrow \rho + p, \tag{11}$$

we rewrite the continuity and the Euler equation.

The second step is the interpretation of the active gravitational mass density i.e., the density that sources the gravitational field. Hence the following redefinition:

$$\rho_g \rightarrow \rho + 3p \tag{12}$$

which is related to the trace of the energy-momentum tensor, will become the source of the Poisson equation.

NEWTONIAN, RELATIVISTIC, AND NEO-NEWTONIAN FORMULATIONS FOR STATIC CONFIGURATIONS: STARS

Very simple, spherically symmetric, and static geometries are perfect for studying astrophysical objects like stars. In this section we start reviewing the Newtonian and general relativistic formulations. They will be useful for a comparison with the neo-Newtonian approach to stellar hydrodynamical equilibrium we present at the end of the section.

The Newtonian hydrostatic equilibrium equations are widely derived in the literature [1]. They read

$$\frac{dp}{dr} = -\frac{G\rho M(r)}{r^2} \tag{13}$$

and

$$\frac{dM(r)}{dr} = 4\pi r^2 \rho. \tag{14}$$

In principle, the system (13, 14) can be solved numerically since an appropriate EoS $p \equiv p(\rho)$ is provided.

On the other hand, there is a similar set of equations when general relativity is adopted. They are known as the Tolman–Oppeinheimer–Volkoff (TOV) equation,

$$\frac{dp}{dr} = -\frac{G\rho M(r)}{r^2}\frac{\left(1+\frac{p}{\rho}\right)\left(1+\frac{4\pi r^3 p}{M(r)}\right)}{\left(1-\frac{2GM(r)}{r}\right)} \tag{15}$$

Only with $p \ll \rho$, Eq. (13) cannot be obtained because of the Schwartzs child-like correction in the denominator (15). After restoring units the correct Newtonian limit occurs for c→∞.

The TOV equation is coupled to the mass definition

$$\frac{dM(r)}{dr} = 4\pi r^2 \rho, \tag{16}$$

which has the same form as its Newtonian counterpart (14).

Therefore, the full equilibrium configuration for stellar systems is obtained by solving the system (15) sourced by the differential relation (16).

Now we derive the neo-Newtonian correspondence for the hydrostatic equilibrium condition. In addition to the modified Poisson equation (8) one finds

$$\frac{1}{\rho_i}\frac{dp}{dr} = -\frac{G}{r^2}\int_0^r 4\pi r'^2(\rho+3p)dr'. \tag{17}$$

Equation (17) with the proper neo-Newtonian identifications for inertial/passive-gravitational mass densities results in

$$\frac{dp}{dr} = -\frac{G(\rho+p)}{r^2}\tilde{M}(r) \tag{18}$$

where one also uses the relation

$$\frac{d\tilde{M}(r)}{dr} = 4\pi r^2(\rho+3p). \tag{19}$$

Note that the above relation differs from the standard definition for the mass of an object,

$$M = \int_0^R 4\pi r^2 \rho dr. \tag{20}$$

Hence, Eqs. (18) and (19) represent the neo-Newtonian version of the TOV system (16) and (15).

NUMERICAL RESULTS

We will solve now the differential equations for the internal structure of stars. Our goal is to compare the solution for the Newtonian formulation (13) and (14); the full relativistic TOV equations which is (15) sourced by the differential relation (16) and the neo-Newtonian equations (18) and (19).

White dwarfs are objects in which the simple Newtonian formalism works quite well. Therefore, the comparison we propose here seems to be meaningless if using such Newtonian stars. On the other hand, neutron stars are perfect laboratories for testing the relativistic corrections incorporated by the TOV equations.

After specifying the EoS p(ρ) for the stellar interior, the solution of the equilibrium equations will also depends on specifying the central stellar pressure $p(r=0) = p_0$ value one obtains the total mass and radius of the star. Unfortunately the inner composition of neutron stars are not well understood and the correct neutron star EoS is still unknown. The typical densities found in such objects are of order or greater than the nuclear saturation density where the common way to obtain the EoS of nuclear matter is via microscopic many-body calculations based on phenomenological relativistic mean-field theories and nucleon–nucleon potentials. Usually, either variational or Monte-Carlo techniques are used (see [16, 17] for a review).

One possibility is that precise observations of spin rate, mass, and radius of many different stars can lead to the reconstruction of the neutron star properties, i.e., one could indirectly obtain p(ρ). Remarkably, recent observations of very massive neutron star, with mass as high as 1.97±0.04$M_\odot$ [18], may place upper limits to thermodynamics quantities like energy density, pressure, baryon number density, and chemical potential [19].

There are in the literature many proposals for the nuclear EoS of neutron stars [20–25]. Also, some fits provide a unified EoS for the neutron star interior [26]. Here, since our focus relies on the gravity approach we will adopt simple configurations usually adopted as a first contact with neutron stars physics and which can be found in standard textbooks.

Let us firstly adopt a pure neutron fluid using the Fermi gas model as the Oppenheimer–Volkoff model [9]. This configuration is, strictly speaking, unrealistic since actual neutron stars contain small fractions of protons and electrons which avoid the neutron decay via weak interactions. Also, a pure neutron configuration fails in counting for the nucleon–nucleon interactions which are important to the energy density. However, we use it as a first analysis of neutron stars. A simple introduction to such EoS can be found in Refs. [1, 27]. It corresponds to a polytrope $p \sim \rho^\gamma$ with γ=5/3. We use in this work

$$\frac{\bar{\rho}(\bar{p})}{c^2} = \bar{K}^{-1}\,\bar{p}^{3/5}, \tag{21}$$

where the coefficient of (21) is fitted in Ref. [27] as $\bar{K}=1.914$. The bar over the density and pressure makes them dimensionless quantities via the transformations $p = \epsilon_0\bar{p}$ and $\rho = \epsilon_0\bar{\rho}$, where $\epsilon_0 = 1.603 \times 10^{38}$ ergs/cm3.

For this EoS we plot the mass–radius diagram in Fig. 1. It shows the stable configurations for different values of the central pressure p(r=0)=p_0. Low-mass/large-radius solutions correspond to small starting values of p_0. We plot with the solid line the TOV solution. The neo-Newtonian (solid-red) and the Newtonian (short-dashed) solutions are also shown. Concerning the TOV solution and the Newtonian one, there is a reliable qualitative agreement of these results. As expected, the relativistic solution presents a maximum (here $M_{max} \sim 0.95M\odot$ at $R \sim 8\,\text{km}$) while for the Newtonian case the larger the mass, the smaller the radius.

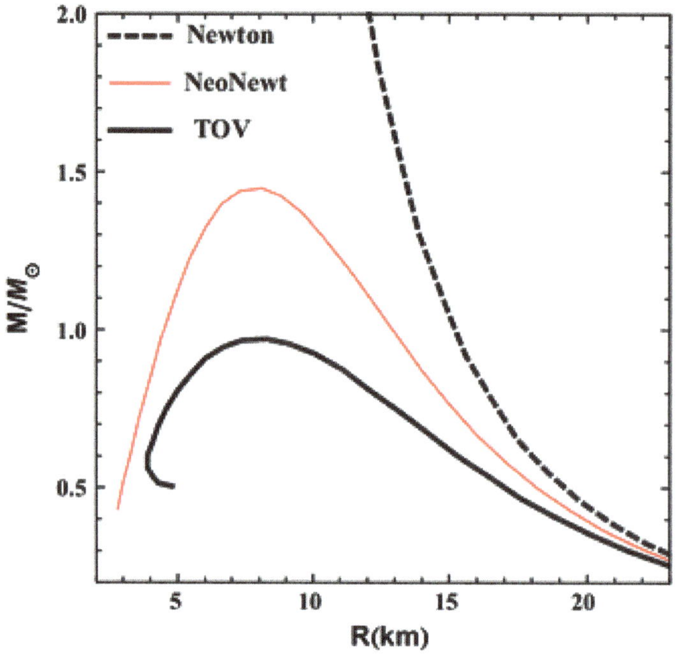

Figure 1: Mass–radius contours for a pure neutron star (without interactions) using a Fermi gas equation of state.

The quantitative equivalence between the TOV and the neo-Newtonian models does not occur. More massive configurations are allowed in the neo-Newtonian context. However, a remarkable result is that the neo-Newtonian solution also presents a maximum mass which is a typical relativistic aspect. This shows that it can be used as a first approximation to the problem.

Performing a quantitative comparison between the Newtonian and the neo-Newtonian cases, although the existence of the maximum, there is, indeed, the indication that by adding the pressure to the Newtonian formalism one obtains more compact stars corroborating the relativistic intuition that pressure "amplifies" gravity effects.

In Fig. 1 we have calculated the mass of the neo-Newtonian case using the definition (20) which is the same as the other cases and therefore allows us a proper comparison of the results. However, within the neo-Newtonian formulation there is some freedom in defining the meaning of mass. For example, the quantity M~M~ (19) could be used instead. Also, as discussed previously, one has to take care with the definitions of inertial, passive-gravitational and active gravitational masses. Hence, being the actual definition of mass in the neo-Newtonian formalism a tricky issue, we plot in Fig. 2 the mass diagrams with a comparison of the possible interpretations of mass within the neo-Newtonian framework. The curve m0m0 is the same as shown previously in Fig. 1. The quantity M~M~ is represented by the long dashed line in these plots. We call it m3m3, i.e., $m_3 \sim \int(\rho + 3p)r^2 dr$. Also, we display the case $m_1 \sim \int(\rho + p)r^2 dr$.

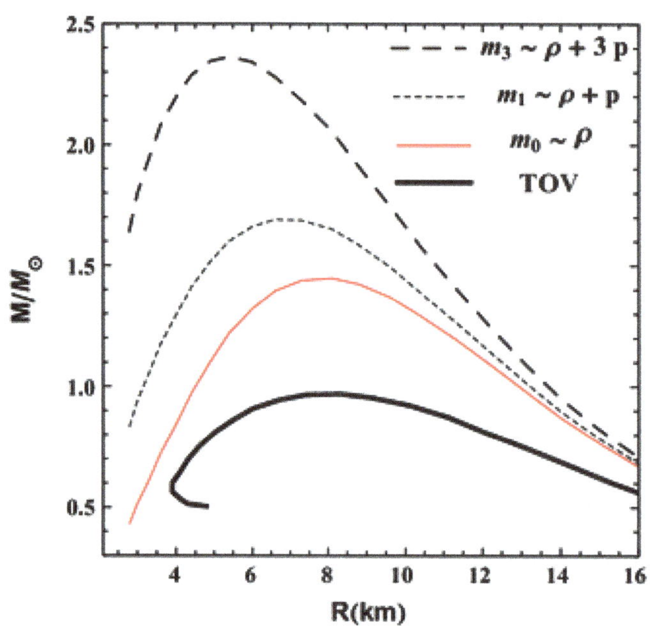

Figure 2: Mass–radius contours for a pure neutron star (without interactions) using a Fermi gas equation of state.

Now we repeat the above comparison between the three approaches but using a more realistic EoS. We assume a pure neutron star but with nucleon–nucleon interactions. We follow the Prakash method for building such an interacting Fermi gas model EoS [28]. This equation assumes the polytropic form

$$\frac{\rho(\bar{p})}{c^2} = (\kappa_0 \epsilon_0)^{-1/2} \bar{p}^{1/2} \qquad (22)$$

where now $\epsilon_0 = m_n^4 c^5 / 3\pi^3 \hbar^3$ and the nuclear (in) compressibility is κ_0=363MeV.

Solving again the equilibrium equations, we show in Fig. 3 the mass–radius contours. Here, the Newtonian equations do not apply. Together with the TOV (black solid) and the standard neo-Newtonian (red solid line) result, we add here the other options for calculating the mass in the neo-Newtonian case (dashed lines). Here again, independently on how we compute the neo-Newtonian mass, it does not agree with the full relativistic solution.

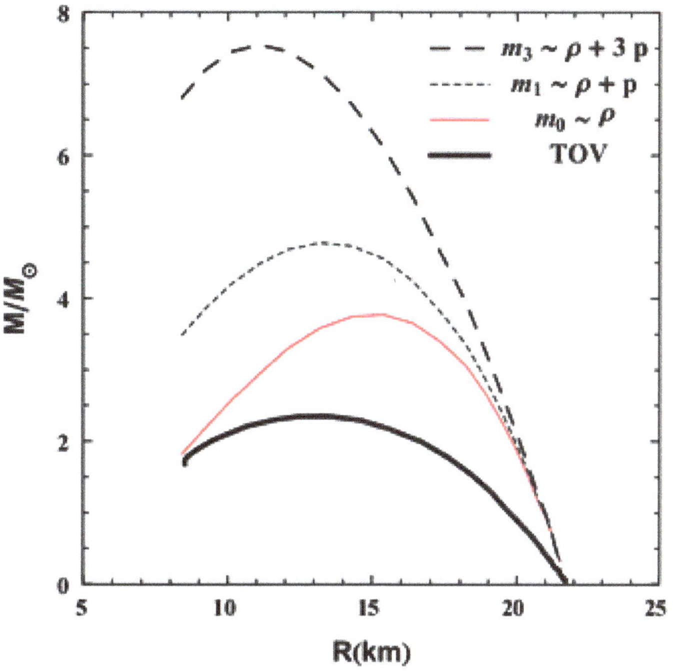

Figure 3: Mass–radius contours for a pure neutron star (with nucleon–nucleon interactions) using a Fermi gas equation of state.

The effect of the nucleons' interactions is quite remarkable since the allowed maximum masses are now much larger than the situation in Fig. 1. The TOV maximum mass now $M \sim 2.3\, M_\odot$ (at $R = 13.5\,\text{km}$) is a reflection of the large value of the nuclear (in)compressibility κ_0=363MeV used in this analysis.

In any of the cases studied here, the neo-Newtonian model was not able to correctly, i.e., quantitatively, describe the general relativistic predictions. Therefore, although this formalism represents a reliable tool for studying cosmology, the same does not seem to happen for stellar configurations.

CONCLUSIONS

It is a remarkable fact that the general relativity predictions for expanding backgrounds can be mimicked by simple Newtonian models. The dynamics of an expanding Newtonian sphere is the same as the relativistic Friedmann–Lemaitre–Robertson–Walker universe dominated by matter, i.e., the Einstein–de Sitter universe. Although the Newtonian cosmology fails in describing epochs of the universe where the pressure is relevant, the neo-Newtonian cosmology was developed to fill this gap.

The question we addressed in this work was how to understand whether such exact correspondence between the neo-Newtonian model and the full relativistic theory remains the same for static configurations. Hence, neutron stars seem to be the perfect laboratory for testing our proposal.

After reviewing the standard Newtonian and the TOV equations, we developed the neo-Newtonian version for the hydrostatic equilibrium in static spherical configurations. The neo-Newtoniam formalism for stars has a structure which looks like the TOV one but without the Schwartzschild-like correction in the denominator of Eq. (15).

We probed both the Newtonian, the neo-Newtonian, and the relativistic approaches by calculating their equilibrium configurations for some specific neutron stars equation of states. Firstly we used the Fermi gas approximation for a pure neutron star (the classical Oppenheimer–Volkoff) model. The results were shown in Figs. 1 and 2. In general, a clear difference between the Newtonian and the relativistic theory is the existence of a maximum mass in the latter. As one can see in Fig. 1, although the neo-Newtonian result is not the same as the relativistic one, it presents a maximum mass. This result is a remarkable aspect of the neo-Newtonian formalism. However, the neo-Newtonian value for the maximum mass $M_{nN} \sim 1.45 M_\odot$ (see Fig. 1) is much larger than we could expect for this EoS from simple analytical arguments [29]. Therefore, the neo-Newtonian method seems to somehow overestimate the stellar maximum mass.

We also used a more realistic EoS for neutron stars where the interaction nucleon–nucleon is allowed. Concerning the comparison between the fluid models used here, the same qualitative results obtained in the first analysis are kept.

It is therefore shown that the neo-Newtonian hydrodynamics formulated to match GR predictions for expanding background reproduces qualitatively the relativistic effects when applied to spherically symmetric configurations but does not fit quantitatively the GR results in this situation.

ACKNOWLEDGMENTS

We are thankful to J. A. de Freitas Pacheco for a critical reading and suggestions on the manuscript. We thank CNPq (Brazil) and FAPES (Brazil) for partial financial support. I.G. Salako also thanks ICTP/IMSP (Benin) for partial support.

Appendix A: About the stellar virial Theorem

Here, we provide a complementary discussion of stellar configurations by studying the proper modifications to the virial theorem. We keep here the standard assumptions like hydrostatic equilibrium and that stars are made up of ideal gases.

The total internal energy of a star can be written as

$$E_i = \int_0^{M_T} u\mathrm{d}M = \int_0^{M_T} \frac{3}{2}\frac{KT}{\mu}\mathrm{d}M = \int_0^{M_T} \frac{3}{2}\frac{p}{\rho}\mathrm{d}M. \tag{23}$$

where we have used the ideal gas EoS, $P = \rho N_A KT/\mu$. The above result forms one of contributions to the virial theorem. The other one comes from the hydrostatic equilibrium. For example, if we multiply (13) by $4\pi r^3$ and integrate over the entire star we find

$$\int_0^R 12\pi r^2 p\mathrm{d}r = -E_g, \tag{24}$$

where, as usual, $E_g = -\int_0^R GM(r)\rho 4\pi r\mathrm{d}r$. Since $\mathrm{d}M = 4\pi r^2\rho\mathrm{d}r$ and having in mind (23), it results in the standard result [11],

$$2E_i + E_g = 0. \tag{25}$$

Now, let us follow the same steps but introducing the neo-Newtonian concepts. First of all, note that since the computation of the internal energy EiEi involves microscopic/ thermodynamic physics this quantity will not be changed if we work with either modified kinematical or gravitational concepts. Hence, we will assume that the internal energy (23) remains the same.

The neo-Newtonian formalism involves a different definition for the mass. It is worth noting that the definition of $\mathrm{d}\tilde{M}$ (19) can be integrated and written as

$$\tilde{M} - M = \int_0^R 12\pi r^2 p\mathrm{d}r = 2\frac{E_i}{c^2}, \tag{26}$$

where MM obeys to (14). Note that we have recovered the proper units by including the speed of light cc. Taking the limit $c \to \infty$ we see that $\tilde{M} = M$ and therefore the Newtonian result is recovered.

We take now the neo-Newtonian hydrostatic equilibrium (18) and proceed as above multiplying it by 4πr3dr4πr3dr and integrating over the entire star. We find a very similar virial configuration,

$$2E_i + \tilde{E}_g = 0, \tag{27}$$

but the modified gravitational energy $\tilde{E}_g$ here re

$$\tilde{E}_g = E_g^{\rho\rho} + 4E_g^{\rho p} + 3E_g^{pp}. \tag{28}$$

The quantities $E^{\rho p}$ and E^{pp} are the new contributions to the virial equilibrium due to the pressure effects. The contributions to (28) are defined here using a simplified notation

$$E_g^{XY} = -\int_0^R \frac{G}{r}\left(\frac{4\pi r^3}{3}\right) XY 4\pi r^2\mathrm{d}r, \tag{29}$$

where X and Y can be either the density ρ or the pressure pp.

The combination X=Y=ρ results in the standard Newtonian gravitational energy as used in (24). The quantity EρpgEgρp results from adopting X=ρ and Y=p. There is also a pure pressure square term contribution EppEpp, which is calculated adopting X=Y=p.

REFERENCES

1. S. Weinberg, *Gravitation and Cosmology: Principles and Applications of the General Theory of Relativity* (Wiley, New York, 1972)
2. Brans, R. Dicke, Phys. Rev. 124, 925 (1961)
3. de Felice, S. Tsujikawa, Living Rev. Relat. 13, 3 (2010)
4. T. Clifton, P.G. Ferreira, A. Padilla, C. Skordis, Phys. Rep. 513(1), 1 (2012)
5. E.A. Milne, Q. J. Math. 5, 64 (1934)
6. E.A. Milne, W.H. McCrea, Q. J. Math. 5, 73 (1934)
7. W.H. McCrea, Proc. R. Soc. Lond. 206, 562 (1951)
8. E.R. Harrison, Ann. Phys. (N. Y.) 35, 437 (1965)
9. J.R. Oppenheimer, G.M. Volkoff, Phys. Rev. 55(4), 374 (1939)
10. R.C. Tolman, Phys. Rev. 55(4), 364 (1939)
11. S. Chandrasekhar, *An Introduction to the Study of Stellar Structure* (Dover, New York, 1967)
12. P.H. Nguyen, J.F. Pedraza, Phys. Rev. D 88, 064020 (2013)
13. Enrico Barausse, Luis Lehner, Phys. Rev. D 88, 024029 (2013)
14. J.A.S. Lima, V. Zanchin, R. Brandenberger, Mon. Not. R. Astron. Soc. 291, L1 (1997)
15. R.R.R. Reis, Phys. Rev. D67, 087301 (2003); Erratum-ibid D68, 089901(2003)
16. D. Page, S. Reddy, Ann. Rev. Nucl. Part. Sci. 56, 327 (2006)

17. J.M. Lattimer, Ann. Rev. Nucl. Part. Sci. 62, 485 (2013)
18. P.B. Demorest, T. Pennucci, S.M. Ransom, M.S.E. Roberts, J.W.T. Hessels, Nature 467, 1081 (2010)
19. K. Hebeler, J.M. Lattimer, C.J. Pethick, A. Schwenk, Astrophys. J. 773, 11 (2013)
20. Chodos et al., Phys. Rev. D 9, 3741 (1974)
21. M. Prakash, T.L. Ainsworth, J.M. Lattimer, Phys. Rev. Lett. 61, 2518 (1988)
22. F. Douchin, P. Haensel, A&A 380, 151 (2001)
23. Akmal, V.R. Pandharipande, D.G. Ravenhall, Phys. Rev. C 58, 1804 (1998)
24. Friedman, V.R. Pandharipande, Nucl. Phys. A 361, 502 (1981)
25. M. Baldo, I. Bombaci, G.F. Burgio, A&A 328, 274 (1997)
26. A.Y. Potekhin, A.F. Fantina, N. Chamel, J.M. Pearson, S. Goriely, Astron. Astrophys. 560, A48 (2013)
27. R.R. Silbar, S. Reddy, Am. J. Phys. 72, 892 (2004); Erratum-ibid. 73, 286 (2005)
28. M. Prakash, Equation of state, in *The Nuclear Equation of State*, ed. by A. Ausari, L. Satpathy (World Scientific, Singapore, 1996)
29. M. Nauenberg, G. Chapline Jr, ApJ 179, 277 (1973)

CHAPTER 15

EULERIAN-LAGRANGIAN FORMULATION FOR COMPRESSIBLE NAVIER-STOKES EQUATIONS

Carlos Cartes and Orazio Descalzi

INTRODUCTION

The Eulerian-Lagrangian formulation of the (inviscid) Euler dynamics in terms of advected Weber-Clebsch potentials (Lamb, 1932), was extended by Constantin to cover the viscous Navier-Stokes dynamics (Constantin, 2001). Numerical studies (Ohkitani & Constantin, 2003), of this formulation of the Navier-Stokes equations concluded that the diffusive Lagrangian map becomes non-invertible under time evolution and requires resetting for its calculation. They proposed the observed sharp increase of the frequency of resettings as a new diagnostic of vortex reconnection. In previous work we were able (Cartes et al., 2007; 2009) to complement these results, using an approach that is based on a generalised set of equations of motion for the Weber-Clebsch potentials, that turned out to depend on a parameter τ, which has the unit of time for the Navier-Stokes case. Also to extend our formulation to magnetohydrodynamics, and thereby obtain a new diagnostic for magnetic reconnection. In this work we present a generalisation of the Weber-Clebsch variables in order to describe the compressible Navier-Stokes dynamics. Our main result is a good agreement between the dynamics for the velocity and density fields that come from the dynamics of Weber-Clebsch variables and direct numerical simulations of the compressible Navier-Stokes equations. We first present the inviscid Eulerian-Lagrangian theory, then Constantin's extension to viscous fluids and derive our equations of motion for the Weber-Clebsch potentials that describe the compressible Navier-Stokes dynamics. Then, performing direct numerical simulations of the Taylor-Green vortex, we check that our formulation reproduces the compressible dynamics.

EULERIAN-LAGRANGIAN THEORY

Euler equations and Clebsch variables

Let us consider the incompressible Euler equations with constant density, fixed to one, for the velocity field u

$$\partial_t \mathbf{u} + \mathbf{u} \cdot \nabla \mathbf{u} = -\nabla p$$
$$\nabla \cdot \mathbf{u} = 0, \tag{1}$$

here p is the pressure field. Now the equations for evolution of the vorticity $\omega = \nabla \times u$ field

$$D_t \omega = \omega \cdot \nabla \mathbf{u}, \tag{2}$$

where D_t is the convective derivative

$$D_t = \partial_t + \mathbf{u} \cdot \nabla. \tag{3}$$

A well known consequence of this equation is the preservation of vorticity lines (Helmholtz's theorem)

Here we introduce Clebsch variables (Lamb, 1932). They can be considered as a representation of vorticity lines. In fact from this transformation, which defines the velocity field in terms of scalar variables (λ, μ, φ)

$$\mathbf{u} = \lambda \nabla \mu - \nabla \phi, \tag{4}$$

we can write the vorticity field as

$$\omega = \nabla \times \mathbf{u} = \nabla \lambda \times \nabla \mu. \tag{5}$$

Vorticity lines r(s) are defined as the solutions of

$$\frac{d\mathbf{r}}{ds} = \omega(\mathbf{r}(s)), \tag{6}$$

which admits integrals

Citation: Carlos Cartes and Orazio Descalzi (2011). Eulerian-Lagrangian Formulation for Compressible Navier-Stokes Equations, Hydrodynamics - Optimizing Methods and Tools, Prof. Harry Schulz (Ed.), InTech, DOI: 10.5772/26963.

$$\lambda(\mathbf{r}(s)) = \text{const.}$$
$$\mu(\mathbf{r}(s)) = \text{const.} \tag{7}$$

In other words the intersections of surfaces λ =const. and μ =const. are the vorticity lines. If vorticity lines follow Euler equations and are preserved, then the fields λ and μ follow the fluid.

Clebsch variables can also be used to find a variational principle for Euler equations. We can write a Lagrangian density for Euler equations

$$\mathcal{L} = \frac{|\mathbf{u}|^2}{2} + \lambda\partial_t\mu, \tag{8}$$

and the variations of L in function of the fields λ, μ and φ give us the system equations

$$\frac{\delta\mathcal{L}}{\delta\mu} = -D_t\lambda = 0$$
$$\frac{\delta\mathcal{L}}{\delta\lambda} = D_t\mu = 0$$
$$\frac{\delta\mathcal{L}}{\delta\phi} = \nabla\cdot\mathbf{u} = 0. \tag{9}$$

From this system and the identity

$$[\nabla, D_t] \equiv (\nabla\mathbf{u})\cdot\nabla, \tag{10}$$

we can obtain the evolution equation for u

$$D_t\mathbf{u} = -\nabla\left(D_t\phi + \frac{1}{2}\mathbf{u}^2\right) \tag{11}$$

WEBER TRANSFORMATION

Let us note a^i as the initial coordinate (at t = 0) of a fluid element and X^i (a, t) its position at time t and note $A^i(x, t)$ the inverse application: $a^i \equiv A^i(X^i(a, t), t)$. At time t Eulerian coordinates are by definition the variables $x^i = X^i(a, t)$ then the Lagrangian velocity of a fluid element is

$$\tilde{u}^i(a,t) = \frac{\partial X^i}{\partial t}(a,t) \tag{12}$$

and its acceleration

$$\frac{\partial\tilde{u}^i}{\partial t}(a,t) = \frac{\partial^2 X^i}{\partial t^2}(a,t) \tag{13}$$

Newton equations for the fluid element are

$$\frac{\partial^2 X^i}{\partial t^2}(a,t) = F_X^i(a,t) \tag{14}$$

where the forces $F_X^i(a,t)$ are given by

$$F_X^i(a,t) = -\frac{\partial p}{\partial x^i}(X(a,t),t) \tag{15}$$

and p (X (a, t)) is the pressure field in Eulerian coordinates. Therefore the movement equations for the fluid elements are

$$\frac{\partial^2 X^i}{\partial t^2}(a,t) = -\frac{\partial p}{\partial x^i}(X(a,t),t) \tag{16}$$

For an incompressible fluid, the transformation matrix, between Lagrangian and Eulerian coordinates, verifies

$$\det\left(\frac{\partial X^i}{\partial a^j}\right) = 1, \tag{17}$$

this value is fixed from the relation between the volume elements in the two coordinate systems. We also note that this transformation is always invertible.

From Eq. (16) we perform a coordinate transformation for the derivatives of p using

$$\frac{\partial}{\partial x^i} = \frac{\partial A^j}{\partial x^i}\frac{\partial}{\partial a^j}$$
$$\frac{\partial}{\partial a^i} = \frac{\partial X^j}{\partial a^i}\frac{\partial}{\partial x^j} \tag{18}$$

$$\frac{\partial^2 X^i}{\partial t^2}(a,t) = -\frac{\partial A^j}{\partial x^i}\frac{\partial\tilde{p}}{\partial a^j}(a,t) \tag{19}$$

where $\tilde{p}(a,t)$ is the pressure field in Lagrangian coordinates.

We multiply Eq. (19) with the inverse coordinate transformation $\frac{\partial X^i}{\partial a^j}$ in order to obtain

$$\frac{\partial^2 X^i}{\partial t^2}(a,t)\frac{\partial X^i}{\partial a^j}(a,t) = -\frac{\partial\tilde{p}}{\partial a^j}(a,t) \tag{20}$$

which is the Lagrangian form for the dynamic equations. The left hand side of this equation can be written as

$$\frac{\partial^2 X^i}{\partial t^2}(a,t)\frac{\partial X^i}{\partial a^j}(a,t) = \frac{\partial}{\partial t}\left(\frac{\partial X^i}{\partial t}(a,t)\frac{\partial X^i}{\partial a^j}(a,t)\right) - \frac{1}{2}\frac{\partial}{\partial a^j}\left|\frac{\partial X^i}{\partial t}(a,t)\right|^2 \tag{21}$$

and Eq. (20) becomes

$$\frac{\partial}{\partial t}\left(\frac{\partial X^i}{\partial t}(a,t)\frac{\partial X^i}{\partial a^j}(a,t)\right) = -\frac{\partial\tilde{q}}{\partial a^j}(a,t) \tag{22}$$

where the term $\tilde{q}(a,t)$ is given by

$$\tilde{q}(a,t) = \tilde{p}(a,t) - \frac{1}{2}\left|\frac{\partial X^i}{\partial t}(a,t)\right|^2 \tag{23}$$

Now let us integrate Eq. (22) over t, maintaining a^i fixed

$$\left[\frac{\partial X^i}{\partial t}(a,t)\frac{\partial X^i}{\partial a^j}(a,t)\right]_0^t = \frac{\partial X^i}{\partial t}(a,t)\frac{\partial X^i}{\partial a^j}(a,t) - \tilde{u}_0^j(a)$$

to obtain

$$\frac{\partial X^i}{\partial t}(a,t)\frac{\partial X^i}{\partial a^j}(a,t) - \tilde{u}_0^j(a) = -\frac{\partial\tilde{\phi}}{\partial a^j}(a,t) \tag{24}$$

where $\tilde{\phi}$ is written as

$$\tilde{\phi}(a,t) = \int_0^t\left(\tilde{p}(a,s) - \frac{1}{2}\left|\frac{\partial X^i}{\partial t}(a,s)\right|^2\right)ds \tag{25}$$

This equation system (24) is called Weber transformation (Lamb, 1932). Now we perform a coordinate transformation identifying $\tilde{\mu}^i(a,t) = a^i$ and the initial velocity

$\tilde{\lambda}^i(a,t) = \tilde{u}_0^i(a)$ we obtain the evolution equations for the fields in Lagrangian coordinates

$$\frac{\partial \tilde{\lambda}^i}{\partial t}(a,t) = 0$$
$$\frac{\partial \tilde{\mu}^i}{\partial t}(a,t) = 0$$
$$\frac{\partial \tilde{\phi}}{\partial t}(a,t) = \tilde{p}(a,t) - \frac{1}{2}\left|\frac{\partial X^i}{\partial t}(a,t)\right|^2 \tag{27}$$

If we now go to the Eulerian coordinates, identifying $\mu^i(x,t) = A^i(x,t)$ and $\lambda^i(x,t) = u_0^i(\mu(x,t))$, we obtain the Weber-Clebsch transformation

$$u^i(x,t) = \sum_{j=1}^{3} \lambda^j \frac{\partial \mu^j}{\partial x^i} - \frac{\partial \phi}{\partial x^i} \tag{28}$$

Using the convective derivative, the dynamic equations for the Clebsch variables Eq. (27) can be written in Eulerian coordinates as

$$D_t\lambda^i(x,t) = 0$$
$$D_t\mu^i(x,t) = 0$$
$$D_t\phi(x,t) = p(x,t) - \frac{1}{2}\left|u^i(x,t)\right|^2 \tag{29}$$

The Weber-Clebsch transformation Eq. (28) and its evolution laws Eq. (29) are very similar to Clebsch variables Eq. (4) and the system (9). An important difference is the number of potential pairs.

If we use Clebsch variables Eq. (4) to represent the velocity field u

$$\mathbf{u} = \lambda\nabla\mu - \nabla\phi, \tag{30}$$

we have the problem that u is restricted to fields with mean helicity

$$h = \int_V \mathbf{u}\cdot\omega d^3x \tag{31}$$

of value zero (Grossmann, 1975). In fact, writing h in terms of Clebsch variables

$$h = \int_V (\lambda\nabla\mu - \nabla\phi)\cdot(\nabla\lambda\times\nabla\mu)\,d^3x = -\int_V \nabla\phi\cdot(\nabla\lambda\times\nabla\mu)\,d^3x \tag{32}$$

the term $\lambda\nabla\mu$ is perpendicular to $\nabla\lambda \times \nabla\mu$ and then their scalar product is zero. For the other terms, we integrate by parts

$$\int_V \nabla\phi\cdot(\nabla\lambda\times\nabla\mu)\,d^3x = \oint_{\partial V} \phi\,(\nabla\lambda\times\nabla\mu)\cdot d\mathbf{s} - \int_V \phi\nabla\cdot(\nabla\lambda\times\nabla\mu) \tag{33}$$

but in a periodic domain the first term in the right hand side is zero. We also know that $\nabla\cdot\omega = 0$ and therefore we have

$$\nabla\cdot(\nabla\lambda\times\nabla\mu) = 0 \tag{34}$$

and we finally get

$$h = \int_V \mathbf{u}\cdot\omega d^3x = 0. \tag{35}$$

If we consider now two pairs of Clebsch variables, for each component of the velocity field, we have

$$u^j = \sum_{i=1}^{2} \lambda^i \frac{\partial \mu^i}{\partial x^j} - \frac{\partial \phi}{\partial x^j}, \tag{36}$$

and we arrive to a system of equations of second degree in its unknowns, this system does not have an analytic solution and we don't have a systematic way to find λ^i and μ^i for an arbitrary velocity field u.

If we use now the same number of pairs as spatial variables (three in this case), we get the Weber-Clebsch transformation

$$\mathbf{u} = \sum_{i=1}^{3} \lambda^i\nabla\mu^i - \nabla\phi, \tag{37}$$

with this representation we can write an arbitrary velocity field defining, at t = 0:

$$\lambda^i(x,0) = u^i(x,0)$$
$$\mu^i(x,0) = x^i$$
$$\phi(x,0) = 0 \tag{38}$$

which is completely equivalent to the Weber transformation.

Constantin's formulation of Navier-Stokes equations

Here we will recall Constantin's extension for the Eulerian-Lagrangian formulation of Navier-Stokes equations.

The departing point (Constantin, 2001), is the expression for the Eulerian velocity u = (u^1, u^2, u^3) from the Weber-Clebsch transformation

$$u^i = \sum_{m=1}^{3} \lambda^m \frac{\partial \mu^m}{\partial x^i} - \frac{\partial \phi}{\partial x^i}. \tag{39}$$

The fields in this equation admit the same interpretation as in the Weber transformation: λ m are the Lagrangian velocity components, μ^m are the Lagrangian coordinates and φ fixes the incompressibility condition for the velocity field.

In a way similar to the Weber transformation, we have the Lagrangian coordinates $a^i = \mu^i(x,t)$ and the Eulerian coordinates $x^i = X^i(a,t)$.

If we now consider the first term of the right hand side in Eq. (39) as a coordinate transformation, it is possible to write their derivatives in Lagrangian coordinates, as in Eq. (18)

$$\frac{\partial}{\partial a^i} = \sum_{m=1}^{3} \frac{\partial X^m}{\partial a^i}\frac{\partial}{\partial x^m} \tag{40}$$

In the same way it is possible to write the derivatives of the Eulerian coordinate in terms of Lagrangian coordinates

$$\frac{\partial}{\partial x^i} = \sum_{m=1}^{3} \frac{\partial \mu^m}{\partial x^i} \frac{\partial}{\partial a^m} \tag{41}$$

We also have the relation for the commutators

$$\left[\frac{\partial}{\partial x^i}, \frac{\partial}{\partial x^k}\right] = 0$$
$$\left[\frac{\partial}{\partial a^i}, \frac{\partial}{\partial a^k}\right] = 0.$$

Using relations Eq. (40), Eq. (41) and Eq. (42) we can compute the commutators between ∂_x and ∂_a

$$\left[\frac{\partial}{\partial a^i}, \frac{\partial}{\partial x^k}\right] = \left[\frac{\partial}{\partial a^i}, \frac{\partial \mu^m}{\partial x^k}\frac{\partial}{\partial a^m}\right] = \frac{\partial}{\partial a^i}\left(\frac{\partial \mu^m}{\partial x^k}\right)\frac{\partial}{\partial a^m}. \tag{43}$$

Introducing the displacement vector $\ell^m = \mu^m - x^m$ which relates the Eulerian position x to the original Lagrangian position μ, we can express the commutator Eq. (43) as

$$\left[\frac{\partial}{\partial a^i}, \frac{\partial}{\partial x^k}\right] = \frac{\partial}{\partial a^i}\left(\frac{\partial \ell^m}{\partial x^k}\right)\frac{\partial}{\partial a^m} = C_{m,k;i}\frac{\partial}{\partial a^m} \tag{44}$$

The term $C_{m,k;i}$ is related to the Christoffel coefficients Γ^m_{ij} of the flat connection in R^3 by the formula

$$\Gamma^m_{ij} = -\frac{\partial X^k}{\partial a^j} C_{m,k;i}. \tag{45}$$

We consider now the diffusive evolution of our fields, with that goal in mind we define the operator

$$\Gamma = \partial_t + \mathbf{u}\cdot\nabla - \nu\triangle, \tag{46}$$

where ν is the viscosity and u is the Eulerian velocity. When the operator Eq. (46) is applied over a vector or a matrix each component is taken in an independent way.

Constantin imposes that the coordinates μ^i are advected and diffused so they follow

$$\Gamma\mu^i = 0 \tag{47}$$

We also need a coordinate transformation that can be invertible at any time t, that condition is always satisfied when the diffusion is zero (ν = 0) and the fluid is incompressible, because the fluid element volume is preserved by the coordinate transformation, and therefore

$$\mathrm{Det}(\nabla\mu) = 1, \tag{48}$$

where

$$(\nabla\mu)_{jk} = \frac{\partial \mu^j}{\partial x^k} \tag{49}$$

In order to get the evolution for the λ^i fields we apply D_t on Eq. (39), and using the relation

$$\left[D_t, \frac{\partial}{\partial x^i}\right] = -\frac{\partial u^l}{\partial x^i}\frac{\partial}{\partial x^l} \tag{50}$$

we obtain

$$D_t u^i = \sum_{m=1}^{3}\left(D_t\lambda^m \frac{\partial \mu^m}{\partial x^i} + \lambda^m \frac{\partial}{\partial x^i} D_t \mu^m\right) - \frac{\partial}{\partial x^i}\left(D_t\phi + \frac{1}{2}\mathbf{u}^2\right) \tag{51}$$

We also have, from Navier-Stokes equations:

$$D_t u^i = \nu\triangle u^i - \frac{\partial}{\partial x^i} p \tag{52}$$

We compute the term $\triangle u^i$ with the transformation Eq. (39), using Eq. (47) and regrouping the terms we obtain

$$\frac{\partial}{\partial x^i}\left(\Gamma\phi + \frac{1}{2}\mathbf{u}^2 - p\right) = \sum_{m=1}^{3}\left(\Gamma\lambda^m \frac{\partial \mu^m}{\partial x^i} - 2\nu\frac{\partial \lambda^m}{\partial x^k}\frac{\partial}{\partial x^k}\frac{\partial \mu^m}{\partial x^i}\right) \tag{53}$$

Now we split the φ field to obtain the pressure equation

$$\Gamma\phi + \frac{1}{2}\mathbf{u}^2 - p = c \tag{54}$$

To obtain the λ^l dynamics we have to invert the transformation matrix

$$\frac{\partial \mu^m}{\partial x^i} \tag{55}$$

in Eq. (53) if the determinant of the transformation matrix follows

$$\det\left(\frac{\partial \mu^m}{\partial x^i}\right) = 0. \tag{56}$$

then it will be impossible to perform a coordinate transformation. Therefore, if the matrix is invertible, the λ^l dynamics is written as

$$\Gamma\lambda^l = 2\nu\frac{\partial \lambda^m}{\partial x^k} C_{m,k;l} \tag{57}$$

We have to remark that the dynamics of u is completely described by Eq. (47), (57) and the incompressibility condition for u, thus Eq. (54) becomes an identity

GENERALISATION OF CONSTANTIN'S FORMULATION

We begin with the Weber-Clebsch transformation for the velocity field u

$$\mathbf{u} = \sum_{i=1}^{3}\lambda^i\nabla\mu^i - \nabla\phi \tag{58}$$

and perform a variation on the Weber-Clebsch transformation Eq. (58) to obtain the relation (Cartes et al., 2007)

$$\delta\mathbf{u} = \sum_{i=1}^{3}\left(\delta\lambda^i\nabla\mu^i - \delta\mu^i\nabla\lambda^i\right) - \nabla\left(\delta\phi - \sum_{i=1}^{3}\delta\mu^i\lambda^i\right) \tag{59}$$

here δ represents a spatial or temporal variation. In the system (59) it is already possible to see that we have three equations (δu) and six unknowns to find ($\delta\lambda^i$ and $\delta\mu^i$, $\delta\phi$ is fixed by the continuity equation).

In order to write the temporal evolution of u, in terms of Weber-Clebsch potentials, we use the convective derivative D_t and the identity

$$[\nabla, D_t] \equiv (\nabla\mathbf{u})\cdot\nabla \tag{60}$$

We compute now the convective derivative for u, which can be written as a function of the potentials

$$D_t\mathbf{u} = \sum_{i=1}^{3}\left(D_t\lambda^i\nabla\mu^i + \lambda^i\nabla D_t\mu^i - \lambda^i(\nabla\mathbf{u})\cdot\nabla\mu^i\right) - \nabla D_t\phi + (\nabla\mathbf{u})\cdot\nabla\phi \tag{61}$$

For that purpose we write in gradient form

$$\sum_{i=1}^{3}\lambda^i\nabla D_t\mu^i = \sum_{i=1}^{3}\left(\nabla\left(\lambda^i D_t\mu^i\right) - D_t\mu^i\nabla\lambda^i\right) \tag{62}$$

and noting that

$$\sum_{i=1}^{3}\left(\lambda^i(\nabla\mathbf{u})\cdot\nabla\mu^i\right) - (\nabla\mathbf{u})\cdot\nabla\phi = \frac{1}{2}\nabla\mathbf{u}^2 \tag{63}$$

Finally we regroup the gradients

$$D_t\mathbf{u} = \sum_{i=1}^{3}\left(D_t\lambda^i\nabla\mu^i - D_t\mu^i\nabla\lambda^i\right) - \nabla\left(D_t\phi + \frac{1}{2}\mathbf{u}^2 - \sum_{i=1}^{3}D_t\mu^i\lambda^i\right) \tag{64}$$

We must note that this expression is very similar to Eq. (59), the only difference is given by the term $\frac{1}{2}\mathbf{u}^2$, that comes from the commutator between the gradient and the convective derivative.

General formulation for the compressible Navier-Stokes Equations

We now consider the compressible Navier-Stokes equations with a general forcing term f

$$D_t\mathbf{u} = -\nabla w + \mathbf{f}[\mathbf{u}, x, t]$$
$$\partial_t\rho = -\nabla\cdot(\rho\mathbf{u}). \tag{65}$$

where w is the enthalpy and ρ the density field.

For this work we will consider, for simplicity and without loss of generality, a barotropic fluid, then the relation for the enthalpy w is

$$w = \frac{(\rho - 1)}{Ma^2} \tag{66}$$

where Ma is the Mach number for a flow of density $\rho_0 = 1$ and velocity u ~ 1. In this approximation we suppose the density field ρ is very near to the uniformity state and consequently the Mach number is small.

The usual compressible Navier-Stokes equations are obtained when the forcing term f is the viscous dissipation

$$\mathbf{f} = \nu\triangle\mathbf{u}. \tag{67}$$

The idea is to find the evolution equations, in the most general way, for the potentials Eq. (58), now we replace Dt , in the equations of motion Eq. (65), by its expression Eq. (64) and we define

$$D_t\lambda^i = L^i[\lambda, \mu]$$
$$D_t\mu^i = M^i[\lambda, \mu] \tag{68}$$

To wit we made the separation in Eq. (64) and Eq. (65) between gradient and non-gradient terms

$$D_t\phi + \frac{1}{2}|\mathbf{u}|^2 - \sum_{i=1}^{3}M^i\lambda^i = w + G \tag{69}$$

$$\sum_{i=1}^{3}\left(L^i\nabla\mu^i - M^i\nabla\lambda^i\right) = \mathbf{f} - \nabla G \tag{70}$$

here G is an arbitrary gauge function, which comes from the fact that the separation in gradient and non-gradient terms is not unique.

The equation system (70) has 3 linear equations and 6 unknowns L^i, M^i . In order to solve this system with f we must remark that, when ν = 0, the fields λ^i and μ^i follow Euler dynamics

$$D_t\lambda^i = 0$$
$$D_t\mu^i = 0 \tag{71}$$

In order to obtain evolution equations in the same way as (Constantin, 2001) we look for advection diffusion equations. With that goal in mind we introduce $\widetilde{L^i}$ and $\widetilde{M^i}$, defined by

$$D_t\lambda^i = L^i[\lambda, \mu] = \nu\triangle\lambda^i + \widetilde{L^i}[\lambda, \mu]$$
$$D_t\mu^i = M^i[\lambda, \mu] = \nu\triangle\mu^i + \widetilde{M^i}[\lambda, \mu] \tag{72}$$

The terms $\widetilde{L^i}$ and $\widetilde{M^i}$ must verify

$$\sum_{i=1}^{3}\left(\widetilde{L^i}\nabla\mu^i - \widetilde{M^i}\nabla\lambda^i\right) = \tilde{\mathbf{f}} - \nabla\tilde{G} \tag{73}$$

where $\tilde{G}$ is an arbitrary scalar function, linked to the old function G, by the relation G = $\tilde{G} - \nu\triangle\phi + \nu\lambda^i\triangle\mu^i$, and

$$\tilde{\mathbf{f}} = 2\nu\sum_{i=1}^{3}\sum_{\alpha=1}^{3}\partial_\alpha\lambda^i\partial_\alpha\nabla\mu^i. \tag{74}$$

The process used to obtain $\widetilde{L^i}$ and $\widetilde{M^i}$ consists in solving the linear system (73).

Moore-Penrose solution

As the system (73) is under determined, we must impose additional restrictions to solve it. The most straightforward way is to force the coefficients $\widetilde{M^i} = 0$ as in Constantin's formulation then we will have 3 equations for the 3 unknowns. Another, more general, method relies in the imposition of additional conditions on the solution's length. For that purpose we use the Moore-Penrose algorithm (Ben-Israel & Greville, 1974; Moore, 1920; Penrose, 1955), which produces 3 additional conditions that allow us to solve this more general system (73).

For the under determined case, the Moore-Penrose general solution consists in finding the solution to the linear system (73) with the imposition that the norm

$$\sum_{i=1}^{3}\left(\widetilde{L^i}\widetilde{L^i}+\tau^{-2}\widetilde{M^i}\widetilde{M^i}\right) \quad (75)$$

is minimal

The constant τ is introduced here because λ^i and μ^i have different dimensions. In fact, the Weber-Clebsch transformation Eq. (58) means that the dimensions for λ^i and μ^i are

$$\begin{aligned}\left[\lambda^i\right] &= \frac{L}{T}\\ \left[\mu^i\right] &= L\end{aligned} \quad (76)$$

because the product λ∇μ has the same dimensions as the velocity and the fields μ i have the dimensions of L they are the Lagrangian coordinates of the system. Then, from equations (72), it is straightforward that the dimensions of $\widetilde{L^i}$ and $\widetilde{M^i}$ i are

$$\begin{aligned}\widetilde{L^i} &= \frac{L}{T^2}\\ \widetilde{M^i} &= \frac{L}{T}\end{aligned} \quad (77)$$

and the parameter τ in Eq. (75) has the dimension of time.

The Moore-Penrose general solution which minimises the norm Eq. (75) (Cartes et al., 2007), is given by Eq. (78) and Eq. (79)

$$\widetilde{L^i} = \nabla\mu^i\cdot\mathbb{H}^{-1}\cdot\left(\tilde{\mathbf{f}}-\nabla\tilde{G}\right) \quad (78)$$

$$\widetilde{M^i} = -\tau^2\,\nabla\lambda^i\cdot\mathbb{H}^{-1}\cdot\left(\tilde{\mathbf{f}}-\nabla\tilde{G}\right) \quad (79)$$

where H represents the squared symmetric matrix

$$\mathbb{H}_{\alpha\beta} \equiv \sum_{i=1}^{3}\left(\tau^2\partial_\alpha\lambda^i\partial_\beta\lambda^i+\partial_\alpha\mu^i\partial_\beta\mu^i\right) \quad (80)$$

and the arbitrary function $\tilde{G}$ is given by

$$\tilde{G} = \Delta^{-1}\nabla\cdot\tilde{\mathbf{f}} \quad (81)$$

in order to minimise the general norm Eq. (82)

$$S[\tilde{G}] = \frac{1}{2}\int_\Omega\left(\tilde{\mathbf{f}}-\nabla\tilde{G}\right)\cdot\mathbb{H}^{-1}\cdot\left(\tilde{\mathbf{f}}-\nabla\tilde{G}\right)W(x^\alpha)d^dx. \quad (82)$$

with the objective to achieve numerical stability in our simulations. Replacing these solutions $\widetilde{L^i}$ and $\widetilde{M^i}$ i in Eq. (72) we arrive to the explicit evolution equations

$$\begin{aligned}D_t\lambda^i &= \nu\Delta\lambda^i+\nabla\mu^i\cdot\mathbb{H}^{-1}\cdot\left(\tilde{\mathbf{f}}-\nabla\tilde{G}\right)\\ D_t\mu^i &= \nu\Delta\mu^i+-\tau^2\nabla\lambda^i\cdot\mathbb{H}^{-1}\cdot\left(\tilde{\mathbf{f}}-\nabla\tilde{G}\right)\end{aligned} \quad (83)$$

Comparison of the invertibility conditions

Constantin's method will have problems when the determinant det(∇μ) = 0 which is the case in a manifold of codimension 1. In three dimensional space the generic situation becomes that, for any point in the space (x^1 , x^2 , x^3), there is a time t ∗ for which the determinant becomes zero.

$$\mathbb{H} = \nabla(\boldsymbol{\mu})\cdot\nabla(\boldsymbol{\mu})^T+\tau^2\nabla(\boldsymbol{\lambda})\cdot\nabla(\boldsymbol{\lambda})^T \quad (84)$$

which corresponds to isolated points in a manifold of codimension 4 in space-time. In consequence the condition det(∇μ) = 0 will arrive more frequently because of its lower codimension, than the condition with a higher codimension, for det(H) and τ → 0 is a singular limit.

Resettings

As we saw, when the determinant det(H) is zero the Weber-Clebsch potential evolution equations (83) are no longer defined.

In order to avoid this situation, we follow (Ohkitani & Constantin, 2003) and we perform a resetting. More precisely, when the spatial minimum of the determinant

$$\mathrm{Min}\left(\det(\mathbb{H})\right) \leq \epsilon^2 \quad (85)$$

where ε is a pre defined lower limit. We reset the fields in the following way

$$\begin{aligned}\mu_p^i &= 0\\ \lambda^i &= u^i(t_0)\\ \phi &= 0,\end{aligned} \quad (86)$$

where $u^i(t_0)$ are the components of the velocity field obtained from Eq. (58) in the instant t_0, when the resetting is performed, and the fields λ_i , μ_i and φ_i are generated in the same way as in section 4.2. It was already pointed (Cartes et al., 2007; Ohkitani & Constantin, 2003), that the vanishing of det(H) is related to intense particle diffusion that takes place near reconnection of vorticity lines in the case of incompressible fluids, that means the spatial position of the minima of det(H) are the places where the reconnections take place.

NUMERICAL RESULTS

In this section we will show the results from numerical simulations of our formulation for compressible Navier-Stokes equations. We used pseudo-spectral methods because they are easy to implement and their high precision. The technical details of the implementation are described in section 6.

Taylor-Green flow

The Taylor-Green flow is an standard flow used in the study of turbulence (Taylor & Green, 1937). Its advantages are the existence of numerous studies, see for instance (Brachet et al., 1983) and references therein, which allow us to perform comparisons, at the same time we can economise memory and computation resources by using its symmetries (Cartes et al., 2007). The initial Taylor-Green condition is:

$$\begin{aligned}u^1 &= \sin x\cos y\cos z\\ u^2 &= -\cos x\sin y\cos z\\ u^3 &= 0.\end{aligned} \quad (87)$$

As the length and the initial velocity are of order 1, the Reynolds number is defined as R = 1/ν.

Periodic Field Generation

Periodic fields are generated from the Weber-Clebsch representation Eq. (58) as

$$\mu^i = x^i + \mu^i_p, \tag{88}$$

we also impose that μ^i_p and the other fields λ^i and φ in Eq. (58) are periodic. In order to generate an arbitrary velocity field u we can use

$$\mu^i_p = 0$$
$$\lambda^i = u^i$$
$$\phi = 0. \tag{89}$$

We note that the non-periodic part of μ^i in Eq. (88) is made in a way that μ^i gradients are periodic

The initial ρ is given by imposing the incompressibility condition over w at t = 0

$$w_0 = -\Delta^{-1}\nabla \cdot (\mathbf{u} \cdot \nabla \mathbf{u}) \tag{90}$$

and the relation Eq. (66).

Simulation Results

The following simulations were made using a spatial resolution of 128^3 points, a Reynolds number of 200 and a Mach number of 0.3. We will compare the velocity field which comes from simulations made with the Weber-Clebsch potentials with the velocity field that comes from a direct numerical simulation of Navier-Stokes equations. Navier-Stokes equations are integrated using standard pseudo-spectral methods (Gottlieb & Orszag, 1977). The temporal scheme is Adams-Bashforth of order 2 (for details see section 6.3).

In order to characterise and measure the precision of our algorithm for the Weber-Clebsch potentials, we compute the associated enstrophy which is defined as

$$\Omega(t) = \sum_k k^2 E(k,t) \tag{91}$$

where E(k, t) is the energy spectrum and can be described from the velocity field in Fourier space ^u(k,t) as

$$E(k,t) = \frac{1}{2} \sum_{k-\frac{\Delta k}{2}<|\mathbf{k}'|<k+\frac{\Delta k}{2}} |\hat{\mathbf{u}}(\mathbf{k}',t)|^2 \tag{92}$$

Then E(k, t) is obtained as the mean value over spherical shells with thickness Δk = 1. This enstrophy is computed from the velocity field which comes from the Weber-Clebsch and direct Navier-Stokes simulations

Fig. (1) shows the temporal evolution of the enstrophy for different values of the parameter τ. We found good agreement between our formulation and the direct Navier-Stokes simulations. The spatial mean of the quantity ²/2, which represents the density field, can be seen in Fig. (2).

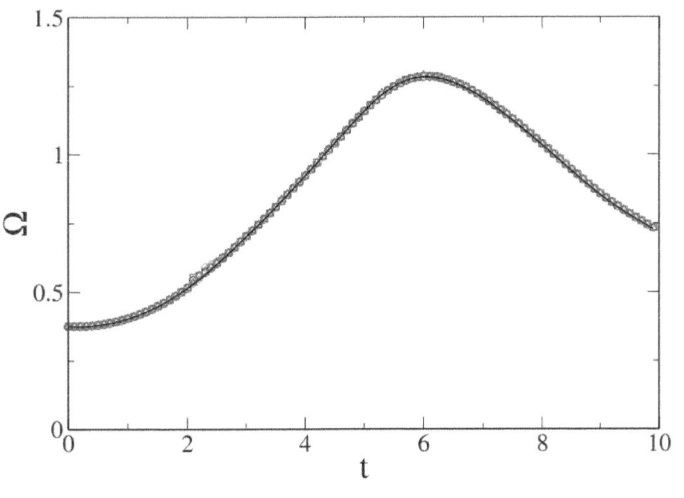

Figure.1: Temporal evolution of the enstrophy Ω for a Reynolds number of 200 and a Mach number of Ma = 0.3 with τ = 0, 0.01, 0.1 and 1 (∘, , ♦ and △), the continuous line represents the direct compressible Navier-Stokes simulation.

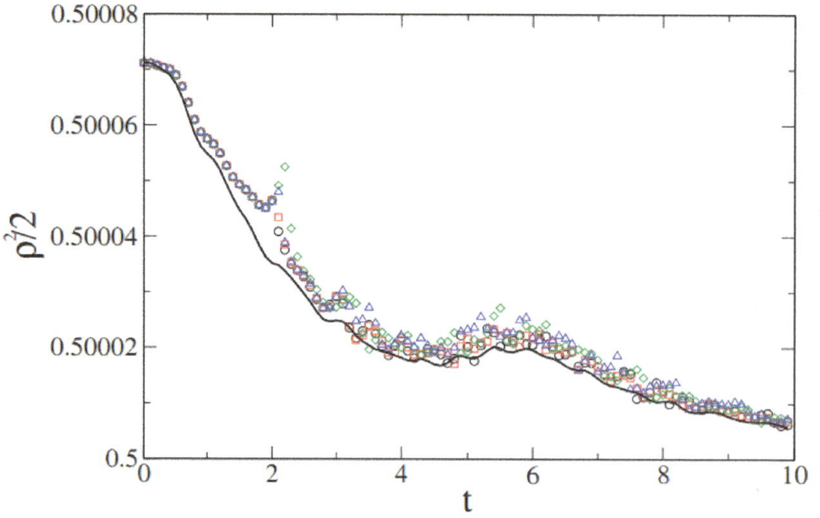

Figure. 2. Temporal evolution of the quantity $\rho^2/2$ for a Reynolds number of 200 and a Mach number of Ma = 0.3 with $\tau = 0, 0.01, 0.1$ and 1 (°, , ♦ and Δ), the continuous line represents the direct compressible Navier-Stokes simulation.

As our λ and μ fields evolved in time we had to reset them to be able to continue the simulation as the coordinate transformation becomes non-invertible. The temporal evolution of the interval between resettings is characterised by

$$\triangle t_j = t_j - t_{j-1} \tag{93}$$

where t_j is the resetting time, we fixed the value for the lower limit of det(H) as $\epsilon^2 = 0.01$, is shown in Fig. (3). We can see that, for a given time, the interval is a growing function of τ. However the shape of Δt is well preserved even when the range of τ goes through several orders of magnitude

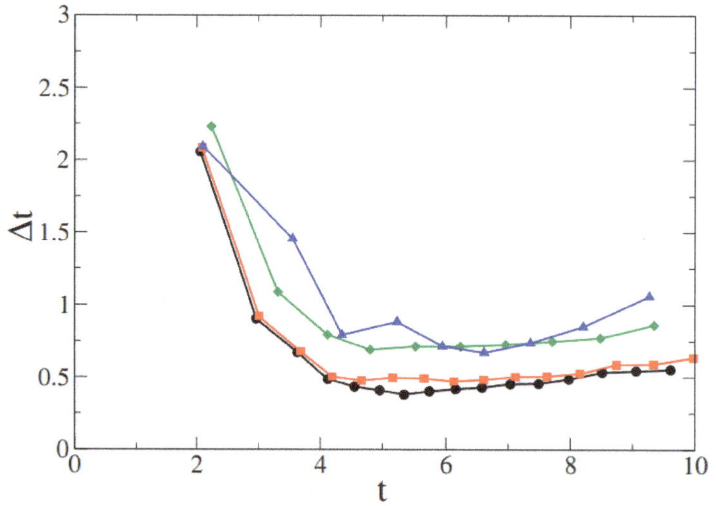

Figure 3: Temporal evolution of the interval between resettings Δt_j versus the resetting time t_j for a Reynolds number of 200 and a Mach number of Ma = 0.3 with $\tau = 0, 0.01, 0.1$ and 1 (°, , ♦ and Δ).

CONCLUSIONS AND PERSPECTIVES

We arrived to a good agreement between the derived generalised equations of motion for the Weber-Clebsch potentials that implying that the velocity field follows the compressible Navier-Stokes equations. These new equations were shown to depend on a parameter with the dimension of time, τ. Direct numerical simulations of the Taylor-Green vortex were performed in order to validate this new formulation. This Eulerian-Lagrangian formulation of compressible Navier-Stokes equations, allows us to study in detail the reconnection process, the turbulence generated by such process and the sound generated by those moving fluids using for example the two antiparallel vortex approach (Virk et al., 1995). This subject is known as aeroacoustics (Lighthill, 1952), which is relevant for aerodynamic noise production, and is a key issue in the design of air planes, turbines, etc. We arrived to a good agreement between the derived generalised equations of motion for the Weber-Clebsch potentials that implying that the velocity field follows the compressible Navier-Stokes equations. These new equations were shown to depend on a parameter with the dimension of time, τ. Direct numerical simulations of the Taylor-Green vortex were performed in order to validate this new formulation. This Eulerian-Lagrangian formulation of compressible Navier-Stokes equations, allows us to study in detail the reconnection process, the turbulence generated by such process and the sound generated by those moving fluids using for example the two antiparallel vortex approach (Virk et al., 1995). This subject is known as aeroacoustics (Lighthill, 1952), which is relevant for aerodynamic noise production, and is a key issue in the design of air planes, turbines, etc.

ACKNOWLEDGMENTS

The authors wish to thank Marc–Etienne Brachet and Jaime Cisternas for valuable comments. We acknowledge the financial support of FONDECYT (Projects No. 3110028 and No. 1110360) and Universidad de los Andes through FAI initiatives.

REFERENCES

1. Ben-Israel, A. & Greville, T. N. E. (1974). Generalized Inverses: Theory and Applications, Wiley-Interscience [John Wiley & Sons], New York. (reprinted by Robert E. Krieger Publishing Co. Inc., Huntington, NY, 1980.).
2. Brachet, M. E., Meiron, D. I., Orszag, S. A., Nickel, B. G., Morf, R. H. & Frisch, U. (1983).Small–scale structure of the Taylor–Green vortex, J. Fluid Mech. 130: 411–452.
3. Cartes, C., Bustamante, M.-D. & Brachet, M.-E. (2007). Generalized Eulerian-Lagrangian description of Navier-Stokes dynamics.
4. Cartes, C., Bustamante, M.-D., Pouquet, A. & Brachet, M.-E. (2009). Capturing reconnection phenomena using generalized Eulerian–Lagrangian description in Navier–Stokes and resistive MHD, Fluid Dynamics Research 41.
5. Constantin, P. (2001). An Eulerian–Lagrangian approach to the Navier–Stokes equations, Commun. Math. Phys. 216: 663–686.
6. Cooley, J. W. & Tukey, J. W. (1965). An algorithm for the machine calculation of complex Fourier series, Math. Comput. 19: 297–301.
7. Gottlieb, D. & Orszag, S. A. (1977). Numerical Analysis of Spectral Methods, SIAM, Philadelphia.Grossmann, S. (1975). An order-parameter field theory for turbulent fluctuations, Physical Review A 11(6): 2165–2172.
8. Lamb, H. (1932). Hydrodynamics, Cambridge University Press, Cambridge.
9. Lighthill, M. J. (1952). On sound generated aerodynamically. i. general theory, Proceedings of The Royal Society A: Mathematical, Physical and Engineering Sciences 211: 564–587.
10. Moore, E. H. (1920). On the reciprocal of the general algebraic matrix, Bulletin of the American Mathematical Society 26: 394–395.
11. Ohkitani, K. & Constantin, P. (2003). Numerical study of the Eulerian–Lagrangian formulation of the Navier–Stokes equations, Physics of Fluids 15(10): 3251–3254.
12. Penrose, R. (1955). A generalized inverse for matrices, Proceedings of the Cambridge Philosophical Society 51: 406–413.
13. Taylor, G. I. & Green, A. E. (1937). Mechanism of the production of small eddies from large ones, Proc. Roy. Soc. Lond. A 158: 499–521.
14. Virk, D., Hussain, F. & Kerr, R. M. (1995). Compressible vortex reconnection, Journal of Fluid Mechanics 304.

CHAPTER 16

DYNAMICALLY INCOMPRESSIBLE FLOW

Christopher Depcik and Sudarshan Loya

INTRODUCTION

Quite often, researchers model a flow as dynamically incompressible without realizing it. This version of the governing equations has been employed to model exhaust aftertreatment devices since the initial work of Vardi and Biller in 1968 (Vardi & Biller, 1968). The small channels in these devices, along with a relatively low flow rate of exhaust gases coming from the engine, promote laminar flow with a speed of approximately one to ten meters per second. This speed is well below the compressible threshold of around 100 m/s or a Mach number of 0.3. As a result, the chemical species equations can be decoupled from the energy equation promoting a computationally faster and easier to program numerical model. While this assumption is indeed valid in this example, only a few researchers have directly stated that the gas is being modeled as dynamically incompressible (Byrne & Norbury, 1993; Depcik et al., 2010). In fact, when this concept is mentioned, reviews of the main author's submitted work in this realm often come back confused as to its meaning. The reviewers wonder how a gas, which is inherently compressible, can be modeled as incompressible. The key wording for this type of situation is "dynamically" incompressible. To clarify this situation, this chapter provides a thorough investigation into this modeling phenomenon. The efforts begin by explaining the threshold by which a gas can be treated as dynamically incompressible. Dynamic incompressibility differs from incompressibility (ρ = constant) by the fact that the density of the gas is considered as being approximately constant ($\rho \approx$ constant). This small change in equal to approximately equal sign has a large bearing on the results. From this assumption, the governing equations of flow are re-derived adding in this assumption in order to provide the proper fundamental versions of these equations for modeling. Of significant importance, when the Law of Conservation of Energy is formulated, an Energy Equation paradox ensues. In particular, two apparently equally valid versions of this equation are found. From this result, this work provides a unique insight into this paradox and indicates the correct description. Moreover, the governing equation of chemical species is included in this chapter, as it not often presented in fluid mechanics books; however, it is important for modeling chemically reactive flow, such as the situation with catalytic exhaust aftertreatment devices.

After describing the governing equations, the failure of this approximation is presented in order for the reader to understand when a model can provide inaccurate results. Finally, this chapter documents a specific example that can lead to highly erroneous results if a modeler does not comprehend the influence of the dynamically incompressible assumption on reacting flows. As a result, this chapter will provide a helpful tool for others when they begin their usage of this version of the fluid mechanics equations.

LAW OF CONSERVATION OF MASS

Conservation of mass is derived from a differential control volume as indicated in Fig. 1. The differential amount of mass (δm) given within the control volume is a function of the product of the fluid density (ρ) and volume (V):

$$\delta m = \rho \delta \mathbb{V} \tag{1}$$

with the differential volume indicated as:

$$\delta \mathbb{V} = dxdydz \tag{2}$$

Citation: Depcik and Sudarshan Loya (2012). Dynamically Incompressible Flow, Advanced Methods for Practical Applications in Fluid Mechanics, Prof. Steven Jones (Ed.), InTech, DOI: 10.5772/27183

It is important to note that all derivations that follow in this chapter, the differential control volume does not change with respect to time.

At each side of this control volume, a mass flux enters and exits the respective control surfaces as a function of local velocity conditions where u, v, and w represent the velocity in the x, y, and z directions respectively. The net mass flux is represented in the x-direction as shown in Fig. 1 as:

$$\text{Net Mass Flux}(x) = \left[\left(\rho u + \frac{\partial(\rho u)}{\partial x}dx\right) - \rho u\right]dydz = \frac{\partial(\rho u)}{\partial x}dxdydz \tag{3}$$

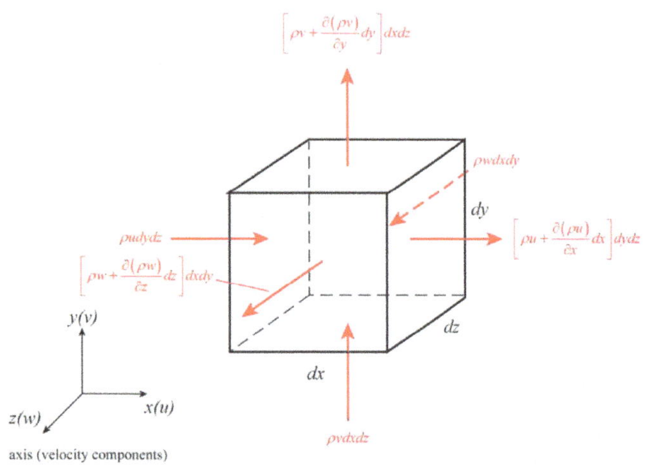

Figure 1: Mass fluxes through a differential control volume.

In other words, the mass flux changes over the differential control surface in each direction are written as a function of its respective derivative in that direction. Hence, the change in mass flux in the x-direction in the previous equation is equal to:

$$\Delta(\rho u) = \frac{\partial(\rho u)}{\partial x}dx \tag{4}$$

with partial derivatives indicated because each variable is a function of all three dimensions. The change in slope of the x-direction mass flux is multiplied over the distance which the slope changes in order to calculate the difference in mass flux in that respective direction.

In the absence of nuclear reactions, conservation of mass states that there will not be a change in mass encompassing the control volume. Hence, the difference of mass within the control volume as a function of time is balanced by the net flow of mass through the control volume during a certain amount of time:

$$\frac{\delta m}{\delta t} + \sum \text{Net Mass Flux} = 0 \tag{5}$$

Written in differential format, this equation becomes:

$$\frac{\partial \rho}{\partial t}\delta V + \left[\frac{\partial(\rho u)}{\partial x}dx\right]dydz + \left[\frac{\partial(\rho v)}{\partial y}dy\right]dxdz + \left[\frac{\partial(\rho w)}{\partial z}dz\right]dxdy = 0 \tag{6}$$

Dividing through by the differential control volume, the conservation of mass is described as:

$$\frac{\partial \rho}{\partial t} + \frac{\partial(\rho u)}{\partial x} + \frac{\partial(\rho v)}{\partial y} + \frac{\partial(\rho w)}{\partial z} = 0 \tag{7}$$

This can be written in vector format for simplicity as:

$$\frac{\partial \rho}{\partial t} + \nabla \cdot (\rho \mathbf{V}) = 0 \tag{8}$$

where the vector V is equal to [u v w];

This version of the conservation of mass is known as the conservative formulation as all variables are embedded within the partial derivatives. However, it is often represented in non-conservative format by first expanding the derivatives:

$$\frac{\partial \rho}{\partial t} + \rho\frac{\partial u}{\partial x} + u\frac{\partial \rho}{\partial x} + \rho\frac{\partial v}{\partial y} + v\frac{\partial \rho}{\partial y} + \rho\frac{\partial w}{\partial z} + w\frac{\partial \rho}{\partial z} = 0 \tag{9}$$

and incorporating the definition of divergence in velocity along with the material derivative:

$$\frac{D\rho}{Dt} + \rho\nabla \cdot \mathbf{V} = 0 \tag{10}$$

Both versions of the conservation of mass have their place in numerical analysis for incompressible and compressible flows. In this next section, the criterion for incompressible flow is described.

MACH NUMBER CRITERION

A fluid can either be considered incompressible or compressible depending on the interrelationship between pressure, density, velocity and temperature. Liquids and solids are nearly always incompressible as their density remains relatively constant independent of variations in these variables. An exact incompressible fluid has the properties of an infinite sound speed, where a pressure disturbance is felt everywhere within the fluid at each instant in time. Gases are compressible since their density changes as a function of pressure and temperature; often related through the ideal gas law. However, a fluid can be considered dynamically incompressible when the speed of sound (aka low-amplitude pressure waves) is significantly faster than the velocity of the working fluid. In this case, the pressure and temperature are not directly related to the density of the medium and a simplification of the governing equations can occur (Kee et al., 2003). In the study of fluid dynamics, aerodynamicists define a non-dimensional parameter after Ernst Mach that relates the fluid velocity to the speed of sound:

$$M = \frac{\mathbf{V}}{a} \tag{11}$$

This parameter helps normalize different working fluids and relate their effects to different flow conditions. As a result, here it will help designate the conditions of incompressible flow independent of the working fluid. The conservation of mass, equation (8), assuming incompressible flow, results in the following governing equation:

$$\nabla \cdot (\rho \mathbf{V}) = 0 \tag{12}$$

By explaining in one-dimension for simplicity, the advection component now equals:

$$\nabla \cdot (\rho \mathbf{V}) \rightarrow \frac{\partial(\rho u)}{\partial x} = 0 \tag{13}$$

and further expanding the derivative in non-conservative format recovers:

$$\frac{\partial(\rho u)}{\partial x} = \rho\frac{\partial u}{\partial x} + u\frac{\partial \rho}{\partial x} = 0 \tag{14}$$

For density to be constant, the second term must be small compared to the first term:

$$u\frac{\partial \rho}{\partial x} \ll \rho\frac{\partial u}{\partial x} \Rightarrow \frac{\delta\rho}{\rho} \ll \frac{\delta u}{u} \tag{15}$$

A process of similarity can be used to compare both differential components in equation (15), given small changes in their respective components. This equation indicates that compressibility needs to be considered only when velocity variations are responsible for density variations. The speed of sound (a) is given by the following thermodynamic relationship under isentropic situations:

$$a^2 = \left(\frac{\partial p}{\partial \rho}\right)_s \approx \frac{\delta p}{\delta \rho} \tag{16}$$

which is then converted similarly to equation (15) into small changes of the respective variables.

$$\delta p \approx -\rho u \delta u \tag{17}$$

Pressure is then related to the velocity by using Bernoulli's equation for irrotational, onedimensional flow when elevation changes are negligible:

$$\frac{u^2}{a^2} = \mathrm{M}^2 \ll 1 \tag{18}$$

Experimental observations indicate that the following condition can typically be assumed to indicate the limit of dynamically incompressible flow:

$$\mathrm{M} \leq 0.3 \tag{19}$$

Therefore, when the Mach number is relatively small, it is reasonable to utilize the dynamic incompressibility assumption for the governing equations of fluid dynamics. From this result, the governing equation of continuity is simplified from equation (10) to:

$$\nabla \cdot \mathbf{V} = 0 \tag{20}$$

The divergence of velocity in this equation $(\nabla)\cdot\mathrm{V}$ is the time rate change of volume of a moving fluid element, per unit volume. Equation (20) means that the change in volume (volume dilation) of the fluid element is zero. This result is consistent with equation (1); i.e., if density and mass are constant, the volume is required to be constant. Written out explicitly using directional variables, the divergence of velocity is given as:

$$\nabla \cdot \mathbf{V} = \frac{\partial u}{\partial x} + \frac{\partial v}{\partial y} + \frac{\partial w}{\partial z} = 0 \tag{21}$$

LAW OF CONSERVATION OF MOMENTUM

Consider a fluid element subjected to body forces caused by gravitation, and to surface forces like pressure and shear stresses caused by fluid friction as shown in Fig. 2. From Newton's second law of motion, the time rate of change of momentum of a body (mV) equals the net force (F) exerted on it.

$$\mathbf{F} = \frac{d}{dt}(m\mathbf{V}) \tag{22}$$

where the left hand side represents the forces exerted on the body and the right hand side is the time rate of change and net flow of momentum within the control volume.

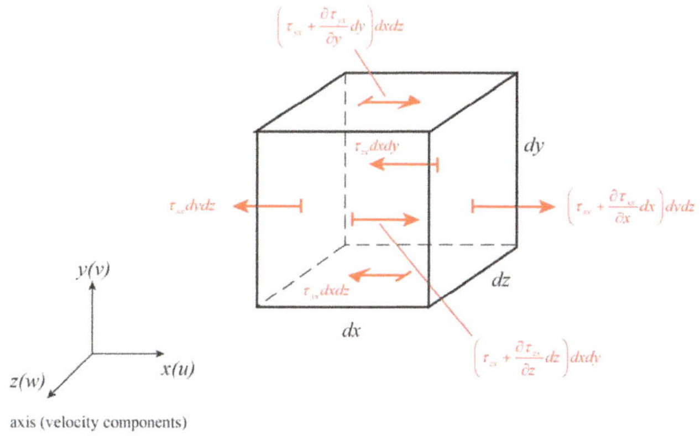

Figure 2: Forces acting on a control volume, with only x-direction illustrated.

In the absence of external forces, the time rate of change of momentum inside the control volume would be balanced by the momentum flux entering the control volume during a certain amount of time (analogous to the conservation of mass):

$$\frac{\delta(m\mathbf{V})}{\delta t} + \sum \text{Net Momentum Flux} = 0 \tag{23}$$

Since the mass entering the control volume on each side brings along with it a respective amount of momentum from all three directions, the momentum flux can be derived in a similar manner as the net mass flux as:

$$\text{Net Momentum Flux}(x) = \left[\left(\rho u \mathbf{V} + \frac{\partial(\rho u \mathbf{V})}{\partial x}dx\right) - \rho u \mathbf{V}\right]dydz = \frac{\partial(\rho u \mathbf{V})}{\partial x}dxdydz \tag{24}$$

For example, the x-momentum equation would equal, in the absence of external forces:

$$\frac{\partial(\rho u)}{\partial t}\delta\mathbb{V} + \left[\frac{\partial(\rho uu)}{\partial x}dx\right]dydz + \left[\frac{\partial(\rho uv)}{\partial y}dy\right]dxdz + \left[\frac{\partial(\rho uw)}{\partial z}dz\right]dxdy = 0 \tag{25}$$

utilizing equations (1) and (23).

Hence, the total change of momentum inside the control volume for all three-dimensions via the right hand side of equation (22) equals:

$$\frac{d}{dt}(m\mathbf{V}) \to \frac{\partial(\rho\mathbf{V})}{\partial t}\delta\mathbb{V} + (\nabla\cdot\rho\mathbf{V}\mathbf{V})\delta\mathbf{S} \tag{26}$$

where δS is a vector indicating the control surface in the respective direction:

$$\delta S_x = dydz \ ; \ \delta S_y = dxdz \ ; \ \delta S_z = dxdy \tag{27}$$

The forces on this fluid element are broken up into those that act at a distance, caused by force fields acting through space proportional to the control volume (body), and shear and strain forces (surface) that include both normal and tangential components proportional to the area:

$$\mathbf{F} = \mathbf{F}_B + \mathbf{F}_S \tag{28}$$

where F_B are body forces and FS are surface forces in the above equation. Body forces in this case include only gravity:

$$\mathbf{F}_B = (\rho\delta\mathbb{V})\mathbf{g} \tag{29}$$

where g is a vector oriented in the direction of the solved equation. Other body forces do exist like electromagnetic forces; however, these forces are beyond the scope of this chapter. The surface forces act on the boundary of the control volume and are applied by external stresses on the sides of the element. Similar to the net momentum flux, a net stress component can be derived as indicated via Fig. 2 and equation (24)

$$\text{Net Stress}(x) = \frac{\partial\tau_{xx}}{\partial x}dxdydz + \frac{\partial\tau_{yx}}{\partial y}dydxdz + \frac{\partial\tau_{zx}}{\partial z}dzdxdy \tag{30}$$

where the first subscript on the stress component (τ) indicates the surface orientation by providing the direction of its outward normal and the second subscript indicates the direction of the force component. In the absence of shear and strain, flow can and will happen because of the normal compressive stress exerted by the working fluid on the boundaries. This is called the hydrostatic pressure condition, as pressure force always acts normal and against the control surface in each direction. Hence, the stress tensor can be represented as:

$$\boldsymbol{\tau} = -p\delta_{ij} + \begin{bmatrix} \tau_{xx} & \tau_{xy} & \tau_{xz} \\ \tau_{yx} & \tau_{yy} & \tau_{yz} \\ \tau_{zx} & \tau_{zy} & \tau_{zz} \end{bmatrix} = -p\delta_{ij} + \boldsymbol{\tau}' \tag{31}$$

with the complete surface force represented as:

$$\delta\mathbf{F}_S = \boldsymbol{\tau}\cdot\delta\mathbf{S} \tag{32}$$

The rest of the stress tensor in equation (31) is calculated by considering the motion of a fluid element as it moves through a flow field. Fluids can undergo a number of different phenomena, as indicated in Fig. 3:

- Translation – linear movement from one location to another
- Rotation – the sides of the element may change as a function of pure rotation
- Angular deformation – distortion of the element by converting perpendicular planes to non-perpendicular planes
- Linear deformation – a change in shape without a change in orientation

Volume dilatation – rate of change of volume per unit volume

Of these different influences, the last three relate to the strain that the fluid element undergoes. Only a brief summary of these components is presented in this section and the readers may wish to refer to the following references for a full derivation (Anderson, 1995; Schlichting & Gersten, 2000; Fox et al., 2004; White, 2003; Panton, 2005; White, 2006).

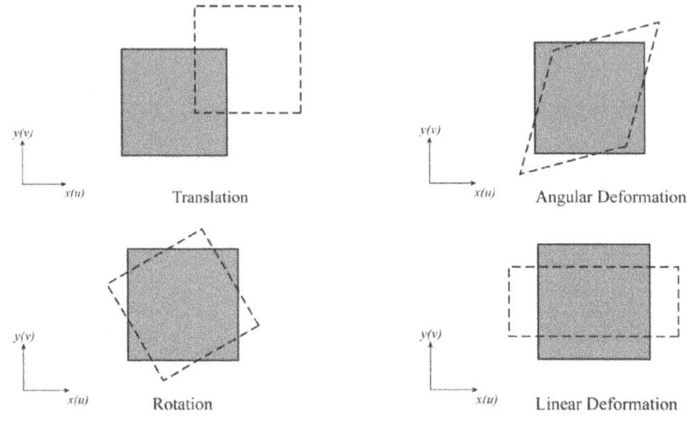

Figure 3: Impact of fluid motion on fluid element as it flows through a flow field in twodimensions.

Angular deformation occurs when the fluid element experiences a rate of deformation given by the change in velocity over the distance computed. In order to formulate a mathematical interpretation of this phenomenon, two assumptions are often made. The first relates to the assumption of a Newtonian Fluid, where the stress at a point is linearly dependent on the rates of strain (deformation) of the fluid. The second involves classifying the fluid as isotropic where the properties of the fluid are independent of direction of orientation. As a result, the components in the stress tensor that are a function of angular deformation are:

$$\tau_{yz} = \tau_{zy} = \mu\left(\frac{\partial w}{\partial y} + \frac{\partial v}{\partial z}\right); \ \tau_{xy} = \tau_{yx} = \mu\left(\frac{\partial u}{\partial y} + \frac{\partial v}{\partial x}\right); \ \tau_{zx} = \tau_{xz} = \mu\left(\frac{\partial u}{\partial z} + \frac{\partial w}{\partial x}\right) \tag{33}$$

where μ is a constant of proportionality that can be measured experimentally; also known as the dynamic viscosity.

Because of velocity gradients, the fluid element can deform as it moves. Linear deformation happens when the flow causes strain by stretching or shrinking the shape of the fluid element. Because of the isotropic condition previously specified, it was found that the linear coefficient of proportionality is equal to twice the constant of proportionality of angular deformation:

$$\tau_{xx,a} = 2\mu\frac{\partial u}{\partial x}; \ \tau_{yy,a} = 2\mu\frac{\partial v}{\partial y}; \ \tau_{zz,a} = 2\mu\frac{\partial w}{\partial z} \tag{34}$$

The fluid can also undergo a volumetric dilatation that equals the rate of change of the volume per unit volume. This component can be thought as the elasticity or compressibility of the working fluid, and while volume dilation and linear deformation are similar concepts, they are not the same. For example, a fluid element can undergo linear deformation (change in shape) but not necessarily volume dilation (change in volume). Similar to linear deformation, volume dilation does not involve any angular deflection and is purely linear in nature. Of interest, even though the fluid volume might only change in the x-direction, it does impact the other momentum equations. This is because the time-change component of the other momentum equations via equation (26) includes the density of the working fluid. Hence, even though the volume is changing only in the x-direction, it will influence the other governing equations through the density dependence. Moreover, because the isotropic condition is applied, the deformation is independent of the coordinate axis under which it is expressed. Therefore, the strain will impact all dimensions; hence, the normal component will be equal among all three dimensions:

$$\tau_{xx,b} = \tau_{yy,b} = \tau_{zz,b} = \lambda\left(\frac{\partial u}{\partial x} + \frac{\partial v}{\partial y} + \frac{\partial w}{\partial z}\right) = \lambda(\nabla \cdot \mathbf{V}) \tag{35}$$

The coefficient λ was originally considered independent of μ and is often called the second coefficient of viscosity in order to help differentiate volume dilation from angular deformation. In linear elasticity according to a Newtonian description, this variable is called Lamé's constant. For estimation of this variable, consult Gad-el-Hak for an understanding of the assumptions present in its calculation (Gad-el-Hak, 1995). At this point, all phenomena have been developed and the momentum equations can be written out explicitly. The x-momentum equation expands to:

$$\begin{aligned}\frac{\partial(\rho u)}{\partial t}\delta\mathbb{V} + \left[\frac{\partial(\rho uu)}{\partial x}dx\right]dydz + \left[\frac{\partial(\rho uv)}{\partial y}dy\right]dxdz + \left[\frac{\partial(\rho uw)}{\partial z}dz\right]dxdy = \\ (\rho\delta\mathbb{V})g_x - \left(\frac{\partial p}{\partial x}dx\right)dydz + \left(\frac{\partial \tau_{xx}}{\partial x}dx\right)dydz + \left(\frac{\partial \tau_{yx}}{\partial y}dy\right)dxdz + \left(\frac{\partial \tau_{zx}}{\partial z}dz\right)dydz\end{aligned} \tag{36}$$

and dividing through by the volume while making the assumption of incompressibility recovers:

$$\rho\left[\frac{\partial u}{\partial t} + u\frac{\partial u}{\partial x} + v\frac{\partial u}{\partial y} + w\frac{\partial u}{\partial z} + u\left(\underbrace{\frac{\partial u}{\partial x} + \frac{\partial v}{\partial y} + \frac{\partial w}{\partial z}}_{\nabla\cdot\mathbf{V}=0}\right)\right] = \rho g_x - \frac{\partial p}{\partial x} + \frac{\partial \tau_{xx}}{\partial x} + \frac{\partial \tau_{yx}}{\partial y} + \frac{\partial \tau_{zx}}{\partial z} \tag{37}$$

Simplifying and incorporating the stress components equals:

$$\rho\frac{Du}{Dt} = \rho g_x + \frac{\partial}{\partial x}\left[-p + 2\mu\frac{\partial u}{\partial x}\right] + \frac{\partial}{\partial y}\mu\left(\frac{\partial u}{\partial y} + \frac{\partial v}{\partial x}\right) + \frac{\partial}{\partial z}\mu\left(\frac{\partial u}{\partial z} + \frac{\partial w}{\partial x}\right) \tag{38}$$

Note that Lamé's constant does not appear in the above equation because the velocity gradient via equation (20) is approximately equal to zero. Therefore, the influence of Lame's constant (λ) is ignored as it is multiplied by a negligible term; e.g. viscous stress is much smaller in magnitude than other flow parameters, like pressure, which results in the multiplication of a relatively small term by a negligible term. For a Newtonian fluid, viscosity depends on temperature and pressure. In the case of a dynamically incompressible fluid, as discussed later, the change of temperature and pressure across the region must be relatively small in order for the assumption of dynamic incompressibility to hold. Hence, there is a negligible change in viscosity as a function of these parameters and the assumption of constant viscosity is valid.

$$\rho\frac{Du}{Dt} = -\frac{\partial p}{\partial x} + \mu\left[\underbrace{\frac{\partial^2 u}{\partial x^2} + \frac{\partial^2 u}{\partial y^2} + \frac{\partial^2 u}{\partial z^2}}_{\nabla^2 u}\right] + \mu\left[\frac{\partial^2 u}{\partial x^2} + \frac{\partial^2 v}{\partial y\partial x} + \frac{\partial^2 w}{\partial z\partial x}\right] + \rho g_x \tag{39}$$

By collecting the derivative in the second to last term on the right hand side, this results in:

$$\rho\frac{Du}{Dt} = -\frac{\partial p}{\partial x} + \mu\nabla^2 u + \mu\frac{\partial}{\partial x}\left[\underbrace{\frac{\partial u}{\partial x} + \frac{\partial v}{\partial y} + \frac{\partial w}{\partial z}}_{\nabla\cdot\mathbf{V}=0}\right] + \rho g_x \tag{40}$$

Hence, this term disappears since the derivative of the velocity gradient is approximately equal to zero from dynamic incompressibility. Therefore, the final x-momentum governing equation for a dynamically incompressible flow equals:

$$\rho\frac{Du}{Dt} = -\frac{\partial p}{\partial x} + \mu\nabla^2 u + \rho g_x \tag{41}$$

Similarly, the momentum equations for dynamic incompressible flow in the y and z directions can be obtained as:

$$\rho\frac{Dv}{Dt}=-\frac{\partial p}{\partial y}+\mu\nabla^2 v+\rho g_y \tag{42}$$

$$\rho\frac{Dw}{Dt}=-\frac{\partial p}{\partial z}+\mu\nabla^2 w+\rho g_z \tag{43}$$

LAW OF CONSERVATION OF ENERGY

The physical principle governing the law of conservation of energy is that the total energy of the system must be conserved. Similar to the last two sections, consider a small fluid element moving with the fluid flow as in Fig. 4. The rate of change of total energy inside the fluid element (E) will be equal to the addition of the net heat flux (Q) into the element and rate of work done (W) on the fluid element due to body and shear forces. In mathematical form, this is represented as:

$$\frac{d}{dt}(mE)=\frac{dQ}{dt}-\frac{dW}{dt} \tag{44}$$

where heat transfer is defined as positive into the control volume and work is defined as positive out of the control volume. The total energy inside the fluid element is a function of its internal energy, potential and kinetic energy due to the translational motion of the fluid element:

$$E=e+\tfrac{1}{2}\tilde{V}^2+\mathbf{g} \tag{45}$$

where the effective kinetic energy velocity incorporates all three components of direction:

$$\tilde{V}^2=u^2+v^2+w^2 \tag{46}$$

In the absence of heat transfer and work, the time rate of change of energy inside the control volume would be balanced by the energy flux entering the control volume during a certain amount of time (analogous to the conservation of mass):

$$\frac{\delta(mE)}{\delta t}+\sum\text{Net Energy Flux}=0 \tag{47}$$

Since the mass entering the control volume on each side brings along with it a respective amount of energy from all three directions, the energy flux can be derived in a similar manner as the net mass flux as:

$$\text{Net Energy Flux}(x)=\left[\left(\rho uE+\frac{\partial(\rho uE)}{\partial x}dx\right)-\rho uE\right]dydz=\frac{\partial(\rho uE)}{\partial x}dxdydz \tag{48}$$

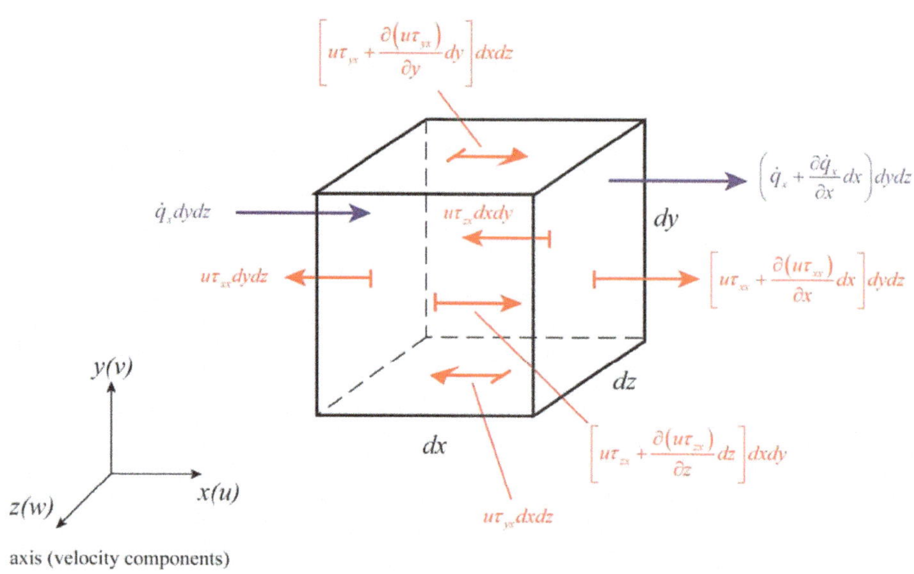

Figure 4: Energy fluxes through an infinitesimally small fluid element with only the fluxes in the x-direction illustrated.

Hence, written in differential format, the left hand side of equation (44) becomes:

$$\frac{d}{dt}(mE) \rightarrow \frac{\partial(\rho E)}{\partial t}\delta\mathbb{V} + \left[\frac{\partial(\rho u E)}{\partial x}dx\right]dydz + \left[\frac{\partial(\rho v E)}{\partial y}dy\right]dxdz + \left[\frac{\partial(\rho w E)}{\partial z}dz\right]dxdy \tag{49}$$

The net heat flux, as indicated in Fig. 4, incorporates a negative sign in order to account for the definition that heat transfer into the control volume is positive:

$$\text{Net Heat Flux}(x) = -\left[\left(\dot{q}_x + \frac{\partial \dot{q}_x}{\partial x}dx\right) - \dot{q}_x\right]dydz = -\frac{\partial \dot{q}_x}{\partial x}dxdydz \tag{50}$$

Based on phenomenological evidence, Fourier found that the heat transfer rate in the direction per unit area perpendicular to the direction of transfer is linearly proportional to the temperature gradient in this direction. Furthermore, through the isotropic condition previously mentioned, all three directions incorporate the same proportionality constant, k, in determination of the heat flux:

$$\dot{q}_x = -k\frac{\partial T}{\partial x};\ \dot{q}_y = -k\frac{\partial T}{\partial y};\ \dot{q}_z = -k\frac{\partial T}{\partial z} \tag{51}$$

This constant, referred hereto as thermal conductivity with a value that is a function of time, is analogous in its concept to viscosity in the momentum equation. Note that if the derivative of temperature is negative, heat transfer is positive as it is propagating in the forward direction (moving from hot to cold). Hence, the total net heat flux is equal to:

$$\frac{dQ}{dt} \rightarrow \left\{\frac{\partial}{\partial x}\left(k\frac{\partial T}{\partial x}\right) + \frac{\partial}{\partial y}\left(k\frac{\partial T}{\partial y}\right) + \frac{\partial}{\partial z}\left(k\frac{\partial T}{\partial z}\right)\right\}dxdydz \tag{52}$$

The work done on a fluid element is a function of the forces acting on this element. As indicated in the momentum equation, there are two forces (body and surface) evident. The rate of work done by a force is the product of this force and the component of velocity in the direction of the force. In order words, the time rate of work is equal to the force done over a certain distance as a function of time. Hence, the work done by the body force is represented as:

$$\text{Body Work} = (\rho\delta\mathbb{V})\mathbf{g}\cdot\mathbf{V} \tag{53}$$

The rate of work done by the surface forces are the product of the stresses with the component of velocity in the corresponding direction. For example, the rate of work in the xdirection is equal to:

$$\text{Net Work}(x) = -\frac{\partial(u\tau_{xx})}{\partial x}dxdydz - \frac{\partial(u\tau_{yx})}{\partial y}dydxdz - \frac{\partial(u\tau_{zx})}{\partial z}dzdxdy \tag{54}$$

including a negative sign as the force on the control volume is opposing the direction of the flow.

Including the work done by pressure and surface forces in all three directions, the total rate of work done is equal to:

$$\frac{dW}{dt} = -\left\{\frac{\partial(u\tau_{xx} + v\tau_{xy} + w\tau_{xz})}{\partial x} + \frac{\partial(u\tau_{yx} + v\tau_{yy} + w\tau_{yz})}{\partial y} + \frac{\partial(u\tau_{zx} + v\tau_{zy} + w\tau_{zz})}{\partial z}\right\}dxdydz \tag{55}$$

And the governing equation for the conservation of energy, incorporating equations (49), (52) and (55) into equation (44) while dividing by the volume, equals:

$$\begin{aligned}\frac{\partial(\rho E)}{\partial t} + \frac{\partial(\rho u E)}{\partial x} + \frac{\partial(\rho v E)}{\partial y} + \frac{\partial(\rho w E)}{\partial z} = \\ \frac{\partial}{\partial x}\left(k\frac{\partial T}{\partial x}\right) + \frac{\partial}{\partial y}\left(k\frac{\partial T}{\partial y}\right) + \frac{\partial}{\partial z}\left(k\frac{\partial T}{\partial z}\right) + \\ \frac{\partial(u\tau_{xx} + v\tau_{xy} + w\tau_{xz})}{\partial x} + \frac{\partial(u\tau_{yx} + v\tau_{yy} + w\tau_{yz})}{\partial y} + \frac{\partial(u\tau_{zx} + v\tau_{zy} + w\tau_{zz})}{\partial z} + \rho(\mathbf{g}\cdot\mathbf{V})\end{aligned} \tag{56}$$

which can be represented in vector format as:

$$\frac{\partial(\rho E)}{\partial t} + \nabla\cdot(\rho\mathbf{V}E) = \nabla\cdot(k\nabla T) - \nabla(p\mathbf{V}) + \nabla(\mathbf{V}\cdot\boldsymbol{\tau}') + \rho(\mathbf{g}\cdot\mathbf{V}) \tag{57}$$

with the pressure term separated out from the stress tensor as doing so will become important in a later section.

As a result, the energy equation governing fluid flow consists of two energy sources. Work through the body forces accelerates the fluid and increases its kinetic energy, while thermal energy conduction (heat flux) increases its internal energy (Panton, 2005). This equation can be further simplified by expanding the left hand side:

$$\rho\frac{\partial E}{\partial t}+\rho\mathbf{V}\cdot\nabla E+E\left[\underbrace{\frac{\partial\rho}{\partial t}+\nabla\cdot(\rho\mathbf{V})}_{0}\right]=\nabla\cdot(k\nabla T)-\nabla(p\mathbf{V})+\nabla(\mathbf{V}\cdot\boldsymbol{\tau}')+\rho(\mathbf{g}\cdot\mathbf{V}) \tag{58}$$

and incorporating the conservation of mass via equation (8) in order to recover:

$$\rho\frac{DE}{Dt}=\nabla\cdot(k\nabla T)-\nabla(p\mathbf{V})+\nabla(\mathbf{V}\cdot\boldsymbol{\tau}')+\rho(\mathbf{g}\cdot\mathbf{V}) \tag{59}$$

Of importance, when the mechanical work component is subtracted from the total energy equation, the remaining part is called the thermal energy equation. As discussed earlier, mechanical work is equal to product of force and velocity. All the forces acting on the body are described via the momentum equation. Hence, in order to obtain the mechanical energy equation in the x-direction, one can multiply the momentum equation (37) (before simplifying this equation further) by the respective velocity component u as follows:

$$\rho\frac{D(u^2/2)}{Dt}=\left[-u\frac{\partial p}{\partial x}+u\frac{\partial\tau_{xx}}{\partial x}+u\frac{\partial\tau_{yx}}{\partial y}+u\frac{\partial\tau_{zx}}{\partial z}+\rho u g_x\right] \tag{60}$$

Similarly, the mechanical energy equations for the y and z directions can be found as:

$$\rho\frac{D(v^2/2)}{Dt}=\left[-v\frac{\partial p}{\partial y}+v\frac{\partial\tau_{xy}}{\partial x}+v\frac{\partial\tau_{yy}}{\partial y}+v\frac{\partial\tau_{zy}}{\partial z}+\rho v g_y\right] \tag{61}$$

$$\rho\frac{D(w^2/2)}{Dt}=\left[-w\frac{\partial p}{\partial z}+w\frac{\partial\tau_{xz}}{\partial x}+w\frac{\partial\tau_{yz}}{\partial y}+w\frac{\partial\tau_{zz}}{\partial z}+\rho w g_z\right] \tag{62}$$

By adding these equations, the total mechanical energy is obtained:

$$\begin{aligned}\rho\frac{D\left(\frac{\tilde{V}^2}{2}\right)}{Dt}&=-\left(u\frac{\partial p}{\partial x}+v\frac{\partial p}{\partial y}+w\frac{\partial p}{\partial z}\right)+u\left(\frac{\partial\tau_{xx}}{\partial x}+\frac{\partial\tau_{yx}}{\partial y}+\frac{\partial\tau_{zx}}{\partial z}\right)\\&+v\left(\frac{\partial\tau_{xy}}{\partial x}+\frac{\partial\tau_{yy}}{\partial y}+\frac{\partial\tau_{zy}}{\partial z}\right)+w\left(\frac{\partial\tau_{xz}}{\partial x}+\frac{\partial\tau_{yz}}{\partial y}+\frac{\partial\tau_{zz}}{\partial z}\right)\\&+\underbrace{\rho u g_x+\rho v g_y+\rho w g_z}_{\rho(\mathbf{g}\cdot\mathbf{V})}\end{aligned} \tag{63}$$

which in vector format equals:

$$\rho\frac{D}{Dt}\left(\frac{\tilde{V}^2}{2}\right)=-\mathbf{V}\cdot\nabla p+\mathbf{V}\cdot\nabla\cdot\boldsymbol{\tau}'+\rho(\mathbf{g}\cdot\mathbf{V}) \tag{64}$$

When equation (45) is used in equation (59) and the derivitives are expanded on the righthand side, the result is:

$$\rho\frac{D(e+\mathbf{g})}{Dt}+\rho\frac{D}{Dt}\left(\frac{\tilde{V}^2}{2}\right)=\nabla\cdot(k\nabla T)-p\nabla\cdot\mathbf{V}-\mathbf{V}\cdot\nabla p+\mathbf{V}\cdot\nabla\cdot\boldsymbol{\tau}'+\boldsymbol{\tau}':\nabla\mathbf{V}+\rho(\mathbf{g}\cdot\mathbf{V}) \tag{65}$$

The total mechanical energy from equation (64) can now be subtracted from the total energy equation, equation (65), in order to recover the thermal energy equation. Assuming that gravity has a negligible influence, the result is:

$$\rho\frac{De}{Dt}=\nabla\cdot(k\nabla T)-p\nabla\cdot\mathbf{V}+\boldsymbol{\tau}':\nabla\mathbf{V} \tag{66}$$

In order to utilize the thermal energy equation for modeling purposes, it is customary to convert it into temperature as the dependant variable (White, 2003). This conversion can be accomplished using either of two different methods as illustrated in the next section. As a result, an energy equation paradox results.

Energy Equation Paradox

Internal energy is a thermodynamic property that can be expressed by two fundamental properties of state. Here it is expressed in terms of temperature and specific volume:

$$de = \underbrace{\left(\frac{\partial e}{\partial T}\right)\Bigg|_v}_{c_v} dT + \left(\frac{\partial e}{\partial v}\right)\Bigg|_T dv \tag{67}$$

Note that the first term on the right hand side is the definition of the constant volume specific heat. Substituting equation (67) in equation (66) finds:

$$\rho c_v \frac{DT}{Dt} + \rho\left(\frac{\partial e}{\partial v}\right)\Bigg|_T \frac{Dv}{Dt} = \nabla\cdot(k\nabla T) - p\nabla\cdot\mathbf{V} + \boldsymbol{\tau}' : \nabla\mathbf{V} \tag{68}$$

However, the specific volume is reciprocal of density and, therefore equation (68) becomes:

$$\rho c_v \frac{DT}{Dt} + \left(\frac{\partial e}{\partial v}\right)\Bigg|_T \left(-\frac{1}{\rho}\right)\frac{D\rho}{Dt} = \nabla\cdot(k\nabla T) - p\nabla\cdot\mathbf{V} + \boldsymbol{\tau}' : \nabla\mathbf{V} \tag{69}$$

Alternatively, internal energy can be expressed in terms of enthalpy as:

$$e = h - p/\rho \tag{70}$$

Written in derivative format, equation (70) becomes:

$$de = dh - dp/\rho + p\left(d\rho/\rho^2\right) \tag{71}$$

Since enthalpy is a thermodynamic property, it can additionally be expressed by two fundamental state variables; here, it is expressed as a function of temperature and pressure:

$$dh = \underbrace{\left(\frac{\partial h}{\partial T}\right)_p}_{c_p} dT + \left(\frac{\partial h}{\partial p}\right)_T dp \tag{72}$$

where the first term on the right hand side is the definition of the constant pressure specific heat of a fluid.

Using the basic law of thermodynamics along with Maxwell's relation,

$$\left(\frac{\partial h}{\partial p}\right)_T = \left[v + T\left(\frac{\partial s}{\partial p}\right)_T\right] = \left[v + T\left(\frac{\partial v}{\partial T}\right)_p\right] = \frac{1}{\rho}\left[1 - \frac{T}{\rho}\left(\frac{\partial \rho}{\partial T}\right)_p\right] \tag{73}$$

And substituting equations (72) and (73) into equation (71), one can write the internal energy as a function of the constant pressure specific heat:

$$de = c_p dT + \frac{1}{\rho}\left[1 - \frac{T}{\rho}\underbrace{\left(\frac{\partial \rho}{\partial T}\right)_p}_{\beta}\right] dp - \frac{dp}{\rho} + \frac{p}{\rho^2} d\rho \tag{74}$$

Note that the derivative of density with respect to temperature is the definition of the thermal expansion coefficient ([1]). In particular, a substance will expand with an increase in energy (heating) and contract with a decrease (cooling) with this dimensional response expressed as the coefficient of thermal expansion. Furthermore, incorporating (74) into equation (66):

$$\rho\left[c_p \frac{DT}{Dt} - \frac{T\beta}{\rho}\frac{Dp}{Dt} + \frac{p}{\rho^2}\frac{D\rho}{Dt}\right] = \nabla\cdot(k\nabla T) - p\nabla\cdot\mathbf{V} + \boldsymbol{\tau}' : \nabla\mathbf{V} \tag{75}$$

The terms in equation (75) can be rearranged and regrouped to produce:

$$\rho c_p \frac{DT}{Dt} = T\beta \frac{Dp}{Dt} - p\left[\underbrace{\frac{1}{\rho}\frac{D\rho}{Dt} + (\nabla \cdot \mathbf{V})}_{0}\right] + \boldsymbol{\tau}' : \nabla\mathbf{V} + \nabla \cdot (k\nabla T) \quad (76)$$

Note that the expression in square brackets on the right hand side is the continuity equation for dynamically incompressible flow in non-conservative format and must therefore be equal to zero. As a result, the final thermal equation following this procedure equals:

$$\rho c_p \frac{DT}{Dt} = T\beta \frac{Dp}{Dt} + \boldsymbol{\tau}' : \nabla\mathbf{V} + \nabla \cdot (k\nabla T) \quad (77)$$

As a result, equation (69) and equation (77) both represent the thermal energy equation utilizing temperature as the dependant variable. If one applies the assumption of dynamic incompressibility to both equations, the results are:

$$\rho c_v \frac{DT}{Dt} = \nabla \cdot (k\nabla T) \quad (78)$$

$$\rho c_p \frac{DT}{Dt} = \nabla \cdot (k\nabla T) \quad (79)$$

as the terms that were eliminated are negligible for an incompressible fluid. In case of truly incompressible liquids and solids, the difference between the specific heat at constant volume and constant pressure vanishes. Hence, equation (78) and (79) reduce to a single equation:

$$\rho c \frac{DT}{Dt} = \nabla \cdot (k\nabla T) \quad (80\ )$$

However, because the specific heats of gases have distinct values equations (78) and (79) lead to a unique problem called the Energy Equation Paradox (White, 2003; Panton, 2005). Both equation (78) and equation (79) cannot be simultaneously valid for the dynamically incompressible flow of gases. In particular, equation (78) implies that advection of internal energy is balanced by heat conduction, while equation (79) implies that enthalpy advection is balanced by heat conduction. In order to explain this paradox and obtain a solution, one must first understand the assumption of dynamic incompressibility, its conditions and application in the governing equations.

Any gaseous flow is assumed to be incompressible if the velocity is less than a Mach number of 0.3 and there is not a large local change in temperature and pressure. For such a flow condition, density is assumed to be constant (ρ) = constant and the divergence of velocity is set equal to zero (∇)= V 0 . However, in actuality, there is a negligible change in density (ρ) ≈ constant and divergence of velocity is not quite zero (∇)· ≈ V 0 . As a result, when the temperature gradient is not large, the conduction and advection of gases are relatively small and nearly of same magnitude as the divergence of velocity. Hence, in the thermal energy equation, any term containing the divergence of velocity or substantial derivative of pressure, temperature or density cannot be set immediately to zero without a thorough review.

Viscous Dissipation

Viscous dissipation is always positive and acts to create internal energy (Panton, 2005). This change in energy is irreversible and it is written as a dyadic product of two tensors, shear stress and gradient of velocity $(\boldsymbol{\tau}' : \nabla\mathbf{V})$ resulting in scalar work (Kee et al., 2003). Viscous dissipation describes rate of work for shape change at constant volume. For dynamically incompressible flow, change in shape at constant volume is negligible as density is assumed to be constant; hence, viscous dissipation is relatively small (Kee et al., 2003). Moreover, viscous dissipation becomes important when the fluid is highly viscous or turbulent (Kreith et al., 2010). In this case, the fluid is a gas with low viscosity under low Mach number situations. Therefore, the change in internal energy due to viscous dissipation will not influence the internal energy significantly and, subsequently, the temperature. As a result, it can be neglected in both equations (69) and (77).

Substantial Derivative of Pressure

The substantial derivative of pressure factor $[T\beta(Dp/Dt)]$ includes the thermal expansion coefficient, β. For an ideal gas, this coefficient is the reciprocal of temperature; hence, it cancels out the temperature component leaving just the material pressure derivative (Kundu & Cohen, 2010). For incompressible flow, thermodynamics properties (like $k,\ c_p,\ \mu$) are often considered constant. Although they fundamentally change with temperature, one of the overriding assumptions for dynamic incompressibility is that there is not a substantial temperature change. If this assumption does not hold,the flow must be treated as compressible (Panton, 2005; Depcik et al., 2010). This assumption allows decoupling of the continuity and momentum equations from the thermal energy equation. Hence, all three velocities and pressure can be solved without needing to compute the temperature simultaneously. Therefore, the velocity field and pressure are unaffected by thermal changes in incompressible flow, since they are derived from the mass and momentum equations. Thus, pressure is represented as a force and not as a property influencing temperature. If the pressure increases or decreases across the incompressible flow region, the level of all pressures increases or decreases respectively. As a result, the change in pressure across the flow is negligible and it can be eliminated from equation (60).

Substantial Derivative of Density

For dynamically incompressible flow, the change in density is negligible. Moreover, any change in internal energy corresponding to a change in volume is marginal because the change in volume itself is small, as indicated in equation (21). Since this component is a product of two trivial terms, it is neglected in equation (69).

Pressure Times Divergence of Veloacity

Although the divergence of velocity may be small, pressure across the flow is significant in magnitude (Kee et al., 2003). Hence, this term is not inconsequential as the product is on the same order of scale as that of conduction: $p(\nabla \cdot \mathbf{V})$. As a result, applying the previous discussions and the influence of this component, the final thermal energy equation is obtained as a function of constant volume or constant pressure specific heats:

Pressure Times Divergence of Velocity

Although the divergence of velocity may be small, pressure across the flow is significant in magnitude (Kee et al., 2003). Hence, this term is not inconsequential as the product is on the same order of scale as that of conduction: $p(\nabla \cdot \mathbf{V})$. As a result, applying the previous discussions and the influence of this component, the final thermal energy equation is obtained as a function of constant volume or constant pressure specific heats:

$$\rho c_v \frac{DT}{Dt} = \nabla \cdot (k\nabla T) - p(\nabla \cdot \mathbf{V}) \tag{81}$$

$$\rho c_p \frac{DT}{Dt} = \nabla \cdot (k\nabla T) \tag{82}$$

In order to validate the previous discussion, the above two equations must generate the same results. Hence, the following derivation proves the methodology and explains the paradox.

For dynamic incompressibility, the change in density of a particle is negligible and therefore density is assumed constant. In mathematical form, this is represented as:

$$\frac{D\rho}{Dt} = 0 = \frac{1}{\rho}\frac{D\rho}{Dt} = \frac{1}{\rho}\frac{\partial \rho}{\partial p}\bigg|_T \frac{Dp}{Dt} + \frac{1}{\rho}\frac{\partial \rho}{\partial T}\bigg|_p \frac{DT}{Dt} \tag{83}$$

Since density is governed by the thermodynamic equation of state, it can be represented as a function of temperature and pressure (Panton, 2005). In addition, equation (83) can be written as:

$$\frac{1}{\rho}\frac{D\rho}{Dt} = \alpha \frac{Dp}{Dt} - \beta \frac{DT}{Dt} = 0 \tag{84}$$

where the isothermal compressibility ‘┘’ and thermal expansion coefficient, ┘, are the characteristic of fluid and hence cannot be set to zero (Turcotte & Schubert, 2002; Graetzel & Infelta, 2002). However, if the change in pressure and change in temperature are sufficiently small, then equation (84) is equal to zero.

Incorporating equation (84) into equation (21):

$$-(\nabla \cdot \mathbf{V}) = \alpha\left(\frac{Dp}{Dt}\right) - \beta\left(\frac{DT}{Dt}\right) \tag{85}$$

As a result, equation (81) is modified to:

$$\rho c_v \frac{DT}{Dt} = \nabla \cdot (k\nabla T) + p\left[\alpha\left(\frac{Dp}{Dt}\right) - \beta\left(\frac{DT}{Dt}\right)\right] \tag{86}$$

For an ideal gas, isothermal compressibility ‘a’ is the reciprocal of pressure and the thermal expansion coefficient, β, is the reciprocal of temperature (Roy, 2001; Honerkamp, 2002; Kundu & Cohen, 2010). Incorporating these simplifications into equation (86) results in:

$$\rho\left(c_v + \frac{\beta}{\rho\alpha}\right)\frac{DT}{Dt} = \nabla \cdot (k\nabla T) + \frac{Dp}{Dt} \tag{87}$$

As explained earlier, the change in pressure across the flow is negligible and the substantial derivative of pressure in this equation is set to zero. Moreover, the specific heats of gases are related by the gas constant as:

$$c_p = c_v + R \Rightarrow c_v + \frac{p}{\rho T} \tag{88}$$

Using the property of isothermal compressibility and thermal expansion coefficient for ideal gases, equation (88) converts to:

$$c_p = c_v + \frac{p}{\rho T} = c_v + \frac{\beta}{\rho\alpha} \tag{89}$$

When equation (89) is used in equation (87), the result is identical to equation (79):

$$\rho c_p \frac{DT}{Dt} = \nabla \cdot (k\nabla T) \tag{79}$$

This alternative development of equation (79) resolves the energy equation paradox; for dynamically incompressible flow, the advection of enthalpy is balanced by conduction.

LAW OF CONSERVATION OF SPECIES

The law of conservation of species follows the same principles as the law of conservation of mass. At each side of the control volume indicated in Fig. 5, a species flux enters and exits the respective control surfaces as a function of local velocity conditions where u_A, v_A, and w_A represent the total species A velocity in the x, y, and z directions, respectively. The net species flux is represented in the x-direction as shown in Fig. 5 as

$$\text{Net Species Flux}(x) = \left[\left(\rho_A u_A + \frac{\partial(\rho_A u_A)}{\partial x}dx\right) - \rho_A u_A\right]dydz = \frac{\partial(\rho_A u_A)}{\partial x}dxdydz \tag{90}$$

where ρ_A is the density of the individual species considered.

Because reactions of chemical species may occur, the difference of mass of the species within the control volume as a function of time is balanced by the net flow of species through the control volume during a certain amount of time while including a local production or destruction rate:

$$\frac{\delta m_A}{\delta t} + \sum \text{Net Species Flux} = \dot{\omega}_A \delta \mathbb{V} \tag{91}$$

where the differential mass is written as:

$$\delta m_A = \rho_A \delta \mathbb{V} \tag{92}$$

and the chemical reaction rate has the units of kg $m^{-3}\ s^{-1}$, consumes species A when negative, and acts over the entire control volume. In differential format, equation (91) becomes:

$$\frac{\partial\rho_A}{\partial t}\delta\mathbb{V}+\left[\frac{\partial(\rho_A u_A)}{\partial x}dx\right]dydz+\left[\frac{\partial(\rho_A v_A)}{\partial y}dy\right]dxdz+\left[\frac{\partial(\rho_A w_A)}{\partial z}dz\right]dxdy=\dot{\omega}_A dxdydz \quad (93)$$

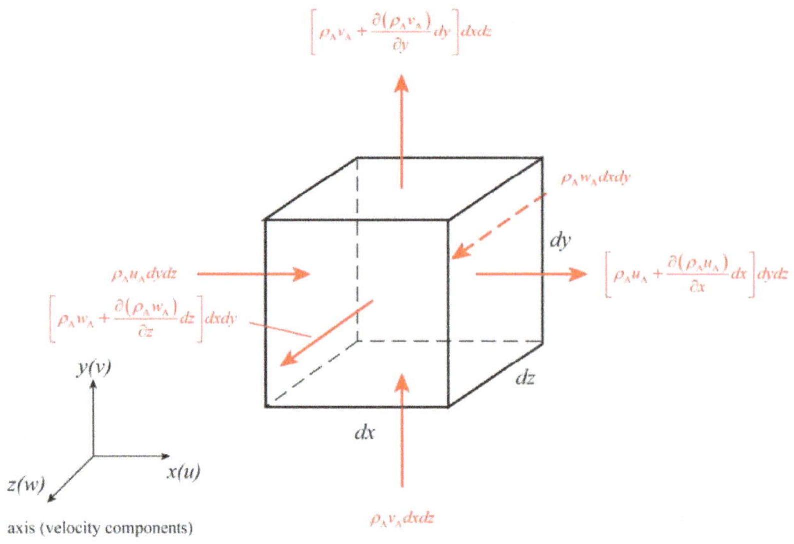

Figure 5: Species fluxes through a differential control volume.

When equation (93) is divided through by $\delta\mathbb{V}=\delta x\delta y\delta z$, the differential control volume, the conservation of species A is described as:

$$\frac{\partial\rho_A}{\partial t}+\frac{\partial(\rho_A u_A)}{\partial x}+\frac{\partial(\rho_A v_A)}{\partial y}+\frac{\partial(\rho_A w_A)}{\partial z}=\dot{\omega}_A \quad (94)$$

This equation can be written in vector format for simplicity as:

$$\frac{\partial\rho_A}{\partial t}+\nabla\cdot(\rho_A\mathbf{V}_A)=\dot{\omega}_A \quad (95)$$

where the vector V_A is equal to $[u_A\ v_A\ w_A]$.

In this equation, the mass flux rate of individual species is the product of the specific density of that species and velocity in the given direction. However, the velocity of individual species depends not only on the bulk velocity of the flow, but also on concentration gradients. If there is a difference in concentration of species at various points across the flow, the species will move from regions of high concentrations to that of low concentrations. This motion is analogous to the phenomena of heat conduction from high temperatures to low temperatures. The velocity induced by a concentration gradient is called the diffusion velocity.

Consider a multicomponent system with different species in the mixture moving with different velocities in different directions. As shown in Fig. 6, V represents the mass averaged bulk velocity of the flow $(V = u\mathbf{i} + v\mathbf{j} + w\mathbf{k})$ and $\mathbf{V}^*$ represents the molar averaged flow velocity. Both V and V* will differ in magnitude and direction as they contain unique weighing factors. Letting VA be the velocity of species A, this value is independent on the molar weight or mass of the species A as it represents only that species. The difference between velocity of individual species and the mass averaged velocity is called the mass diffusion velocity of that species. In this case, it is υA. Similarly, the difference between individual species velocity and the molar averaged velocity of the flow is called the molar diffusion velocity υ_A^*.

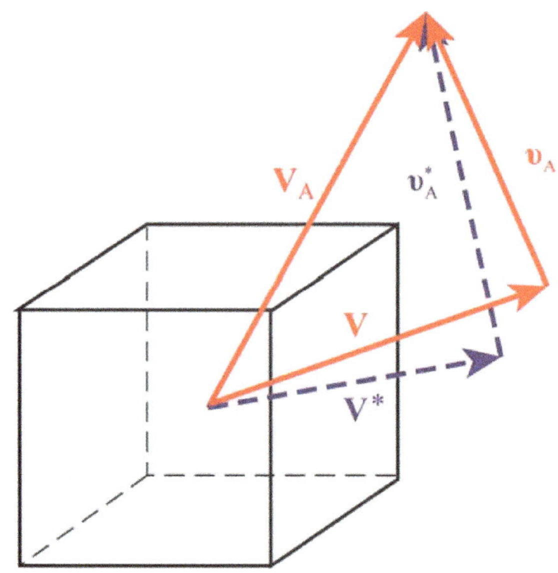

Figure 6: Diagram of mass and molar averaged bulk velocity vectors.

Therefore, the velocity of individual species A is given as:

$$\mathbf{V}_A=\mathbf{V}+\boldsymbol{\upsilon}_A \quad (96)$$

Hence, the conservation of species equals:

$$\frac{\partial\rho_A}{\partial t}+\nabla\cdot(\rho_A\mathbf{V})+\nabla\cdot(\rho_A\boldsymbol{\upsilon}_A)=\dot{\omega}_A \quad (97)$$

Adolf Fick investigated the diffusion velocity in the above equation in detail. He explained that υA could be expressed using diffusion coefficients (δ). The use of this coefficient in the mass flux expression was later called Fick's law of diffusion and is written in mathematical format as:

$$\rho_A \mathbf{\upsilon}_A = -\rho_A \delta_{Am} \nabla(\ln Y_A) \tag{98}$$

where the mass fraction is given by

$$Y_A = \frac{\rho_A}{\rho} \tag{99}$$

with the m subscript on the diffusion coefficient indicating the value of species A with respect to the entire mixture as described later.

Substituting this expression back into Fick's law of diffusion:

$$\rho_A \mathbf{\upsilon}_A = -\rho_A \delta_{Am} \nabla\left(\ln \frac{\rho_A}{\rho}\right) \tag{100}$$

solving for the logarithmic derivative,

(e.g., $\frac{\partial[\ln(\rho_A/\rho)]}{\partial x} = \frac{\rho_A}{\rho}\frac{\partial(\rho/\rho_A)}{\partial x}$),

$$\rho_A \mathbf{\upsilon}_A = -\rho \delta_{Am} \nabla\left(\frac{\rho_A}{\rho}\right) \tag{101}$$

and substituting equations (99) and (101) into the conservation of species, equation (97), recovers:

$$\frac{\partial(\rho Y_A)}{\partial t} + \nabla \cdot (\rho Y_A \mathbf{V}) - \nabla \cdot (\rho \delta_{Am} \nabla Y_A) = \dot{\omega}_A \tag{102}$$

The first two components on the left hand side of equation (102) can be expanded to obtain,

$$\rho \frac{\partial Y_A}{\partial t} + \rho \mathbf{V} \nabla \cdot Y_A + Y_A \left[\underbrace{\frac{\partial \rho}{\partial t} + \nabla \cdot (\rho \mathbf{V})}_{0}\right] - \nabla \cdot (\rho \delta_{Am} \nabla Y_A) = \dot{\omega}_A \tag{103}$$

and conservation of mass requires the term in square brackets of equation (103) to be zero so that:

$$\rho \frac{\partial Y_A}{\partial t} + \rho \mathbf{V} \nabla \cdot Y_A - \nabla \cdot (\rho \delta_{Am} \nabla Y_A) = \dot{\omega}_A \tag{104}$$

This equation is further modified as per the assumptions of dynamic incompressibility. Since the mass fraction of the species is related to the molar concentration of the species by:

$$Y_A = \frac{\bar{C}_A W_A}{\rho} \tag{105}$$

where W_A is the molecular weight of species A, the conservation of species equation for dynamic incompressible flow is:

$$\frac{\partial \bar{C}_A}{\partial t} + \mathbf{V} \nabla \cdot \bar{C}_A - \nabla \cdot \left(\delta_{Am} \nabla \bar{C}_A\right) = \frac{\dot{\omega}_A}{W_A} \tag{106}$$

Fick's law expresses the diffusion velocities in terms of diffusion coefficients. However, calculation of these coefficients has always been subject to interpretation in the literature. The standard method of calculating these values is a three-step process. First, the diffusion coefficient for one species is calculated as in a binary mixture consisting of two gases with one gas is held as the base. The binary diffusion coefficient is then calculated for all other gases in the mixture, keeping the base gas constant. Then, the diffusion coefficient of the base gas is calculated for the mixture from all of the calculated binary diffusion coefficients. This method is then repeated for all the gases in the mixture keeping one gas as the base each time. Many ways have been proposed to calculate the binary diffusion coefficient in the mentioned procedure (Perry & Green, 1984; Cussler, 1997; Kee et al., 2003; Kuo, 2005). After studying the referenced literature, the authors feel that binary diffusion between two species (A and B here) is best calculated by the following (Cussler, 1997; Kee et al., 2003):

$$\delta_{AB} = \frac{0.00186T^{1.5}\left(\frac{1}{W_A} + \frac{1}{W_B}\right)^{0.5}}{p\sigma_{AB}^2\Omega} \quad (107)$$

Equation (107) assumes all gases to be non polar, and although values of σ (Collision diameter of the molecule) and Ω (dimensionless energy integral based on temperature and Boltzmann constant) are not available for all gases. However, this equation provides for a relatively high accuracy, within an eight percent error range with experimental data (Cussler, 1997). This accuracy is appreciable as it is the closest any equation predicting binary diffusion coefficient can get. Even the commercial software program, transport CHEMKIN, supports this same equation for calculating these values (Reaction Design, 2003). From these binary coefficients, the mixture averaged diffusion coefficients can be calculated. The most accurate method to calculate these coefficients is to use a full multicomponent system that involves inverting an L by L matrix, where L is number of species (Wilke, 1950; Perry & Green, 1984; Cussler, 1997; Kee et al., 2003). However, this method is computationally expensive and not required in most numerical models. As a result, most researchers use approximate formulas as follows.

When the mass diffusion velocity is given as a function of mass fractions,

$$\mathbf{v}_A = \frac{-1}{Y_A}\delta_{Am}\nabla Y_A \quad (108)$$

The mixture-averaged diffusion coefficient is represented as:

$$\frac{1}{\delta_{Am}} = \sum_{j\neq A}^{L}\frac{X_A}{\delta_{Aj}} + \frac{X_A}{1-Y_j}\sum_{j\neq A}^{L}\frac{Y_j}{\delta_{Aj}} \quad (109)$$

However, when the mass diffusion velocity is written using mole fractions (or molar concentrations),

$$\mathbf{v}_A = \frac{-1}{X_A}\delta'_{Am}\nabla X_A \quad (110)$$

The mixture-averaged diffusion coefficient is now written as:

$$\delta'_{Am} = \frac{1-Y_A}{\sum_{j\neq A}^{L}\frac{X_j}{\delta_{Aj}}} \quad (111)$$

The reason for the two different methodologies in calculating the diffusion coefficient relates to the solution of the mole or mass fraction version of the governing equation of chemical species. In specific, since mole and mass fractions are related by the molecular weight of the species, diffusion coefficients must be calculated differently in order to take this weighting factor into account. For more information on this topic, please consult the efforts of Kee et al. (Kee et al., 2003).

FAILURE OF DYNAMIC INCOMPRESSIBILITY

As discussed in the Mach Number Criterion section, when the pressure and temperature are not directly related to the density of the medium, one may simplify the governing equations via dynamic incompressibility (Kee et al., 2003). For catalytic exhaust aftertreatment modeling, a previous effort by the first author shows that traditional catalyst model equations assume dynamic incompressibility in order to simplify the solution technique by decoupling the species and energy equations (Depcik & Assanis, 2005). This model, represented by Figure 7, simulates an open channel interacting with the surface of the catalyst (washcoat impregnated with catalytic materials) by virtue of literature derived source terms, similar to equation (91), that model the interactions within the boundary layer. Use of dynamic incompressibility for simulating the gas in the open channel increases the computational speed of the model in order to make it suitable for transient regulatory tests and kinetic constant optimization.

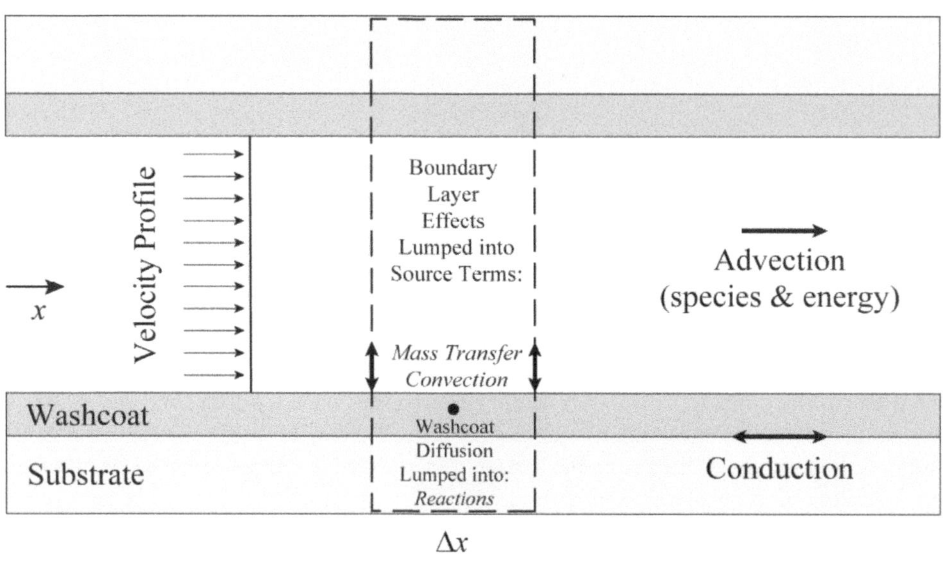

Figure 7: One-dimensional catalytic exhaust aftertreatment model description.

Since the laminar channels of a monolithic catalyst do not contribute significantly to the pressure drop, the pressure throughout the catalyst is nearly constant. However, when a temperature profile exists, according to the ideal gas law the density must change along the monolith. This was stated previously by Byrne and Norbury (Byrne & Norbury, 1993) where they assert that the variation of gas density with temperature suggests that the flow should be treated as compressible for aftertreatment modeling; the assertion was further verified by Depcik et al. (Depcik et al., 2010). Compressibility leads to variations in velocity of the gas and a decreasing residence time of the gas in the catalyst.

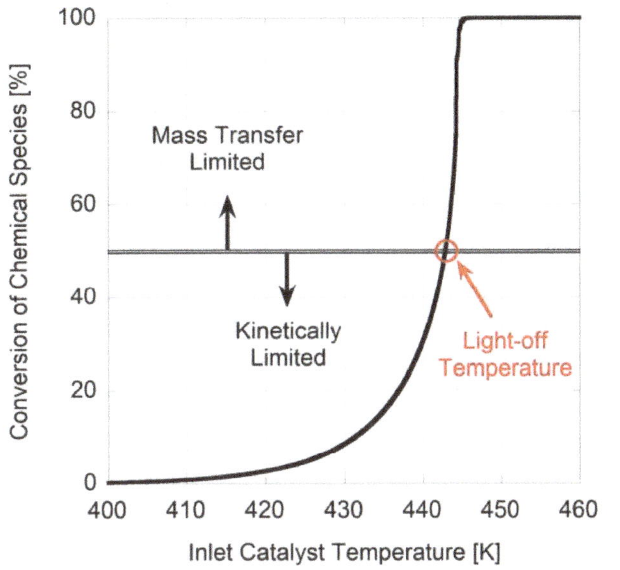

Figure 8: Example of light-off temperature of a catalyst as indicated by the 50% conversion of the chemical species of interest.

One significant goal of the catalyst model is to simulate a "light-off event" when the catalyst transitions from kinetic to mass transfer regimes as illustrated in Figure 8. In other words, when the temperature is low, the reactions taking place on the surface are slow and there is little conversion of species across the catalyst (chemical kinetic limited). As the temperature increases, the reactions increase in magnitude until they hit a certain activation energy that causes an exponential jump in the magnitude of the conversion (50% conversion of inlet species is typically considered the light-off point). As a result, during a light-off experiment the temperature of the catalyst can change dramatically if a significant amount of exothermic reactions takes place. Hence, from a modeling standpoint, it becomes erroneous to employ dynamic incompressibility at this point. However, many researchers neglect this fact and continue to simulate the gas dynamics in this manner. After light-off of a catalyst, it is commonplace to assume the reaction is now mass transfer limited and not kinetic limited. Hence, once entering the catalyst, all species will be converted. From a regulatory standpoint, modeling the catalyst after light-off is virtually unnecessary because most emissions occur in the cold start phase when temperatures and reaction speed is low (Koltsakis & Stamatelos, 1997). Therefore, during the times the model is most needed, pressure and temperature through the catalyst are relatively uniform; i.e. the use of dynamic incompressibility is valid. As a result, it is important to understand that the assumption of dynamic incompressibility fails under situations that involve a significant pressure drop or temperature change. This failure occurs because the fluid is no longer held constant to the ideal gas law and the density remains independent of changes in pressure or temperature. However, by understanding the application of the model within certain parameters, it is possible to utilize dynamic incompressibility in a safe manner.

REACTION RATE USAGE

One item that requires further discussion within the catalytic exhaust aftertreatment community is the use of mole fractions for reactions rates; e.g. the right hand side of equation (104). It must be stressed that reactions rates depend on concentration in principle. In the general case when the operation pressure can differ, the more suitable choice for formulating reactions rates is via concentration rate expressions because they inherently take into account changes in density. For example, consider a case of catalytic reactor working at one atmosphere. If reaction rates are expressed in terms of mole fractions, the reaction rate does not respond to the change of pressure. One obtains the same production rates even for a vanishing pressure. When the pressure is lowered, the subsequent production of species is slower because of a less dense mixture. Thus, a model not limited to a singular pressure must be based on concentration and take into account density variations; i.e. similar temperatures at different pressures would result in different densities but could have the same mole fractions. In other words, a model based on concentration can work for different operational pressures; whereas, a model based on mole fractions is fixed for one pressure. Since pressure changes along the monolith are small and most automotive catalysts work at similar pressures, the use of mole fraction-based expressions does not cause significant problems, but it is important for the reader to understand the implication of doing so. If the reaction rate is written using concentration-based reaction expressions, the results will be significantly different because density is constant and does not respond to the temperature change, as discussed in the previous section; hence, reaction rates will be unique at each cell as axial temperature fluctuations change the calculated rate. For a more in-depth discussion of this topic with respect to catalytic exhaust aftertreatment devices, please consult reference (Depcik et al., 2010). For simplicity, an example is given here for homogeneous (gas phase) reactions in order to demonstrate the concept. The right hand side of the compressible version of the chemical species equation (104), when homogeneous reactions occur, should be written in terms of the gas concentration

$$\dot{\omega}_{A} = A\exp\left(\frac{-E}{R_{u}T}\right)C_{CO} \tag{112}$$

where equation (112) expresses a linear reaction rate dependant on the species carbon monoxide according to the traditional Arrhenius rate expression.

However, when the dynamic incompressible version is used via equation (106)

$$\frac{\dot{\omega}_{A}}{W_{A}} = A\exp\left(\frac{-E}{R_{u}T}\right)X_{CO} \tag{113}$$

mole fractions must be used. The pre-exponential components (A) in both expressions can be equated at specific pressures and temperatures via density.

CONCLUSION

Dynamically incompressible flow is a convenient simplification of the governing equations of fluid mechanics in order to solve for flow conditions below a Mach number of 0.3. This assumption decouples the equations, because the fluid properties are no longer subject to the constraint of the ideal gas law. Consequently, the equations can be solved in a reduced computational time. However, when pressure and temperature change significantly across the computational domain, the assumption of dynamically incompressible flow falters. In order to aid in the development of the reader, this chapter provided the proper background into this computational methodology and illustrated the correct tactic of computing reaction rates for the governing equation of chemical species.

REFERENCES

1. Anderson, J. J. (1995). Computational Fluid Dynamics: The Basics with Applications (1st Ed), McGraw-Hill, 0070016852, New York.
2. Byrne, H. and J. Norbury (1993). Mathematical Modelling of Catalytic Converters. Math. Engng. Ind. vol. 4, no. 1, pp. 27-48.
3. Cussler, E. L. (1997). Diffusion: Mass Transfer in Fluid Systems (2nd Ed), Cambridge University Press, 0521450780, New York.
4. Depcik, C. and D. Assanis (2005). One-Dimensional Automotive Catalyst Modeling. Progress in Energy and Combustion Science, vol. 31, no. 4, pp. 308-369.
5. Depcik, C., A. Kobiera, et al. (2010). Influence of Density Variation on One-Dimensional Modeling of Exhaust Assisted Catalytic Fuel Reforming. Heat Transfer Engineering, vol. 31, no. 13, pp. 1098-1113.
6. Fox, R., A. McDonald, et al. (2004). Introduction to Fluid Mechanics (6th Ed), John Wiley & Sons, Inc., 0471202312, Hoboken, New Jersey.
7. Gad-el-Hak, M. (1995). Questions in Fluid Mechanics: Stokes' Hypothesis for a Newtonian, Isotropic Fluid. Journal of Fluids Engineering, vol. 117, no. 1, p. 3-5.
8. Graetzel, M. and P. Infelta (2002). The Bases of Chemical Thermodynamics: Vol. 1, Universal Publishers, 1581127723.
9. Honerkamp, J. (2002). Statistical Physics: An Advanced Approach with Applications (2nd), Springer, 3540430202, Berlin.
10. Kee, R. J., M. E. Coltrin, et al. (2003). Chemically Reacting Flow: Theory and Practice (1st Ed), John Wiley & Sons, Inc., 0472361793, Hoboken, New Jersey.
11. Koltsakis, G. C. and A. M. Stamatelos (1997). Catalytic Automotive Exhaust Aftertreatment. Progress in Energy and Combustion Science, vol. 23, no. 1, p. 1-39.
12. Kreith, F., R. M. Manglik, et al. (2010). Principles of Heat Transfer (7th Ed), Cengage Learning, Inc., 0495667706.
13. Kundu, P. K. and I. Cohen (2010). Fluid Mechanics (4th Ed), Academic Press, 0123813999, New York.
14. Kuo, K. K. (2005). Principles of Combustion (2nd Ed), John Wiley & Sons, Inc., 0471046892, Hoboken, New Jersey.
15. Panton, R. L. (2005). Incompressible Flow (3rd Ed), John Wiley & Sons, Inc., 047126122X, Hoboken, New Jersey.
16. Perry, R. and D. Green (1984). Perry's Chemical Engineering Handbook (6th Ed), McGraw-Hill, Inc., 0070494797, New York.
17. Reaction Design (2003). Transport Core Utility Manual. Available from: http://eccentric.mae.cornell.edu/~laniu/MAE643/CHEMKIN3.7/transport.pdf
18. Roy, S. K. (2001). Thermal Physics and Statistical Mechanics, New Age International Publishers, 81-224-1302-1, New Delhi.
19. Schlichting, H. and K. Gersten (2000). Boundary Layer Theory (8th Ed), Springer, 3540662707, Berlin.
20. Turcotte, D. L. and G. Schubert (2002). Geodynamics (2nd), Cambridge University Press, 0521661862, New York.
21. Vardi, J. and W. F. Biller (1968). Thermal Behavior of an Exhaust Gas Catalytic Converter. Industrial and Engineering Chemistry - Process Design and Development, vol. 7, no. 1, p. 83-90.
22. White, F. (2003). Fluid Mechanics (5th Ed), McGraw-Hill, 0072402172, Boston.
23. White, F. (2006). Viscous Fluid Flow (3rd Ed), McGraw-Hill, 0072402318, New York.
24. Wilke, C. R. (1950). Diffusional Properties of Multicomponent Gases. Chemical Engineering Science, vol. 46, no. 2, p. 95-104.

CHAPTER 17

ANALYSIS OF FLOW PARAMETERS OF A NEWTONIAN FLUID THROUGH A CYLINDRICAL COLLAPSIBLE TUBE

Caroline W. Kanyiri, Mathew Kinyanjui and Kang'ethe Giterere

Keywords: Flow parameters, Collapsible tube, Steady flow, Newtonian fluid

BACKGROUND

Fluid is a type of matter which undergoes continuous deformation when some external force is applied. It is said to undergo deformation if the distance between any two neighboring molecules change. A fluid is said to be Newtonian if it obeys the Newton's law of viscosity which states that the shear stress is proportional to the velocity gradient. The viscosity does not change with the rate of flow.

A collapsible tube is any tube with sufficiently flexible walls that it can elastically accommodate deformation to a highly noncircular cross section when the external pressure exceeds the internal pressure.

The study of flow through collapsible tubes is of utmost importance in biological studies as well as in industries. Vessel collapse is seen in the veins, such as in the veins of a hand raised above the level of the heart or in the jugular vein when a person is standing upright. In the arteries, collapse occurs when high external pressures are applied, such as when an artery is compressed by a sphygmomanometer cuff during blood pressure measurement. Similarly, in the industry, collapse may be experienced during cementing operations, trapped fluid expansion, or well evacuation, among many others. Most oilfield tubulars also experience collapse.

Citation: Kanyiri C.W., Kinyanjui M. & Giterere K., Analysis of flow parameters of a Newtonian fluid through a cylindrical collapsible tube. doi:10.1186/2193-1801-3-566.

Literature Review

Flow through collapsible tubes has been extensively studied in the laboratory. Pioneering work on collapsible tubes, explain that the veins play an important role in controlling the output of the heart. This control function of the veins is a passive one, and is as a result of their ability to collapse and inflate. Several experimental studies have been done on flexible tubing. For instance Bertram (1986) did an experimental study on collapsible and elastic tubes with finite-length and the upstream and downstream ends held open. He concluded that when the external pressure exceeded the fluid pressure by a sufficiently large amount, the tube buckled non-axi symmetrically, leading to a nonlinear relation between pressure-drop and flow rate.

Bertram et al. (1990) explained that at sufficiently large Reynolds numbers, the system produces self-excited oscillations. Jensen and Heil (2003) also realized that asymmetrically collapsed vessels readily develop flow-induced, self excited oscillations. Physiological examples include wheezing during forced expiration and the development of korottkoff sounds during blood pressure measurement. Earlier works by Luo and Pedley (1998) who investigated the effect of wall inertia on the self-excited oscillations in a collapsible channel flow show that tension-induced instabilities were the main cause of the self-excited oscillations. Makinde (2005) further described the fluid dynamics of a collapsible tube using a mathematical model. He observed that the fluid axial velocity profile was parabolic with maximum value at centerline. He also noted that fluid axial velocity generally decreases with an increase in tube contraction due to the strong influence of the negative transmural pressure owing to marked reduction of rigidity. Marzo et al. (2005) studied three-dimensional collapse of a steady flow through finite-length elastic tubes numerically. Three-dimensional solid elements were used for the elastic wall, allowing the wall thickness to be specified. Previous findings by Hazel and Heil (2003) for thinner-walled tubing were confirmed and also he showed the existence of significant differences if a thick-walled tube is used. Andrew et al. (2008) described the role of venous valves in pressure shielding. A one-dimensional mathematical model of a collapsible tube, with the facility to introduce valves at any position, was used. It was found out that a valve decreased the dynamic pressures applied to a vein when gravity is applied by a considerable amount. Emilie and Patrice (2010) developed a simple and effective numerical physiological tool to help clinicians and researchers in the understanding of flow phenomena. One-dimensional Runge–Kutta discontinuous Galerkin (RK-DG) method coupled with lumped parameter models for the boundary conditions was used. It was noted that the efficiency of muscular calf pump is strongly dependent on the valves pathology and the walking frequency. Eleuterio and Annunziato (2013) formulated a one-dimensional time-dependent non-linear mathematical model for physiological fluid flow in collapsible tubes with discontinuous material properties. He observed that although the solution algorithm dealt with idealized cases, it is uniquely well-suited for assessing the performance of numerical methods intended for simulating more general situations.

This research study has presented a one dimensional mathematical model of fluid flow through collapsible tube. From the Literature review above, a comprehensive study considering the flow parameters such as longitudinal tension and volumetric flow rate and their effects on the cross sectional area of a collapsible tube, flow velocity and internal pressure of fluid in a collapsible tube has not been done. This study therefore aimed at coming up with a more comprehensive model of flow through collapsible tubes thus expanding the understanding of fluid flow through collapsible tubes.

The continuity equation

This equation arises from the fact that matter is neither created nor destroyed. The rate at which mass enters a system is equal to the rate at which mass leaves the system. The differential form for a general continuity equation is given by;

$$\frac{\partial \rho}{\partial t} + \vec{\nabla}\left(\rho \vec{u}\right) = 0 \tag{1}$$

where ρ is the fluid density and u⃗ $\vec{u}$ is the fluid's velocity.

For incompressible fluid flow, ρ is assumed to be a constant and hence equation (1) reduces to;

$$\nabla \vec{u} = 0 \tag{2}$$

Equation (2) means that the divergence of velocity is zero.

In Cartesian co-ordinate form and considering a one dimensional fluid flow equation (2) is expressed as;

$$\frac{\partial u}{\partial x} = 0 \tag{3}$$

The volumetric flow rate Q is given by area multiplied by velocity per unit time, therefore equation (3) becomes;

$$\frac{\partial (Au)}{\partial x} = \frac{\partial Q}{\partial x} = 0 \tag{4}$$

Equation (4) is derived from the fact that mass is always conserved in fluid systems regardless of the pipeline complexity or direction of flow. The volumetric flow rate Q is constant but area and velocity of the fluid flow are variable so that if A decreases, u increases and vice versa.

Equation of conservation of momentum

The equation of conservation of momentum is derived from Newton's second law of motion, which states that the time rate change of momentum of a body matter is equal to the net external forces applied to the body. The momentum equation can be expressed as;

$$\frac{du}{dt} + u\frac{du}{dx} = -\frac{1}{\rho}\frac{dp}{dx} - Ru + g \tag{5}$$

where R>0 is a friction factor and g is the gravitational acceleration when the tube is held vertically.

For steady laminar fluid flow, equation (5) is given by,

$$\rho u \frac{\partial u}{\partial x} = -\frac{\partial p}{\partial x} - f_L \frac{\mu u}{D_e}\left(\frac{s}{A}\right), \tag{6}$$

where s the peripheral length and f_l is the skin-friction coefficient for laminar flow.

The Tube law

The tube law relates the transmural pressure (internal pressure –external pressure) to the cross-sectional area of a collapsible tube. It is given by,

$$p - p_E = \phi\left(\frac{A}{A_0}\right) \tag{7}$$

Putting into consideration the longitudinal tension in the tube law equation (7) becomes;

$$p - p_E = \phi\left(\frac{A}{A_o}\right) - T\frac{\partial^2 A}{\partial x^2}, \tag{8}$$

Where $\phi\left(\frac{A}{A_0}\right) = K_{PE}\left(\alpha^n - \alpha^{-\frac{3}{2}}\right)$, $\alpha = \frac{A}{A_0}$ and K_{PE} is the combined stiffness, which represents the overall stiffness of the tube, whether collapsed or distended.

Final set of equations

Making the substitution Q = Au, equation (6) becomes;

$$\rho \frac{Q}{A}\frac{\partial \frac{Q}{A}}{\partial x} = -\frac{\partial p}{\partial x} - f_l \frac{\mu}{D_e}\frac{Q}{A}\frac{s}{A} \tag{9}$$

Which reduces to

$$\frac{\partial p}{\partial x} = \rho\frac{Q^2}{A^3}\frac{\partial A}{\partial x} - f_l \frac{\mu}{D_e}\frac{Q}{A}\frac{s}{A} \tag{10}$$

Thus the final set of equations for a steady laminar fluid flow through a cylindrical collapsible tube are;

$$\frac{\partial p}{\partial x} = \rho\frac{Q^2}{A^3}\frac{\partial A}{\partial x} - f_l \frac{\mu}{D_e}\frac{Q}{A}\frac{s}{A} \tag{11}$$

And

$$p - p_E = K_{PE}\left(\alpha^n - \alpha^{-\frac{3}{2}}\right) - T\frac{\partial^2 A}{\partial x^2} \tag{12}$$

METHOD OF SOLUTION

The equations governing the flow problem were written in finite difference form and then reorganized and written in matrix form.

The governing equations describing the steady, incompressible laminar fluid flow through a cylindrical collapsible tube, in finite difference form are given as:

$$p_i = p_{i-1} + \frac{\rho Q^2 (A_i - A_{i-1})}{A_{im}(i)^3} - \Delta x f_l \frac{\mu}{D_e}\frac{Q}{A_{im}(i)}\frac{s_i}{A_{im}(i)} \tag{13}$$

where s_i is the peripheral length and is expressed as $s_i = 2\pi\sqrt{\frac{A_i}{\Pi}}$, $A_{im}(i)$ is the area at pressure nodes and is expressed as $A_{im}(i) = \frac{A_i - A_{i-1}}{2}$ $i = 2 \ldots N$ and D_e is the hydraulic diameter expressed as $D_e = 2\pi r$.

Equation (13) is derived from equation (11), the derivatives have been replaced by their corresponding finite approximations.

Discretizing equation (12) with central differencing of $\frac{\partial^2 A}{\partial x^2}$ yields;

$$T\frac{A_{i+1} - 2A_i + A_{i-1}}{\Delta x^2} = p_E - p_i + K_{PE}\left(\alpha^{10} - \alpha^{-\frac{3}{2}}\right) \tag{14}$$

The right hand side of equation (14) is linearized using the Taylor expansion of the term $K_{PE}\left(\alpha^{10} - \alpha^{-\frac{3}{2}}\right)$ expanded about the point $A_i = c$ to get equation (15) as follows;

$$T\frac{A_{i+1} - 2A_i + A_{i-1}}{\Delta x^2} = p_E - p_i + K_{PE}\left[\left(\frac{c}{A_0}\right)^{10} - \left(\frac{c}{A_0}\right)^{-\frac{3}{2}}\right] + (A_i - c)K_{PE}\left[\frac{10}{c}\left(\frac{c}{A_0}\right)^{10} + \frac{3}{2c}\left(\frac{c}{A_0}\right)^{-\frac{3}{2}}\right] \tag{15}$$

Rearranging equation (15) in order to put the like terms together yields;

$$T\frac{A_{i+1} + A_{i-1}}{\Delta x^2} + A_i\left[\frac{-2T}{\Delta x^2} + K_{PE}\left(\frac{10}{c}\left(\frac{c}{A_0}\right)^{10} + \frac{3}{2c}\left(\frac{c}{A_0}\right)^{-\frac{3}{2}}\right)\right] = p_E - p_i + K_{PE}\left[\left(\frac{c}{A_0}\right)^{10} - \left(\frac{c}{A_0}\right)^{-\frac{3}{2}}\right] - K_{PE}\left[10\left(\frac{c}{A_0}\right)^{10} + \frac{3}{2}\left(\frac{c}{A_0}\right)^{-\frac{3}{2}}\right] \tag{16}$$

Equation (16) is subject to the following boundary conditions;

$A_1 = A_0$, $A_N = A_0$

where A_0 is the area at the inlet.

Equation (16) is represented in matrix form and the coefficient matrix is tridiagonal.

The matrix system is of the form $B\ \vec{A} = \vec{R}$ where B is the tridiagonal matrix.

To write the equation in matrix form, let β represent $K_{PE}\left(\frac{10}{c}\left(\frac{c}{A_0}\right)^{10} + \frac{3}{2c}\left(\frac{c}{A_0}\right)^{-\frac{3}{2}}\right) - \frac{2T}{\Delta x^2}$ and α represent $p_E + K_{PE}\left[\left(\frac{c}{A_0}\right)^{10} - \left(\frac{c}{A_0}\right)^{-\frac{3}{2}}\right] - K_{PE}\left[10\left(\frac{c}{A_0}\right)^{10} + \frac{3}{2}\left(\frac{c}{A_0}\right)^{-\frac{3}{2}}\right]$.

The matrix system of the form $B\ \vec{A} = \vec{R}$ is as shown below;

$$\begin{pmatrix} \beta & \frac{T}{\Delta x^2} & 0 & \cdot & \cdot & \cdot & \cdot \\ \frac{T}{\Delta x^2} & \beta & \frac{T}{\Delta x^2} & 0 & \cdot & \cdot & \cdot \\ 0 & \frac{T}{\Delta x^2} & \beta & \frac{T}{\Delta x^2} & 0 & \cdot & \cdot \\ \vdots & \vdots & \vdots & \vdots & \vdots & \vdots & \vdots \\ \cdot & \cdot & \cdot & 0 & \frac{T}{\Delta x^2} & \beta & \frac{T}{\Delta x^2} \\ \cdot & \cdot & \cdot & \cdot & 0 & \frac{T}{\Delta x^2} & \beta \end{pmatrix} \begin{pmatrix} A_2 \\ A_3 \\ A_4 \\ A_5 \\ \vdots \\ A_{N-3} \\ A_{N-2} \\ A_{N-1} \end{pmatrix} = \begin{pmatrix} -p_2 + \alpha - T\frac{A_0}{\Delta x^2} \\ -p_3 + \alpha \\ -p_4 + \alpha \\ \vdots \\ -p_{N-2} + \alpha \\ -p_{N-1} + \alpha - T\frac{A_0}{\Delta x^2} \end{pmatrix}$$

The tridiagonal matrix B is given by;

$$\begin{pmatrix} K_{PE}\begin{pmatrix}\frac{10}{c}\left(\frac{c}{A_0}\right)^{10} + \\ \frac{3}{2c}\left(\frac{c}{A_0}\right)^{-\frac{3}{2}}\end{pmatrix} - \frac{2T}{\Delta x^2} & \frac{T}{\Delta x^2} & 0 & \cdot & \cdot & \cdot \\ \frac{T}{\Delta x^2} & K_{PE}\begin{pmatrix}\frac{10}{c}\left(\frac{c}{A_0}\right)^{10} + \\ \frac{3}{2c}\left(\frac{c}{A_0}\right)^{-\frac{3}{2}}\end{pmatrix} - \frac{2T}{\Delta x^2} & \frac{T}{\Delta x^2} & 0 & \cdot & \cdot \\ 0 & \frac{T}{\Delta x^2} & K_{PE}\begin{pmatrix}\frac{10}{c}\left(\frac{c}{A_0}\right)^{10} + \\ \frac{3}{2c}\left(\frac{c}{A_0}\right)^{-\frac{3}{2}}\end{pmatrix} - \frac{2T}{\Delta x^2} & \frac{T}{\Delta x^2} & 0 & \cdot \\ \vdots & \vdots & \vdots & \vdots & \vdots & \vdots \\ \cdot & \cdot & 0 & \frac{T}{\Delta x^2} & K_{PE}\begin{pmatrix}\frac{10}{c}\left(\frac{c}{A_0}\right)^{10} + \\ \frac{3}{2c}\left(\frac{c}{A_0}\right)^{-\frac{3}{2}}\end{pmatrix} - \frac{2T}{\Delta x^2} & \frac{T}{\Delta x^2} \\ \cdot & \cdot & \cdot & 0 & \frac{T}{\Delta x^2} & K_{PE}\begin{pmatrix}\frac{10}{c}\left(\frac{c}{A_0}\right)^{10} + \\ \frac{3}{2c}\left(\frac{c}{A_0}\right)^{-\frac{3}{2}}\end{pmatrix} - \frac{2T}{\Delta x^2} \end{pmatrix}$$

And vector R is given by:

$$\vec{R} = \begin{pmatrix} -p_2 + p_E + k_{PE}\left(\left(\frac{c}{A_0}\right)^{10} - \left(\frac{c}{A_0}\right)^{-\frac{3}{2}}\right) - k_{PE}\left[10\left(\frac{c}{A_0}\right)^{10} + \frac{3}{2}\left(\frac{c}{A_0}\right)^{-\frac{3}{2}}\right] - \frac{T}{\Delta x^2}A_0 \\ -p_4 + p_E + k_{PE}\left(\left(\frac{c}{A_0}\right)^{10} - \left(\frac{c}{A_0}\right)^{-\frac{3}{2}}\right) - k_{PE}\left[10\left(\frac{c}{A_0}\right)^{10} + \frac{3}{2}\left(\frac{c}{A_0}\right)^{-\frac{3}{2}}\right] \\ -p_4 + p_E + k_{PE}\left(\left(\frac{c}{A_0}\right)^{10} - \left(\frac{c}{A_0}\right)^{-\frac{3}{2}}\right) - k_{PE}\left[10\left(\frac{c}{A_0}\right)^{10} + \frac{3}{2}\left(\frac{c}{A_0}\right)^{-\frac{3}{2}}\right] \\ \vdots \\ -p_{N-2} + p_E + k_{PE}\left(\left(\frac{c}{A_0}\right)^{10} - \left(\frac{c}{A_0}\right)^{-\frac{3}{2}}\right) - k_{PE}\left[10\left(\frac{c}{A_0}\right)^{10} + \frac{3}{2}\left(\frac{c}{A_0}\right)^{-\frac{3}{2}}\right] \\ -p_{N-1} + p_E + k_{PE}\left(\left(\frac{c}{A_0}\right)^{10} - \left(\frac{c}{A_0}\right)^{-\frac{3}{2}}\right) - k_{PE}\left[10\left(\frac{c}{A_0}\right)^{10} + \frac{3}{2}\left(\frac{c}{A_0}\right)^{-\frac{3}{2}}\right] - \frac{T}{\Delta x^2}A_0 \end{pmatrix}$$

Results and Discussion

The tridiagonal matrix obtained along with vector R were used to obtain the following graphs using MATLAB program.

Effect of varying Longitudinal Tension on the cross sectional area of the collapsible tube

In order to determine the effect of longitudinal tension on the cross sectional area of the collapsible tube, other parameters were held constant while the longitudinal tension was varied. This resulted to three curves that were plotted on the same axes as shown in Figure 1.

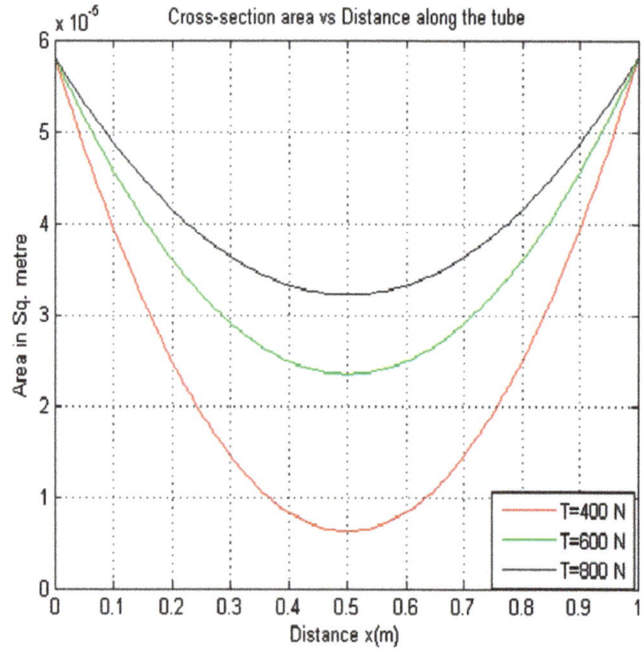

Figure 1: Cross sectional area versus distance with $K_{PE} = 1.21 \times 10^{-5}$ ρ **= 1.0 × 10**^{3} Pe = 4.00×10^3 r = 4.3×10^{-3} Q = 5×10^{-6}.

From Figure 1 it is observed that when the longitudinal tension was increased from 400 N to 800 N holding the other parameters constant, the cross sectional area increased from 6.206×10^{-6} to 3.215×10^{-5}square meters. This can be explained by the reduction of the tube's tendency to collapse as the longitudinal tension increases which consequently leads to decrease in collapse hence increase in the cross sectional area.

Effect of varying Longitudinal Tension on the flow velocity

The longitudinal tension was varied while other parameters were held constant. This resulted to three curves that were plotted on the same axes as shown in Figure 2 below.

From Figure 2, it is observed that as the longitudinal tension increases from 400 N to 800 N, the flow velocity decreases from 0.8051 to 0.1555 m/s. The flow velocity decreases when the longitudinal tension increases because of the already increased cross sectional area. This happens in order to maintain a constant discharge.

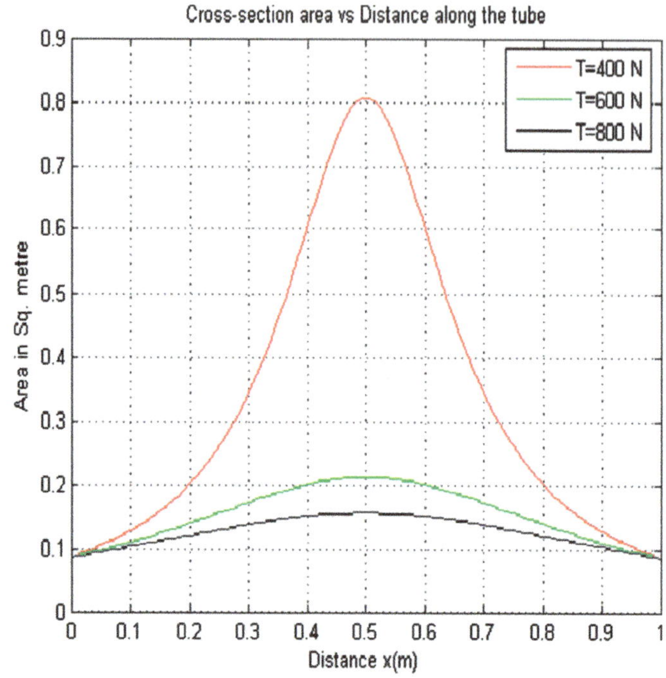

Figure 2: Flow Velocity versus distance with $K_{PE} = 1.21 \times 10^{-5}$ ρ = **1.0 × 10**^{3}Pe = 4.00×10^3r = 4.3×10^{-3}Q = 5×10^{-6}.

Effect of varying Longitudinal Tension on the internal pressure

In order to determine the effect of longitudinal tension on the internal pressure, other parameters were held constant while the longitudinal tension was varied. This resulted to three curves that were plotted on the same axes as shown in Figure 3.

From Figure 3 it is observed that as the longitudinal tension increases the internal pressure increases. When the longitudinal tension is 400 N the internal pressure is 3679 Pascals and when the longitudinal tension is increased to 800 N, the internal pressure increases to 3992 Pascals. The increase in internal pressure as the longitudinal tension increases is due to the decrease in flow velocity. From Bernoulli principle, the sum of pressure energy at any part plus the kinetic energy per unit volume plus the potential energy per unit volume at that point is always constant and therefore a decrease in flow velocity leads to an increase in pressure.

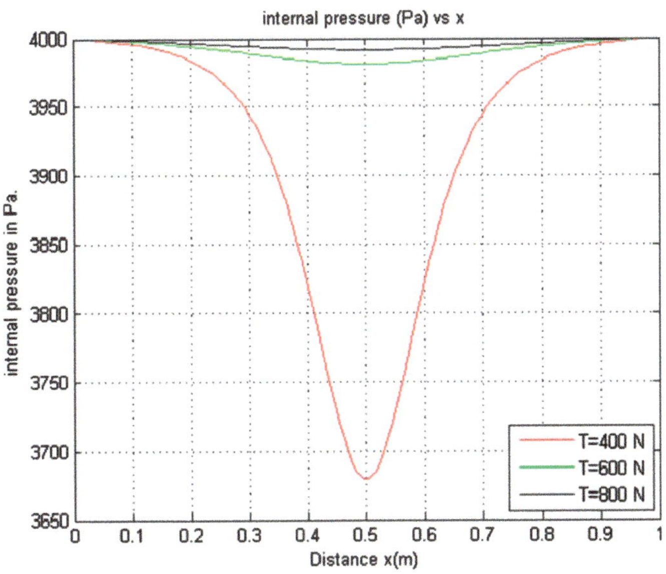

Figure 3: Internal Pressure versus distance with $K_{PE} = 1.21 \times 10^{-5}$ $\boldsymbol{\rho = 1.0 \times 10^3}$ $Pe = 4.00 \times 10^3$ $r = 4.3 \times 10^{-3}$ $Q = 5 \times 10^{-6}$.

Effect of varying volumetric flow rate on the cross sectional area

To investigate the effects of varying volumetric flow rate on the cross sectional area of a collapsible tube, the other parameters were held constant while the volumetric flow rate was varied. The curves obtained were plotted on the same axes as shown in Figure 4 below.

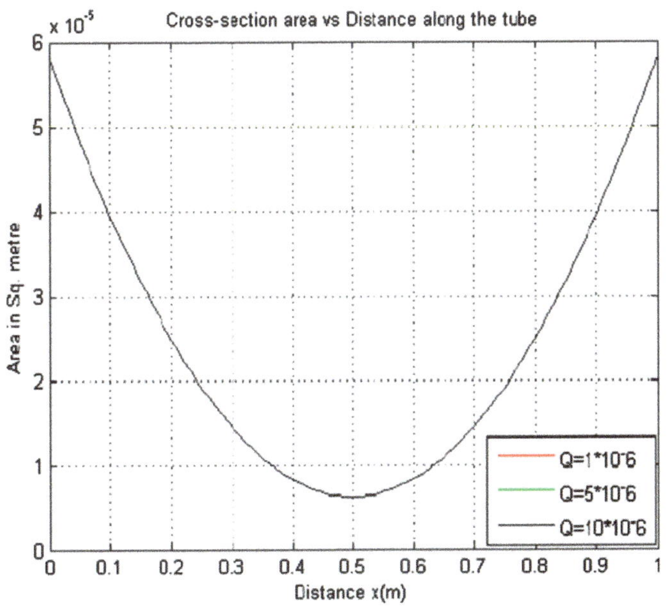

Figure 4: Cross sectional area versus distance for $T = 4.0 \times 10^2$ $K_{PE} = 1.21 \times 10^{-5}$ $\boldsymbol{\rho = 1.0 \times 10^3}$ $Pe = 4.00 \times 10^3$ $r = 4.3 \times 10^{-3}$.

From Figure 4 it is observed that change in the volumetric flow rate does not affect the cross sectional area meaning that the cross sectional area is largely independent of the flow rate. The cross sectional area remains as 6.206×10^{-6} m/s as the volumetric flow rate changes. This is because as the flow rate increases, the pressure drop increases in order to maintain the steady flow rate.

Effect of varying volumetric flow rate on the flow velocity

The volumetric flow rate was varied while the other parameters were held constant. The curves obtained were then plotted on the same axes as shown in Figure 5 below.

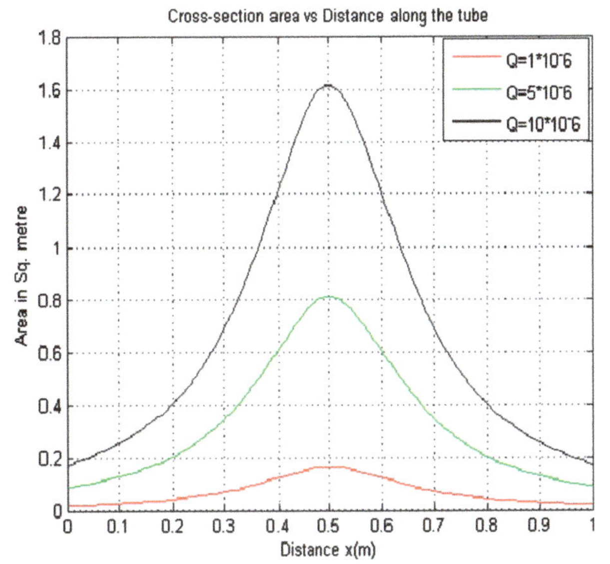

Figure 5: Flow velocity versus distance for $T = 4.0 \times 10^2$ $Kp_E = 1.21 \times 10^{-5}$ $\boldsymbol{\rho = 1.0 \times 10^3}$ $Pe = 4.00 \times 10^3$ $r = 4.3 \times 10^{-3}$.

From Figure 5 it is observed that as volumetric flow rate increases, the flow velocity also increases. When the volumetric flow rate increases from 1×10^{-6} to 10×10^{-6}, the flow velocity increases from 0.1611 m/s to 1.611 m/s. This is because the volumetric flow rate is directly proportional to the flow velocity for a given cross sectional area. In this case the cross sectional area is constant hence an increase in volumetric flow rate translates to an increase in the flow velocity.

Effects of varying volumetric flow rate on the internal pressure

To investigate the effects of varying the volumetric flow rate on the internal pressure of a collapsible tube, different values of volumetric flow rate were used to plot curves while the other parameters were held constant. The curves obtained were plotted on the same axes as shown in Figure 6 below.

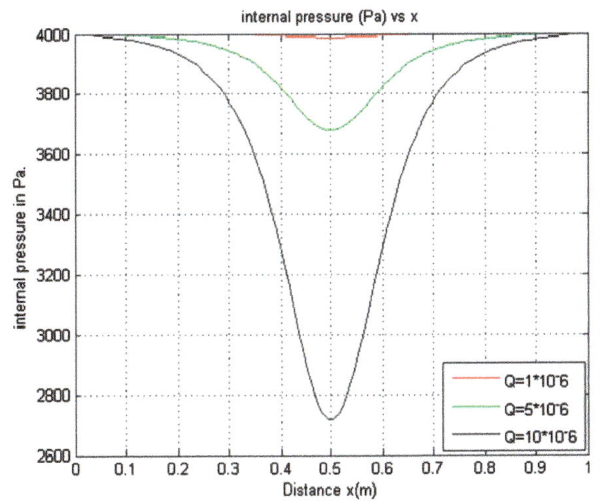

Figure 6: Internal pressure versus distance for $T = 4.0 \times 10^2$ $Kp_E = 1.21 \times 10^{-5}$ $\boldsymbol{\rho = 1.0 \times 10^3}$ $Pe = 4.00 \times 10^3$ $r = 4.3 \times 10^{-3}$.

From Figure 6, it is noted that as volumetric flow rate increases, the internal pressure decreases. As the volumetric flow rate increases from 1×10^{-6} to 10×10^{-6}, the internal pressure decreases from 3987 Pascals to 2718 Pascals. This is as a result of the already increased flow velocity. From Bernoulli principle an increase in flow velocity leads to a decrease in pressure.

DISCUSSION

The objective of this study was to do an analysis of the flow parameters of a Newtonian fluid through a cylindrical collapsible tube. The flow parameters considered are longitudinal tension and volumetric flow rate. The effects of these flow parameters on the cross sectional area of a collapsible tube, flow velocity and internal pressure have been analyzed.

Longitudinal tension describes how tight the tube is pulled out when attached at the edges while the volumetric flow rate refers to the volume of fluid which passes through a given surface per unit time. The values of the cross sectional area, flow velocity and internal pressure are taken at the point where the collapse of the tube is mostly felt, that being midway along the length of the tube. As fluid flows through the elastic tube there is collision between molecules hence a decrease in kinetic energy. Pressure energy is converted into kinetic energy to maintain the flow velocity since the cross sectional area is the same. This leads to a decrease in internal pressure. Since the external pressure remains constant, it exceeds the internal pressure causing the tube to collapse. The collapse leads to a decrease in cross sectional area and consequently the flow velocity increases in order to maintain a constant flow rate. An increase in velocity of the fluid leads to an increase in the collision between the molecules hence greater loss in kinetic energy. This causes the internal pressure to decrease even more. In addition, according to Bernoulli principle, an increase in fluid velocity leads to a decrease in pressure.

CONCLUSION

The extent to which the tube collapses is dependent on the longitudinal tension. An increase in this parameter reduces the tube's tendency to collapse and therefore leads to an increase in the cross sectional area of the tube. Consequently, the flow velocity decreases and the internal pressure increases. It is therefore noted that longitudinal tension is directly proportional to both the cross sectional area and internal pressure and inversely proportional to the flow velocity.

The volumetric flow rate is largely independent of the cross sectional area and therefore any change in the discharge affects the flow velocity and consequently the internal pressure. The volumetric flow rate is directly proportional to the flow velocity and inversely proportional to the internal pressure.

REFERENCES

1. Andrew JN, Patricia VL, David RH: The role of venous valves in pressure shielding. J Am Soc Mass Spectrom 2008, 7: 8.
2. Bertram CD: Unstable equilibrium behavior in collapsible tubes. J Biomech 1986, 19: 61-69. 10.1016/0021-9290(86)90109-0
3. Bertram CD, Raymond CJ, Pedley TJ: Mapping of instabilities for flow through collapsible tubes of differing lengths. J Fluid Struct 1990, 4: 125-153. 10.1016/0889-9746(90)90058-D
4. Eleuterio FT, Annunziato S: Flow in collapsible tubes with discontinuous mechanical properties: mathematical model and exact solutions. Commun Comput Phys 2013, 13: 361-385.
5. Emilie M, Patrice F: Accurate modeling of unsteady flows in collapsible tubes. Comput Methods Biomech Biomed Engin 2010, 13: 279-290. 10.1080/10255840903190726
6. Hazel A, Heil M: Steady finite-Reynolds number flows in three dimensional collapsible tubes. J Fluid Mech 2003, 486: 79-103.
7. Jensen OE, Heil M: High frequency self-excited oscillations in a collapsible channel flow. J Fluid Mech 2003, 481: 235-268.
8. Luo XY, Pedley TJ: The effects of wall inertia on flow in a two dimensional collapsible channel.J Fluid Mech 1998, 363: 253-280.
9. Makinde OD: Collapsible tube flow: a mathematical model. Rom J Phys 2005, 50: 493-506.
10. Marzo A, Luo XY, Bertram CD: Three-dimensional collapse and steady flow in thick-walled flexible tubes. J Fluid Struct 2005, 20: 817-835. 10.1016/j.jfluidstructs.2005.03.008

CHAPTER 18

ON RADIATION EFFECTS ON HYDROMAGNETIC NEWTONIAN LIQUID FLOW DUE TO AN EXPONENTIAL STRETCHING SHEET

PK Kameswaran, M Narayana, P Sibanda and G Makanda

INTRODUCTION

The study of laminar boundary layer flow over a stretching sheet has received considerable attention in the recent past due to its immense application in industry, for example, in extrusion processes such as the polymer extrusion from a dye and wire drawing. Other engineering applications of the stretching sheet problem include polymer sheet extrusion from a dye, drawing, tinning and annealing of copper wires, glass fiber and paper production, the cooling of a metallic plate in a cooling bath and so on. There has been tremendous amount of work on the stretching sheet problem in the past several decades (see Crane [1], Gupta and Gupta [2], Grubka and Bobba [3], Dutta and Gupta [4], Siddappa and Abel [5], Chen and Char [6], Lahaet al. [7], Chakrabarti and Gupta [8], Anderson et al. [9], Siddheshwar and Mahabaleswar [10], Abel and Mahesha [11], Abel et al. [12] and the references therein).

The above studies concern the linear stretching sheet problem but most of the practical situations involve a non-linear stretching sheet such as an exponential one. With this in mind, several authors have considered the velocity of the sheet to vary exponentially with the distance from the slit. Elbashbeshy [13] was among the first to study the exponentially stretching sheet problem. He considered a perforated sheet and examined the effect of wall mass suction on the flow and heat transfer over an exponentially stretching surface. Using a suitable similarity transformation, he transformed the momentum equation into a non-linear Riccati type equation and solved it iteratively. Ishak [14] studied the MHD boundary layer flow due to an exponentially stretching sheet with radiation effect. He found that the local heat transfer rate at the surface decreased with increasing values of the magnetic and radiation parameters. The flow and heat transfer from an exponentially stretching surface was considered by Magyari and Keller [15]. They examined the heat and mass transfer characteristics and compared with the well-known results of the power-law models. Sanjayanand and Khan [16] studied the heat and mass transfer in a viscoelastic boundary layer flow over an exponentially stretching sheet. They found that the viscoelastic parameter enhances the thermal boundary layer thickness. The effect of viscous dissipation on the mixed convection heat transfer from an exponentially stretching surface was studied by Partha et al. [17]. They observed a rapid growth in the non-dimensional skin friction coefficient with the mixed convection parameter. The influence of thermal radiation on the boundary layer flow due to an exponentially stretching sheet is studied by Sajid and Hayat [18]. Khan [19] presented an elegant solution of the viscoelastic boundary layer flow over an exponentially stretching sheet in terms of Whittaker's function.

The characteristics desired of the final product in an extrusion process depend on the rate of stretching and cooling. Hence, it is very important to have a controlled cooling environment where the flow over the stretching sheet can be regulated by external agencies like a magnetic field. An exponential variation of a magnetic field is used, among other applications, to determine the diamagnetic susceptibility of plasma. Steenbeck [20] determined the diamagnetic susceptibility of a cylindrical plasma for axial magnetic fields with various gas pressure and magnetic field strengths. Tonks [21] studied the effects of a magnetic field in the plasma of

Citation: PK Kameswaran, M Narayana, P Sibanda, G Makand and On radiation effects on hydromagnetic Newtonian liquid flow due to an exponential stretching sheet, DOI: 10.1186/1687-2770-2012-105.

an arc. Pavlov [22] considered the magnetohydrodynamic flow of an incompressible viscous fluid over a linearly stretching surface. Sarpakaya [23] extended Pavlov's work to non-Newtonian fluids. Subsequent studies by Andersson [24], Lawrence and Rao [25], Abel et al. [26], Cortell [27] concerned the magnetohydrodynamic flow of viscoelastic liquids over a stretching sheet. Radiation effects on MHD flow past an exponentially accelerated isothermal vertical plate with uniform mass diffusion in the presence of a heat source was studied by Reddy et al. [28]. They observed that the velocity decreases with an increase in the magnetic parameter due to a resistive drag force which tends to resist the fluid flow and thus reduces the velocity. The boundary layer thickness was also found to decrease with an increase in the magnetic parameter.

Most of the earlier work neglected radiation effects. If the polymer extrusion process is placed in a thermally controlled environment, radiation could become important. As with magnetohydrodynamics, careful control of thermal radiative heat transfer has an effect on the characteristics of the final product. Many researchers have considered the effect of thermal radiation on flows over stretching sheets. Studies by Raptis [29], Raptis and Perdikis [30] address the effect of radiation in various situations. Siddheshwar and Mahabaleswar [10] studied the effects of radiation and heat source on MHD flow of a viscoelastic liquid and heat transfer over a stretching sheet. Bidin and Nazar [31] studied the effects of numerical solution of the boundary layer flow over an exponentially stretching sheet with thermal radiation. They observed that the temperature profiles and the thermal boundary layer thickness increase slightly with an increase in the Eckert number. They also showed that an increase in Pr causes a decrease in temperature profiles and the thermal boundary layer thickness. Physically, if Princreases, the thermal diffusivity decreases, and these phenomena lead to the decreasing of energy ability that reduces the thermal boundary layer. Elbashbeshy and Dimian [32] analyzed boundary layer flow in the presence of radiation effect and heat transfer over the wedge with a viscous coefficient. Thermal radiation effects on hydro-magnetic flow due to an exponentially stretching sheet were studied by Reddy and Reddy [33]. They found that as radiation increases, the temperature profiles and thermal boundary layer thickness also increase. They also observed that the temperature profiles and thermal boundary layer thickness increase slightly with an increase in the Eckert number. Raptis et al. [34] studied the effect of thermal radiation on the magnetohydrodynamic flow of a viscous fluid past semi-infinite stationary plate and Hayat et al. [35] extended the analysis for the second grade fluid.

In addition to a magnetic field and thermal radiation, one has to consider the viscous dissipation effects due to frictional heating between fluid layers. The effect of viscous dissipation in natural convection processes has been studied by Gebhart [36] and Gebhart and Mollendorf [37]. They observed that the effect of viscous dissipation is predominant in vigorous natural convection and mixed convection processes. They also showed the existence of a similarity solution for the external flow over an infinite vertical surface with an exponential variation of surface temperature. Vajravelu and Hadjinicalaou [38] studied the heat transfer characteristics over a stretching surface with viscous dissipation in the presence of internal heat generation or absorption.

In this paper, we investigate the effects of various physical and fluid parameters such as the magnetic parameter, radiation parameter and viscous dissipation parameter on the flow and heat transfer characteristics of an exponentially stretching sheet. The momentum, energy and concentration equations are coupled and nonlinear. By using suitable similarity variables, these equations are converted into coupled ordinary differential equations and are solved analytically and numerically by using the Runge-Kutta-Fehlberg and Newton-Raphson schemes.

MATHEMATICAL FORMULATION

Consider the two-dimensional magnetohydrodynamic flow of a Newtonian fluid over a stretching sheet. The origin of the system is located at the slit from which the sheet is drawn. The x-axis is taken along the continuous stretching surface and points in the direction of motion. The y-axis is perpendicular to the plate. The sheet velocity is assumed to vary as an exponential function of the distance x from the slit. The temperature and concentration far away from the fluid are assumed to be T∞T_∞ and C_∞ respectively as shown in Figure 1. The sheet-ambient temperature and concentration differences are also assumed to be exponential functions of the distance x from the slit. A variable magnetic field of strength B(x) is applied normally to the sheet. Under the usual boundary layer approximation, subject to radiation and viscous dissipation effects, the equations governing the momentum, heat and mass transports can be written as

$$\frac{\partial u}{\partial x} + \frac{\partial v}{\partial y} = 0, \tag{1}$$

$$u\frac{\partial u}{\partial x} + v\frac{\partial u}{\partial y} = \nu\frac{\partial^2 u}{\partial y^2} - \frac{\sigma B^2}{\rho}u, \tag{2}$$

$$u\frac{\partial T}{\partial x} + v\frac{\partial T}{\partial y} = \alpha\frac{\partial^2 T}{\partial y^2} - \frac{1}{\rho C_p}\frac{\partial q_r}{\partial y} + \frac{\sigma B^2}{\rho C_p}u^2 + \frac{\nu}{C_p}\left(\frac{\partial u}{\partial y}\right)^2, \tag{3}$$

$$u\frac{\partial C}{\partial x} + v\frac{\partial C}{\partial y} = D\frac{\partial^2 C}{\partial y^2} \tag{4}$$

where u, v are the velocity components in the x, y directions respectively, ν is the kinematic viscosity, ρ is the density, σ is the electrical conductivity of the fluid, T is the temperature, C is the concentration, $\alpha = k/\rho C_p$ is the thermal diffusivity, k is the thermal conductivity, C_p is the specific heat at constant pressure, qrq_r is the radiative heat flux, and D is the species diffusivity.

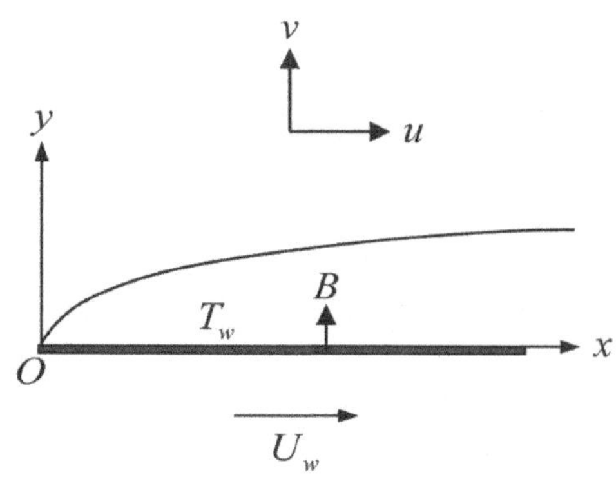

Figure 1: Schematics of the problem.

The boundary conditions for Equations (1)-(4) have the form

$$u = U_w = U_0 e^{\frac{x}{L}}, \qquad v = 0, \qquad T = T_w = T_\infty + T_0 e^{\frac{2x}{L}},$$

$$C = C_w = C_\infty + C_0 e^{\frac{2x}{L}} \quad \text{at } y = 0,$$

$$u \to 0, \qquad T \to T_\infty, \qquad C \to C_\infty \quad \text{as } y \to \infty. \tag{5}$$

Here the subscripts w, ∞ refer to the surface and ambient conditions respectively, T_0, C_0 are positive constants, U_0 is the characteristic velocity, and L is the characteristic length.

To facilitate a similarity solution, the magnetic field $B(x)$ is assumed to be of the form

$$B(x) = B_0 e^{\frac{x}{2L}}, \tag{6}$$

where B_0 is a constant. It is also assumed that the fluid is weakly electrically conducting so that the induced magnetic field is negligible. Following Rosseland's approximation, the radiative heat flux q_r is modeled as

$$q_r = -\frac{4\sigma^*}{3k^*}\frac{\partial T^4}{\partial y}, \tag{7}$$

where σ^* is the Stefan-Boltzman constant, k^* is the mean absorption coefficient. Assuming that the temperature differences within the flow are sufficiently small such that T^4 may be expressed as a linear function of temperature $T^4 \equiv 4T_\infty^3 T - 3T_\infty^4$, we have

$$\frac{\partial q_r}{\partial y} = -\frac{16\sigma^* T_\infty^3}{3k^*}\frac{\partial^2 T}{\partial y^2}. \tag{8}$$

Continuity Equation (1) is satisfied by introducing a stream function ψ such that

$$u = \frac{\partial \psi}{\partial y}, \qquad v = -\frac{\partial \psi}{\partial x}. \tag{9}$$

The following similarity variables are used:

$$\left.\begin{aligned} &u = U_0 e^{\frac{x}{L}} f_\eta(\eta), \qquad v = -\left(\frac{\nu U_0}{2L}\right)^{\frac{1}{2}} e^{\frac{x}{2L}}\left[f(\eta) + \eta f_\eta(\eta)\right], \\ &T = T_\infty + T_0 e^{\frac{2x}{L}} \theta(\eta), \qquad C = C_\infty + C_0 e^{\frac{2x}{L}} \phi(\eta), \\ &\eta = \left(\frac{U_0}{2\nu L}\right)^{\frac{1}{2}} y e^{\frac{x}{2L}}, \end{aligned}\right\} \tag{10}$$

where η is the similarity variable, f(η) is the dimensionless stream function, θ(η) is the dimensionless temperature, and ϕ(η) is the dimensionless concentration.

On using Equations (6), (8) and (10), Equations (2)-(5) transform into the following two-point boundary value problem:

$$f_{\eta\eta\eta} + f f_{\eta\eta} - 2f_\eta^2 - M f_\eta = 0, \tag{11}$$

$$\left(1 + \frac{4}{3}K\right)\theta_{\eta\eta} + Pr(f\theta_\eta - 4f_\eta\theta) + GbPr\left(Mf_\eta^2 + f_{\eta\eta}^2\right) = 0, \tag{12}$$

$$\phi_{\eta\eta} + Sc(f\phi_\eta - 4f_\eta\phi) = 0, \tag{13}$$

$$f(0) = 0, \qquad f_\eta(0) = 1, \qquad f_\eta(\infty) \to 0, \tag{14}$$

$$\theta(0) = 1, \qquad \theta(\infty) \to 0, \tag{15}$$

$$\phi(0) = 1, \qquad \phi(\infty) \to 0. \tag{16}$$

The non-dimensional constants appearing in Equations (11)-(13) are the magnetic parameterM, the radiation parameter K, the Prandtl number Pr, the Gebhart number Gb, and the Schmidt number Sc respectively defined as

$$M = \frac{2\sigma B_0^2 L}{\rho U_0}, \qquad K = \frac{4\sigma^* T_\infty^3}{k^* k}, \qquad Pr = \frac{\rho \nu C_p}{k},$$

$$Gb = \frac{U_0^2}{C_p T_0}, \qquad Sc = \frac{\nu}{D}.$$

SKIN FRICTION, HEAT AND MASS TRANSFER COEFFICIENTS

The parameters of engineering interest in heat and mass transport problems are the skin friction coefficient C_f, the local Nusselt number Nu_x, and the local Sherwood number Sh_x. These parameters respectively characterize the surface drag, wall heat and mass transfer rates.

The shearing stress at the surface of the wall τ_w is given by

$$\tau_w = -\mu\left[\frac{\partial u}{\partial y}\right]_{y=0} = -\frac{\mu U_0}{L}\sqrt{\frac{Re}{2}}e^{\frac{3x}{2L}}f''(0), \tag{17}$$

where μ is the coefficient of viscosity and $Re = \frac{LU_0}{\nu}$ is the Reynolds number. The skin friction coefficient is defined as

$$C_f = \frac{2\tau_w}{\rho u_w^2}, \tag{18}$$

and using Equation (17) in Equation (18), we obtain

$$\frac{C_f\sqrt{Re_x/2}}{\sqrt{x/L}} = -f''(0). \tag{19}$$

The heat transfer rate at the surface flux at the wall is given by

$$q_w = -k\left[\frac{\partial T}{\partial y}\right]_{y=0} = \frac{-k(T_w - T_\infty)}{L}\sqrt{\frac{Re}{2}}e^{\frac{x}{2L}}\theta'(0), \tag{20}$$

where k is thermal conductivity of the fluid. The Nusselt number is defined as

$$Nu_x = \frac{x}{k}\frac{q_w}{T_w - T_\infty}. \tag{21}$$

Using Equation (20) in Equation (21), the dimensionless wall heat transfer rate is obtained as follows:

$$\frac{Nu_x}{\sqrt{x/L}\sqrt{Re_x/2}} = -\theta'(0). \tag{22}$$

The mass flux at the surface of the wall is given by

$$J_w = -D\left[\frac{\partial C}{\partial y}\right]_{y=0} = \frac{-D(C_w - C_\infty)}{L}\sqrt{\frac{Re}{2}}e^{\frac{x}{2L}}\phi'(0), \tag{23}$$

and the Sherwood is defined as

$$Sh_x = \frac{x}{D}\frac{J_w}{C_w - C_\infty}. \tag{24}$$

Using (23) in (24), the dimensionless wall mass transfer rate is obtained as

$$\frac{Sh_x}{\sqrt{x/L}\sqrt{Re_x/2}} = -\phi'(0). \tag{25}$$

In Equations (19), (22) and (25), RexRe_x represents the local Reynolds number and it is defined as $Re_x = \frac{xU_w}{\nu}$.

ANALYTICAL SOLUTION

Solution of Momentum Equation

The momentum boundary layer equation is partially decoupled from the energy and species equations. Integrating Equation (11) with η once over to the interval [0,η], we obtain

$$f_{\eta\eta} + f f_\eta = -s + \int_0^\eta \left[3f_\eta^2 + Mf_\eta\right] d\eta, \tag{26}$$

where $s = -f_{\eta\eta}(0)$. Letting $\eta \to \infty$; we obtain

$$s = \int_0^\infty \left[3f_\eta^2 + Mf_\eta\right] d\eta. \tag{27}$$

Integrating Equation (26) once again, we obtain

$$f_\eta + \frac{1}{2}f_\eta^2 = 1 - s\eta + \int_0^\eta \left[\int_0^{\eta_1} \left[3f_{\eta_2}^2 + Mf_{\eta_2}\right] d\eta_2\right] d\eta_1. \tag{28}$$

The solution procedure of Equation (28) can be reduced to the sequential solutions of the Riccati type equation of the form

$$f_\eta^{(n)} + \frac{1}{2}f^{(n)2} = RHS\left[f_\eta^{(n-1)}\right] \tag{29}$$

This iteration algorithm has to be solved by substituting suitable zero-order approximations $f_\eta^{(0)}(\eta)$ for $f_\eta(\eta)$ into the right-hand side of Equation (28). We assume a zero-order approximation as

$$f_\eta^{(0)}(\eta) = e^{-s_0\eta}, \tag{30}$$

which satisfies the condition at infinity. Integrating Equation (30) with respect to η and using the condition $f_\eta^{(0)}(0) = 0$ we get

$$f_\eta^{(0)}(\eta) = \frac{1 - e^{-s_0\eta}}{s_0}. \tag{31}$$

Using the above solution in Equation (27), the approximate value of s can be obtained as

$$s_0 = \sqrt{\frac{3}{2} + M}, \qquad f_{\eta\eta}^{(0)}(0) = -s_0. \tag{32}$$

Now substituting all the derivatives of zero-order approximation $f_\eta^{(0)}(\eta)$ into the right-hand side of Equation (28), we obtain the equation for first-order iteration $f_\eta^{(1)}$ as follows:

$$f_\eta^{(1)} + \frac{1}{2}f^{(1)2} = 1 + \frac{3}{4s_0^2}\left[e^{-2s_0\eta} - 1\right] + \frac{M}{s_0^2}\left[e^{-s_0\eta} - 1\right] \tag{33}$$

Further, we assume that the first-order iterate of f satisfies the boundary conditions on f as given in (14). The above non-linear Riccati type equation can be solved in terms of a confluent hypergeometric Whittaker function as discussed by Khan [19]. However, we restrict ourselves to the zero-order solution, and similarly, to heat and mass transport equations.

4.2 Solution of heat transfer equation

Using the zero-order approximations of f and f_η and further introducing a new variable

$$\xi = -\frac{Pr}{s_0^2}e^{-s_0\eta}, \tag{34}$$

Equation (12) and the thermal boundary conditions (15) take the form

$$\left(1 + \frac{4}{3}K\right)\xi\theta_{\xi\xi} + \left(1 + \frac{4}{3}K - Pr^* - \xi\right)\theta_\xi + 4\theta = -\frac{Gb}{Pr}s_0^2\left(M + s_0^2\right)\xi, \tag{35}$$

$$\theta\left(-Pr^*\right) = 1, \qquad \theta\left(0^-\right) \to 0, \tag{36}$$

where $Pr^* = Pr/s_0^2$ is the modified Prandtl number. The solution of Equation (35) is assumed in the form of

$\theta(\xi) = \theta_c(\xi) + \theta_p(\xi),$

where $\theta_c(\xi)$ is the complementary solution and $\theta_p(\xi)$ is the particular solution. The complementary solution of Equation (35) is obtained in terms of confluent hypergeometric function in the following form:

$$\theta_c(\xi) = C_0\xi^{\alpha}M\left[\alpha-4,\alpha+1,\frac{-\xi}{1+\frac{4}{3}K}\right], \tag{37}$$

where

$$M[a,b,z] = \sum_{r=0}^{\infty}\frac{a(a+1)\cdots(a+r-1)}{b(b+1)\cdots(b+r-1)}\frac{z}{r!}$$

is Kummer's function (see Abramowitz and Stegun [39]) and

$$\alpha = \frac{Pr^*}{1+\frac{4}{3}K}$$

The particular solution is obtained as

$$\theta_p(\xi) = a_0\xi^2 + a_1\xi^3 + a_2\xi^4, \tag{38}$$

where

$$a_0 = -\left(\frac{Gb}{Pr^*}\right)\frac{M+s_0^2}{4(1+\frac{4}{3}K)-2Pr^*},$$

$$a_1 = \frac{-2a_0}{9(1+\frac{4}{3}K)-3Pr^*},$$

$$a_2 = \frac{-a_1}{16(1+\frac{4}{3}K)-4Pr^*}.$$

Now, the complete solution can be written as

$$\theta(\xi) = \theta_c(\xi) + \theta_p(\xi). \tag{39}$$

Making use of the boundary conditions (36) and rewriting the solution in terms of the variableη, we get

$$\theta(\eta) = C_1\frac{e^{-s_0\eta\alpha}M[\alpha-4,\alpha+1,-\alpha e^{-s_0\eta}]}{M[\alpha-4,\alpha+1,-\alpha]} + a_0Pr^{*2}e^{-2s_0\eta} - a_1Pr^{*3}e^{-3s_0\eta} + a_2Pr^{*4}e^{-4s_0\eta}, \tag{40}$$

where

$$C_1 = 1 - a_0Pr^{*2} + a_1Pr^{*3} - a_2Pr^{*4}.$$

Solution of mass transfer equation

Using the zero-order approximation of f and f_η and further introducing a new variable

$$\zeta = -\frac{Sc}{s_0^2}e^{-s_0\eta}, \tag{41}$$

Equation (13) and the thermal boundary conditions in (16) take the form

$$\zeta\phi_{\zeta\zeta} + (1-Sc^*-\zeta)\phi_\zeta + 4\phi = 0, \tag{42}$$

$$\phi(-Sc^*) = 1, \qquad \phi(0^-) \to 0, \tag{43}$$

where $Sc^* = Sc/s_0^2$ is the modified Schmidt number. Following the solution procedure discussed in the case of the energy equation, the solution of Equation (42) is obtained in terms of confluent hypergeometric function as

$$\phi(\eta) = \frac{e^{-s_0Sc^*\eta}M[Sc^*-4,Sc^*+1,-Sc^*e^{-s_0\eta}]}{M[Sc^*-4,Sc^*+1,-Sc^*]}. \tag{44}$$

SOLUTION PROCEDURE

The set of non-linear differential Equations (11)-(13) subject to the boundary conditions (14)-(16) were solved numerically using an efficient Runge-Kutta-Fehlberg method with a shooting technique, which is described in Pal and Shivakumara [40]. The most important step in this method is to choose an appropriate finite value of $\eta\to\infty$. In order to determine $\eta\to\infty$ for the boundary value problem described by Equations (11)-(13), we start with initial guess values for a particular set of physical parameters to obtain $f''(0)$, $\theta'(0)$ and $\phi'(0)$. The solution procedure is repeated with another large value of $\eta\to\infty\eta\to\infty$ until two successive values of $f''(0)$, $\theta'(0)$ and $\phi'(0)$ differ only by a specified significant digit. The value of η may change for a different set of physical parameters. Once the appropriate value of η is determined, the coupled boundary value problem given by Equations (11)-(13) is solved numerically using the method of superposition. In this method, third-order non-linear Equation (11), second-order Equations (12) and (13) have been reduced to five ordinary differential equations as follows:

$$f_1' = f_2, \qquad f_2' = f_3,$$
$$f_3' = 2f_2^2 + Mf_2 - f_1f_3,$$
$$f_4' = f_5,$$
$$f_5' = -\frac{Pr}{(1+\frac{4}{3}K)}\left[f_1f_5 - 4f_2f_4 + Gb(Mf_2^2 + f_3^2)\right],$$
$$f_6' = f_7,$$
$$f_7' = Sc(4f_2f_6 - f_1f_7), \tag{45}$$

where

$$f_1 = f, \quad f_2 = f', \quad f_3 = f'', \quad f_4 = \theta, \quad f_5 = \theta', \quad f_6 = \phi, \quad f_7 = \phi', \tag{46}$$

and a prime denotes differentiation with respect to η. The boundary conditions now become

$$f_1 = 0, \quad f_2 = 1, \quad f_3 = s_1, \quad f_4 = 1,$$
$$f_5 = s_2, \quad f_6 = 1, \quad f_7 = s_3 \quad \text{at } \eta \to 0, \tag{47}$$

$$f_2 = 0, \quad f_4 = 0, \quad f_6 = 0 \quad \text{as } \eta \to \infty, \tag{48}$$

where s1s_1, s2s_2 and s3s_3 are determined such that $f_2(\infty)=0$, $f_4(\infty)=0$ and $f_6(\infty)=0$. Thus, to solve this system, we require six initial conditions. However, since we have only three initial conditions for f and two initial conditions for θ and ϕ, the conditions f″(0), θ′(0), ϕ′(0) are to be determined by the shooting method using the initial guess values s1s_1, s2s_2 and s3 s_3 until the conditions $f_2(\infty)=0$, $f_3(\infty)=0$ and $f_5(\infty)=0$ are satisfied. In this paper, we employed the shooting technique with the Runge-Kutta-Fehlberg scheme to determine two more unknowns in order to convert the boundary value problem to an initial value problem. Once all the six initial conditions were determined, the resulting differential equations were integrated using an initial value solver. For this purpose, the fifth-order Runge-Kutta-Fehlberg integration scheme was used.

RESULTS AND DISCUSSION

Analytical and numerical solutions were obtained for the effects of radiation and viscous dissipation for the MHD flow over an exponentially stretching sheet. Similarity transformations were used to transform the governing partial differential equations of flow, heat and mass transfer into a system of non-linear ordinary differential equations. The zero-order approximate solution for the dimensionless stream function f has been obtained analytically. Solutions of the energy and species equations were obtained in terms of confluent hypergeometric functions. The accuracy of the method was established by comparing the analytical solution with the numerical solution obtained by a shooting method together with Runge-Kutta-Fehlberg and Newton-Raphson schemes. The skin friction, heat and mass transfer coefficients are tabulated in Tables 1-3. The effects of magnetic, radiation and viscous dissipation parameters on the velocity f′(η), temperature θ(η) and concentration ϕ(η) profiles are shown in Figures 2-6.

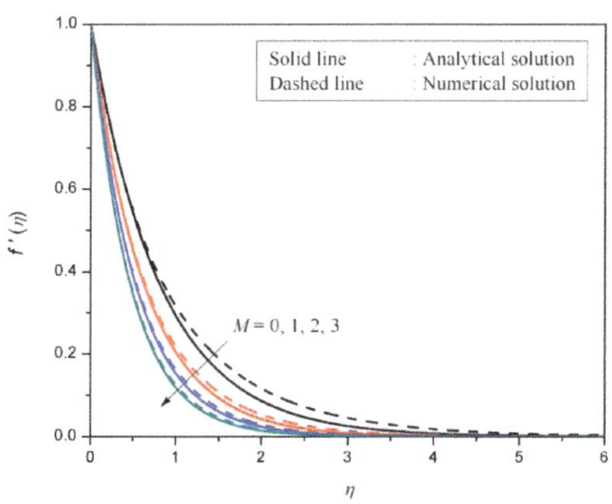

Figure 2: Effect of the magnetic parameter (M) on velocity profile forPr=7$Pr = 7$,Gb=0.2$Gb = 0.2$,K=0.5$K = 0.5$,Sc=1$Sc = 1$.

Table 1: A comparison of−f″(0)obtained by the analytical method with the shooting technique for different values ofM

M	−f″(0)	
	Analytical	**Numerical**
0	1.22474	1.281809
1	1.58114	1.629178
2	1.87083	1.912620
3	2.12132	2.158736
5	2.54951	2.581130
10	3.39116	3.41529

Table 1 provides values of the skin friction coefficient for different values of the magnetic parameter M. Increasing values of M result in considerable opposition to the flow in the form of a Lorenz drag which enhances the values of the skin friction coefficient. Table 2 highlights the effect of the magnetic field, radiation and dissipation on the dimensionless wall temperature gradient. It is evident that all the three parameters reduce the values of the wall temperature gradient. Table 3 shows that the increase in Schmidt numbers leads to the increase in the dimensionless wall concentration gradient, while the opposite trend is observed in the case of the magnetic parameter. The results confirm a good agreement between analytical and numerical results.

Table 2: A comparison of−θ′(0)obtained by the analytical method with the shooting technique for different values ofM,GbandKfor fixed values ofP_r=7

M	*K*	*Gb*	−θ′(0)	
			Analytical	**Numerical**
0	0.5	0.2	3.82684	3.822508
1			3.48576	3.483155
2			3.19181	3.191131
3			2.92781	2.928577
1	0	0.2	4.56379	4.556219
	0.5		3.48576	3.483155
	1		2.90556	2.905805
	2		2.25649	2.260503
	3		1.88314	1.889815
1	0.5	0	3.94905	3.946604
		0.1	3.71740	3.714879
		0.2	3.48576	3.483155
		0.5	2.79082	2.787982
		1	1.63260	1.629360

Table 3: A comparison of−ϕ′(0)$-\phi'(0)$obtained by the analytical method with the shooting technique for different values ofM,Sc

M	*Sc*	−ϕ′(0)	
		Analytical	**Numerical**
0	1	1.79791	1.805684
1		1.69115	1.699309
2		1.60312	1.611410
3		1.52781	1.535984
1	1	1.69115	1.699309
	2	2.58672	2.589044
	5	4.34813	4.344825
	10	6.32456	6.318568

The skin friction coefficients are shown in Table 4 for different values of the magnetic parameter in the absence of the physical parameters (i.e., $P_r=S_c=K=G_b=0$). We observe that skin friction coefficient increases with an increase in the magnetic parameter. It is interesting to note that the value of the wall skin-friction coefficient in the non-magnetic (M=0 $M=0$) and magnetic (M=1) cases are in good agreement with the results presented by Reddy and Reddy [33].

Table 4: A comparison of $-f''(0)$for different values ofMfor fixed values of $Pr = Sc = K = Gb = 0$

M	−f″(0)	
	Reddy and Reddy [33]	**Present**
0	1.28213	1.28181
1	1.62918	1.62918
2	-	1.91262
3	-	2.15874
4	-	2.37937

Figure 2 shows the variation of the velocity profile against the magnetic parameter. We notice that the effect of the magnetic parameter is to reduce the velocity of the fluid in the boundary layer region. This is due to an increase in the Lorentz force, similar to Darcy's drag observed in the case of flow through a porous medium. This adverse force is responsible for slowing down the motion of the fluid in the boundary layer region. These results are similar to the results obtained by Reddy and Reddy [33].

The variation of the temperature distribution with the magnetic parameter is shown in Figure 3. The thermal boundary layer thickness increases with increasing values of the magnetic parameter. The opposing force introduced in the form of the Lorentz drag contributes in increasing the frictional heating between the fluid layers, and hence energy is released in the form of heat. This results in thickening of the thermal boundary layer.

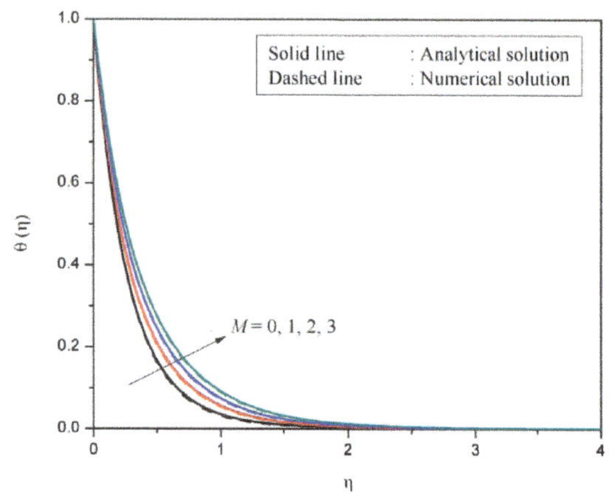

Figure 3: Effect of Mon temperature profile for $P_r=7,G_b=0.2,K=0.5,S_c=1$.

The effect of the magnetic parameter on the concentration profile is shown in Figure 4. It is observed that increases in the values in M result in thickening of the species boundary layer.

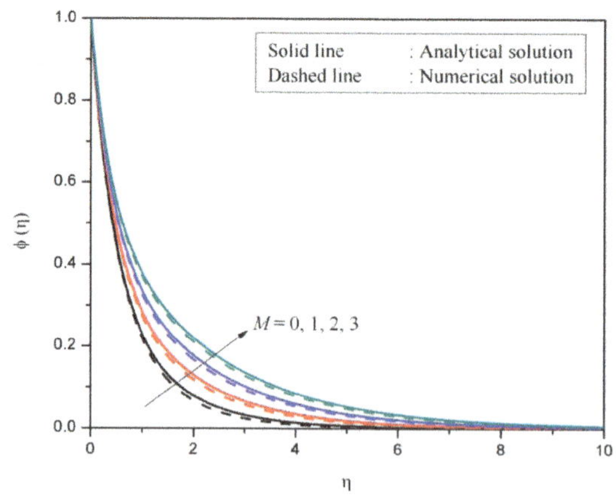

Figure 4: Effect ofMon concentration profile for$P_r=7,G_b=0.2,K=0.5,S_c=1$.

influence of the thermal radiation parameter K on temperature is shown in Figure 5. It is clear that thermal radiation enhances the temperature in the boundary layer region. Thus radiation should be kept at its minimum in order to facilitate better cooling environment. The radiation parameter K defines the relative contribution of conduction heat transfer to thermal radiation transfer.

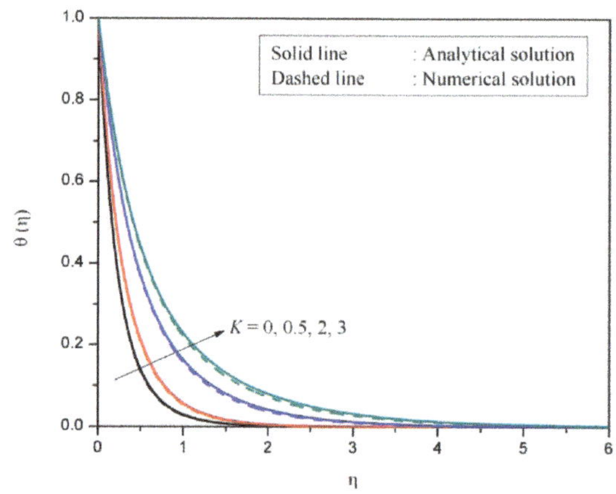

Figure 5: Effect of the radiation parameter on temperature profile for$G_b=0.2,M=1,P_r=7,S_c=1$.

The effect of the Gebhart number Gb on the heat transfer is shown in Figure 6. It is clear that the temperature in the boundary layer region increases with an increase in the viscous dissipation parameter. We also note that since the energy equation is partially decoupled from the momentum and species conservation equations, the parameters affecting the energy equation, namely, the Prandtl number, the radiation parameter and the Gebhart number, do not alter velocity and concentration profiles. We also observe a good agreement between the analytical and numerical solutions through Figures 2-6.

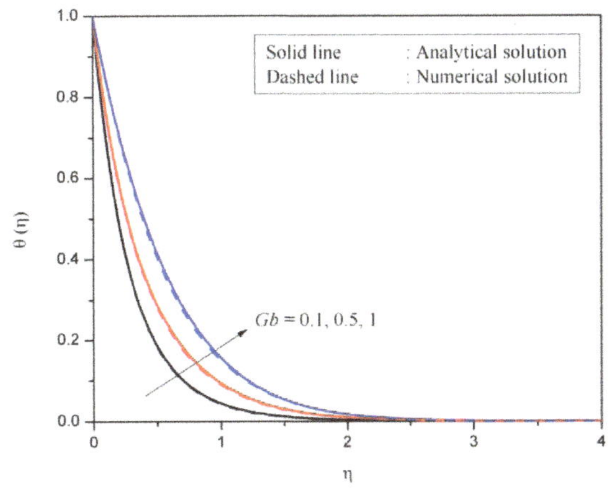

Figure 6: Effect of viscous dissipation on temperature profile for P_r=7,M=1,K=0.5,S_c=1.

CONCLUSIONS

The problem of hydromagnetic Newtonian liquid flow due to an exponentially stretching sheet in the presence of radiation and viscous dissipation effects has been analyzed. Exact solutions were found in terms of hypergeometric functions, and a comparison of analytical and numerical results was shown. We found that the effect of the magnetic parameter is to reduce the velocity of the fluid in the boundary layer region. It was also observed that the increase in values of Mresults in thickening of the species boundary layer. The combined and individual effects of the magnetic parameter M, the radiation parameter K, and the viscous dissipation parameter Gbare to increase the heat transfer rates. Under some limiting conditions when the parameters Pr,Sc, K, Gb are zero, the current results agree well with available results in the literature.

ACKNOWLEDGEMENTS

The authors are grateful to the National Research Foundation (NRF) and the University of KwaZulu-Natal for financial support.

AUTHORS' CONTRIBUTIONS

The problem was conceived in discussions by the authors. PKK and MN carried out the analytical and numerical computations, while GM and PS participated in the design of the study and drafted the manuscript.

REFERENCES

1. Crane LJ: Flow past a stretching plate. Z. Angew. Math. Phys. 1970, 21: 645-647. 10.1007/BF01587695
2. Gupta PS, Gupta AS: Heat and mass transfer on a stretching sheet with suction or blowing. Can. J. Chem. Eng. 1977, 55: 744-746. 10.1002/cjce.5450550619
3. Grubka LG, Bobba KM: Heat transfer characteristics of a continuous stretching surface with variable temperature. J. Heat Transf. 1985, 107: 248-250. 10.1115/1.3247387
4. Dutta BK, Gupta AS: Cooling of a stretching sheet in a various flow. Ind. Eng. Chem. Res. 1987, 26: 333-336. 10.1021/ie00062a025
5. Siddappa B, Abel MS: Non-Newtonian flow past a stretching plate. Z. Angew. Math. Phys. 1985, 36: 47-54. 10.1007/BF00949032
6. Chen CK, Char MI: Heat transfer of a continuous stretching surface with suction or blowing. J. Math. Anal. Appl. 1988, 135: 568-580. 10.1016/0022-247X(88)90172-2
7. Laha MK, Gupta PS, Gupta AS: Heat transfer characteristics of the flow of an incompressible viscous fluid over a stretching sheet. Wärme-Stoffübertrag. 1989, 24: 151-153.
8. Chakarabarti A, Gupta AS: Hydromagentic flow and heat transfer over a stretching sheet. Q. Appl. Math. 1979, 37: 73-78.
9. Andersson HI, Bech KH, Dandapat BS: Magnetohydrodynamic flow of a power-law fluid over a stretching sheet. Int. J. Non-Linear Mech. 1992, 27: 929-936. 10.1016/0020-7462(92)90045-9
10. Siddheshwar PG, Mahabaleswar US: Effects of radiation and heat source on MHD flow of a viscoelastic liquid and heat transfer over a stretching sheet. Int. J. Non-Linear Mech. 2005, 40: 807-820. 10.1016/j.ijnonlinmec.2004.04.006
11. Abel MS, Mahesha N: Heat transfer in MHD viscoelastic fluid flow over a stretching sheet with variable thermal conductivity, non-uniform heat source and radiation. Appl. Math. Model. 2008, 32: 1965-1983. 10.1016/j.apm.2007.06.038
12. Abel MS, Siddheshwar PG, Mahesha N: Effects of thermal buoyancy and variable thermal conductivity on the MHD flow and heat transfer in a power-law fluid past a vertical stretching sheet in the presence of a non-uniform heat source.Int. J. Non-Linear Mech. 2009, 44: 1-12. 10.1016/j.ijnonlinmec.2008.08.002
13. Elbashbeshy EMA: Heat transfer over an exponentially stretching continuous surface with suction. Arch. Mech. 2001, 53(6):643-651.
14. Ishak A: MHD boundary layer flow due to an exponentially stretching sheet with radiation effect. Sains Malays. 2011, 40(4):391-395.
15. Magyari E, Keller B: Heat and mass transfer in the boundary layers on an exponentially stretching continuous surface. J. Phys. D, Appl. Phys. 1999, 32: 577-585. 10.1088/0022-3727/32/5/012
16. Sanjayanand E, Khan SK: On heat and mass transfer in a viscoelastic boundary layer flow over an exponentially stretching sheet. Int. J. Therm. Sci. 2006, 45: 819-828. 10.1016/j.ijthermalsci.2005.11.002
17. Partha MK, Murthy PVSN, Rajasekhar GP: Effect of viscous dissipation on the mixed convection heat transfer from an exponentially stretching surface. Heat Mass Transf. 2005, 41: 360-366. 10.1007/s00231-004-0552-2
18. Sajid M, Hayat T: Influence of thermal radiation on the boundary layer flow due to an exponentially stretching

sheet. Int. Commun. Heat Mass Transf. 2008, 35: 347-356. 10.1016/j.icheatmasstransfer.2007.08.006

19. Khan SK: Boundary layer viscoelastic fluid flow over an exponentially stretching sheet. Int. J. Appl. Mech. Eng. 2006, 11(2):321-335.

20. Steenbeck M: Wissen. Veroff. Siemens 1936., 15(2): Article ID 1

21. Tonks L: Theory of magnetic effects in the plasma of an arc. Phys. Rev. 1939, 56: 360-373. 10.1103/PhysRev.56.360

22. Pavlov KB: Magnetohydromagnetic flow of an incompressible viscous fluid caused by deformation of a surface. Magn. Gidrodin. 1974, 4: 146-147.

23. Sarpakaya T: Flow of non-Newtonian fluids in a magnetic field. AIChE J. 1961, 7: 324-328. 10.1002/aic.690070231

24. Andersson HI: MHD flow of a viscoelastic fluid past a stretching surface. Acta Mech. 1992, 95: 227-230. 10.1007/BF01170814

25. Lawrence PS, Rao BN: The non-uniqueness of the MHD flow of a viscoelastic fluid past a stretching sheet. Acta Mech. 1995, 112: 223-228. 10.1007/BF01177490

26. Abel MS, Joshi A, Sonth RM: Heat transfer in MHD viscoelastic fluid flow over a stretching surface. Z. Angew. Math. Mech. 2001, 81: 691-698.

27. Cortell R: Flow and heat transfer of an electrically conducing fluid of second grade over a stretching sheet subject to suction and to a transverse magnetic field. Int. J. Heat Mass Transf. 2006, 49: 1851-1856. 10.1016/j.ijheatmasstransfer.2005.11.013

28. Reddy PBA, Reddy NB, Suneetha S: Radiation effects on MHD flow past an exponentially accelerated isothermal vertical plate with uniform mass diffusion in the presence of heat source. J. Appl. Fluid Mech. 2012, 5(3):119-126.

29. Raptis A: Radiation and viscoelastic flow. Int. Commun. Heat Mass Transf. 1999, 26(6):889-895. 10.1016/S0735-1933(99)00077-9

30. Raptis A, Perdikis C: Viscoelastic flow by the presence of radiation. Z. Angew. Math. Mech. 1998, 78: 277-279. 10.1002/(SICI)1521-4001(199804)78:4<277::AID-ZAMM277>3.0.CO;2-F

31. Bidin B, Nazar R: Numerical solution of the boundary layer flow over an exponentially stretching sheet with thermal radiation. Eur. J. Sci. Res. 2009, 33(4):710-717.

32. Elbashbeshy EMA, Dimian MF: Effect of radiation on the flow and heat transfer over a wedge with variable viscosity. Appl. Math. Comput. 2002, 132: 445-454. 10.1016/S0096-3003(01)00205-3

33. Reddy PBA, Reddy NB: Thermal radiation effects on hydromagnetic flow due to an exponentially stretching sheet. Int. J. Appl. Math. Comput. 2011, 3(4):300-306.

34. Raptis A, Perdikis C, Takhar HS: Effect of thermal radiation on MHD flow. Int. J. Heat Mass Transf. 2004, 153: 645-649.

35. Hayat T, Abbas Z, Sajid M, Asghar S: The influence of thermal radiation on MHD flow of a second grade fluid. Int. J. Heat Mass Transf. 2007, 50: 931-941. 10.1016/j.ijheatmasstransfer.2006.08.014

36. Gebhart B: Effect of viscous dissipation in natural convection. J. Fluid Mech. 1962, 14: 225-232. 10.1017/S0022112062001196

37. Gebhart B, Mollendorf J: Viscous dissipation in external natural convection flows.J. Fluid Mech. 1969, 38: 97-107. 10.1017/S0022112069000061

38. Vajravelu K, Hadjinicolaou A: Heat transfer in a viscous fluid over a stretching sheet with viscous dissipation and internal heat generation. Int. Commun. Heat Mass Transf. 1993, 20: 417-430. 10.1016/0735-1933(93)90026-R

39. Abramowitz M, Stegun LA National Bureau of Standards, AMS 55. Handbook of Mathematical Functions 1972.

40. Pal D, Shivakumara IS: Mixed convection heat transfer from a vertical plate embedded in a sparsely packed porous medium. Int. J. Appl. Mech. Eng. 2006, 11(4):929-939.

CHAPTER 19

INCOMPRESSIBLE NON-NEWTONIAN FLUID FLOWS

Quoc-Hung Nguyen and Ngoc-Diep Nguyen

INTRODUCTION

A non-Newtonian fluid is a fluid whose flow properties differ in many ways from those of Newtonian fluids. Most commonly the viscosity of non-Newtonian fluids is not independent of shear rate or shear rate history. In practice, many fluid materials exhibits non-Newtonian fluid behavior such as: salt solutions, molten, ketchup, custard, toothpaste, starch suspensions, paint, blood, and shampoo etc. In a Newtonian fluid, the relation between the shear stress and the shear rate is linear, passing through the origin, the constant of proportionality being the coefficient of viscosity. In a non-Newtonian fluid, the relation between the shear stress and the shear rate is different, and can even be time-dependent. Therefore a constant coefficient of viscosity cannot be defined. In the previous parts of this book, the mechanics of Newtonian fluid have been mentioned. In this chapter, the common rheological models of non-Newtonian fluids are introduced and several approaches concerned with non-Newtonian fluid flows are considered. In addition, the solution of common non-Newtonian fluid flows in a circular pipe, annular and rectangular duct are presented.

Citation: Quoc-Hung Nguyen and Ngoc-Diep Nguyen (2012). Incompressible Non-Newtonian Fluid Flows, DOI: 10.5772/26091.

CLASSIFICATION OF NON-NEWTONIAN FLUI

As above mentioned, a non-Newtonian fluid is one whose flow curve (shear stress versus shear rate) is nonlinear or does not pass through the origin, i.e. where the apparent viscosity, shear stress divided by shear rate, is not constant at a given temperature and pressure but is dependent on flow conditions such as flow geometry, shear rate, etc. and sometimes even on the kinematic history of the fluid element under consideration. Such materials may be conveniently grouped into three general classes:

1. fluids for which the rate of shear at any point is determined only by the value of the shear stress at that point at that instant; these fluids are variously known as 'time independent' , ' purely viscous' , 'inelastic' or 'generalized Newtonian fluids');
2. more complex fluids for which the relation between shear stress and shear rate depends, in addition, upon the duration of shearing and their kinematic history; they are called 'time-dependent fluids', and finally,
3. substances exhibiting characteristics of both ideal fluids and elastic solids and showing partial elastic recovery, after deformation; these are categorized as 'viscoplastic fluids'.

Among the three groups, the time independent Non-Newtonian fluids are the most popular and easiest to handle in analysis. In this chapter, only this group of Non-Newtonian fluids are considered.

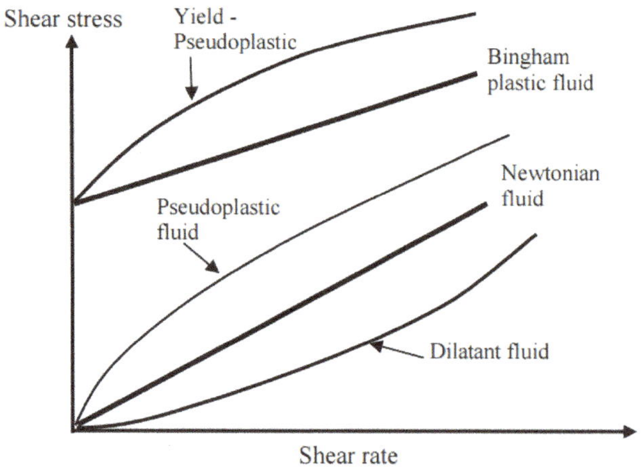

Figure 1: Types of time-independent non-Newtonian fluid.

In simple shear, the flow behaviour of this class of materials may be described by the following constitutive relation,

$$\tau_{yx} = f(\dot{\gamma}_{yx}) \tag{1}$$

This equation implies that the value of $\dot{\gamma}_{yx}$ at any point within the sheared fluid is determined only by the current value of shear stress at that point or vice versa. Depending upon the form of the function in equation (1), these fluids may be further subdivided into three types: shear-thinning or pseudoplastic, shear-thickening or dilatant and viscoplastic.

Shear-Thinning Or Pseudo-Plastic Fluids

The most common type of time-independent non-Newtonian fluid behaviour observed is Pseudo-plasticity or shear-thinning, characterized by an apparent viscosity which decreases with increasing shear rate. Both at very low and at very high shear rates, most shearthinning polymer solutions and melts exhibit Newtonian behaviour, i.e., shear stress–shear rate plots become straight lines and on a linear scale will pass through origin. The resulting values of the apparent viscosity at very low and high shear rates are known as the zero shear viscosity, μ_0, and the infinite shear viscosity μ_∞, respectively. Thus, the apparent viscosity of a shear-thinning fluid decreases from μ_0 to μ_∞ with increasing shear rate. Many mathematical expressions of varying complexity and form have been proposed in the literature to model shear-thinning characteristics; some of these are straightforward attempts at curve fitting, giving empirical relationships for the shear stress (or apparent viscosity)–shear rate curves for example, while others have some theoretical basis in statistical mechanics – as an extension of the application of the kinetic theory to the liquid state or the theory of rate processes, etc. Only a selection of the more widely used viscosity models is given here; more complete descriptions of such models are available in many books (Bird et al ., 1987 ; Carreau et al ., 1997) and in a review paper (Bird, 1976).

i. The power-law model

The relationship between shear stress and shear rate for this type of fluid can be mathematically expressed as follows:

$$\tau_{yx} = K(\dot{\gamma}_{yx})^n \tag{2}$$

So the apparent viscosity for the so-called power-law fluid is thus given by:

$$\mu = K(\dot{\gamma}_{yx})^{n-1} \tag{3}$$

For

n < 1, the fluid exhibits shear-thinning properties

n = 1, the fluid shows Newtonian behaviour

n > 1, the fluid shows shear-thickening behaviour

In these equations, K and n are two empirical curve-fitting parameters and are known as the fluid consistency coefficient and the flow behaviour index respectively. For a shear thinning fluid, the index may have any value between 0 and 1. The smaller the value of n, the greater is the degree of shear-thinning. For a shear-thickening fluid, the index n will be greater than unity. When n=1, equations (3) becomes the constitutive equation of Newtonian fluid. Although the power-law model offers the simplest representation of shear-thinning behaviour, it does have a number of limitations. Generally, it applies over only a limited range of shear rates and therefore the fitted values of K and n will depend on the range of shear rates considered. Furthermore, it does not predict the zero and infinite shear viscosities. Finally, it should be noted that the dimensions of the flow consistency coefficient, K, depend on the numerical value of n and therefore the K values must not be compared when the n values differ. On the other hand, the value of K can be viewed as the value of apparent viscosity at the shear rate of unity and will therefore depend on the time unit (e.g. second, minute or hour) employed. Despite these limitations, this is perhaps the most widely used model in the literature dealing with process engineering applications. Table 1 provides a compilation of the power-law constants (K and n) for a variety of substances.

ii. The Carreau viscosity equation

When there are significant deviations from the power-law model at very high and very low shear rates, it is necessary to use a model which takes account of the limiting values of viscosities μ_0 and μ_∞ . Based on the molecular network considerations, Carreau (1972) put forward the following viscosity model.

$$\frac{\mu - \mu_\infty}{\mu_0 - \mu_\infty} = [1 + (\lambda \dot{\gamma}_{yx})^2]^{(n-1)/2} \tag{4}$$

where n (< 1) and λ are two curve-fitting parameters. This model can describe shear thinning behaviour over wide ranges of shear rates but only at the expense of the added complexity of four parameters. This model predicts Newtonian fluid behaviour $\mu = \mu_0$ when either n = 1 or λ =0 or both.

Table 1: Typical values of power-law constants for a few systems

System	Temperature (K)	n	M (Pa s^n)
Agro- and food-related products			
Ammonium alginate solution (3.37%)	297	0.5	13
Apple butter	-	0.15	200
Apple sauce	300	0.3–0.45	12–22
Apricot puree	300	0.3–0.4	5–20
Banana puree	293–315	0.33–0.5	4–10
Carrot puree	298	0.25	25
Chicken (minced)	296	0.10	900
Chocolate	303	0.5	0.7
Guava puree	296.5	0.5	40
Human blood	300	0.9	0.004
Mango pulp	300–340	0.3	3–10
Marshmallow cream	-	0.4	560
Mayonnaise	298	0.6	5–100
Papaya puree	300	0.5	10
Peach puree	300	0.38	1–5
Peanut butter	-	0.07	500
Pear puree	300	0.4–0.5	1–5
Plum puree	287	0.35	30–80
Tomato concentrate (5.8% solid)	305	0.6	0.22
Tomato ketchup	295	0.24	33
Tomato paste	-	0.5	15
Whipped desert toppings	-	0.12	400
Yoghurt	293	0.5–0.6	25
Polymer melts			
High density polyethylene (HDPE)	453-493	0.6	3.75–6.2 x 10^3
High impact polystyrene	443-483	0.20	3.5–7.5 x 10^4
Polystyrene	463-498	0.25	1.5–4.5 x 10^4
Polypropylene	453-473	0.40	4.5–7 x 10^3
Low density polyethylene (LDPE)	433-473	0.45	4.3–9.4 x 10^3
Nylon	493-508	0.65	1.8–2.6 x 10^3
Polymethylmethyacrylate (PMMA)	493–533	0.25	2.5–9 x 10^4
Polycarbonate	553-593	0.65–0.8	1–8.5 x 10^3
Personal care products			
Nail polish	298	0.86	750
Mascara	298	0.24	200
Toothpaste	298	0.28	120
Sunscreen lotions	298	0.28	75
Ponds cold cream	298	0.45	25
Oil of Olay	298	0.22	25

iii. The Cross viscosity equation

Another four parameter model which has gained wide acceptance is due to Cross (1965) which, in simple shear, is written as:

$$\frac{\mu - \mu_\infty}{\mu_0 - \mu_\infty} = \frac{1}{1 + k(\dot{\gamma}_{yx})^n} \qquad (5)$$

Here $(<1)$and k are two fitting parameters whereas μ_0 and μ_∞ are the limiting values of the apparent viscosity at low and high shear rates, respectively. This model reduces to the Newtonian fluid behaviour as k $\rightarrow$ 0. Similarly, when, $\mu << \mu_0$ and $\mu >> \mu_\infty$, it reduces to the familiar power-law model, equation (3). Though initially Cross (1965) suggested that a constant value of n =2/3 was adequate to approximate the viscosity data for many systems, it is now thought that treating the index, n, as an adjustable parameter offers considerable improvement over the use of the constant value of n (Barnes et al. , 1989).

iv. The Ellis fluid model

When the deviations from the power-law model are significant only at low shear rates, it is more appropriate to use the Ellis model. The three viscosity equations presented so far are examples of the form of equation (1). The three-constant Ellis model is an illustration of the inverse form. In simple shear, the apparent viscosity of an Ellis model fluid is given by:

$$\mu = \frac{\mu_0}{1 + (\tau_{yx} / \tau_{1/2})^{\alpha - 1}} \qquad (6)$$

In this equation, μ_0 is the zero shear viscosity and the remaining two constants α (> 1) and $\tau_{1/2}$ are adjustable parameters. While the index α is a measure of the degree of shear thinning behaviour (the greater the value of α , greater is the extent of shear-thinning), $\tau_{1/2}$ represents the value of shear stress at which the apparent viscosity has dropped to half its zero shear value. Equation (6) predicts Newtonian fluid behaviour in the limit of $\tau_{1/2} \rightarrow$ ¥. This form of equation has advantages in permitting easy calculation of velocity profiles from a known stress distribution, but renders the reverse operation tedious and cumbersome. It can easily be seen that in the intermediate range of shear stress (or shear rate), $(\tau_{yx} / \tau_{1/2})^{\alpha-1} >> 1$, and equation (6) reduces to equation (3) with $n = (1/\alpha)$ and $m = (\mu_0 \tau_{1/2}^{\alpha-1})^{1/\alpha}$.

Viscoplastic Fluid Behaviour

This type of fluid behaviour is characterized by the existence of a yield stress (τ_0) which must be exceeded before the fluid will deform or flow. Conversely, such a material will deform elastically (or flow en masse like a rigid body) when the externally applied stress is smaller than the yield stress. Once the magnitude of the external stress has exceeded the value of the yield stress, the flow curve may be linear or non-linear but will not pass through origin (Figure 1). Hence, in the absence of surface tension effects, such a material will not level out under gravity to form an absolutely flat free surface. One can, however, explain this kind of fluid behaviour by postulating that the substance at rest consists of three-dimensional structures of sufficient rigidity to resist any external stress less than τ_0. For stress levels greater than τ_0, however, the structure breaks down and the substance behaves like a viscous material. In some cases, the build-up and breakdown of structure has been found to be reversible, i.e., the substance may regain its initial value of the yield stress.

Strictly speaking, it is virtually impossible to ascertain whether any real material has a true yield stress or not, but nevertheless the concept of a yield stress has proved to be convenient in practice because some materials closely approximate to this type of flow behaviour, e.g. see Barnes and Walters (1985) , Astarita (1990) , Schurz (1990) and Evans (1992) . Many workers in this field view the yield stress in terms of the transition from a solid-like (high viscosity) to a liquid-like (low viscosity) state which occurs abruptly over an extremely narrow range of shear rates or shear stress (Uhlherr et al ., 2005). It is not uncommon for the two values of viscosity to differ from each other by several orders of magnitude. The answer to the question whether a fluid has a yield stress or not seems to be related to the choice of a time scale of observation. Common examples of viscoplastic fluid behaviour include particulate suspensions, emulsions, foodstuffs, blood and drilling mud, etc. (Barnes, 1999).

Over the years, many empirical expressions have been proposed as a result of straightforward curve-fitting exercises. A model based on sound theory is yet to emerge. Three commonly used models for viscoplastic fluids are: Bingham plastic model, HerschelBulkley model and Casson model.

i. The Bingham plastic model

This is the simplest equation describing the flow behaviour of a fluid with a yield stress and, in steady one-dimensional shear, it is described by

$$\tau_{yx} = \tau_0 + \mu(\dot{\gamma}_{yx}) \text{ for } |\tau_{yx}| > |\tau_0|$$

$$\dot{\gamma}_{yx} = 0 \text{ for } |\tau_{yx}| < |\tau_0| \tag{7}$$

Often, the two model parameters τ_0 and μ are treated as curve-fitting constants irrespective of whether or not the fluid possesses a true yield stress.

ii. The Herschel-Bulkley fluid model

A simple generalization of the Bingham plastic model to embrace the non-linear flow curve (for $\tau_{yx} > \tau_0$) is the three constant Herschel–Bulkley fluid model. In one dimensional steady shearing motion, the model is written as:

$$\tau_{yx} = \tau_0 + K(\dot{\gamma}_{yx})^n \text{ for } |\tau_{yx}| > |\tau_0|$$

$$\dot{\gamma}_{yx} = 0 \text{ for } |\tau_{yx}| < |\tau_0| \tag{8}$$

It is again noted that, the dimensions of K depend upon the value of n. With the use of the third parameter, this model provides a somewhat better fit to some experimental data.

iii. The Casson fluid model

Many foodstuffs and biological materials, especially blood, are well described by this two constant model as:

$$|\tau_{yx}|^{1/2} = |\tau_0|^{1/2} + (\mu / \dot{\gamma}_{yx})^{1/2} \text{ for } |\tau_{yx}| > |\tau_0|$$

$$\dot{\gamma}_{yx} = 0 \text{ for } |\tau_{yx}| < |\tau_0| \tag{9}$$

This model has often been used for describing the steady shear stress–shear rate behaviour of blood, yoghurt, tomato purée, molten chocolate, etc. The flow behaviour of some particulate suspensions also closely approximates to this type of behaviour. The comparative performance of these three as well as several other models for viscoplastic behaviour has been thoroughly evaluated in an extensive review paper by Bird et al . (1983) and a through discussion on the existence, measurement and implications of yield stress has been provided by Barnes (1999).

Shear-thickening or Dilatant Fluid Behaviour

Dilatant fluids are similar to pseudoplastic systems in that they show no yield stress but their apparent viscosity increases with increasing shear rate; thus these fluids are also called shear-thickening. This type of fluid behaviour was originally observed in concentrated suspensions and a possible explanation for their dilatant behaviour is as follows: At rest, the voidage is minimum and the liquid present is sufficient to fill the void space. At low shear rates, the liquid lubricates the motion of each particle past others and the resulting stresses are consequently small. At high shear rates, on the other hand, the material expands or dilates slightly (as also observed in the transport of sand dunes) so that there is no longer sufficient liquid to fill the increased void space and prevent direct solid–solid contacts which result in increased friction and higher shear stresses. This mechanism causes the apparent viscosity to rise rapidly with increasing rate of shear. The term dilatant has also been used for all other fluids which exhibit increasing apparent viscosity with increasing rate of shear. Many of these, such as starch pastes, are not true suspensions and show no dilation on shearing. The above explanation therefore is not applicable but nevertheless such materials are still commonly referred to as dilatant fluids.

Of the time-independent fluids, this sub-class has received very little attention; consequently very few reliable data are available. Until recently, dilatant fluid behaviour was considered to be much less widespread in the chemical and processing industries. However, with the recent growing interest in the handling and processing of systems with high solids loadings, it is no longer so, as is evidenced by the number of recent review articles on this subject (Barnes et al., 1987; Barnes, 1989; Goddard and Bashir, 1990). Typical examples of materials exhibiting dilatant behaviour include concentrated suspensions of china clay, titanium dioxide (Metzner and Whitlock, 1958) and of corn fl our in water (Griskey et al., 1985). The limited information reported so far suggests that the apparent viscosity–shear rate data often result in linear plots on double logarithmic coordinates over a limited shear rate range and the flow behaviour may be represented by the power-law model, with the flow behaviour index, n, greater than unity, i.e.,

$$\mu = K(\dot{\gamma}_{xy})^{n-1} \tag{10}$$

One can readily see that for n > 1, equation (10) predicts increasing viscosity with increasing shear rate. The dilatant behaviour may be observed in moderately concentrated suspensions at high shear rates, and yet, the same suspension may exhibit pseudoplastic behaviour at lower shear rates. This section is concluded by Table 2 providing a list of materials displaying a spectrum of non-Newtonian flow characteristics in diverse applications to reinforce idea yet again of the ubiquitous nature of such flow behaviour.

Table 2: Non-Newtonian characteristics of some common materials

Practical fluid	Characteristics	Consequence of non-Newtonian behaviour
Toothpaste	Bingham Plastic	Stays on brush and behaves more liquid-like while brushing
Drilling muds	Bingham Plastic	Good lubrication properties and ability to convey debris
Non-drip paints	Thixotropic	Thick in the tin, thin on the brush
Wallpaper paste	Pseudoplastic and Viscoelastic	Good spreadability and adhesive properties
Egg white	Visco-elastic	Easy air dispersion (whipping)
Molten polymers	Visco-elastic	Thread-forming properties
' Bouncing Putty '	Visco-elastic	Will flow if stretched slowly, but will bounce (or shatter) if hit sharply
Wet cement aggregates	Dilatant and thixotropic	Permit tamping operations in which small impulses produce almost complete settlement
Printing inks	Pseudoplastic	Spread easily in high speed machines yet do not run excessively at low speeds
Waxy crude oils	Viscoplastic and Thixotropic	Flows readily in a pipe, but difficult to restart the flow

RABINOWITSCH-MOONEY EQUATION

Consider a one-directional flow of fluid through a circular tube with radius R, Figure 2. The volumetric flow rate through an annular element of area perpendicular to the flow and of width δr is given by

$$\delta Q = 2\pi r\delta r.v_x \tag{11}$$

and, consequently, the flow rate through the whole tube is

$$Q = 2\pi\int_0^R rv_x dr \tag{12}$$

Integrating by parts gives

$$Q = 2\pi\left\{\left[\frac{r^2 v_x}{2}\right]_0^{r_i} + \int_0^{r_i}\frac{r^2}{2}\left(-\frac{dv_x}{dr}\right)dr\right\} \tag{13}$$

Provided there is no slip at the tube wall, the first term in equation (13) vanishes. Equation (13) then can be written as

$$Q = \pi\int_0^R r^2(-\dot{\gamma})dr \tag{14}$$

If the fluid is time-independent and homogeneous, the shear stress is a function of shear rate only. The inverse is that the shear rate $\dot{\gamma}'$, is a function of shear stress τ_{rx} only and the variation of τ_{rx} with r is known from the following well-known equation:

$$\frac{\tau_{rx}}{\tau_w} = \frac{r}{R} \tag{15}$$

where $\dot{\gamma}'$ is interpreted as a function of t instead of r

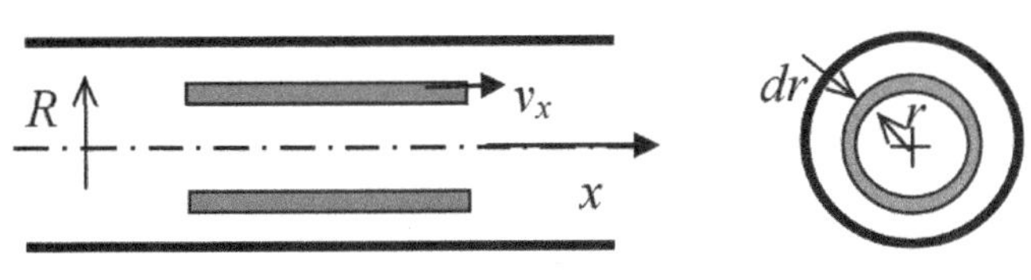

Figure 2: Geometric presentation of MR fluid in o circular tube

Writing equation (16) in terms of the flow characteristic gives

$$\frac{8u}{D} = \frac{4Q}{\pi R^3} = \frac{4}{\tau_w^3}\int_0^{\tau_w}\tau^2(-\dot{\gamma})d\tau$$

where u is the average velocity of the fluid flow and D is the diameter of the tube. For flow in a pipe or tube the shear rate is negative so the integral in equation (17) is positive. For a given relationship between t and $\dot{\gamma}$, the value of the integral depends only on the value of τ_w. Thus, for a non-Newtonian fluid, as well as for a Newtonian fluid, the flow characteristic 8u/D is a unique function of the wall shear stress τ_w.

The shear rate $\dot{\gamma}$ can be extracted from equation (17) by differentiating with respect to t . Moreover, if a definite integral is differentiated w.r.t. the upper limit (τ_w), the result is the integrand evaluated at the upper limit. It is convenient first to multiply equation (17) by 3 w t throughout, then differentiating w.r.t. τ_w gives

$$3\tau_w^2 \frac{8u}{D} + \tau_w^3 \frac{d(8u/D)}{d\tau_w} = 4\tau_w^2(-\dot{\gamma})_w \quad (18)$$

Rearranging equation (18) gives the wall shear rate $\dot{\gamma}_w$ as

$$-\dot{\gamma}_w = \frac{8u}{D}\left[\frac{3}{4} + \frac{1}{4}\frac{\tau_w}{(8u/D)}\frac{d(8u/D)}{d\tau_w}\right] \quad (19)$$

Making use of the relationship dx/x = dlnx, equation (19) can be written as

$$-\dot{\gamma}_w = \frac{8u}{D}\left[\frac{3}{4} + \frac{1}{4}\frac{d\ln(8u/D)}{d\ln\tau_w}\right] \quad (20)$$

As the wall shear rate $\dot{\gamma}_{wN}$ & for a Newtonian fluid in laminar flow is equal to (-8u/D), equation (20) can be expressed as

$$\dot{\gamma}_w = \dot{\gamma}_{wN}\left[\frac{3}{4} + \frac{1}{4}\frac{d\ln(8u/D)}{d\ln\tau_w}\right]$$

Equations (20) and (21) are forms of the Rabinowitsch-Mooney equation. It shows that the wall shear rate for a non-Newtonian fluid can be calculated from the value for a Newtonian fluid having the same flow rate in the same pipe, the correction factor being the quantity in the square brackets. The derivative can be estimated by plotting ln(8u/D) against ln τ_w and measuring the gradient. Alternatively the gradient may be calculated from the (finite) differences between values of ln(8du/D) and ln τ_w . Thus the flow curve τ_w against $\dot{\gamma}_w$ can be determined. The measurements required and the calculation procedure are as follows.

1. Measure Q at various values of $\Delta P_f / L$, preferably eliminating end effects.
2. Calculate τ_ω from the pressure drop measurements and the corresponding values of the flow characteristic $(8du/D = 4Q/\pi R^3)$ from the flow rate measurements.
3. Plot $\ln(8du/D)$ against ln τ_w and measure the gradient at various points on the curve. Alternatively, calculate the gradient from the differences between the successive values of these quantities.
4. Calculate the true wall shear rate from equation (20) with the derivative determined in step 3. In general, the plot of $\ln(8du/D)$ against ln τ_w will not be a straight line and the gradient must be evaluated at the appropriate points on the curve.

Example 1: The flow rate-pressure drop measurements shown in Table 3 were made in a horizontal tube having an internal diameter D = 6 mm, the pressure drop being measured between two tapings 2.0m apart. The density of the fluid, r, was 870 kg/m^3. Determine the wall shear stress-flow characteristic curve and the shear stress-true shear rate curve for this material.

Table 3: The results are shown in Table 4

Pressure drop (bar)	Mass flow rate x 10^3 (kg/s)
0.384	0.0864
0.519	0.463
0.716	1.37
0.965	2.76
1.16	4.13
1.29	5.20
1.46	6.78
1.60	8.15

Table 4

τ_ω (Pa)	$8du/D$	gradient n'	$(3n'+1)/4n'$	$-\dot{\gamma}_\omega$ (s^{-1})
28.8	4.68	0.157	2.34	11.0
38.9	25.1	0.232	1.83	45.9
53.7	74.3	0.375	1.42	106
72.4	150	0.439	1.32	197
87.0	224	0.475	1.28	286
96.8	282	0.475	1.28	360
110	367	0.475	1.28	469
120	442	0.475	1.28	564

CALCULATION OF FLOW RATE-PRESSURE DROP RELATIONSHIP FOR LAMINAR FLOW USING T –G DATA

Flow rate-pressure drop calculations for laminar non-Newtonian flow in pipes may be made in various ways depending on the type of flow information available. When the flow data are in the form of flow rate and pressure gradient measured in a tubular viscometer or in a pilot scale pipeline, direct scale-up can be done as described in Section 5. When the data are in the form of shear stress-shear rate values (tabular or graphical), the flow rate can be calculated directly using equation (17), where D is the diameter of the pipe to be used and w t is the wall shear stress corresponding to the specified pressure gradient. Whether obtained with a rotational instrument or with a tubular viscometer, the data provide the relationship between t and $\dot{\gamma}$. Numerical evaluation of the integral in equation (17) can be done using selected pairs of values of t and $\dot{\gamma}$ ranging from 0 to τ_w.

If the t - $\dot{\gamma}$, relationship can be accurately represented by a simple algebraic expression, such as the power law, over the required range then this may be used to substitute for $\dot{\gamma}$, in equation (17), allowing the integral to be evaluated analytically. Both these methods are illustrated in the following example

Example 2: Using the viscometric data given in Table 5 calculate the average velocity for the material flowing through a pipe of diameter 37mm when the pressure gradient is 1.1kPa/m.

Table 5.

$\dot{\gamma}(s^{-1})$	τ (*Pa*)	μ_a (*Pa* s)
0.00911	0.0417	4.58
0.0911	0.175	1.95
0.911	0.708	0.777
9.111	2.82	0.310
91.11	11.22	0.123
102.3	12.03	0.118

Calculations

The wall shear stress is given by

$$\tau_w = \frac{D\Delta P}{4L} = \frac{(37 \times 10^{-3}\text{m})(1100 \text{ Pa/m})}{4} = 10.18 \text{ Pa}$$

the flow characteristic

$$\frac{8u}{D} = \frac{4}{\tau_w^3}\int_0^{\tau_w} \tau^2(-\dot{\gamma})d\tau$$

It is necessary to evaluate the integral from t = 0 to t = 10.18Pa. This can be done by calculating $\tau^2\dot{\gamma}$ for each of the values given in the table and plotting $\tau^2\dot{\gamma}$ against t. The area under the curve between t = 0 and t = 10.18Pa can then be measured. An alternative, which will be used here, is to use a numerical method such as Simpson's rule. This requires values at equal intervals of t. Dividing the range of integration into six strips and interpolating the data allows Table 6 to be constructed.

Table 6:

τ (*Pa*)	$\dot{\gamma}$	$\tau^2\dot{\gamma}$
0.00	0.0	0.00 (Centerline)
1.70	3.91	11.24
3.39	12.41	142.8
5.09	24.38	631.0
6.78	39.39	1812
8.48	57.14	4108
10.18	77.43	8016 (pipe wall)

By Simpson's rule

$$\int_0^{10.18} \tau^2\dot{\gamma}d\tau \approx \frac{10.18/6}{3}[0+8016+4(11.24+613+4108)+2(142.8+1812)] = 17490 \text{ Pa}^3\text{s}^{-1}$$

From equation (17)

$$u = \frac{(37 \times 10^{-3}\text{m})(17490 \text{ Pa}^3\text{s}^{-1})}{2(10.18 \text{ Pa})^3} = 0.307 \text{ m/s}$$

The above is the general method but in this case the viscometric data can be well represented by $\tau = 0.749\dot{\gamma}^{0.60}$ *Pa*, thus $\dot{\gamma} = 1.62\tau^{1.667}$ s^{-1} This allows the integral in equation (17) to be evaluated analytically.

$$\int_0^{\tau_w} \tau^2\dot{\gamma}d\tau = 1.62\int_0^{10.18} \tau^{3.667}d\tau = 17510 \; Pa^3s^{-1}$$

This agrees with the value found by numerical integration and would give the same value for u. Note that the values of the apparent viscosity μ_0 were not used; they were provided to show that the fluid is strongly shear thinning. If the data were available as values of μ_0 at corresponding values of $\dot{\gamma}$, then t should be calculated as their product. The table of values of $\tau^2\dot{\gamma}$ (Table 6) illustrates the fact that flow in the centre makes a small contribution to the total flow: flow in the outer parts of the pipe is most significant. As mentioned previously, the minus sign in equation (17) reflects the fact that the shear rate is negative for flow in a pipe. In the above calculations, the absolute values of $\dot{\gamma}$, and t have been used and the minus sign has therefore been dropped.

WALL SHEAR STRESS-FLOW CHARACTERISTIC CURVES AND SCALE-UP FOR LAMINAR FLOW

When data are available in the form of the flow rate-pressure gradient relationship obtained in a small diameter tube, direct scale-up for flow in larger pipes can be done. It is not necessary to determine the t - $\dot{\gamma}$ curve with the true value of $\dot{\gamma}$ calculated from the

Rabinowitsch-Mooney equation (Equation (20)). Equation (17) shows that the flow characteristic is a unique function of the wall shear stress for a particular fluid:

$$\frac{8u}{D} = \frac{4}{\tau_w^3}\int_0^{\tau_w} \tau^2(-\dot{\gamma})d\tau$$

In the case of a Newtonian fluid, substituting $\dot{\gamma} = -\tau/\mu$ into the above equation and evaluating the integral gives

$$\frac{8u}{D} = \frac{\tau_w}{\mu} \qquad (22)$$

Recall that the wall shear rate for a Newtonian fluid in laminar flow in a tube is equal to -8 /u D . In the case of a non-Newtonian fluid in laminar flow, the flow characteristic is no longer equal to the magnitude of the wall shear rate. However, the flow characteristic is still related uniquely to t_w because the value of the integral, and hence the right hand side of equation (17), is determined by the value of t_w.

If the fluid flows in two pipes having internal diameters D_1 and D_2 with the same value of the wall shear stress in both pipes, then from equation (17) the values of the flow characteristic are equal in both pipes:

$$\frac{8u_1}{D_1} = \frac{8u_2}{D_2} \qquad (23)$$

So the average velocities are related by

$$\frac{u_1}{u_2} = \frac{D_1}{D_2}$$

By substituting for u or by writing the flow characteristic as $-4Q/\pi R^3$, the volumetric flow rates are related by

$$\frac{Q_1}{Q_2} = \left(\frac{D_1}{D_2}\right)^3 \qquad (25)$$

It is important to appreciate that the same value of τ_w requires different values of the pressure gradient in the two pipes. It is convenient to represent the flow behaviour as a graph of τ_w plotted against 8 /u D , as shown in Figure 3. In accordance with the above discussion, all data fit a single line for laminar flow. The graph is steeper for turbulent flow and different lines are found for different pipe diameters. It is noteworthy that the same would be found for Newtonian flow if the data were plotted in this way and the laminar flow line would be a straight line of gradient μ passing through the origin. The plot in Figure 3 is not a true flow curve because the flow characteristic is equal to the magnitude of the wall shear rate only in the case of Newtonian laminar flow.

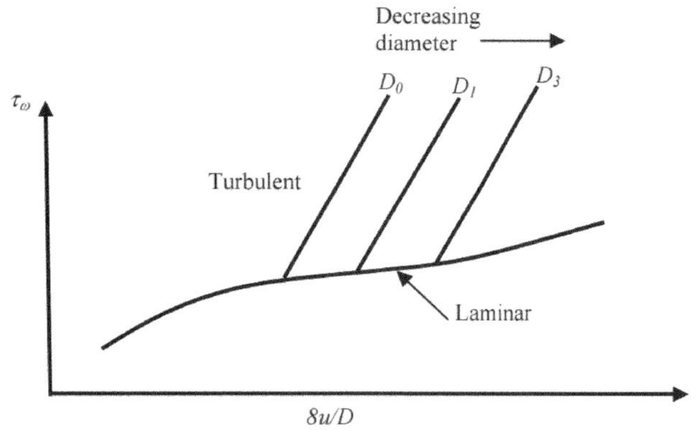

Figure 3: Shear stress at the pipe wall against flow characteristic for a non-Newtonian fluid flowing in a pipe

Given a wall shear stress-flow characteristic curve such as that in Figure 3, the flow ratepressure drop relationship can be found for any diameter of pipe provided the flow remains laminar and is within the range of the graph. For example, if it is required to calculate the pressure drop for flow in a pipe of given diameter at a specified volumetric flow rate, the value of the flow characteristic $(8u/D = 4Q/\pi R^3)$ is calculated and the corresponding value of the wall shear stress w t read from the graph. The pressure gradient, and hence the pressure drop for a given pipe length, can then be calculated.

It is found useful to define two quantifies K' and n' in order to describe the τ_w -flow characteristic curve. If the laminar flow data are plotted on logarithmic axes as in Figure 4, then the gradient of the curve defines the value of n' :

$$n' = \frac{d\ln\tau_w}{d\ln(8u/D)} \tag{26}$$

The equation of the tangent can be written as

In general, both K' and n' have different values at different points along the curve. The values should be found at the point corresponding to the required value of τ_w. In some cases, the curve in Figure 4 will be virtually straight over the range required and a single value may be used for each of K' and n'. Although equation (27) is similar to the equation of a power law fluid, the two must not be confused.

The reason for defining n' in this way can be seen from equation (21) where the inverse of the derivative occurs in the correction factor. Equation (20) can be written in terms of n' as

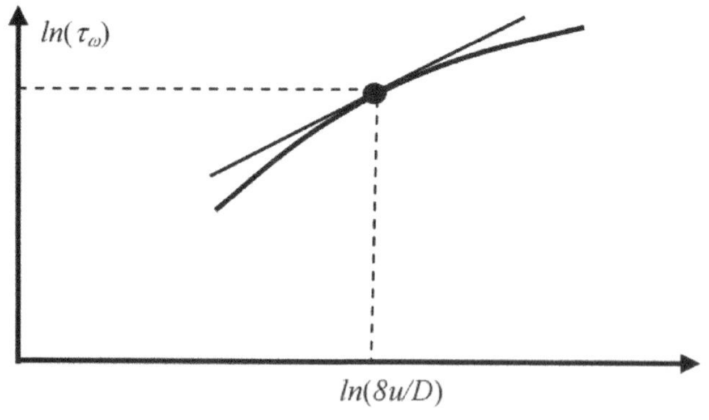

Figure 4: Logarithmic plot of wall shear stress against flow characteristic: the gradient at a point defines n'

$$\dot{\gamma}_\omega = \dot{\gamma}_{\omega N}\left(\frac{3n'+1}{4n'}\right) \tag{28}$$

Equation (28) is helpful in showing how the value of the correction factor in the Rabinowitsch-Mooney equation corresponds to different types of flow behaviour. For a Newtonian fluid, n' = 1 and therefore the correction factor has the value unity. Shear thinning behaviour corresponds to n' < 1 and consequently the correction factor has values greater than unity, showing that the wall shear rate $\dot{\gamma}_w$ is of greater magnitude than the value for Newtonian flow. Similarly, for shear thickening behaviour, $\dot{\gamma}_w$ is of a smaller magnitude than the Newtonian value $\dot{\gamma}_{wN}$. The value correction factor varies from 2.0 for n' = 0.2 to 0.94 for n' = 1.3.

GENERALIZED REYNOLDS NUMBER FOR FLOW IN PIPES

It is recalled that for Newtonian flow in a pipe, the Reynolds number is defined by

$$Re = \frac{\rho u D}{\mu} \tag{29}$$

In the case of non-Newtonian flow, it is necessary to use an appropriate apparent viscosity. Although the apparent viscosity μ_a is defined in the same way as for a Newtonian fluid, it no longer has the same fundamental significance and other, equally valid, definitions of apparent viscosities may be made. In flow in a pipe, where the shear stress varies with radial location, the value of μ_a varies. It is shown that the conditions near the pipe wall that are most important. The value of μ_a evaluated at the wall is given by

$$\mu_a = -\frac{\text{shear stress at wall}}{\text{shear rate at wall}} = \frac{\tau_\omega}{(-dv_x/dr)_\omega} \tag{30}$$

Another definition is based, not on the true shear rate at the wall, but on the flow characteristic. This quantity, which may be called the apparent viscosity for pipe flow, is given by

$$\mu_{ap} = \frac{\text{shear stress at wall}}{\text{flow characteristic}} = \frac{\tau_\omega}{(8u/d_i)} \tag{31}$$

For laminar flow, μ_{ap} has the property that it is the viscosity of a Newtonian fluid having the same flow characteristic as the non-Newtonian fluid when subjected to the same value of wall shear stress. In particular, this corresponds to the same volumetric flow rate for the same pressure gradient in the same pipe. This suggests that μap might be a useful quantity for correlating flow rate-pressure gradient data for non-Newtonian flow in pipes. This is found to be the case and it is on μ_{ap} that a generalized Reynolds number Re' is based

$$\mathrm{Re}' = \frac{\rho u d_i}{\mu_{ap}} \quad (32)$$

Representing the fluid's laminar flow behaviour in terms of K' and n'

$$\tau_\omega = K'\left(\frac{8u}{d_i}\right)^{n'} \quad (33)$$

The pipe flow apparent viscosity, defined by equation 31, is given by

$$\mu_{ap} = \frac{\tau_\omega}{8u/d_i} = K'\left(\frac{8u}{d_i}\right)^{n'-1} \quad (34)$$

Using μ_{ap} in Equation (34), the generalized Reynolds number takes the form

$$Re' = \frac{\rho u^{2-n'} d_i^{n'}}{8^{n'-1} K'} \quad \mathbf{(35)}$$

Use of this generalized Reynolds number was suggested by Metzner and Reed (1955). For Newtonian behaviour, K' = μ and n' = 1 so that the generalized Reynolds number reduces to the normal Reynolds number.

TURBULENT FLOW OF INELASTIC NON-NEWTONIAN FLUIDS IN PIPES AND CIRCULAR DUCTS

Turbulent flow of Newtonian fluids is described in terms of the Fanning friction factor, which is correlated against the Reynolds number with the relative roughness of the pipe wall as a parameter. The same approach is adopted for non-Newtonian flow but the generalized Reynolds number is used. The Fanning friction factor is defined by

$$f = \frac{\tau_\omega}{\frac{1}{2}\rho u^2} \quad (.36)$$

It is straightforward to show that the Fanning friction factor for laminar non-Newtonian flow becomes

$$f = 16/\mathrm{Re}' \quad (37)$$

This is of the same form as equation for Newtonian flow and is one reason or using this form of generalized Reynolds number. Equation (37) provides another way of calculating the pressure gradient for a given flow rate for laminar non-Newtonian flow.

LAMINAR-TURBULENT TRANSITION

Turbulent flow of Newtonian fluids is described in terms of the Fanning friction factor, which is correlated against the Reynolds number with the relative roughness of the pipe wall as a parameter. The same approach is adopted for non-Newtonian flow but the generalized Reynolds number is used. The Fanning friction factor is defined by

$$f = \frac{\tau_\omega}{\frac{1}{2}\rho u^2} \quad (36)$$

It is straightforward to show that the Fanning friction factor for laminar non-Newtonian flow becomes

$$f = 16/\mathrm{Re}' \quad (37)$$

This is of the same form as equation for Newtonian flow and is one reason or using this form of generalized Reynolds number. Equation (37) provides another way of calculating the pressure gradient for a given flow rate for laminar non-Newtonian flow.

Laminar-turbulent Transition

A stability analysis made by Ryan and Johnson (1959) suggests that the transition from laminar to turbulent flow for inelastic non-Newtonian fluids occurs at a critical value of the generalized Reynolds number that depends on the value of n'. The results of this analysis are shown in Figure 5. This relationship has been tested for shear thinning and for Bingham plastic fluids and has been found to be accurate. Over the range of shear thinning behaviour encountered in practice, 0.2 ≤ n' ≤ 1, the critical value of Re' is in the range 2100 ≤ Re' ≤ 2400.

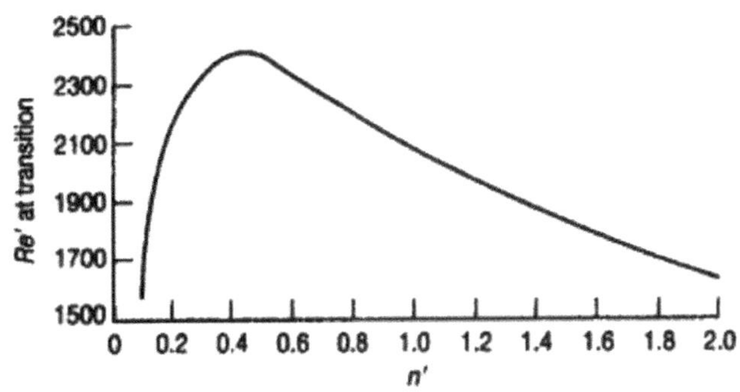

Figiure 5: Variation of the critical value of the Reynolds number with n'

Friction Factors for Turbulent Flow in Smooth Pipes

Experimental results for the Fanning friction factor for turbulent flow of shear thinning fluids in smooth pipes have been correlated by Dodge and Metzner (1959) as a generalized form of the yon Kármán equation:

$$\frac{1}{f^{1/2}} = \frac{4.0}{(n')^{0.75}}\log[f^{1-n'/2}Re'] - \frac{0.40}{(n')^{1.2}} \quad (38)$$

This correlation is shown in Figure 6. The broken lines represent extrapolation of equation (38) for values of n' and Re' beyond those of the measurements made by Dodge and Metzner. More recent studies tend to confirm the findings of Dodge and Metzner but do not significantly extend the range of applicability. Having determined the value of the friction factor f for a specified flow rate and hence Re', the pressure gradient can be calculated in the normal way.

Example 3

A general time-independent non-Newtonian liquid of density 961 kg/m^3 flows steadily with an average velocity of 2.0m/s through a tube 3.048 m long with an inside diameter of 0.0762 m. For these conditions, the pipe flow consistency coefficient K' has a value of 1.48 Pa s$^{0.3}$ and n' a value of 0.3. Calculate the values of the apparent viscosity for pipe flow μ_{ap}, the generalized Reynolds number Re' and the pressure drop across the tube, neglecting end effects.

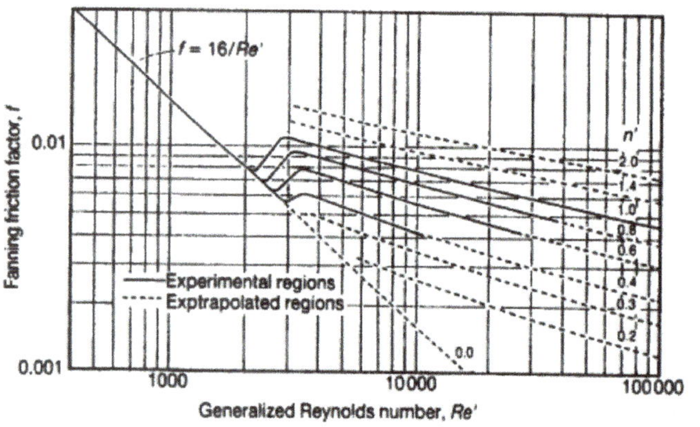

Figure 6: Friction factor chart for purely viscous non-Newtonian fluids.

Calculations

The flow characteristic is given by

$$\frac{8u}{D} = \frac{8(2.0m/s)}{0.0762m} = 210s^{-1}$$

and

$$\left(\frac{8u}{D}\right)^{n'-1} = 210^{(0.3-1.0)} = 0.0237s^{0.7}$$

Hence

$$\mu_{ap} = K'\left(\frac{8u}{D}\right)^{n'-1} = (1.48 \text{ Pa s}^{0.3})(0.0237\ s^{0.7}) = 0.0351 \text{ Pa s}$$

and

$$Re' = \frac{\rho u D}{\mu_{ap}} = \frac{(0.0762 \text{ m})(2.0 \text{ m})(961 \text{ kg/m}^3)}{(0.0351 \text{ Pa s})} = 4178$$

From Figure 6, the Fanning friction factor f has a value 0.0047. Therefore the pressure drop is given by

$$\Delta P_f = 4f\left(\frac{L}{d_i}\right)\frac{\rho u^2}{2} = \frac{2(0.0047)(3.048 \text{ m})(961 \text{ kg/m}^3)(2.0 \text{ m/s})^2}{(0.0762 \text{ m})} = 1445 \text{ Pa}$$

LAMINAR FLOW OF INELASTIC FLUIDS IN NON-CIRCULAR DUCTS

Analytical solutions for the laminar flow of time-independent fluids in non-axisymmetric conduits are not possible. Numerous workers have obtained approximate and/or complete numerical solutions for specific flow geometries including square, rectangular and triangular pipes (Schechter, 1961 ; Wheeler and Wissler, 1965 ; Miller, 1972 ; Mitsuishi and Aoyagi, 1969, 1973). On the other hand, semi-empirical attempts have also been made to develop methods for predicting pressure drop for time-independent fluids in ducts of noncircular cross-section. Perhaps the most systematic and successful friction factor analysis is that provided by Kozicki et al . (1966, 1967) . It is useful to recall here that the equation (19) is a generalized equation for the laminar flow of time-independent fluids in a tube and it can be slightly rearranged as:

$$-\dot{\gamma}_w = f(\tau_w) = \frac{1}{4}\tau_w \frac{d(8u/D)}{d\tau_w} + \frac{3}{4}\left(\frac{8u}{D}\right) \tag{39}$$

Similarly, one can parallel this approach for the fully developed laminar flow of time independent fluids in a thin slit (Figure 7) to derive the following relationship

$$-\dot{\gamma}_w = \left(-\frac{dV_z}{dr}\right)_w = f(\tau_w) = \tau_w \frac{d(u/h)}{d\tau_w} + 2\left(\frac{u}{h}\right) \tag{40}$$

In order to develop a unified treatment for ducts of various cross-sections, it is convenient to introduce the usual hydraulic diameter D_h (defined as four times the area for flow/wetted perimeter) into equations (39) and 40).

$$-\dot{\gamma}_\omega = f(\tau_\omega) = \frac{1}{4}\tau_\omega \frac{d(8u/D_h)}{d\tau_\omega} + \frac{3}{4}\left(\frac{8u}{D_h}\right) \tag{41}$$

For the slit shown in Figure 7, the hydraulic diameter Dh= 4h , and thus equation (39) is rewritten as:

$$-\dot{\gamma}_w == \frac{1}{2}\tau_w \frac{d(8u/D_h)}{d\tau_w} + \left(\frac{8u}{D_h}\right) \tag{42}$$

By noting the similarity between the form of the Rabinowitsch–Mooney equations for the flow of time-independent fluids in circular pipes (equation (41)) and that in between two plates (equation (42)), they suggested that it could be extended to the ducts having a constant cross-section of arbitrary shape as follows:

$$-\dot{\gamma}_w = f(\bar{\tau}_w) = a\bar{\tau}_w \frac{d(8u/D_h)}{d\bar{\tau}_w} + b\left(\frac{8u}{D_h}\right) \tag{43}$$

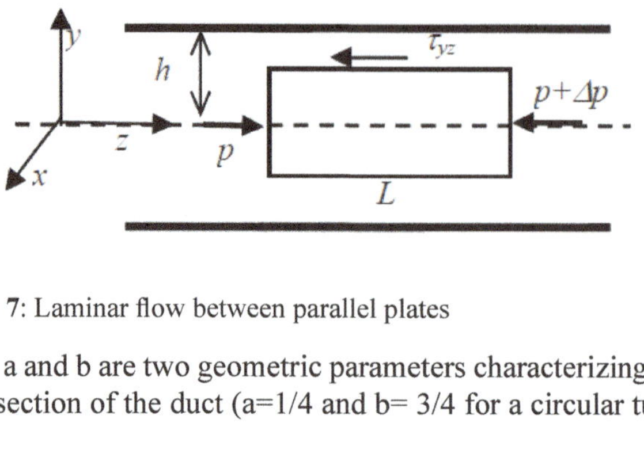

Figure 7: Laminar flow between parallel plates

where a and b are two geometric parameters characterizing the cross-section of the duct (a=1/4 and b= 3/4 for a circular tube, and a =1/2 and b =1 for the slit) and $\bar{\tau}_w$ is the mean value of shear stress at the wall, and is related to the pressure gradient as:

$$\bar{\tau}_w = \frac{D_h}{4}\left(\frac{-\Delta P}{L}\right) \tag{44}$$

For constant values of a and b , equation (42) is an ordinary differential equation of the form (d y/d x) + p(x) y = q(x) which can be integrated to obtain the solution as:

$$y = e^{-\int p(x)dx}\int e^{\int p(x)dx} q(x)dx + C_0 \tag{45}$$

Now identifying $y = (8V/D_h)$ and $x = \bar{\tau}_w$, $p(x) = (b/a\,\bar{\tau}_w)$ and $q(x) = (f(\bar{\tau}_w)/a\,\bar{\tau}_w)$, the solution to equation (43) is given as:

$$\frac{8u}{D_h} = \frac{1}{a}(\bar{\tau}_w)^{-b/a}\int_0^{\bar{\tau}_w} \xi^{(b/a)-1} f(\tau)d\xi \tag{46}$$

Table 7: Values of a and b depending on geometry of the ducts

1. *Concentric annuli*

$\sigma = Ri/Ro$	a	b
0.00	0.2500	0.7500
0.01	0.3768	0.8751
0.03	0.4056	0.9085
0.05	0.4217	0.9263
0.10	0.4455	0.9510
0.20	0.4693	0.9737
0.30	0.4817	0.9847
0.40	0.4890	0.9911
0.50	0.4935	0.9946
0.80	0.4992	0.9994
1.00	0.5000	1.0000

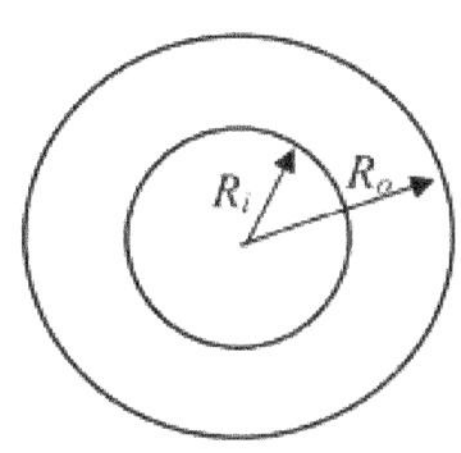

2. *Elliptical ducts*

$\varepsilon = b'/a'$	a	b
0.00	0.3084	0.9253
0.10	0.3018	0.9053
0.20	0.2907	0.8720
0.30	0.2796	0.8389
0.40	0.2702	0.8107
0.50	0.2629	0.7886
0.60	0.2575	0.7725
0.80	0.2515	0.7546
1.00	0.2500	0.7500

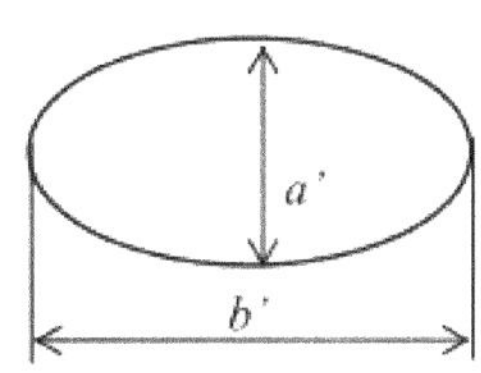

3. *Rectangular ducts*

$e = H/W$	a	b
0.00	0.5000	1.0000
0.25	0.3212	0.8182
0.50	0.2440	0.7276
0.75	0.2178	0.6866
1.00	0.2121	0.6766

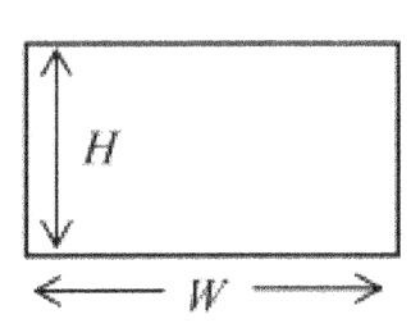

4. *Isosceles triangular ducts*

2α	a	b
10°	0.1547	0.6278
20°	0.1693	0.6332
40°	0.1840	0.6422
60°	0.1875	0.6462
80°	0.1849	0.6438
90°	0.1830	0.6395

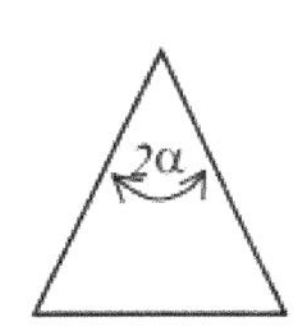

5. *Regular polygonal ducts*

N	a	b
4	0.2121	0.6771
5	0.2245	0.6966
6	0.2316	0.7092
8	0.2391	0.7241

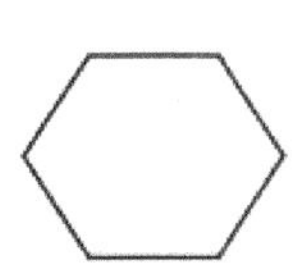

where ξ is a dummy variable of integration. The constant C_0 has been evaluated by using the condition that when $V = 0$, $\bar{\tau}_w = 0$ and therefore, $C_0 = 0$.

$$\bar{\tau}_w = K\left\{\frac{8u}{D_n}\left(b + \frac{a}{n}\right)\right\}^n \tag{47}$$

which can be rewritten in terms of the friction factor, $f = 2\bar{\tau}_w/\rho u^2$ as:

$$f = \frac{16}{Re_g} \tag{48}$$

where the generalized Reynolds number,

$$Re_g = \frac{\rho u^{2-n} D_h^n}{8^{n-1} K\left(b + \frac{a}{n}\right)^n} \tag{49}$$

The main virtue of this approach lies in its simplicity and the fact that the geometric parameters a and b can be deduced from the behaviour of Newtonian fluids in the same flow geometry. Table 7 lists values of a and b for a range of flow geometries commonly encountered in process applications. Kozicki et al.(1966) argued that the friction factor of the turbulent flow in non-circular ducts can be calculated from the following equation

$$\frac{1}{\sqrt{f}} = \frac{4}{n^{0.75}}\log_{10}(\mathrm{Re}_g f^{(2-n)/2}) - \frac{0.4}{n^{1.2}} + 4n^{0.25}\log[\frac{a(a+bn)}{3n+1}]$$

(50)

Note that since for a circular tube, a= 1/4 and b =3/4, equation (50) is consistent with that for circular pipes. The limited data available on turbulent flow in triangular (Irvine Jr, 1988), rectangular (Kostic and Hartnett, 1984) and square ducts (Escudier and Smith, 2001) conforms to equation (48). In the absence of any definite information, Kozicki and Tiu (1988) suggested that the Dodge–Metzner criterion, Re_g £ 2100, can be used for predicting the limit of laminar flow in non-circular ducts.

Some further attempts have been made to simplify and/or improve upon the two geometric parameter method of Kozicki et al . (1966, 1967). Delplace and Leuliet (1995) revisited the definition of the generalized Reynolds number (equation (49)) and argued that while the use of a and b accounts for the non-circular cross-sections of the ducts, but the factor 8n-1 appearing in the denominator is strictly applicable for the flow in circular ducts only. Their reasoning hinges on the fact that for the laminar flow of a Newtonian fluid, the product (f.Re) is a function of the conduit shape only. Thus, they wrote

$$\beta = \frac{48}{(f.Re)} \tag{51}$$

where both the (Fanning) friction factor and the Reynolds number are based on the use of the hydraulic diameter, Dh and the mean velocity of the flow, u. Furthermore, they were able to link the geometric parameters a and b with the new parameter β as follows:

$$a = \frac{1}{1+\beta};\quad b = \frac{\beta}{1+\beta} \tag{52}$$

and finally, the factor of 8^{n-1} in the denominator in equation (49) is replaced $(24/\beta)^{n-1}$. With these modifications, one can use the relationship f =(16/Reg) to estimate the pressure gradient for the laminar flow of a power-law fluid in a non-circular duct for which the value of β is known either from experiments or from numerical results. Therefore, this approach necessitates the knowledge of only one parameter (β) as opposed to the two geometric parameters, namely, a and b in the method of Kozicki et al . (1966) and Kozicki and Tiu (1967), albeit a similar suggestion was also made by Miller (1972) and Liu (1983). Finally, for the limiting case of a circular pipe, evidently β=3 thereby leading to a=(1/4) and b=(3/4) and the two definitions of the Reynolds number coincide, as expected. The values of β for a few standard duct shapes are summarized in Table 5. While in laminar flow, these two methods give almost identical predictions, the applicability of the modified method of Delplace and Leuliet (1995) has not been checked in the transitional and turbulent flow regions. Scant analytical and experimental results suggest that visco-elasticity in a fluid may induce secondary motion in non-circular conduits, even under laminar conditions. However, measurements reported to date indicate that the friction factor–Reynolds number behaviour is little influenced by such secondary flows (Hartnett and Kostic, 1989).

Example 4

A power-law fluid (K = 0.3 Pa.s^n and n = 0.72) of density 1000 kg/m^3 is flowing in a series of ducts of the same flow area but different cross-sections as listed below:

i. concentric annulus with R= 37mm and j=(R/Ri)= 0.40
ii. circular pipe of radius R
iii. rectangular, (H /W) =0.5
iv. elliptical, b'/a'= 0.5

Estimate the pressure gradient required to maintain an average velocity of 1.25m/s in each of these channels. Use the geometric parameter method. Also, calculate the value of the generalized Reynolds number as a guide to the nature of the flow.

Solution

i. For a concentric annulus, j = 0.4
   - From Table 6 we have: a =0.489; b = 0.991
   - The hydraulic diameter, D_h = 2R(1-σ)=0.044m.

$$\mathrm{Re}_g = \frac{\rho u^{2-n} D_h^n}{8^{n-1}K(b+\frac{a}{n})^n} = 579$$

   - Reynolds number,
   - Thus, the flow is laminar and the friction factor is estimated as: f=1/578=0.0276 and

$$-\frac{\Delta P}{L} = \frac{2f\rho u^2}{D_h} = 1963 Pa/m$$

ii. For a circular tube, the area of flow
   - For a circular pipe, a =0.25, b = 0.75, Dh = D = 0.0678m

$$\mathrm{Re}_g = \frac{\rho u^{2-n} D_h^n}{8^{n-1}K(b+\frac{a}{n})^n} = 1070$$

   - Reynolds number,
   - The flow is laminar and the friction factor is estimated as: f=1/1070=0.01495 and

$$-\frac{\Delta P}{L} = \frac{2f\rho u^2}{D_h} = 689 Pa/m$$

iii. For a rectangular duct with H /W = 0.5, H =0.0425 m and W= 0.085 m (for the same area of flow), and from Table 6: a=0.244, b= 0.728
   - Dh = 4HW/2(H+W) = 0.0567m

$$\mathrm{Re}_g = \frac{\rho u^{2-n} D_h^n}{8^{n-1}K(b+\frac{a}{n})^n} = 960$$

   - Reynolds number

f=1/960=0.0167 and $-\frac{\Delta P}{L} = \frac{2f\rho u^2}{D_h} = 919 Pa/m$

   -

iv. elliptical, b'/a'= 0.5.

- From Table 6 : a=0.2629, b= 0.7886
- D_h = = 0.0607m - Reynolds number, Re 953
- f =0.0168 and $-\frac{\Delta P}{L} = 864 Pa/m$

REFERENCES

1. Astarita, G., The Engineering Reality of the Yield Stress, J. Rheol. 34 (1990) 275
2. Barnes , H.A. and Walters , K. , The yield stress myth, Rheol. Acta 24 (1985) 323 .
3. Barnes , H.A. , Edwards , M.F. and Woodcock , L.V. , Applications of computer simulations to dense suspension rheology, Chem. Eng. Sci. 42 (1987) 591 .
4. Barnes , H.A. , Hutton , J.F. and Walters , K. , An Introduction to Rheology , Elsevier, Amsterdam (1989) .
5. Barnes , H.A., The yield stress - a review, J. Non-Newt. Fluid Mech. 81 (1999) 133.
6. Bird , R.B. , Useful non-Newtonian models, Annu. Rev. Fluid Mech. 8 (1976) 13.
7. Bird , R.B. , Dai , G.C. and Yarusso , B.J. , The rheology and flow of viscoplastic materials, Rev. Chem. Eng. 1 (1983) 1 .
8. Bird , R.B. , Armstrong , R.C. and Hassager , O. , Dynamics of polymeric liquids. Fluid dynamics, 2nd edition, Wiley , New York (1987) .
9. Cross , M.M. , Rheology of non-Newtonian fluids: A new flow equation for pseudoplastic systems, J. Colloid Sci. 20 (1965) 417.
10. Carreau , P.J. , Rheological equations from molecular network theories, Trans. Soc. Rheol. 16 (1972) 99.
11. Carreau , P.J. , Dekee , D. and Chhabra , R.P., Rheology of polymeric systems: Principles and applications , (1997)
12. Delplace , F. and Leuliet , J. C. , Generalized Reynolds number for the flow of power law fluids in cylindrical ducts of arbitrary cross-section, Chem. Eng. Journal, 56, 33-37.,Chem. Eng. J. 56 (1995) 33.
13. Dodge, D. W. and Metzner, A. B., Turbulent flow of non-Newtonian systems, AIChE Journal. 5 (1959) 189-204.
14. Escudier , M.P. and Smith , S. , Non-Newtonian liquids through a square duct, Proc. Royal Soc. 457 (2001) 911 .
15. Evans , I. D. , On the nature of the yield stress, J. Rheol. 36 (1992) 1313 .
16. Goddard , J.D. and Bashir , Y. , Recent developments in structured continua II (Chapter 2) , Longman , London (1990) .
17. Griskey, R.G., Nechrebecki, D.G., Notheis , P.J. and Balmer , R.T., Rheological and pipelineflow dehavior of corn starch dispersions, J. Rheol. 29(1985) 349
18. Hartnett , J.P. and Kostic , M. , Heat transfer to Newtonian and non- Newtonian fluids in rectangular duct, Adv. Heat Transf. 19 (1989) 247.
19. Irvine , Jr. , T.F. , A generalized blasius equation for power law liquids, Chem. Eng. Commun. 65 (1988) 39.
20. Johnson , A.T. , Biological Process Engineering , Wiley , New York (1999).
21. Kozicki , W. , Chou , C.H. and Tiu , C. , Non-Newtonian flow in ducts of arbitrary crosssectionalshape, Chem. Eng. Sci. 21 (1966) 665 .
22. Kozicki , W. , and Tiu , C., Non-Newtonian flow through open channels, Chem. Eng. 45 (1967) 127 .
23. Kozicki , W. and Tiu , C. , Encyclopedia of Fluid Mechanics, Vol. 7, Gulf, Houston , p. 199 (1988).
24. Kostic , M. and Hartnett , J.P. , Predicting turbulent friction factors of non- Newtonian fluids in noncircular ducts, Int. Comm. Heat Mass Transf. 11 (1984) 345 .
25. Metzner , A.B. and Whitlock , M. , Flow behavior of concentrated (dilatant) suspensions, Trans. Soc. Rheol. 2 (1958) 239 .
26. Metzner, A.B. and Reed, J.C., Flow of non-Newtonian fluids correlation of the laminar, transition, and turbulent-flow regimes, AIChE Journal. 1 (1955) 434-40.
27. Miller , C. , Predicting non-Newtonian flow behaviour in ducts of unusual cross-sections Ind. Eng. Chem. Fundam. 11 (1972) 524 .
28. Mitsuishi , N. and Aoyagi , Y. , Non-Newtonian flow in non- circular ducts, Chem. Eng. Sci . 24 (1969) 309 .
29. Mitsuishi , N. and Aoyagi , Y. , Non-Newtonian fluid flow in an eccentric annulus, J. Chem. Eng. Jpn. 6 (1973) 402 .
30. Ryan, N.W. and Johnson, M.M., Transition from laminar to turbulent flow in pipes, AIChEJournal. 5 (1959) 433.
31. Schechter, R.S. , On the steady flow of a non-Newtonian fluid in cylinder ducts, AIChE J. 7(1961) 445 .
32. Schurz , J. , The yield stress - an empirical reality, Rheol. Acta. 29 (1990) 170 .
33. Uhlherr , P.H.T. , Guo , J. , Zhang , X-M. , Zhou , J.Z and Tiu , C. , The shear-induced solidliquidtransition in yield stress materials with chemically different structures, J.Non-Newt. Fluid Mech. 125 (2005) 101 .
34. Wheeler , J.A. and Wissler , E.H. , The Friction factor-Reynolds number relation for a steadyflow of pseudoplastic fluids through rectangular ducts, AIChE J. 11 (1965) 207.

CHAPTER 20

THE STRONG SOLUTIONS FOR A CLASS OF FLUID-PARTICLE INTERACTION NON-NEWTONIAN MODELS

Changjia Wang

INTRODUCTION AND MAIN RESULTS

This paper is concerned about the existence and uniqueness of the weak solutions to the non-Newtonian magneto-micropolar fluid equations in $\Omega \times (0,T)$, which are described by

$$\begin{cases} u_t + (u \cdot \nabla)u - \operatorname{div}(|\mathbf{e}(u)|^{p-2}\mathbf{e}(u)) + \nabla(\pi + \frac{1}{2}|b|^2) = \chi \operatorname{rot} \omega + (b \cdot \nabla)b, \\ \omega_t + (u \cdot \nabla)\omega - \mu \Delta \omega + 2\chi \omega = \chi \operatorname{rot} u, \\ b_t - \lambda \Delta b + (u \cdot \nabla)b - (b \cdot \nabla)u = 0, \\ \operatorname{div} u = \operatorname{div} b = 0, \end{cases} \quad (1)$$

here $\Omega \in R^3$ is an open-bounded domain with Lipschitz boundaries, and the unknowns u, ω, b, π denote the velocity of the fluid, the micro-rotational velocity, magnetic field and hydrostatic pressure, respectively. χ, μ, λ are positive numbers associated with properties of the material: χ is the vortex viscosity, μ is spin viscosity and $\frac{1}{\lambda}$ is the magnetic Reynold. In (1.1), e(u) is the symmetric part of the velocity gradient, i.e.,

$$2\mathbf{e}(u) = \nabla u + (\nabla u)^T.$$

To (1.1) we append the following initial and boundary conditions

$$u(x,0) = u_0(x), \qquad \omega(x,0) = \omega_0(x), \qquad b(x,0) = b_0(x), \quad \forall x \in \Omega, \quad (1.2)$$

$$u(x,t) = \omega(x,t) = b(x,t) = 0, \quad \forall (x,t) \in \Sigma_T, \quad (1.3)$$

where $\Sigma_T = \partial\Omega \times (0,T)$.

The theory of micropolar fluid was first proposed by Eringen [1] in 1966, which enabled us to consider some physical phenomena that cannot be treated by the classical NavierStokes equations for the viscous incompressible fluid, for example, the motion of animal blood, liquid crystals and dilute aqueous polymer solutions, etc.

If p = 2, system (1.1) reduces to the classical magneto-micropolar fluid equations, and there are many earlier results concerning the weak and strong solvability in a bounded domain ∈ R^3. For strong solutions, Galdi and Rionero [2] stated, without proof, the results of existence and uniqueness of strong solutions. Rojas-Medar [3] studied it and established the local in time existence and uniqueness of strong solutions by the spectral Galerkin method. In 1999 Ortega-Torres and Rojas-Medar [4]proved global existence of strong solutions with the small initial values. For weak solutions, Rojas-Medar and Boldrini [5] proved the existence of weak solutions, and in the 2D case, also proved the uniqueness of the weak solutions.

On the other hand, there are few existence results about the non-Newtonian magnetomicropolar fluid, i.e., the $p \neq 2$ case. In a recent work, Gunzburger et al. [6] studied the reduced problem (with both $\omega = 0$ and $\chi = 0$), and gave the global unique solvability of the first initial-boundary value problem in a bounded two or three-dimensional domain. Improved results are proved for the periodic boundary condition case. In

Citation: Changjia Wang, The strong solutions for a class of fluid-particle interaction non-Newtonian models, DOI: 10.1186/1687-2770-2013-182.

this paper, we will prove the global existence and uniqueness of the weak solutions for the full system (1.1)-(1.3) under the condition that $p \geq 5/2$. These results are based on the Galerkin method and a series of uniform estimates, which do not depend on the parameters.

Throughout this work, we use a standard notation $L^p(\Omega)$ (normed $\|\cdot\|_p$) for Lebesgue Lp-spaces, as well as $W^{k,p}(\Omega)$ (normed $\|\cdot\|_{k,p}$) for the usual Sobolev spaces. As usual, $C_0^\infty(\Omega)$ denotes the set of all C∞-functions with the compact support in . Given $T > 0$ and a Banach space X, we denote by $L^q(0,T;X)$ Bochner spaces, which are equipped with the norm·

$$\|\cdot\|_{L^q(0,T;X)} := \left(\int_0^T \|\cdot\|_X^q \, ds\right)^{\frac{1}{q}}.$$

We also introduce the following functional vector spaces:

$$\mathcal{V} \equiv \{u \in C_0^\infty(\Omega), \operatorname{div} u = 0\},$$

$$H \equiv \text{the closure of } \mathcal{V} \text{ in } L^2(\Omega),$$

$$V_p \equiv \text{the closure of } \mathcal{V} \text{ in } W^{1,p}(\Omega).$$

We next introduce the definition of a weak solution for problems (1.1)-(1.3).

We say that (u,ω, b) is a weak solution to problems (1.1)-(1.3) if

$$u \in L^\infty(0,T;V_p), \qquad u_t \in L^2(0,T;H),$$
$$\omega \in L^\infty(0,T;W^{1,2}(\Omega)) \cap L^2(0,T;W^{2,2}(\Omega)),$$
$$b \in L^\infty(0,T;V_2) \cap L^2(0,T;W^{2,2}(\Omega)),$$
$$\omega_t \in L^2(0,T;W^{1,2}(\Omega)), \qquad b_t \in L^2(0,T;H) \tag{1.4}$$

satisfy

$$\langle u_t, \phi\rangle + \left\langle |\mathbf{e}(u)|^{p-2}\mathbf{e}(u), \mathbf{e}(\phi)\right\rangle = \chi\langle \operatorname{rot}\omega, \phi\rangle + \langle (b\cdot\nabla)b, \phi\rangle - \langle (u\cdot\nabla)u, \phi\rangle$$
$$\text{for all } \phi \in V_p, \tag{1.5}$$

$$\langle \omega_t, \psi\rangle + \mu\langle \nabla\omega, \nabla\psi\rangle = \chi\langle \operatorname{rot} u, \psi\rangle - \langle (u\cdot\nabla)\omega, \psi\rangle - 2\chi\langle \omega, \psi\rangle$$
$$\text{for all } \psi \in W_0^{1,2}(\Omega), \tag{1.6}$$

$$\langle b_t, \eta\rangle + \lambda\langle \nabla b, \nabla\eta\rangle = \langle (b\cdot\nabla)u, \eta\rangle - \langle (u\cdot\nabla)b, \eta\rangle$$
$$\text{for all } \eta \in V_2, \tag{1.7}$$

where the symbol· $\langle\cdot,\cdot\rangle$ denotes a generic duality pairing.

The following theorem gives the main results of this paper

Theorem 1.1 Let $\Omega \in R^3$ be an open-bounded domain with a Lipschitz boundary ∂. Assume that $p \geq \frac{5}{2}$, $u_0 \in W^{1,p}(\Omega)$, $\omega_0, b_0 \in W^{1,2}(\Omega)$. *Then, for* $\forall T \in (0, +\infty)$, there exists a unique weak solution to problem (1.1)-(1.3) in the sense of Definition1.1

Remark 1.1 If (1.4)-(1.5) hold, it could be easy to introduce the pressure $\pi \in L^\infty(0,T;L^{p'}(\Omega)), p' = p/(p-1)$. This will be done at the end of Section 3.

For latter use, let us state some useful inequalities

Lemma 1.1 (See [7]) (Korn's inequality) Let $1 < p < \infty$. Then there exists a constant $C_p = C_{p(\Omega)}$ such that

$$C\|v\|_p \leq \|v\|_{-1,p} + \|\nabla v\|_{-1,p}.$$

where $\in R^n$ is open and bounded with a Lipschitz boundary.

Lemma 1.2 (See [8]) (On negative norm) Let $1 < p < \infty$, *and let* $v \in W_0^{1,p}(\Omega)$. Then there exists a constant C such that

$$C\|v\|_p \leq \|v\|_{-1,p} + \|\nabla v\|_{-1,p}.$$

Lemma 1.3 (See [9]) Let $v, w \in W^{1,p}(\Omega)$. *For each* $2 < p < \infty$, there exists a constant C'= C' (p)>0 such that

$$\|\mathbf{e}(v) - \mathbf{e}(w)\|_p^p \leq C' \int_\Omega \left(|\mathbf{e}(v)|^{p-2}\mathbf{e}(v) - |\mathbf{e}(w)|^{p-2}\mathbf{e}(w)\right)(\mathbf{e}(v) - \mathbf{e}(w)). \tag{1.9}$$

with

$$\begin{cases} m \leq 3r/(3-r) & \text{for } r \in [1,3), \\ \text{for any } m < \infty & \text{for } r = 3, \\ m = \infty & \text{for } r > 3. \end{cases}$$

Here, $\tilde{C}(m,r) = 0$ if $u|_{\partial\Omega} = 0$ or $\int_\Omega u\,dx = 0$. We will also apply the so-called multiplicative inequalities

$$\|u\|_q \leq C(q)\|\nabla u\|_2^\alpha \cdot \|u\|_2^{1-\alpha} + \hat{C}(q)\|u\|_2$$

with

$$\alpha = 3\left(\frac{1}{2} - \frac{1}{q}\right) \in [0,1], \quad q \in [2,6].$$

If $u|_{\partial\Omega} = 0$ or $\int_\Omega u\,dx = 0$, then $\hat{C}(q) = 0$.

Finally, the paper is organized as follows. In Section 2 we focus on the derivation of the priori estimates for the smooth solutions. On the bases of these estimates, in Section 3, we get the existence result with the help of the Galerkin method. The aim of Section 4 is to give the uniqueness criterion.

The Priori Estimates

Let (u,ω, b) be a smooth solution to system (1.1)-(1.3). The goal of this section is to derive some priori estimates about it. In all the following sections, we always assume that $p \geq \frac{5}{2}$ holds.

Setting $\phi = u$ in (1.5), $\psi = \omega$ in (1.6), $\eta = b$ in (1.7), and observing that $\langle (u\cdot\nabla)u, u\rangle = \langle (u\cdot\nabla)\omega, \omega\rangle = \langle (u\cdot\nabla)b, b\rangle = 0$ we obtain

$$\frac{1}{2}\frac{d}{dt}\|u\|_2^2 + \|\mathbf{e}(u)\|_p^p = \chi\langle \operatorname{rot}\omega, u\rangle + \langle (b\cdot\nabla)b, u\rangle,$$

$$\frac{1}{2}\frac{d}{dt}\|\omega\|_2^2 + \mu\|\nabla\omega\|_2^2 + 2\chi\|\omega\|_2^2 = \chi\langle \operatorname{rot} u, \omega\rangle,$$

$$\frac{1}{2}\frac{d}{dt}\|b\|_2^2 + \lambda\|\nabla b\|_2^2 = \langle (b\cdot\nabla)u, b\rangle.$$

Adding the identities above, noting that $\langle (b\cdot\nabla)b, u\rangle + \langle (b\cdot\nabla)u, b\rangle = 0$, and Korn's inequality (1.8), we get

$$\frac{1}{2}\frac{d}{dt}(\|u\|_2^2 + \|\omega\|_2^2 + \|b\|_2^2) + (C_p\|\nabla u\|_p^p + \mu\|\nabla\omega\|_2^2 + 2\chi\|\omega\|_2^2 + \lambda\|\nabla b\|_2^2)$$
$$= \chi\langle \mathrm{rot}\,\omega, u\rangle + \chi\langle \mathrm{rot}\,u, \omega\rangle$$
$$\leq C\chi(\|\nabla\omega\|_2 \cdot \|u\|_2 + \|\nabla u\|_2 \cdot \|\omega\|_2)$$
$$\leq \varepsilon(\|\nabla\omega\|_2^2 + \|\nabla u\|_p^p) + C_\varepsilon(\|\omega\|_2^2 + \|u\|_2^2) \quad (\text{for } p \geq 2).$$

After choosing ε properly small, integrating over $(0,t), t \in (0,T]$, the Gronwall's inequality yields that

$$\sup_{t\in(0,T)}(\|u\|_2^2 + \|\omega\|_2^2 + \|b\|_2^2) + \int_0^T(\|\nabla u\|_p^p + \|\nabla\omega\|_2^2 + \|\nabla b\|_2^2) \leq C_1, \tag{2.1}$$

where C_1 is a constant depending on the time T and $\|u_0\|_2, \|w_0\|_2, \|b_0\|_2$.

Next, we derive the higher order estimates for ω and b. Setting $\psi = -\Delta\omega$ in (1.6), we find

$$\frac{1}{2}\frac{d}{dt}\|\nabla\omega\|_2^2 + \mu\|\Delta\omega\|_2^2 = \langle(u\cdot\nabla)\omega, \Delta\omega\rangle + 2\chi\langle\omega, \Delta\omega\rangle - \chi\langle \mathrm{rot}\,u, \Delta\omega\rangle$$
$$\leq |\langle(u\cdot\nabla)\omega, \Delta\omega\rangle| + \varepsilon\|\Delta\omega\|_2^2 + C_\varepsilon(\|\omega\|_2^2 + \|\nabla u\|_2^2) \tag{2.2}$$

for the first term on the right hand side, we compute by the divergence free conditions

$$|\langle(u\cdot\nabla)\omega, \Delta\omega\rangle| = \left|\sum_{ijk}\langle\partial_k u_i\,\partial_i\omega_j, \partial_k\omega_j\rangle\right|$$
$$\leq C\int_\Omega|\nabla u||\nabla\omega|^2 \leq C\|\nabla u\|_p \cdot \|\nabla\omega\|^2_{\frac{2p}{p-1}}$$
$$\leq C\|\nabla u\|_p \cdot (\|\nabla\omega\|_2^{\frac{2p-3}{p}} \cdot \|\Delta\omega\|_2^{\frac{3}{p}} + \|\nabla\omega\|_2^2)$$
$$\leq \varepsilon\|\Delta\omega\|_2^2 + C_\varepsilon\|\nabla u\|_p^{\frac{2p}{2p-3}} \cdot \|\nabla\omega\|_2^2 + C\|\nabla u\|_p \cdot \|\nabla\omega\|_2^2, \tag{2.3}$$

where Hölder's, Young's inequality and (1.9), (1.10) have been used.

Inserting (2.3) into (2.2), choosing $\varepsilon = \frac{\mu}{4}$ and integrating over $(0,t), t \in (0,T]$, we have

$$\frac{1}{2}\|\nabla\omega\|_2^2 + \frac{\mu}{2}\int_0^t\|\Delta\omega\|_2^2$$
$$\leq C\int_0^t(\|\nabla u\|_p^{\frac{2p}{2p-3}} + \|\nabla u\|_p)\cdot\|\nabla\omega\|_2^2 + C\int_0^t(\|\omega\|_2^2 + \|\nabla u\|_2^2) + \frac{1}{2}\|\nabla\omega_0\|_2^2$$
$$\overset{(2.1)}{\leq} C\int_0^t(\|\nabla u\|_p^{\frac{2p}{2p-3}} + \|\nabla u\|_p)\cdot\|\nabla\omega\|_2^2 + \frac{1}{2}\|\nabla\omega_0\|_2^2 + C_1, \tag{2.4}$$

since $p \geq \frac{5}{2}$, we have

$$\frac{2p}{2p-3} \leq p. \tag{2.5}$$

Gronwall's inequality and estimate (2.1) now provide the bound

$$\sup_{t\in(0,T)}\|\nabla\omega\|_2^2 + \int_0^T\|\Delta\omega\|_2^2 \leq C_2, \tag{2.6}$$

where C_2 is a constant depending on the time T, C_1 and $\|\nabla\omega_0\|_2$.

Next, set $\eta = -\Delta b$ in (1.7) to discover

$$\frac{1}{2}\frac{d}{dt}\|\nabla b\|_2^2+\lambda\|\Delta b\|_2^2=\langle(u\cdot\nabla)b,\Delta b\rangle-\langle(b\cdot\nabla)u,\Delta b\rangle$$
$$\leq|\langle(u\cdot\nabla)b,\Delta b\rangle|+|\langle(b\cdot\nabla)u,\Delta b\rangle|. \tag{2.7}$$

Reasoning similar to (2.3), we could find

$$|\langle(u\cdot\nabla)b,\Delta b\rangle|\leq\varepsilon\|\Delta b\|_2^2+C_\varepsilon\|\nabla u\|_p^{\frac{2p}{2p-3}}\cdot\|\nabla b\|_2^2+C\|\nabla u\|_p\cdot\|\nabla b\|_2^2. \tag{2.8}$$

For the second term on the right hand side of (2.7), we compute

$$\begin{aligned}&|\langle(b\cdot\nabla)u,\Delta b\rangle|\\&\leq\varepsilon\|\Delta b\|_2^2+C_\varepsilon\|(b\cdot\nabla)u\|_2^2\\&\leq\varepsilon\|\Delta b\|_2^2+C_\varepsilon\|\nabla u\|_p^2\cdot\|b\|^2_{\frac{2p}{p-2}}\\&\leq\varepsilon\|\Delta b\|_2^2+C_\varepsilon\|\nabla u\|_p^2\cdot\|\nabla b\|^2_{\frac{6p}{5p-6}}\\&\leq\varepsilon\|\Delta b\|_2^2+C_\varepsilon\|\nabla u\|_p^2\cdot\left(\|\nabla b\|_2^{\frac{2(2p-3)}{p}}\cdot\|\Delta b\|_2^{\frac{2(3-p)}{p}}+\|\nabla b\|_2^2\right)\\&\leq\varepsilon\|\Delta b\|_2^2+\delta\|\Delta b\|_2^2+C_{\varepsilon\delta}\|\nabla u\|_p^{\frac{2p}{2p-3}}\cdot\|\nabla b\|_2^2+C_\varepsilon\|\nabla u\|_p^2\cdot\|\nabla b\|_2^2,\end{aligned} \tag{2.9}$$

where we have used Hölder's, Young's inequality and (1.9), (1.10). Choosing ε and δ properly small, inserting (2.8)-(2.9) into (2.7) and integrating it over (0,t), t ∈ (0,T], we find

$$\frac{1}{2}\|\nabla b\|_2^2+\frac{\lambda}{2}\int_0^t\|\Delta b\|_2^2\leq C\int_0^t\left(\|\nabla u\|_p^{\frac{2p}{2p-3}}+\|\nabla u\|_p^2\right)\cdot\|\nabla b\|_2^2+\frac{1}{2}\|\nabla b_0\|_2^2. \tag{2.10}$$

Observing (2.5) and estimate (2.1), then Gronwall's inequality yields

$$\sup_{t\in(0,T)}\|\nabla b\|_2^2+\int_0^T\|\Delta b\|_2^2\leq C_3, \tag{2.11}$$

where C_3 is a constant depending on the time T, C_1 and $\|\nabla b_0\|_2$

Reasoning analogously to (2.6) and (2.11), it is easy to see that identity (1.6) with $\psi=\omega_t$, (1.7) with $\eta=b_t$, with the help of (2.6) and (2.11), guarantee the estimate

$$\int_0^T\|\omega_t\|_2^2\leq C_4,\qquad\int_0^T\|b_t\|_2^2\leq C_5, \tag{2.12}$$

where C_4, C_5 are both constants depending only on the time T and some norm of the initial values

In fact (we here only take b_t as an example), set $\eta=b_t$ in (1.7), we deduce that

$$\|b_t\|_2^2+\frac{\lambda}{2}\frac{d}{dt}\|\nabla b\|_2^2\leq|\langle(b\cdot\nabla)u,b_t\rangle|+|\langle(u\cdot\nabla)b,b_t\rangle|$$
$$\leq\varepsilon\|b_t\|_2^2+C_\varepsilon\left(\|(b\cdot\nabla)u\|_2^2+\|(u\cdot\nabla)b\|_2^2\right). \tag{2.13}$$

Now, we compute, by using Hölder's, Young's inequality and (1.9)(1.10)

$$\begin{aligned}\|(b\cdot\nabla)u\|_2^2 &\le \|\nabla u\|_p^2\cdot\|b\|^2_{\frac{2p}{p-2}}\\ &\le C\big(\|\nabla u\|_p^p+\|b\|^{\frac{2p}{p-2}}_{\frac{2p}{p-2}}\big)\\ &\le C\big(\|\nabla u\|_p^p+\|\nabla b\|^{\frac{2p}{p-2}}_{\frac{6p}{5p-6}}\big)\\ &\le C\|\nabla u\|_p^p+C\big(\|\nabla b\|_2^{\frac{2(2p-3)}{p-2}}\cdot\|\Delta b\|_2^{\frac{2(3-p)}{p-2}}+\|\nabla b\|_2^{\frac{2p}{p-2}}\big)\end{aligned} \tag{2.14}$$

and

$$\begin{aligned}\|(u\cdot\nabla)b\|_2^2 &\le \|u\|^2_{\frac{3p}{3-p}}\cdot\|\nabla b\|^2_{\frac{6p}{5p-6}}\\ &\le C\|\nabla u\|_p^2\cdot\|\nabla b\|^2_{\frac{6p}{5p-6}}\\ &\le C\big(\|\nabla u\|_p^p+\|\nabla b\|^{\frac{2p}{p-2}}_{\frac{6p}{5p-6}}\big)\\ &\le C\|\nabla u\|_p^p+C\big(\|\nabla b\|_2^{\frac{2(2p-3)}{p-2}}\cdot\|\Delta b\|_2^{\frac{2(3-p)}{p-2}}+\|\nabla b\|_2^{\frac{2p}{p-2}}\big).\end{aligned} \tag{2.15}$$

Combining (2.13)-(2.15), by choosing $\varepsilon=\frac{1}{2}$, we arrive at

$$\frac{1}{2}\|b_t\|_2^2+\frac{\lambda}{2}\frac{d}{dt}\|\nabla b\|_2^2\le C\big(\|\nabla u\|_p^p+\|\nabla b\|_2^{\frac{2(2p-3)}{p-2}}\cdot\|\Delta b\|_2^{\frac{2(3-p)}{p-2}}+\|\nabla b\|_2^{\frac{2p}{p-2}}\big),$$

noting that $p\ge\frac{5}{2}$, so $2(3-p)/(p-2)\le 2$ and now estimate (2.1), (2.11) and Gronwall's inequality imply the estimate of b_t in (2.22)

In the following, we will derive the bound for u_t. Setting $\eta=u_t$ in (1.5), we deduce that

$$\begin{aligned}&\|u_t\|_2^2+\frac{1}{p}\frac{d}{dt}\|\mathbf{e}(u)\|_p^p\\ &\le |\langle(b\cdot\nabla)b,u_t\rangle|+|\langle(u\cdot\nabla)u,u_t\rangle|+\chi|\langle\operatorname{rot}\omega,u_t\rangle|\\ &\le \varepsilon\|u_t\|_2^2+C_\varepsilon\big(\|(b\cdot\nabla)b\|_2^2+\|(u\cdot\nabla)u\|_2^2+\|\nabla\omega\|_2^2\big).\end{aligned}$$

Integrating it over $(0,t),\ t\in(0,T]$, by choosing $\varepsilon=\frac{1}{2}$ and Korn's inequality, we have

$$\frac{1}{2}\int_0^t\|u_t\|_2^2+C_p\|\nabla u\|_p^p\le C\int_0^t\big(\|(b\cdot\nabla)b\|_2^2+\|(u\cdot\nabla)u\|_2^2\big)+C_1. \tag{2.16}$$

Now, we compute, by (2.11)

$$\begin{aligned}\int_0^t\|(b\cdot\nabla)b\|_2^2 &\le \int_0^t\|b\|_6^2\cdot\|\nabla b\|_3^2\\ &\le C\int_0^t\|\nabla b\|_2^2\cdot\big(\|\nabla b\|_2\cdot\|\Delta b\|_2+\|\nabla b\|_2^2\big)\\ &\le C\sup_{\tau\in(0,t)}\|\nabla b\|_2^3\int_0^t\|\Delta b\|_2+C\sup_{\tau\in(0,t)}\|\nabla b\|_2^4\cdot t\\ &\le C_3^4(1+T),\end{aligned} \tag{2.17}$$

$$\int_0^t\|(u\cdot\nabla)u\|_2^2\le\int_0^t\|u\|^2_{\frac{2p}{p-2}}\cdot\|\nabla u\|_p^2\le C\int_0^t\|\nabla u\|_p^4. \tag{2.18}$$

Inserting (2.17)-(2.18) into (2.16), by appealing to Korn's inequality, it follows that

$$\|\nabla u\|_p^p + \int_0^t \|u_t\|_2^2 \le C \int_0^t \|\nabla u\|_p^p \cdot \|\nabla u\|_p^{4-p} + C_6 + C\|\nabla u_0\|_p^p, \tag{2.19}$$

where C_6 depends on T and C_3. Now, Gronwall's inequality and (2.1) yield that

$$\sup_{t\in(0,T)} \|\nabla u\|_p + \int_0^T \|u_t\|_2^2 \le C_7, \tag{2.20}$$

where C_7 depends on the time T, C_1, C_3 and $\|\nabla u_0\|_p$.

APPROXIMATE SOLUTIONS AND EXISTENCE RESULT

In this section, we show the existence of a weak solution to the system (1.1)-(1.3) via the Galerkin approximations. For this purpose, we take the set $\{\phi^i\}_{i=1}^\infty$ formed by the eigenvectors ϕ^i, $i = 1, 2, \ldots$, of the Stokes operator and the set $\{\psi^i\}_{i=1}^\infty$ formed by the eigenvectors ψ^i, $i = 1, 2, \ldots$, of the Laplace operator. According to the Appendix of [7], the functions $\{\phi^i\}_{i=1}^\infty$ form a basis in the space $V_2 \cap W^{2,2}(\Omega)$. Setting $R_k = \text{span}\{\phi^1, \phi^2, \ldots, \phi^k\}$ and $S_k = \text{span}\{\psi^1, \psi^2, \ldots, \psi^k\}$, we construct the Galerkin approximations { u^k , ω^k , b^k } being of the form

$$u^k(x,t) = \sum_{i=1}^k a_i^k(t)\phi^i(x); \quad \omega^k(x,t) = \sum_{i=1}^k c_i^k(t)\psi^i(x); \quad b^k(x,t) = \sum_{i=1}^k d_i^k(t)\phi^i(x),$$

where $\boldsymbol{a}^k := (a_1^k, \ldots, a_k^k)$, $\boldsymbol{c}^k := (c_1^k, \ldots, c_k^k)$, $\boldsymbol{d}^k := (d_1^k, \ldots, d_k^k)$ solve the system of ordinary equations

$$\langle u_t^k, \phi\rangle + \langle |\mathbf{e}(u^k)|^{p-2}\mathbf{e}(u^k), \mathbf{e}(\phi)\rangle = \chi\langle \operatorname{rot}\omega^k, \phi\rangle + \langle (b^k\cdot\nabla)b^k, \phi\rangle - \langle (u^k\cdot\nabla)u^k, \phi\rangle$$

for all $\phi \in R_k$, (3.1)

$$\langle \omega_t^k, \psi\rangle + \mu\langle \nabla\omega^k, \nabla\psi\rangle = \chi\langle \operatorname{rot} u^k, \psi\rangle - \langle (u^k\cdot\nabla)\omega^k, \psi\rangle - 2\chi\langle \omega^k, \psi\rangle$$

for all $\psi \in S_k$, (3.2)

$$\langle b_t^k, \phi\rangle + \lambda\langle \nabla b^k, \nabla\phi\rangle = \langle (b^k\cdot\nabla)u^k, \phi\rangle - \langle (u^k\cdot\nabla)b^k, \phi\rangle$$

for all $\phi \in R_k$. (3.3)

Moreover, we require that u^k , ω^k , b^k satisfy the following initial conditions

$$\sup_{t\in(0,T)} \|u^k(t)\|_2^2 = \sup_{t\in(0,T)} \sum_{i=1}^k (a_i^k(t))^2 \le C,$$

$$\sup_{t\in(0,T)} \|\omega^k(t)\|_2^2 = \sup_{t\in(0,T)} \sum_{i=1}^k (c_i^k(t))^2 \le C,$$

$$\sup_{t\in(0,T)} \|b^k(t)\|_2^2 = \sup_{t\in(0,T)} \sum_{i=1}^k (d_i^k(t))^2 \le C$$

with upper bounds C that do not depend on k. Moreover, we have for u^k , ω^k and b^k the same estimates for all norms we have obtained in Section 2. More precisely, we have

$$\begin{aligned}
&\sup_{t\in(0,T)} \|u^k\|_2, \quad \sup_{t\in(0,T)} \|\nabla u^k\|_p, \quad \int_0^T \|u_t\|_2^2 \le C,\\
&\sup_{t\in(0,T)} \|\omega^k\|_{1,2}, \quad \int_0^t \|\omega^k\|_{2,2}^2, \quad \int_0^T \|\omega_t\|_2^2 \le C,\\
&\sup_{t\in(0,T)} \|b^k\|_{1,2}, \quad \int_0^t \|b^k\|_{2,2}^2, \quad \int_0^T \|b_t\|_2^2 \le C
\end{aligned} \tag{3.5}$$

with a constant C that does not depend on k.

Uniform estimates (3.5) imply that there exists a subsequence of $\{u^k\}$, $\{\omega^k\}$ and $\{b^k\}$ (not relabeled) such that

$$u^k \rightharpoonup u, \quad \text{weak-* in } L^\infty(0,T;H)\cap L^\infty(0,T;V_p),$$

$$\omega^k \rightharpoonup \omega, \quad \text{weakly in } L^2(0,T;W^{2,2}(\Omega)) \text{ and weak-* in } L^\infty(0,T;W^{1,2}(\Omega)),$$

$$b^k \rightharpoonup b, \quad \text{weakly in } L^2(0,T;V_2\cap W^{2,2}(\Omega)) \text{ and weak-* in } L^\infty(0,T;V_2),$$

$$u_t^k \rightharpoonup u_t, \qquad b_t^k \rightharpoonup b_t, \quad \text{weakly in } L^2(0,T;H),$$

$$\omega_t^k \rightharpoonup \omega_t, \quad \text{weakly in } L^2(0,T;L^2(\Omega)),$$

$$|\mathbf{e}(u^k)|^{p-2}\mathbf{e}(u^k) \rightharpoonup \Lambda, \quad \text{weakly in } L^{p'}(0,T;L^{p'}(\Omega)),$$

where p'= p/(p –1). Therefore, by making use of the Aubin-Lions lemma (see Lions [15], Theorem1.5.1), we have

$$u^k \to u, \quad \text{strongly in } L^2(0,T;H),$$

$$\omega^k \to \omega, \quad \text{strongly in } L^2(0,T;W^{1,2}(\Omega)),$$

$$b^k \to b, \quad \text{strongly in } L^2(0,T;V_2).$$

With the convergence above, it is easy to pass to the limit as $k\to\infty$ in (3.1)-(3.3) to find

$$\langle u_t,\phi\rangle + \langle \Lambda,\mathbf{e}(\phi)\rangle = \chi\langle \operatorname{rot}\omega,\phi\rangle + \langle (b\cdot\nabla)b,\phi\rangle - \langle (u\cdot\nabla)u,\phi\rangle$$

$$\text{for all } \phi\in V_p, \tag{3.6}$$

$$\langle \omega_t,\psi\rangle + \mu\langle \nabla\omega,\nabla\psi\rangle = \chi\langle \operatorname{rot}u,\psi\rangle - \langle (u\cdot\nabla)\omega,\psi\rangle - 2\chi\langle \omega,\psi\rangle$$

$$\text{for all } \psi\in W^{2,2}(\Omega), \tag{3.7}$$

$$\langle b_t,\phi\rangle + \lambda\langle \nabla b,\nabla\phi\rangle = \langle (b\cdot\nabla)u,\phi\rangle - \langle (u\cdot\nabla)b,\phi\rangle$$

$$\text{for all } \phi\in V_2. \tag{3.8}$$

Next, to complete the existence proof, we need to verify that

$$\Lambda = |\mathbf{e}(u)|^{p-2}\mathbf{e}(u). \tag{3.9}$$

By Lemma 1.3, we have

$$\begin{aligned}
\|\mathbf{e}(u^k)-\mathbf{e}(u)\|_p^p &\le C'\int_\Omega (|\mathbf{e}(u^k)|^{p-2}\mathbf{e}(u^k) - |\mathbf{e}(u)|^{p-2}\mathbf{e}(u))(\mathbf{e}(u^k)-\mathbf{e}(u)) \\
&= C'\int_\Omega |\mathbf{e}(u^k)|^{p-2}\mathbf{e}(u^k)\mathbf{e}(u^k) - \int_\Omega |\mathbf{e}(u)|^{p-2}\mathbf{e}(u)(\mathbf{e}(u^k)-\mathbf{e}(u)) \\
&\quad - \int_\Omega |\mathbf{e}(u^k)|^{p-2}\mathbf{e}(u^k)\mathbf{e}(u) \\
&\overset{(3.1)}{=} \chi\langle \operatorname{rot}\omega^k,u^k\rangle + \langle (b^k\cdot\nabla)b^k,u^k\rangle - \langle u_t^k,u^k\rangle \\
&\quad - \int_\Omega |\mathbf{e}(u)|^{p-2}\mathbf{e}(u)(\mathbf{e}(u^k)-\mathbf{e}(u)) - \int_\Omega |\mathbf{e}(u^k)|^{p-2}\mathbf{e}(u^k)\mathbf{e}(u).
\end{aligned}$$

Considering $\lim_{k\to\infty}$ of this identity together with (3.6) implies that

$$\lim_{k\to\infty}\|\mathbf{e}(u^k)-\mathbf{e}(u)\|_p^p \le 0,$$

and thus (3.9) follows. Having the estimates

$$\|u\|_{L^\infty(0,T;W_0^{1,p}(\Omega))}, \quad \|\omega\|_{L^\infty(0,T;W_0^{1,2}(\Omega))}, \quad \|b\|_{L^\infty(0,T;W_0^{1,2}(\Omega))} \le C, \tag{3.10}$$

we can now introduce the pressure from (1.5). For $t\in(0,T]$, define the functional $W^{-1,p'}(\Omega)$ as

$$\langle F, \xi \rangle := \langle \operatorname{div}(|\mathbf{e}(u)|^{p-2}\mathbf{e}(u)), \xi \rangle + \chi \langle \operatorname{rot} \omega, \xi \rangle - \langle b \times \operatorname{rot} b, \xi \rangle$$
$$- \langle (u \cdot \nabla) u, \xi \rangle - \langle u_t, \xi \rangle.$$

We have

$$\langle F, \xi \rangle = 0, \quad \forall \xi \in V_p, \text{ a.e. } t \in (0, T]$$

By using De Rahm's theorem (see [11], Lemma 2.7), we obtain a function $\pi \in L^{p'}(\Omega)$ such that

$$F = -\nabla \pi, \quad \text{a.e. } t \in (0, T].$$

Moreover, due to estimates (3.10),

$$\|\nabla \pi\|_{W^{-1,p'}(\Omega)} \leq C, \quad \text{a.e. } t \in (0, T].$$

Then, by Lemma 1.2, there is a generic constant C, depending only on the data such that

$$\|\pi\|_{p'} \leq C, \quad \text{a.e. } t \in (0, T].$$

Now, we complete the proof of the existence part of Theorem 1.2

UNIQUENESS CRITERION

Let (u_1, ω_1, b_1) and (u_2, ω_2, b_2) be both solutions of the problem. Then, for their difference $\bar{u} = u_1 - u_2, \bar{\omega} = \omega_1 - \omega_2, \bar{b} = b_1 - b_2$, we have

$$\langle \bar{u}_t, \phi \rangle + \langle |\mathbf{e}(u_1)|^{p-2}\mathbf{e}(u_1) - |\mathbf{e}(u_2)|^{p-2}\mathbf{e}(u_2), \mathbf{e}(\phi) \rangle$$
$$= \chi \langle \operatorname{rot} \bar{\omega}, \phi \rangle + \langle (b_1 \cdot \nabla)\bar{b}, \phi \rangle$$
$$+ \langle (\bar{b} \cdot \nabla) b_2, \phi \rangle - \langle (u_1 \cdot \nabla)\bar{u}, \phi \rangle - \langle (\bar{u} \cdot \nabla) u_2, \phi \rangle \quad \text{for all } \phi \in V_p, \tag{4.1}$$

$$\langle \bar{\omega}_t, \psi \rangle + \mu \langle \nabla \bar{\omega}, \nabla \psi \rangle + 2\chi \langle \bar{\omega}, \psi \rangle = \chi \langle \operatorname{rot} \bar{u}, \psi \rangle - \langle (u_1 \cdot \nabla)\bar{\omega}, \psi \rangle - \langle (\bar{u} \cdot \nabla)\omega_2, \psi \rangle$$
$$\text{for all } \psi \in W_0^{1,2}(\Omega), \tag{4.2}$$

$$\langle \bar{b}_t, \eta \rangle + \lambda \langle \nabla \bar{b}, \nabla \eta \rangle = \langle (b_1 \cdot \nabla)\bar{u}, \eta \rangle + \langle (\bar{b} \cdot \nabla) u_2, \eta \rangle - \langle (u_1 \cdot \nabla)\bar{b}, \eta \rangle - \langle (\bar{u} \cdot \nabla) b_2, \eta \rangle$$
$$\text{for all } \eta \in V_2. \tag{4.3}$$

Taking $\phi = \bar{u}$ in (4.1), by Lemma 1.3 and the fact that $\langle (u_1 \cdot \nabla)\bar{u}, \bar{u} \rangle = 0 \text{ and } p > 2$, we obtain

$$\frac{1}{2}\frac{d}{dt}\|\bar{u}\|_2^2 + C_p \|\nabla \bar{u}\|_p^p \leq \chi \langle \operatorname{rot} \bar{\omega}, \bar{u} \rangle + \langle (\bar{b} \cdot \nabla) b_2, \bar{u} \rangle - \langle (\bar{u} \cdot \nabla) u_2, \bar{u} \rangle + \langle (b_1 \cdot \nabla)\bar{b}, \bar{u} \rangle$$
$$\equiv \sum_{i=1}^{3} J_i + \langle (b_1 \cdot \nabla)\bar{b}, \bar{u} \rangle$$

for each $J_i, i = 1, 2, 3$, it follows from Hölder's, Young's inequality and (1.9), (1.10) that

$$J_1 \leq \chi |\langle \operatorname{rot} \bar{\omega}, \bar{u} \rangle| \leq \chi \|\nabla \bar{\omega}\|_2 \cdot \|\bar{u}\|_2 \leq \varepsilon \|\nabla \bar{\omega}\|_2^2 + C_\varepsilon \|\bar{u}\|_2^2,$$

$$J_2 \leq |\langle (\bar{b} \cdot \nabla) b_2, \bar{u} \rangle| \leq \|\nabla b_2\|_6 \cdot \|\bar{b}\|_3 \cdot \|\bar{u}\|_2 \leq C(\|\Delta b_2\|_2 + \|\nabla b_2\|_2) \cdot \|\nabla \bar{b}\|_2 \cdot \|\bar{u}\|_2$$
$$\leq \varepsilon \|\nabla \bar{b}\|_2^2 + C_\varepsilon (\|\Delta b_2\|_2^2 + \|\nabla b_2\|_2^2) \cdot \|\bar{u}\|_2^2,$$

$$J_3 \leq |\langle (\bar{u} \cdot \nabla) u_2, \bar{u} \rangle| \leq \|\nabla u_2\|_p \cdot \|\bar{u}\|_{\frac{2p}{p-1}}^2 \leq \|\nabla u_2\|_p \cdot \|\bar{u}\|_2^{\frac{2p-3}{p}} \cdot \|\nabla \bar{u}\|_2^{\frac{3}{p}}$$
$$\leq \varepsilon \|\nabla \bar{u}\|_2^2 + C_\varepsilon \|\nabla u_2\|_p^{\frac{2p}{2p-3}} \cdot \|\bar{u}\|_2^2,$$

so, we have

$$\begin{aligned}&\frac{1}{2}\frac{d}{dt}\|\bar{u}\|_2^2 + C_p\|\nabla\bar{u}\|_p^p\\&\le\varepsilon\left(\|\nabla\bar{u}\|_2^2 + \|\nabla\bar{\omega}\|_2^2 + \|\nabla\bar{b}\|_2^2\right)\\&\quad + C_\varepsilon\left(1 + \|\nabla b_2\|_2^2 + \|\Delta b_2\|_2^2 + \|\nabla u_2\|_p^{\frac{2p}{2p-3}}\right)\cdot\|\bar{u}\|_2^2 + \left\langle(b_1\cdot\nabla)\bar{b},\bar{u}\right\rangle.\end{aligned}\tag{4.4}$$

Taking $\psi = \bar{\omega}$ in (4.2) and noting that $\langle(u_1\cdot\nabla)\bar{\omega},\bar{\omega}\rangle = 0$, it follows

$$\begin{aligned}\frac{1}{2}\frac{d}{dt}\|\bar{\omega}\|_2^2 + \mu\|\nabla\bar{\omega}\|_2^2 + 2\chi\|\bar{\omega}\|_2^2 &= \chi\langle\operatorname{rot}\bar{u},\bar{\omega}\rangle - \left\langle(\bar{u}\cdot\nabla)\omega_2,\bar{\omega}\right\rangle\\&\equiv\sum_{i=1}^{2} I_i,\end{aligned}$$

and for I_i, $i = 1,2$,

$$\begin{aligned}I_1 &\le \chi\|\nabla\bar{u}\|_2\cdot\|\bar{\omega}\|_2 \le \varepsilon\|\nabla\bar{u}\|_2^2 + C_\varepsilon\|\bar{\omega}\|_2^2,\\ I_2 &\le \|\nabla\omega_2\|_6\cdot\|\bar{\omega}\|_3\cdot\|\bar{u}\|_2 \le C\left(\|\Delta\omega_2\|_2 + \|\nabla\omega_2\|_2\right)\cdot\|\nabla\bar{\omega}\|_2\cdot\|\bar{u}\|_2\\&\le\varepsilon\|\nabla\bar{\omega}\|_2^2 + C_\varepsilon\left(\|\Delta\omega_2\|_2^2 + \|\nabla\omega_2\|_2^2\right)\cdot\|\bar{u}\|_2^2,\end{aligned}\tag{4.5}$$

Similarly, by taking $\eta = \bar{b}$ in (4.3), reasoning analogous as above, we could get

$$\begin{aligned}\frac{1}{2}\frac{d}{dt}\|\bar{b}\|_2^2 + \lambda\|\nabla\bar{b}\|_2^2 &\le \varepsilon\|\nabla\bar{b}\|_2^2 + C_\varepsilon\left(\|\Delta b_2\|_2^2 + \|\nabla b_2\|_2^2\right)\cdot\|\bar{u}\|_2^2\\&\quad + C_\varepsilon\|\nabla u_2\|_p^{\frac{2p}{2p-3}}\cdot\|\bar{b}\|_2^2 + \left\langle(b_1\cdot\nabla)\bar{u},\bar{b}\right\rangle.\end{aligned}\tag{4.6}$$

Adding (4.4)-(4.6) and observing that $\langle(b_1\cdot\nabla)\bar{b},\bar{u}\rangle + \langle(b_1\cdot\nabla)\bar{u},\bar{b}\rangle = 0$, after choosing ε properly small, we finally get

$$\begin{aligned}&\frac{1}{2}\frac{d}{dt}\left(\|\bar{u}\|_2^2 + \|\bar{\omega}\|_2^2 + \|\bar{b}\|_2^2\right) + C\left(\|\nabla\bar{u}\|_2^2 + \|\nabla\bar{\omega}\|_2^2 + \|\nabla\bar{b}\|_2^2\right)\\&\le CF(t)\cdot\left(\|\bar{u}\|_2^2 + \|\bar{\omega}\|_2^2 + \|\bar{b}\|_2^2\right)\end{aligned}$$

with

$$F(t) = 1 + \|\nabla b_2\|_2^2 + \|\Delta b_2\|_2^2 + \|\nabla u_2\|_p^{\frac{2p}{2p-3}} + \|\nabla\omega_2\|_2^2 + \|\Delta\omega_2\|_2^2.$$

Since $2p/(2p-3)\le p$ for $p\ge\frac{5}{2}$, then Gronwall's inequality and the estimates obtained in Section 2 yield that $\bar{u} = \bar{\omega} = \bar{b} = 0$ for $t\in[0,T]$.

This completes the proof of the theorem.

ACKNOWLEDGEMENTS

The author would like to thank the referees for valuable comments and suggestions for improving this paper. This work was partially supported by the National NSF (Grant No. 10971080) of China.

REFERENCES

1. Eringen AC: Theory of micropolar fluids. J. Math. Mech. 1966, 16: 1-18.
2. Galdi GP, Rionero S: A note on the existence and uniqueness of solutions of the micropolar fluid equations. Int. J. Eng. Sci. 1977, 15: 105-108. 10.1016/0020-7225(77)90025-8
3. Rojas-Medar MA: Magneto-micropolar fluid motion: existence and uniqueness of strong solution. Math. Nachr. 1997, 188: 301-319. 10.1002/mana.19971880116
4. Ortega-Torres EE, Rojas-Medar MA: Magneto-micropolar fluid motion: global existence of strong solutions. Abstr. Appl. Anal. 1999, 4: 109-125. 10.1155/S1085337599000287
5. Rojas-Medar MA, Boldrini JL: Magneto-micropolar fluid motion: existence of weak solutions. Rev. Mat. Complut. 1998, 11: 443-460.

6. Gunzburger MD, Ladyzhenskaya OA, Peterson JS: On the global unique solvability of initial-boundary value problems for the coupled modified Navier-Stokes and Maxwell equations. J. Math. Fluid Mech. 2004, 6: 462-482.
7. Málek J, Něcas J, Rokyta M, Ružička M: Weak and Measure-Valued Solutions to Evolutionary PDEs. Chapman & Hall, London; 1996.
8. Bellout H, Bloom F, Nečas J: Solutions for incompressible non-Newtonian fluids. C. R. Math. Acad. Sci. Paris, Sér. I 1993, 317: 795-800.
9. Ebmeyer C, Urbano JM: Quasi-steady Stokes flow of multiphase fluids with shear-dependent viscosity. Eur. J. Appl. Math. 2007, 18: 417-434. 10.1017/S0956792507006948
10. Lions JL: Quelques méthodes de Résolution de Problèmes aux limites non Linéaires. Dunod, Paris; 1969.
11. Amrouche C, Girault V: Decomposition of vector spaces and application to the Stokes problem in arbitrary dimension. Czechoslov. Math. J. 1994, 44: 109-140.

INDEX

T

V